I0063090

Synthesis and Applications of Biopolymer Composites

Synthesis and Applications of Biopolymer Composites

Special Issue Editors

Ana María Díez-Pascual
Patrizia Cinelli

MDPI • Basel • Beijing • Wuhan • Barcelona • Belgrade

MDPI

Special Issue Editors
Ana María Díez-Pascual
Alcalá University
Spain

Patrizia Cinelli
Pisa University
Italy

Editorial Office
MDPI
St. Alban-Anlage 66
4052 Basel, Switzerland

This is a reprint of articles from the Special Issue published online in the open access journal *International Journal of Molecular Sciences* (ISSN 1422-0067) from 2018 to 2019 (available at: https://www.mdpi.com/journal/ijms/special_issues/Biopolymer_Composites)

For citation purposes, cite each article independently as indicated on the article page online and as indicated below:

LastName, A.A.; LastName, B.B.; LastName, C.C. Article Title. *Journal Name* **Year**, *Article Number*, Page Range.

ISBN 978-3-03921-132-6 (Pbk)
ISBN 978-3-03921-133-3 (PDF)

Cover image courtesy of Patrizia Cinelli.

© 2019 by the authors. Articles in this book are Open Access and distributed under the Creative Commons Attribution (CC BY) license, which allows users to download, copy and build upon published articles, as long as the author and publisher are properly credited, which ensures maximum dissemination and a wider impact of our publications.

The book as a whole is distributed by MDPI under the terms and conditions of the Creative Commons license CC BY-NC-ND.

Contents

About the Special Issue Editors ix

Ana María Díez-Pascual
Synthesis and Applications of Biopolymer Composites
Reprinted from: *Int. J. Mol. Sci.* **2019**, *20*, 2321, doi:10.3390/ijms20092321 1

Maria-Beatrice Coltelli, Patrizia Cinelli, Vito Gigante, Laura Aliotta, Pierfrancesco Morganti, Luca Panariello and Andrea Lazzeri
Chitin Nanofibrils in Poly(Lactic Acid) (PLA) Nanocomposites: Dispersion and Thermo-Mechanical Properties
Reprinted from: *Int. J. Mol. Sci.* **2019**, *20*, 504, doi:10.3390/ijms20030504 8

Moritz Koch, Sofía Doello, Kirstin Gutekunst and Karl Forchhammer
PHB is Produced from Glycogen Turn-over during Nitrogen Starvation in *Synechocystis* sp. PCC 6803
Reprinted from: *Int. J. Mol. Sci.* **2019**, *20*, 1942, doi:10.3390/ijms20081942 28

Laura Aliotta, Vito Gigante, Maria Beatrice Coltelli, Patrizia Cinelli and Andrea Lazzeri
Evaluation of Mechanical and Interfacial Properties of Bio-Composites Based on Poly(Lactic Acid) with Natural Cellulose Fibers
Reprinted from: *Int. J. Mol. Sci.* **2019**, *20*, 960, doi:10.3390/ijms20040960 42

Maria Cristina Righetti, Patrizia Cinelli, Norma Mallegni, Carlo Andrea Massa, Simona Bronco, Andreas Stäbler and Andrea Lazzeri
Thermal, Mechanical, and Rheological Properties of Biocomposites Made of Poly(lactic acid) and Potato Pulp Powder
Reprinted from: *Int. J. Mol. Sci.* **2019**, *20*, 675, doi:10.3390/ijms20030675 56

Yun Zhao, Bei Liu, Hongwei Bi, Jinjun Yang, Wei Li, Hui Liang, Yue Liang, Zhibin Jia, Shuxin Shi and Minfang Chen
The Degradation Properties of MgO Whiskers/PLLA Composite In Vitro
Reprinted from: *Int. J. Mol. Sci.* **2018**, *19.*, 2740, doi:10.3390/ijms19092740 73

Patrizia Cinelli, Maurizia Seggiani, Norma Mallegni, Vito Gigante and Andrea Lazzeri
Processability and Degradability of PHA-Based Composites in Terrestrial Environments
Reprinted from: *Int. J. Mol. Sci.* **2019**, *20*, 284, doi:10.3390/ijms20020284 89

Estefanía Lidón Sánchez-Safont, Alex Arrillaga, Jon Anakabe, Luis Cabedo and Jose Gamez-Perez
Toughness Enhancement of PHBV/TPU/Cellulose Compounds with Reactive Additives for Compostable Injected Parts in Industrial Applications
Reprinted from: *Int. J. Mol. Sci.* **2018**, *19*, 2102, doi:10.3390/ijms19072102 103

J. Vincent Edwards, Krystal Fontenot, Falk Liebner, Nicole Doyle nee Pircher, Alfred D. French and Brian D. Condon
Structure/Function   Analysis   of   Cotton-Based   Peptide-Cellulose   Conjugates: Spatiotemporal/Kinetic Assessment of Protease Aerogels Compared to Nanocrystalline and Paper Cellulose
Reprinted from: *Int. J. Mol. Sci.* **2018**, *19*, 840, doi:10.3390/ijms19030840 123

Sanna Siljander, Pasi Keinänen, Anna Räty, Karthik Ram Ramakrishnan, Sampo Tuukkanen, Vesa Kunnari, Ali Harlin, Jyrki Vuorinen and Mikko Kanerva
Effect of Surfactant Type and Sonication Energy on the Electrical Conductivity Properties of Nanocellulose-CNT Nanocomposite Films
Reprinted from: *Int. J. Mol. Sci.* **2018**, *19*, 1819, doi:10.3390/ijms19061819 . . . . **139**

Elena Manaila, Maria Daniela Stelescu and Gabriela Craciun
Degradation Studies Realized on Natural Rubber and Plasticized Potato Starch Based Eco-Composites Obtained by Peroxide Cross-Linking
Reprinted from: *Int. J. Mol. Sci.* **2018**, *19*, 2862, doi:10.3390/ijms19102862 . . . . **153**

Faraz Muneer, Eva Johansson, Mikael S. Hedenqvist, Tomás S. Plivelic and Ramune Kuktaite
Impact of pH Modification on Protein Polymerization and Structure–Function Relationships in Potato Protein and Wheat Gluten Composites
Reprinted from: *Int. J. Mol. Sci.* **2019**, *20*, 58, doi:10.3390/ijms20010058 . . . . **171**

Qichun Liu, Fang Wang, Zhenggui Gu, Qingyu Ma and Xiao Hu
Exploring the Structural Transformation Mechanism of Chinese and Thailand Silk Fibroin Fibers and Formic-Acid Fabricated Silk Films
Reprinted from: *Int. J. Mol. Sci.* **2018**, *19*, 3309, doi:10.3390/ijms19113309 . . . . **185**

Su-Kyoung Baek, Sujin Kim and Kyung Bin Song
Characterization of *Ecklonia cava* Alginate Films Containing Cinnamon Essential Oils
Reprinted from: *Int. J. Mol. Sci.* **2018**, *19*, 3545, doi:10.3390/ijms19113545 . . . . **201**

Diana Isela Sanchez-Alvarado, Javier Guzmán-Pantoja, Ulises Páramo-García, Alfredo Maciel-Cerda, Reinaldo David Martínez-Orozco and Ricardo Vera-Graziano
Morphological Study of Chitosan/Poly (Vinyl Alcohol) Nanofibers Prepared by Electrospinning, Collected on Reticulated Vitreous Carbon
Reprinted from: *Int. J. Mol. Sci.* **2018**, *19*, 1718, doi:10.3390/ijms19061718 . . . . **214**

Na Liang, Shaoping Sun, Xianfeng Gong, Qiang Li, Pengfei Yan and Fude Cui
Polymeric Micelles Based on Modified Glycol Chitosan for Paclitaxel Delivery: Preparation, Characterization and Evaluation
Reprinted from: *Int. J. Mol. Sci.* **2018**, *19*, 1550, doi:10.3390/ijms19061550 . . . . **226**

Liang Li, Na Liang, Danfeng Wang, Pengfei Yan, Yoshiaki Kawashima, Fude Cui and Shaoping Sun
Amphiphilic Polymeric Micelles Based on Deoxycholic Acid and Folic Acid Modified Chitosan for the Delivery of Paclitaxel
Reprinted from: *Int. J. Mol. Sci.* **2018**, *19*, 3132, doi:10.3390/ijms19103132 . . . . **240**

Shaoyun Chen, Min Xiao, Luyi Sun and Yuezhong Meng
Study on Thermal Decomposition Behaviors of Terpolymers of Carbon Dioxide, Propylene Oxide, and Cyclohexene Oxide
Reprinted from: *Int. J. Mol. Sci.* **2018**, *19*, 3723, doi:10.3390/ijms19123723 . . . . **253**

Dongmei Han, Guiji Chen, Min Xiao, Shuanjin Wang, Shou Chen, Xiaohua Peng and Yuezhong Meng
Biodegradable and Toughened Composite of Poly(Propylene Carbonate)/Thermoplastic Polyurethane (PPC/TPU): Effect of Hydrogen Bonding
Reprinted from: *Int. J. Mol. Sci.* **2018**, *19*, 2032, doi:10.3390/ijms19072032 . . . . **268**

Serena Danti, Luisa Trombi, Alessandra Fusco, Bahareh Azimi, Andrea Lazzeri, Pierfrancesco Morganti, Maria-Beatrice Coltelli and Giovanna Donnarumma
Chitin Nanofibrils and Nanolignin as Functional Agents in Skin Regeneration
Reprinted from: *Int. J. Mol. Sci.* **2019**, *20*, 2669, doi:10.3390/ijms20112669 . . . . **280**

About the Special Issue Editors

Ana María Díez-Pascual graduated with a degree in Chemistry in 2001 (awarded Extraordinary Prize) from Complutense University (Madrid, Spain), where she also carried out her Ph.D. studies (2002–2005) on the dynamic and equilibrium properties of fluid interfaces under the supervision of Prof. Rubio. In 2005, Dr. Diez-Pascual worked at the Max Planck Institute of Colloids and Interfaces (Germany), with Prof. Miller, on the rheological characterization of water-soluble polymers. During 2006–2008, she was a Postdoctoral Researcher at the Physical Chemistry Institute of the RWTH Aachen University (Germany), where she worked on the layer-by-layer assembly of polyelectrolyte multilayers onto thermoresponsive microgels. Dr. Diez-Pascual then moved to the Institute of Polymer Science and Technology (Madrid, Spain) and participated in a Canada–Spain joint project to develop carbon nanotube (CNT)-reinforced epoxy and polyetheretherketone composites for transport applications. Currently, Dr. Diez-Pascual is a Postdoctoral Researcher at Alcala University (Madrid, Spain) where she focuses on the development of polymer/nanofiller systems for biomedical applications. She has participated in 22 research projects (11 international and 10 national, of which 3 have been with private companies and where she has been the principal investigator in 6 of the projects). She has published **102 SCI articles** (97% in Q1 journals), has an **h-index of 39**, and **more than 3000 total citations**. More than 50% of her articles are in journals with an impact factor of ≥ 4.8, such as *J. Mater. Chem, Carbon*, and *J. Phys. Chem. C*. She is the first and corresponding author of 2 invited reviews in *Prog. Mater. Sci.* and a frequent reviewer for journals published by ACS, MDPI, and Elsevier. Dr. Diez-Pascual has published 19 book chapters, 2 monographies, and edited 1 book, and is the first author of an international patent. She has contributed to 61 international conferences (45 oral communications, including 6 by invitation) and has been a member of the organizing committee in 3 workshops and 1 national meeting. She has been invited to present seminars at prestigious international research centers such as Max Planck in Germany, NRC in Canada, and the School of Materials in Manchester, UK). She was awarded the TR35 2012 Prize by the Massachusetts Institute of Technology (MIT) for her innovative work in the field of nanotechnology.

Patrizia Cinelli is Associate Professor of Materials Science and Technology at University of Pisa, where she runs courses on "Applied Chemistry, Materials Science and Technology" and "Biodegradable and Sustainable Polymers". She graduated with a degree in Chemistry in 1995 from the University of Florence, and in 1999, she received her Ph.D. in Chemistry studying biodegradable and sustainable polymers for application in agriculture at Pisa University, where part of the work was also carried out at the United States Department of Agriculture (USDA), Peoria, IL, USA, where three stages of the study (each of 6 months duration) were conducted in 1998, 2000, and 2001. She has worked as Researcher at the Interuniversity Consortium of Materials Science and Technology, Florence, Italy, and at the Institute for Chemical and Physical Processes, Pisa Division, of the National Research Council (CNR), Italy. She was a Visiting Scientist at the University of Almeria, Spain, as well as at INTEMA-CONICET, Mar del Plata, Argentina. Prof. Cinelli has accumulated over 20 years of international experience in materials science and polymer technologies while working within the framework of regional, national, and European projects, from FP5 to the current Horizon 2020. She has participated in 20 European projects and is participating in 3 new Horizon 2020 projects in the Bio-Based Industries, acting as coordinator of the ECOAT project GA 837863. Prof. Cinelli is co-author of over 75 articles in peer-reviewed journals, with a h-index 24, 2025 citations, 11 book chapters, and 8 patents on innovative materials, and participated in over 100 international conferences, having been an invited speaker on 5 occasions. Prof. Cinelli is a reviewer for international journals of materials science (published by MDPI and Elsevier) and an evaluator of EUREKA, as well as National Funds projects in Latvia and Denmark. She has been an organizing committee member for 3 workshops and 3 international conferences. Prof. Cinelli has worked on biodegradable polymers and, in particular, on the processing, production, and characterization of biocomposites based on fibers from renewable resources. She has considerable research experience regarding biodegradable materials and the study of their morphological, thermal, and mechanical characterization, as well as the evaluation of sustainability through biodegradation tests and the evaluation of life cycle assessment.

International Journal of
Molecular Sciences

MDPI

Editorial

Synthesis and Applications of Biopolymer Composites

Ana María Díez-Pascual

Department of Analytical Chemistry, Physical Chemistry and Chemical Engineering, Faculty of Sciences, Institute of Chemistry Research "Andrés M. del Río" (IQAR), University of Alcalá, Ctra. Madrid-Barcelona, Km. 33.6, 28805 Alcalá de Henares, Madrid, Spain; am.diez@uah.es; Tel.: +34-918-856-430

Received: 28 March 2019; Accepted: 9 May 2019; Published: 10 May 2019

In recent years, there has been a growing demand for a clean and pollution-free environment and an evident target to minimizing fossil fuel. Therefore, a lot of attention has been focused on research to replace petroleum-based commodity plastics by biodegradable materials arising from biological and renewable resources. Different biopolymers, polymers produced from natural sources either chemically from a biological material or biosynthesized by living organisms, are also suitable alternatives to address these issues due to their outstanding properties including good barrier performance, biodegradation ability, and low weight. However, they generally present poor mechanical properties, a short fatigue life, low chemical resistance, poor long-term durability, and limited processing capability. In order to overcome these deficiencies and develop advanced materials for a wide range of applications, biopolymers can be reinforced with fillers or nanofillers (with at least one of its dimensions in the nanometer range) to form biocomposites or bionanocomposites. In particular, nanostructures can exhibit higher specific surface areas, surface energy, and density, compared to conventional microfillers, and can lead to materials with new and improved properties due to synergistic effects that are better than those arising from the simple rule of mixtures. Therefore, bionanocomposites are advantageous for a wide range of applications, such as medicine, pharmaceutics, cosmetics, food packaging, agriculture, forestry, electronics, transport, construction, and so forth. This Special Issue, with a collection of 17 research articles, provides selected examples of the most recent advances in the synthesis, characterization, and applications of environment friendly and biodegradable biopolymer composites and nanocomposites.

The most widely used biopolymers for the current development of biocomposites are poly(lactic acid) (PLA), cellulose esters, polyhydroxyalkanoates (PHAs), and starch-based plastics [1,2]. PLA is a fully renewable polymer that is both resorbable in the human body and biodegradable in composting plants. It presents biocidal activity because of its tendency to hydrolyze on the surface, producing lactic acid, and is one of the best alternatives to petroleum-based polymers in the packaging, agricultural, personal care, cosmetic, biomedical, and tissue engineering sectors [1,3,4]. A large number of studies have been devoted to extend its processability and the range of applications by reinforcing it with different nanofillers, including cellulose nanocrystals, chitin nanofibers, metal oxide nanoparticles, or clays [5,6]. In particular, the combination of PLA and chitin nanofibers represents a good opportunity for the preparation of bioplastic materials with improved structural and functional properties due to synergistic effects. However, it is difficult to attain a uniform dispersion of these nanofibers within the PLA matrix at the nanoscale level. In this regard, Coltelli et al. [7] used poly(ethylene glycol) (PEG), a biocompatible polymer, to prepare pre-composites that were subsequently added to PLA in the extruder to obtain transparent nanocomposites. The tensile properties did not show a reinforcing effect of up to 12 wt% chitin loading, albeit the nanocomposites maintained high values of elongation at break (>150%). This methodology is advantageous since it can be applied at an industrial level and does not modify the thermo-mechanical properties of plasticized PLA. This is in contrast to the results found upon the addition of diverse types of cellulose microfibers with different aspect ratios, where the stiffness increased with increased filler loading [8]. Some cellulose microfibers can be used without any compatibilization in order to reduce the final composite cost, increase the stiffness, and simultaneously promote the biodegradability of the materials.

An interesting nanofiller for PLA is potato pulp powder, utilized as a residue of the processing for the production and extraction of starch. It consists mainly of lignocellulosic fibers, starch, and proteins, and the cost of the raw material is low, which makes it very appealing for industrial application. In this regard, Righetti et al. [9] developed PLA/potato pulp biocomposites by extrusion followed by injection molding and characterized them in terms of processability and thermal, mechanical, and rheological properties. To make the processing easier, acetyl tributyl citrate (ATBC), derived from naturally occurring citric acid, was used as plasticizer and calcium carbonate was added in low percentages as an inert filler to facilitate the detachment of the injection-molded specimens. A slight drop in stiffness was found, compared to the neat matrix, together with a small reduction in ductility, since the potato pulp particles act as stress concentration sites and promote crack nucleation. Nonetheless, the lower viscosity of the biocomposites is an advantage for the material processing, which meets the requirements for rigid food packaging applications. The biomedical uses of PLA composites are also of great interest. Continuing the progress in this topic, Zhao et al. [10] developed PLA based-composite films, reinforced with stearic acid-modified MgO whiskers via a solution casting method, and studied their in vitro degradation properties and cytocompatibility. The degradation behaviour of the composites was found to increase with increasing MgO content and was pH-dependent. Furthermore, the cytocompatibility of the composites also increased considerably, which is beneficial for promoting cell proliferation and improving the matrix bioactivity.

PHAs are very interesting biopolymers. They are a family of polyesters of hydroxyalkanoic acids, synthetized by microorganisms in the presence of excess carbon and lack of essential nutrients. PHAs have thermoplastic properties similar to those of polypropylene, good mechanical properties, and excellent biodegradability in various ecosystems [11]. The most common PHAs are the homopolymer poly(3-hydroxybutyrate) (PHB) and its copolyester with hydroxyvaleric acid, poly(3-hydroxybutyrate-co-3-hydroxyvalerate) (PHBV), which are well suited for food packaging [12]. Despite their good properties and excellent biodegradability, their costs are relatively high ($€7–12$/kg) compared to other biopolymers, such as PLA ($€2.5–3$/kg), and has limited their use in the medical and pharmaceutical sectors. A lot of effort has been devoted to incorporate low-value nanomaterials into PHAs in order to reduce the cost of the final products. Thus, Cinelli et al. [13] incorporated waste wood sawdust fibers, a byproduct of the wood industry, into PHB via melt extrusion, using ATBC as plasticizer and $CaCO_3$ as inert filler. The impact resistance of the composites increased notably with increased fiber loading. More importantly, the fibers accelerated the degradation of the polymeric matrix in soil. Hence, these composites are interesting in agriculture or plant nursery. With the aim of improving the performance of PHAs in terms of heat resistance, stiffness, and toughness, cellulose fibers and a thermoplastic polyurethane (TPU) have been added to PHBV [14]. To improve PHBV-cellulose interfacial adhesion, different compatibilizing agents were tested, including hexametylene diisocianate, an epoxy-functionalized styrene-acrylic oligomer, and triglycidyl isocyanurate. The diisocianate displayed the best compatibilization ability, with the uppermost values of elongation at break and toughness. This strategy can aid to solve some of the issues that these materials encounter in common applications.

On the other hand, cellulose is environmentally conscious, low-cost, strong, dimension-stable, non-melting, non-toxic, and can be derivatized to covalently append a wide range of biologically active molecules. In particular, Edwards et al. [15] compared the performance of a sensor designed with a nanocellulose aerogel transducer surface, derived from cotton, with cotton filter paper and nanocrystalline cellulose. X-ray crystallography, Michaelis–Menten enzyme kinetics, and circular dichroism were used to assess the structure/function relations of the peptide-cellulose conjugate conformation to enzyme/substrate binding and turnover rates. The aerogel-based sensor yielded the highest enzyme efficiency, ascribed to the binding of the serine protease to the negatively charged cellulose surface.

The interest towards nanoscale cellulose has increased extraordinarily over the last years, owed to its inherent mechanical properties, which are better than those of the source biomass material [16]. The combination of carbon nanotubes (CNTs) and cellulose results in a conductive nanocomposite network that can be used in a wide range of applications, including supercapacitor electrodes, electromagnetic interference shielding devices, and water and pressure sensors [17]. The properties of nanocellulose-CNT composites are affected by the quality of the CNT dispersion, the amount of defects, and the aspect ratio of the CNTs, as well as the strength of the CNT-cellulose interactions. The key challenge is to achieve a uniform and stable CNT dispersion. To attain such goal, Siljander et al. combined ultrasonication with the addition of surfactants [18] and found that there are a number of parameters that strongly affect the nanocomposite conductivity, such as surfactant type and concentration, sonication energy, and the film processing technique and the best performance was attained with the non-ionic surfactant Triton.

Natural rubber is another interesting non-toxic material derived from a renewable stock that has excellent physical properties and, due to its low price, is the elastomer most used in industry worldwide. Continuing the progress in this topic, Manaila et al. [19] developed environmental-friendly natural rubber/plasticized potato starch composites via peroxide cross-linking in the presence of trimethylolpropane trimethacrylate as a cross-linking co-agent. The influence of starch concentration on the mechanical properties, gel fraction and cross-link density, water uptake, structure, and morphology, before and after thermal degradation and natural ageing of the composites, were investigated. Plasticized starch loading up to 20 wt% was found to have a reinforcing effect on the matrix, and also favored its natural degradation; hence, starch can be considered as an interesting alternative to conventional fillers such as silica and carbon black.

Potato starch is a protein-rich polymeric by-product currently used in animal feed industries. Its combination with wheat gluten, also a protein-rich material, is interesting for the development of bio-based plastics [20]. However, these raw materials are difficult to process since their glass transition temperature (T_g) is close to their thermal degradation temperature. Hence, chemical agents that reduce the T_g and broaden the processing window are required. Chemical additives such as NaOH create basic conditions for the proteins, resulting in changes of their secondary and supramolecular structures that lead to improved functional properties of the processed materials [21]. In particular, increasing the wheat gluten content in the composites was found to decrease the protein solubility and the Young's modulus, albeit enabling the manufacture of films with good properties at a low pressing temperature (i.e., 130 °C), thereby contributing to a lower environmental foot-print due to a reduction of energy use [22]. Silk fibroin is another polymeric protein that has outstanding mechanical properties and a tunable biodegradation rate, due to its variable structures. Different fabrication methods can affect the structural transitions and physical properties of silk fibroin materials. In this sense, Liu et al. [23] investigated the variability of structural, thermal, and mechanical properties of two silk films (Chinese and Thailand B. Mori) regenerated from a formic acid solution, as well as their original fibers, using dynamic mechanical analysis (DMA) and Fourier transform infrared spectrometry (FTIR). Chinese silks were found to display a lower T_g, a higher disorder degree, and better elasticity and mechanical strength. Further, as the calcium chloride content in the initial processing solvent increased, the T_g of the samples decreased while their disorder degree raised. These findings provide useful insight into the development of advanced protein biomaterials with different secondary structures.

Renewable polymeric materials, from vegetable or plant oils, can also be used as reliable starting material to access new products with a wide array of structural and functional variations [24]. Their abundant availability and relatively low cost make them industrially attractive for the plastics industry. Vegetable oils (soybean oil, castor oil, linseed oil, etc.) have been polymerized in the presence of various fillers and fibers, such as clays, inorganic nanoparticles, hemp, flax, jute or kenaf fibers, and so forth, leading to biocomposites that show significant improvements in their mechanical properties and thermal stabilities [25–27]. Cinnamon oil is also a highly interesting additive and it can be extracted from various parts of the cinnamon plant, such as leaf, bark, flower, and root. The major compounds in leaf and bark cinnamon oil are eugenol and trans-cinnamaldehyde, respectively, which present antioxidant and antimicrobial activities and can be added to alginate-based films to fabricate active packaging materials. In this regard, Baek et al. [28] added low amounts (up to 1 wt%) of cinnamon leaf and bark oils to Ecklonia cava alginate, in the presence of $CaCl_2$, as a cross-linking agent. As the content of the oils increased, the tensile strength decreased, while the elongation at break increased. The antioxidant activities of the films with bark cinnamon were higher than those of films with leaf cinnamon. In contrast, the antimicrobial activities against Escherichia coli, Salmonella typhimurium, Staphylococcus aureus, and Listeria monocytogenes were better in the films with leaf cinnamon, corroborating that both types of films can be applied as new active packaging materials.

Another interesting biopolymer is chitosan, often obtained from the exoskeleton of crustaceans. It has very valuable properties, including biocompatibility, biodegradability, and antimicrobial activity. The production of chitosan products is difficult due to its insolubility in organic solvents, its ionic character in solution, and the formation of three-dimensional networks by strong hydrogen bonds [29]. However, a great breakthrough has been done with producing chitosan fibers. To improve fiber formation, Sanchez-Alvarado et al. [30] combined an anionic biodegradable poly(vinyl alcohol) (PVA) using the electrospinning technique. Different chitosan concentrations (0.5, 1, 2, and 3 wt%) were tested and the electrospinning parameters (syringe/collector distance, solution flow, and voltage) were optimized. Furthermore, the fibers were treated with ethanolic NaOH solution to make them chemically stable. On the other hand, the grafting of α-tocopherol succinate to the skeleton of glycol chitosan leads to an amphiphilic polymer that can form micelles suitable for the delivery of paclitaxel [31], a powerful anti-tumor drug extensively used in the clinical treatment of tumors. Micelles loaded with this anticancer agent showed good antitumor activities, in vitro and in vivo, and had advantages over commercially available formulations in terms of lower toxicity levels and a higher tolerated dose. Analogously, hydrophobic deoxycholic acid and folic acid (FA) have been used to modify chitosan, leading to another amphiphilic polymer that was a safe and effective carrier for the intravenous delivery of paclitaxel [32].

An alternative approach to produce biodegradable polymeric materials is the use of CO_2. In this sense, Chen et al. [33] copolymerized CO_2 with other monomers, propylene oxide (PO), and cyclohexene oxide (CHO) to synthesize random copolymers, di-block, and tri-block copolymers. Pyrolysis-gas chromatography/mass spectrometry and thermogravimetric analysis/infrared spectrometry techniques were applied to examine the thermal degradation behaviour of the polymers. The results showed that, in all cases, unzipping was the main degradation mechanism. The random copolymer showed a one-step decomposition with very high degradation temperatures. Hence, random copolymerization of CHO, PO, and CO_2 seems to be a better way to improve the thermal stability of poly(propylene carbonate (PPC)–cyclohexyl carbonate than block copolymerization. Blends of PPC and polyester-based TPU have also been developed via melt compounding [34] and the compatibility, thermal, and mechanical properties, as well as the toughening mechanism of the blends, have been investigated via FTIR, differential scanning calorimetry (DSC), DMA, and tensile and impact tests. For these materials, FTIR revealed strong interfacial adhesion between the polymers, which resulted in more enhanced thermal stability and mechanical properties than the individual polymers. Moreover, the blends with 20 wt% polyurethane exhibited a brittle-ductile transition.

It is also interesting to examine the potential of these biopolymer composites from an application viewpoint, considering their properties and costs, as summarized in Table 1. For instance, the incorporation of about 20 wt% of potato pulp powder to PLA offers the possibility to markedly reduce the cost of PLA-based composites for common applications, like food packaging [9]. However, those based on PHB [13], PHBV [14] or comprising vegetable oils [28] are not cost effective in such applications and would only be used in active packaging, or for biomedical purposes like tissue engineering. On the other hand, rubber [19] and potato starch-based [22] composites are an interesting and relatively cheap alternative to petroleum-based plastics. It is important to note that the international market for biopolymers/bioplastics is still in its infancy. Nevertheless, owing to increasing prices of petrochemical feed stocks for plastics, along with growing environmental considerations, would pave the way for a bright future for these materials, including their biodegradable composites. These materials would be essential to realize and maintain a sustainable productive society that produces waste materials at a rate at which they can be reabsorbed by the environment.

Table 1. Summary of biopolymer composites properties, applications, and costs.

Matrix/Filler	Production Method	Properties	Applications	Cost/Kg(€)	Ref.
PLA/PEG/Chit	Extrusion	Low stiffness/High flexibility	Bone & dental implants food packaging	3.0–4.2	[7]
PLA/Cellulose	Extrusion/injection	Improved rigidity & biodegradability	Packaging, automotive industry, building	2.7–3.1	[8]
PLA/Potato pulp	Extrusion/injection	Low stiffness & ductility, good processability	Food packaging	2.4–2.7	[9]
PLA/MgO	Solution casting	Improved stability and bioactivity	Medical implants, tissue engineering, orthopedic devices	2.8–3.3	[10]
PHB/wood sawdust fibers	Extrusion	Improved degradation in soil	Agriculture or plant nursery	5.6–7.0	[13]
PHBV/TPU/cellulose	Extrusion/injection	Balanced heat resistance, stiffness, and toughness.	Food packaging tissue engineering	8.2–9.8	[14]
Nanocellulose/CNT	Cast molding	Good electrical conductivity	Supercapacitor, sensors	2.4–11.5	[18]
Rubber/potato starch	Roller mixing	Accelerated thermal ageing	Vibration isolators, shock mounts, electrical components	1.7–1.9	[19]
Potato starch/wheat gluten	Compression molding	Improved maximum stress & extensibility	Development of bio-based plastics	0.8–1.2	[22]
Alginate/cinnamon oil	Solution casting	Good antibacterial activity	Active packaging materials	7.3–8.2	[28]
PVA/Chitosan	Electrospinning	Good chemical stability	Drug delivery food packaging	1.5–1.8	[30]
PPC/TPU	Melt compounding	Good thermal stability & stiffness	Electronicpackaging applications	4.1–5.0	[34]

What should we expect for the next years? It is clear that the field of biopolymer composites will continue growing with the incorporation of new nanofillers and the development of complex hybrid materials to be applied in a wider range of fields. For instance, a market study by Helmut Kaiser Consultancy has reported that the availability of bioplastics during the last decade has the potential to reduce the petroleum consumption for plastic by 15%–20% by 2025 [35]. The global bioplastics production capacity is set to increase from around 2.1 million tonnes in 2019 to 2.6 million tonnes in 2023. PLA and PHAs are driving this growth. The market is growing rapidly, since a large number of companies are entering it with newer innovations and applications in packaging, food services, agriculture, automotive, electronics, household appliances, and consumer goods. Europe is the largest bioplastic market, owed to limited crude oil reserves. The applications responsible for higher market growth are food and beverage packing, catering products, and bags. The research on biopolymer composites, although still in its initial stage, has shown their great potential to replace conventional composites based on petroleum derived plastics. However, some challenges should be tackled in the future, including appropriate fatigue life, improved capability forfiber forming and drawing, strict control of degradation time, longer operation life time, and comparable strength values to those of advanced composites (i.e., carbon fiber reinforced polymers, CFRPs) for high performance applications, to mention a few. In addition, to fully attain their potential at an industrial level, increased investment capital and well-developed government incentives are desired.

Acknowledgments: Ana M. Díez-Pascual wishes to knowledge the Ministerio de Economía y Competitividad (MINECO) for a "Ramón y Cajal" Research Fellowship co-financed by the EU.

Conflicts of Interest: The author declares no conflict of interest.

References

1. Yu, L.; Dean, K.; Li, L. Polymer blends and composites from renewable resources. *Prog. Polym. Sci.* **2006**, *31*, 576–602. [CrossRef]

2. Díez-Pascual, A.M.; Díez-Vicente, A.L. ZnO-Reinforced Poly(3-hydroxybutyrate-co-3-hydroxyvalerate) Bionanocomposites with Antimicrobial Function for Food Packaging. *ACS Appl. Mater. Interfaces* **2014**, *6*, 9822–9834. [CrossRef]

3. Naffakh, M.; Díez-Pascual, A.M.; Marco, C. Polymer blend nanocomposites based on poly(ʟ-lactic acid), polypropylene and WS2 inorganic nanotubes. *RSC Adv.* **2016**, *6*, 40033–40044. [CrossRef]

4. Saini, P.; Arora, M.; Kumar, M.N.V.R. Poly(lactic acid) blends in biomedical applications. *Adv. Drug Deliv. Rev.* **2016**, *107*, 47–59. [CrossRef]

5. Jonoobi, M.; Harun, J.; Mathew, A.P.; Oksman, K. Mechanical properties of cellulose nanofiber (CNF) reinforced polylactic acid (PLA) prepared by twin screw extrusion. *Compos. Sci. Technol.* **2010**, *70*, 1742–1747. [CrossRef]

6. Nakagaito, A.N.; Kanzawa, S.; Takagi, H. Polylactic Acid Reinforced with Mixed Cellulose and Chitin Nanofibers—Effect of Mixture Ratio on the Mechanical Properties of Composites. *J. Compos. Sci.* **2018**, *2*, 36. [CrossRef]

7. Coltelli, M.-B.; Cinelli, P.; Gigante, V.; Aliotta, L.; Morganti, P.; Panariello, L.; Lazzeri, A. Chitin Nanofibrils in Poly(Lactic Acid) (PLA) Nanocomposites: Dispersion and Thermo-Mechanical Properties. *Int. J. Mol. Sci.* **2019**, *20*, 504. [CrossRef]

8. Aliotta, L.; Gigante, V.; Coltelli, M.B.; Cinelli, P.; Lazzeri, A. Evaluation of Mechanical and Interfacial Properties of Bio-Composites Based on Poly(Lactic Acid) with Natural Cellulose Fibers. *Int. J. Mol. Sci.* **2019**, *20*, 960. [CrossRef]

9. Righetti, M.C.; Cinelli, P.; Mallegni, N.; Massa, C.A.; Bronco, S.; Stäbler, A.; Lazzeri, A. Thermal, Mechanical, and Rheological Properties of Biocomposites Made of Poly(lactic acid) and Potato Pulp Powder. *Int. J. Mol. Sci.* **2019**, *20*, 675. [CrossRef]

10. Zhao, Y.; Liu, B.; Bi, H.; Yang, J.; Li, W.; Liang, H.; Liang, Y.; Jia, Z.; Shi, S.; Chen, M. The Degradation Properties of MgO Whiskers/PLLA Composite In Vitro. *Int. J. Mol. Sci.* **2018**, *19*, 2740. [CrossRef] [PubMed]

11. Bugnicourt, E.; Cinelli, P.; Lazzeri, A.; Alvarez, V. Polyhydroxyalkanoate (PHA): Review of synthesis, characteristics, processing and potential applications in packaging. *Express Polym. Lett.* **2014**, *8*, 791–808. [CrossRef]

12. Díez-Pascual, A.M.; Díez-Vicente, A.L. Poly(3-hydroxybutyrate)/ZnOBionanocomposites with Improved Mechanical, Barrier and Antibacterial Properties. *Int. J. Mol. Sci.* **2014**, *15*, 10950–10973. [CrossRef] [PubMed]

13. Cinelli, P.; Seggiani, M.; Mallegni, N.; Gigante, V.; Lazzeri, A. Processability and Degradability of PHA-Based Composites in Terrestrial Environments. *Int. J. Mol. Sci.* **2019**, *20*, 284. [CrossRef]

14. Sánchez-Safont, E.L.; Arrillaga, A.; Anakabe, J.; Cabedo, L.; Gamez-Perez, J. Toughness Enhancement of PHBV/TPU/Cellulose Compounds with Reactive Additives for Compostable Injected Parts in Industrial Applications. *Int. J. Mol. Sci.* **2018**, *19*, 2102. [CrossRef] [PubMed]

15. Edwards, J.V.; Fontenot, K.; Liebner, F.; Pircher, N.D.; French, A.D.; Condon, B.D. Structure/Function Analysis of Cotton-Based Peptide-Cellulose Conjugates: Spatiotemporal/Kinetic Assessment of Protease Aerogels Compared to Nanocrystalline and Paper Cellulose. *Int. J. Mol. Sci.* **2018**, *19*, 840. [CrossRef]

16. Hoeng, F.; Denneulin, A.; Bras, J. Use of nanocellulose in printed electronics: A review. *Nanoscale* **2016**, *8*, 13131–13154. [CrossRef] [PubMed]

17. Lee, T.W.; Lee, S.E.; Jeong, Y.G. Carbon nanotube/cellulose papers with high performance in electric heating and electromagnetic interference shielding. *Compos. Sci. Technol.* **2016**, *131*, 77–87. [CrossRef]

18. Siljander, S.; Keinänen, P.; Räty, A.; Ramakrishnan, K.R.; Tuukkanen, S.; Kunnari, V.; Harlin, A.; Vuorinen, J.; Kanerva, M. Effect of Surfactant Type and Sonication Energy on the Electrical Conductivity Properties of Nanocellulose-CNT Nanocomposite Films. *Int. J. Mol. Sci.* **2018**, *19*, 1819. [CrossRef]

19. Manaila, E.; Stelescu, M.D.; Craciun, G. Degradation Studies Realized on Natural Rubber and Plasticized Potato Starch Based Eco-Composites Obtained by Peroxide Cross-Linking. *Int. J. Mol. Sci.* **2018**, *19*, 2862. [CrossRef]
20. Waglay, A.; Karboune, S.; Alli, I. Potato protein isolates: Recovery and characterization of their properties. *Food Chem.* **2014**, *142*, 373–382. [CrossRef]
21. Kowalczyk, D.; Baraniak, B. Effects of plasticizers, pH and heating of film-forming solution on the properties of pea protein isolate films. *J. Food Eng.* **2011**, *105*, 295–305. [CrossRef]
22. Muneer, F.; Johansson, E.; Hedenqvist, M.S.; Plivelic, T.S.; Kuktaite, R. Impact of pH Modification on Protein Polymerization and Structure–Function Relationships in Potato Protein and Wheat Gluten Composites. *Int. J. Mol. Sci.* **2019**, *20*, 58. [CrossRef] [PubMed]
23. Liu, Q.; Wang, F.; Gu, Z.; Ma, Q.; Hu, X. Exploring the Structural Transformation Mechanism of Chinese and Thailand Silk Fibroin Fibers and Formic-Acid Fabricated Silk Films. *Int. J. Mol. Sci.* **2018**, *19*, 3309. [CrossRef] [PubMed]
24. Lligadas, G.; Ronda, J.C.; Galià, M.; Cádiz, V. Renewable polymeric materials from vegetable oils: A perspective. *Mater. Today* **2013**, *16*, 337–343. [CrossRef]
25. Díez-Pascual, A.M.; Díez-Vicente, A.L. Wound Healing Bionanocomposites Based on Castor Oil Polymeric Films Reinforced with Chitosan-Modified ZnO Nanoparticles. *Biomacromolecules* **2015**, *16*, 2631–2644. [CrossRef] [PubMed]
26. Díez-Pascual, A.M.; Díez-Vicente, A.L. Epoxidized Soybean Oil/ZnOBiocomposites for Soft Tissue Applications: Preparation and Characterization. *ACS Appl. Mater. Interfaces* **2014**, *6*, 17277–17288. [CrossRef]
27. Díez-Pascual, A.M.; Díez-Vicente, A.L. Development of Linseed Oil/TiO$_2$ Green Nanocomposites as Antimicrobial Coatings. *J. Mater. Chem. B* **2015**, *3*, 4458–4471. [CrossRef]
28. Baek, S.-K.; Kim, S.; Song, K.B. Characterization of Ecklonia cava Alginate Films Containing Cinnamon Essential Oils. *Int. J. Mol. Sci.* **2018**, *19*, 3545. [CrossRef]
29. Homayoni, H.; Hosseini Ravandi, S.A.; Valizadeh, M. Electrospinning of chitosan nanofibers: Processing optimization. *Carbohydr. Polym.* **2009**, *77*, 656–661. [CrossRef]
30. Sanchez-Alvarado, D.I.; Guzmán-Pantoja, J.; Páramo-García, U.; Maciel-Cerda, A.; Martínez-Orozco, R.D.; Vera-Graziano, R. Morphological Study of Chitosan/Poly (Vinyl Alcohol) Nanofibers Prepared by Electrospinning, Collected on Reticulated Vitreous Carbon. *Int. J. Mol. Sci.* **2018**, *19*, 1718. [CrossRef]
31. Liang, N.; Sun, S.; Gong, X.; Li, Q.; Yan, P.; Cui, F. Polymeric Micelles Based on Modified Glycol Chitosan for Paclitaxel Delivery: Preparation, Characterization and Evaluation. *Int. J. Mol. Sci.* **2018**, *19*, 1550. [CrossRef] [PubMed]
32. Li, L.; Liang, N.; Wang, D.; Yan, P.; Kawashima, Y.; Cui, F.; Sun, S. Amphiphilic Polymeric Micelles Based on Deoxycholic Acid and Folic Acid Modified Chitosan for the Delivery of Paclitaxel. *Int. J. Mol. Sci.* **2018**, *19*, 3132. [CrossRef] [PubMed]
33. Chen, S.; Xiao, M.; Sun, L.; Meng, Y. Study on Thermal Decomposition Behaviors of Terpolymers of Carbon Dioxide, Propylene Oxide, and Cyclohexene Oxide. *Int. J. Mol. Sci.* **2018**, *19*, 3723. [CrossRef]
34. Han, D.; Chen, G.; Xiao, M.; Wang, S.; Chen, S.; Peng, X.; Meng, Y. Biodegradable and Toughened Composite of Poly(Propylene Carbonate)/Thermoplastic Polyurethane (PPC/TPU): Effect of Hydrogen Bonding. *Int. J. Mol. Sci.* **2018**, *19*, 2032. [CrossRef]
35. Farmer, N. *Trends in Packaging of Food, Beverages and Other Fast-Moving Consumer Goods*; Woodhead Publishing: Cambridge, UK, 2013; p. 255.

© 2019 by the author. Licensee MDPI, Basel, Switzerland. This article is an open access article distributed under the terms and conditions of the Creative Commons Attribution (CC BY) license (http://creativecommons.org/licenses/by/4.0/).

International Journal of
Molecular Sciences

MDPI

Article

Chitin Nanofibrils in Poly(Lactic Acid) (PLA) Nanocomposites: Dispersion and Thermo-Mechanical Properties

Maria-Beatrice Coltelli [1,2,*], **Patrizia Cinelli** [1,2], **Vito Gigante** [1,2], **Laura Aliotta** [1,2], **Pierfrancesco Morganti** [3,4], **Luca Panariello** [1,2] **and Andrea Lazzeri** [1,2,*]

[1] Department of Civil and Industrial Engineering, University of Pisa, Via Diotisalvi 2, 56122 Pisa, Italy; patrizia.cinelli@unipi.it (P.C.); vito.gigante@dici.unipi.it (V.G.); laura.aliotta@dici.unipi.it (L.A.); luca.panariello@ing.unipi.it (L.P.)

[2] National InterUniversity Consortium of Materials Science and Technology (INSTM), Via Giusti 9, 50121 Florence, Italy

[3] Skin Pharmacology and Dermatology Unit, Campania University "Luigi Vanvitelli", 80100 Naples, Italy; pierfrancesco.morganti@mavicosmetics.it

[4] MAVI SUD, Aprilia (LT), 04011 Aprilia, Italy

* Correspondence: maria.beatrice.coltelli@unipi.it (M.-B.C.); andrea.lazzeri@unipi.it (A.L.); Tel.: +39-050-2217-856 (M.-B.C.)

Received: 14 December 2018; Accepted: 18 January 2019; Published: 24 January 2019

Abstract: Chitin-nanofibrils are obtained in water suspension at low concentration, as nanoparticles normally are, to avoid their aggregation. The addition of the fibrils in molten PLA during extrusion is thus difficult and disadvantageous. In the present paper, the use of poly(ethylene glycol) (PEG) is proposed to prepare a solid pre-composite by water evaporation. The pre-composite is then added to PLA in the extruder to obtain transparent nanocomposites. The amount of PEG and chitin nanofibrils was varied in the nanocomposites to compare the reinforcement due to nanofibrils and plasticization due to the presence of PEG, as well as for extrapolating, where possible, the properties of reinforcement due to chitin nanofibrils exclusively. Thermal and morphological properties of nanocomposites were also investigated. This study concluded that chitin nanofibrils, added as reinforcing filler up to 12% by weight, do not alter the properties of the PLA based material; hence, this additive can be used in bioplastic items mainly exploiting its intrinsic anti-microbial and skin regenerating properties.

Keywords: chitin nanofibrils; poly(lactic acid); nanocomposites

---

1. Introduction

Chitin is the second most abundant biopolymer on earth, having global reserves of 100 billion tons [1]. Waste from the seafood industry is a great source of chitin, because it represents the matrix of the hierarchically structured, fiber-based composite constituting the exoskeleton of crustaceans [2]. On a global level, about 6 mega tons of crustacean shells are discarded per year [3]. For this reason, many researchers evidenced the possibility of exploiting this resource [4] to obtain valuable materials such as chitosan and its derivatives [5], chitin nanofibrils [6–8], inorganics such as calcium carbonate, and molecules such as acetic acid and pyrrole [9].

Chitin nanofibrils (CNs), consisting of colloidal nano-rods, represent the crystalline fraction of the chitin extracted from sea food waste; it was reported that they show anti-microbial properties and favor cells regeneration [10–12].

Poly(lactic acid) (PLA) is a fully renewable polymer compatible with the human body, plants and the environment, being both resorbable in the human body and biodegradable in composting plants [13]. It is reported to have biocidal activity thanks to its tendency to hydrolyze on the surface, producing lactic acid, which exerts a slight anti-microbial activity [14]. For all these reasons PLA is one of the best alternatives to petrol-based polymers in the packaging, agricultural, personal care, cosmetic, biomedical and tissue engineering sectors [15–17]. The modulation of PLA properties to make it suitable for several processing techniques and final applications is an attractive topic of current research about nanomaterials [18]. The possibility of reinforcing PLA using nanofibers was investigated by considering several nano-reinforcements [19] such as cellulose nanocrystals or nanofibers [20,21], sometimes also combined with nanofillers with different aspect ratios, like clays, to modulate barrier properties [22]. Nanocellulose did not show anti-microbial and cell regenerative properties.

In contrast, the combination of PLA and CNs to obtain bionanocomposites can represent a good opportunity for the preparation of bioplastic materials with improved structural and functional properties [23]. Moreover, it is reported that chitin nanofibers require less energy intensive treatments in their production, compared to cellulose nanofibers [24], and that the combination with cellulose nanofibers to obtained nanocomposites can be synergistic [25].

The dispersion of CNs at a nanometric level in PLA was attempted by several techniques. The preliminary modification of chitin nanofibrils was identified as a good strategy to improve its stability in suspension, as observed by Araki, that prepared sterically stabilized chitin nanowhiskers by surface grafting monomethoxy poly(ethylene glycol) (mPEG) [26]. mPEG grafting has been previously reported to effectively improve the dispersion stability of cellulose nanowhiskers [27,28]. By following this strategy based on CNs chemical modification, Zhang et al. [29] acetylated chitin nanofibrils to improve their dispersion in poly (lactic acid) (PLA), but the mechanical properties of composites prepared by solvent casting (not in the molten state) were not improved. Moreover, the chemical modification of diluted suspension can be complex, in view of an industrialization of the composite's production. On the other hand, extrusion is a conventional melt polymer-processing technique, and it would be preferentially adopted for nanocomposite processing for industrial applications. Rizvi et al. [30], working in a twin-screw laboratory extruder, used more traditional compatibilizing agents to improve the adhesion between nanofibrils and the PLA matrix, investigating the effect of different contents of chitin nano-fibrils in melt-blended PLA/chitin and PLA/CN composites. In particular, maleic anhydride (MA) was grafted to PLA at about 2% by weight, producing a modified polymer (PLA-g-MA). Tensile tests showed an effective increase in Young's modulus in nano-composites compared to pure PLA. In particular, it was observed that the Young's modulus underwent an increase of up to 5% chitin content, while higher quantities did not produce any further reinforcement effect. Herrera et al. noticed that only 1% CNs was enough to observe an increase in Young's Modulus. [31]. The tensile strength decreased by increasing the nano-chitin content, but in any case, lower values than those of pure PLA were observed. This decreasing trend was attributed to the hydrolysis of PLA during the preparation, since aqueous solutions of CNs were added in the melt mixer to avoid agglomeration, and therefore, as the amount of reinforcement used increased, the introduced moisture increased accordingly, leading to an increasing impact of hydrolysis. This paper showed the good potential of CNs to improve mechanical performance, but at the same time, the difficulty of dispersing them at the nano-metric level and homogeneously in the PLA matrix. Guan and Naguib [32] investigated PLA/CN nano-composites using MA as a compatibilizing agent and N,N-dimethylacetamide (DMAc) as a dispersing agent. The latter was added to improve the dispersion of nano-fibrils in the composite. The CNs were re-dispersed in DMAc by mechanical stirring to prepare a nano-structured suspension. Moreover, PLA was prepared with grafted anhydride groups (PLA-g-MA) using 2 wt% of MA and 0.5 wt% peroxide as a radical initiator to improve the fiber-matrix adhesion. Tensile test results demonstrated that the presence of both DMAc and PLA-g-MA, without CNs, led to a decrease of the Young's modulus and of tensile strength; however, DMAc caused a significant worsening of the mechanical properties, explained by the authors by considering the degradation generated in the PLA through chain scission during the extrusion. It is also important to note that as the amount of nano-filler increased, two contrasting effects occurred: a reinforcing effect of chitin, which increased stiffness and strength at the expense of ductility; and a negative impact on the mechanical properties which was attributable to the presence of DMAc, which caused the degradation of the matrix. This article evidenced the need to use a dispersing agent, looking for one that does not negatively impact the mechanical performance of PLA. The use of known plasticizers of PLA in PLA-based nanocomposites has been studied by Herrera et al. [31,33,34], who developed triethyl citrate (TEC) in water/alcohol suspensions, fed in the extruder, and investigated the properties of composites containing 1% of CNs. They reported that this amount was enough to modify some key properties, such as anti-fungal activity and antistatic behavior puncture strength, and to improve mechanical properties thanks to the very high surface-to-volume ratio of the nanofibrils. Interestingly, the authors investigated the thermal properties of nanocomposites by DSC, and concluded that the simultaneous presence of CNs and TEC in the explored composition range did not provoke significant changes, except for the evident decrease of the glass transition due to the plasticizing effect of TEC. The effect of the plasticizer content to determine the synergic effect of the plasticizer as a dispersing and toughening aid with a minimum impact on the properties of PLA was considered an interesting topic to further investigate by the authors [31]. This molecule, as well as other citrates, due to the presence of ester groups, showed a very high affinity for PLA matrix [35–38], thus strongly favoring the nanodispersion of CNs in the matrix and the formation of an extended CN/PLA interface.

Poly(ethylene glycol) (PEG) was also successfully used to disperse CNs [39–41] in PLA. More recently, Li et al. [42], aiming at obtaining rigid nanocomposites, used high molecular weight poly(ethylene oxide) (PEO) or PEG and investigated the flexural and impact properties of composites with NC in the range 10–40%, observing a reinforcing effect of CN.

In all these studies, the exigence of effectively dispersing the CNs in PLA led to the necessity of using both a plasticizer and chitin nanofibrils.

Interestingly, Nakagaito et al. [43] used only water as a dispersion medium, which was removed by filtration and drying. Although PLA is insoluble in water, aqueous suspensions can be obtained by using PLA short fibers or particles, that can be easily mixed with cellulose nanofibers in water suspension. After dewatering, the mixture forms paper-like sheets that can be laminated and compression molded. Hence, nanocomposites were obtained by compression molding the filtrates. Static tensile test and dynamic mechanical analysis were performed to evaluate the reinforcement as a function of nanofiber content. Chitin nanofibers delivered reinforcement similar to that of cellulose nanofibers, being especially effective at up to 70 wt% fiber load. The ultimate tensile modulus and strength reached 7.7 GPa and 110 MPa, respectively, at a nanofiber content of 70 wt%. This work evidenced the interesting potential of CNs as a reinforcing agent in a pure PLA matrix. This methodology, however, cannot claim to reach full nanoscaled homogeneity because of the anisotropy flows typical of compression molding process. Subsequently, Li [42] compared PEG and PEO dispersion with this methodology, using a laboratory twin screw extruder. In this case, the Modulus measured by DMTA resultedin 7.6 GPa for the water method (a value in agreement with the one reported by Nakagaito et al. [43]), and 6,5 and 6,0 GPa for PEG and PEO respectively. The lower value obtained by using PEG and PEO can be attributed to their plasticizing effect.

Interestingly, very recently, Shanshina et al. [44] used an ionic liquid-based approach to co-dissolve PLA and CNs and produce PLA fibers containing up to 27% by weight of CNs. The fibers showed improved strength with respect to the fibers obtained by pure PLA. However, processes considering ionic liquid technology, which may be promising for future applications in several fields, are not yet well diffused in the industry.

By an analysis of the literature, it is evident that the combination of reinforcement with the necessity of using dispersing agents is an interesting topic of current research. The use of PEG resulted an effective overall limit of CN agglomeration, but a systematic study about the possibility of modulating nanocomposite properties, considering low amount of CNs, was never carried out. Hence, in this attempt, after the preparation of nanocomposites, their tensile and thermal properties as a function of the content of CNs and PEG were measured and discussed.

Int. J. Mol. Sci. **2019**, *20*, 504

2. Results

2.1. Dispersion of Chitin Nanofibrils in Plasticized Pla

In order to study chitin nanofibrils morphology, water suspensions at 2% by weight of chitin nanofibrils were diluted 1:100, and then one drop of diluted suspension was deposited onto a glass window. By using a field emission scanning electron microscope (FESEM), it was possible to determine the shape of chitin nanofibrils (Figure 1): average length of 20 µm and an average width of 90 nm.

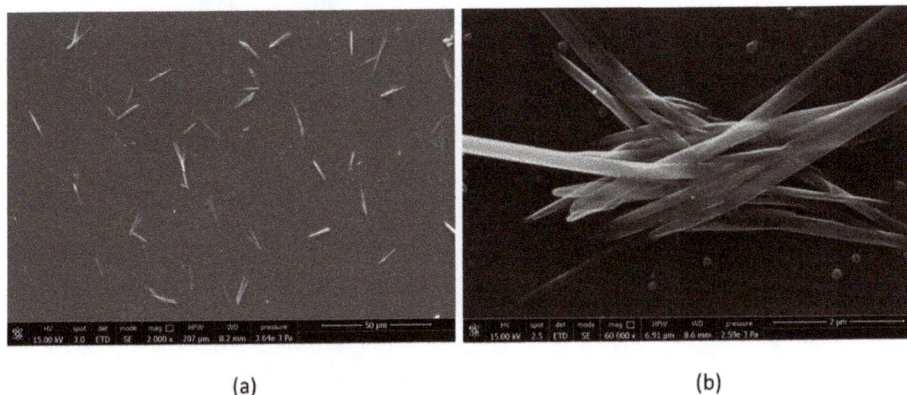

(a)   (b)

Figure 1. FESEM (field emission scanning electron microscopy) micrographs obtained from 2% by weight water suspension of CN diluted 1:100 and deposited on glass. (**a**) 2000× magnification; (**b**) 60,000× magnification.

When the 2% by weight suspension was dried without any previous dilution, and characterized by Scanning Electron Microscopy (SEM), the chitin nanofibrils formed agglomerates (Figure 2). In Figure 2a, it is evident that the drying of nanosized chitin produced large flakes in which the fibers are agglomerated to form a compact structure like that of a sheet of paper. In fact, in between different nanofibrils, having a very high surface to volume ratio, the formation of a high number of hydrogen bonds is thermodynamically favored, and this reasonably results in this compact structure. The structure of a nano-fibrous disordered assembly is clearly observable at the edges of the flakes (Figure 2b).

With a concentration of 1:10 of the CN suspension, PEGs having 400, 1500, 4000, 6000 and 8000 as molecular weight were added in weight ratio 1:1 to CN to obtain five different pre-composites that were dried. The obtained materials were solid, except for the one obtained with PEG 400, that showed a pasty consistency. The solid pre-composites were characterized by SEM and showed, from a morphological point of view, a complex fibrous nano-structure (Figure 2c–f), indicating that PEG constituted the matrix in which chitin nano-fibrils were immersed. Hence, it seemed reasonable that a PEG polymer was present in between the CNs, thus avoiding the formation of compact agglomerates.

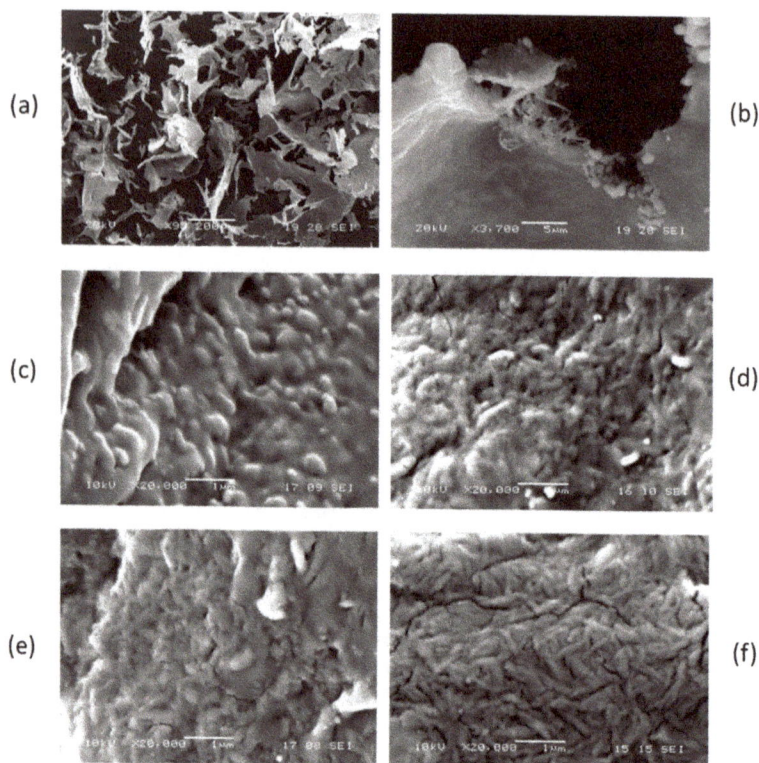

Figure 2. SEM micrographs of (**a**) dried CN suspension flakes; (**b**) magnification of flake edge; (**c**) pre-composite based on PEG 1500; (**d**) pre-composite based on PEG 4000; (**e**) pre-composite based on PEG 6000; (**f**) pre-composite based on PEG 8000.

With the purpose of investigating the interactions between PEG and chitin nanofibrils, some specific infrared characterizations and thermogravimetric tests were performed on the sample obtained by adding to chitin nanofibrils 2% of PEG 8000. The highest molecular weight was selected because, with respect to the same quantity of lower molecular weight PEG, it corresponded to a lower number of macromolecules, and consequently, demonstrated less efficient interactions with the CN. Thus, if the PEG 8000 is able to interact with the CNs, this effect will be stronger in samples with a lower molecular weight.

Regarding the infrared spectrum, CN showed characteristic amide I and Amide II bands at 1618 and 1550 cm^{-1} respectively. The Amide I band is typical of α-chitin [45], as it is split into two components at 1660 and 1630 cm^{-1}. This double band was attributed to the influence of hydrogen bonding or the presence of an enol form of the amide moiety [46–48]). Interestingly, it was found that the infrared bands typical of chitin resulted in the spectrum of the sample containing PEG8000, despite of only 2% of PEG8000 being present. This is particularly evident in the spectrum part where PEG 8000 (Figure 3) did not show any absorption bands, such as in the region 500–800 cm^{-1} and in the region 1500–1800 cm^{-1}. The presence of PEG, that probably at least partially interposes between CNs, thus favors the hypothesis of more complex interactions in between CNs.

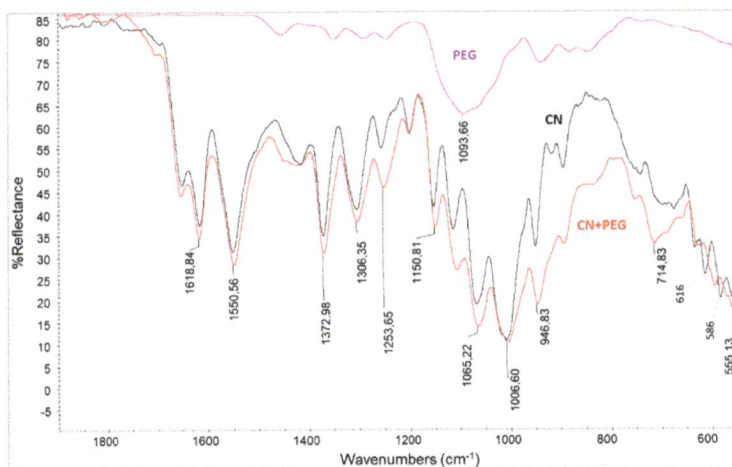

Figure 3. infrared ATR spectra of PEG 8000, CN and mixture CN/PEG8000 98:2. The spectrum of pure PEG is reported with a reduced reflectance intensity to allow a better visualization of the spectra.

The thermo-gravimetric analysis of pure CN in nitrogen flow showed that its main degradation step is at 349 °C (−21.2% by weight), and a second evident mass loss at 394 °C (−9.5%) is present. These values were calculated through the analysis of the derivative curve of the thermogram. A slight loss of mass can be observed also below 140 °C, but it accounts for a reduced loss of mass of the pure chitin (−2.6%). The mass loss observed below 100 °C was attributed to removal of humidity (−4%). The final residue was 58.1%. In the presence of 2% of PEG the thermogravimetry trend was completely changed (Figure 4). The main mass loss at 329 °C was −57.5%, and the final residue presented as 17.4% by weight. Hence, the presence of PEG induced a more efficient thermal degradation of CN, probably because it interposes between the CNs, avoiding the formation of compact agglomerations. This type of assembly is instead typical of pure CN, where inter-fibrils hydrogen bonding is predominant, resulting in the formation of a considerable quantity of carbonaceous residue after thermal degradation in nitrogen atmosphere. Similar results were obtained by Cheng et al., that considered the addition of PEG 1000 to cellulose nanofibrils [49].

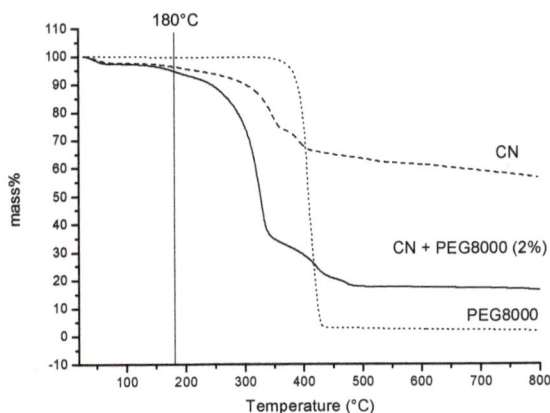

Figure 4. thermogravimetric curves of CN, PEG and CN + PEG (2%). The adopted extrusion temperature is marked with a dash line.

As the morphology of the different composites was very similar, only PEG400 liquid and PEG8000 solid (having the lowest and highest molecular weights respectively) -based samples were considered for preparing PLA based composites in a mini-extruder. The composition of the different composites, reported in Table 1, was selected by considering not only the two different molecular weights, but also with the aim of investigating the composite's properties as a function of PEG concentration and chitin nanofibrils concentration. As in the pre-composite, the weight ratio of PEG and CN was 1:1; for this reason, some PEG was added to obtain the desired content. For the P1Low2NC, a specific pre-composite was prepared with a PEG: CN weight ratio of 1:2.

Table 1. Composition of composites obtained by PLA, chitin nano-fibrils and PEG.

Samples	PEG (%)	MW of PEG	CN
P	—	—	—
P2CN* [a]	—	—	2
P10low	10	400	0
P10low2CN	10	400	2
P10high2CN	10	8000	2
P1low2CN	1	400	2
P5low2CN	5	400	2
P10low5CN	10	400	5
P10low12CN	10	400	12

[a] P2CN* is a sample of Chitin nanofibrils powder obtained by simply drying the CN suspension, without using PEG.

The extrusion was followed by the injection molding of specimens, that resulted in a transparent com pound, which was in agreement with the achievement of a nano-scaled dispersion [50,51]. As shown in Figure 5, P10low2CN and P10high 2CN resulted in a transparent compound, and also P5low2CN, but with the presence of some visible particles.

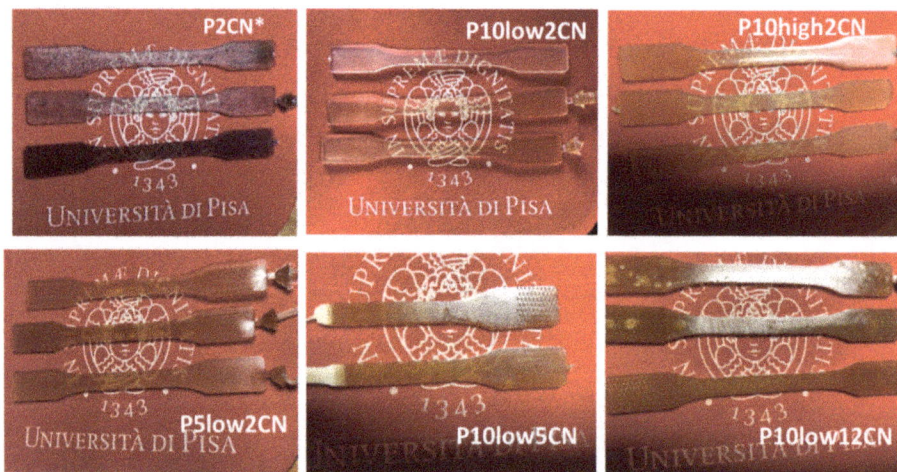

Figure 5. specimens obtained by injection moulding: comparison regarding transparency.

P10low5CN resulted in a transparent but brownish compound, whereas the sample containing 12% of CN (P10low12CN), resulted in a brown compound with a reduced transparency. In general, high transparency and colorlessness was achieved by decreasing the content of CN and increasing the content of PEG. In fact, both these characteristics positively affect the CN nano dispersion. The role of PEG as a dispersing agent is evident. In fact, the P2CN* composite was prepared by adding dried chitin nanofibrils without using PEG, and in this case, the resulting specimens were darker and not transparent because of the presence of visible particles. In this case, the CN formed aggregates that could not be well dispersed in the melt PLA during extrusion.

The samples were then characterized by SEM microscopy by preparing cryofractured surfaces from tensile specimens. From the micrographs in Figure 6, it is possible to see that the PLA containing only PEG400 (P10low sample) present a very high homogeneity apart from the presence of some round domains attributable to the presence of a second phase of PEG. The presence of very big aggregates, i.e., with diameters higher than 10 microns, could be observed in the sample obtained without preparing a pre-composite (P2CN*). In other composites where the chitin nanofibrils were dispersed by the addition of the pre-composite, it was not possible to reveal the presence of chitin nanofibrils aggregates, although several cryofractured surfaces were examined. The micrographs results were similar to those of the sample obtained in absence chitin nanofibrils, with submicrometric droplets being attributable to PEG. This result indicated a very good dispersion of chitin nanofibrils in the composite.

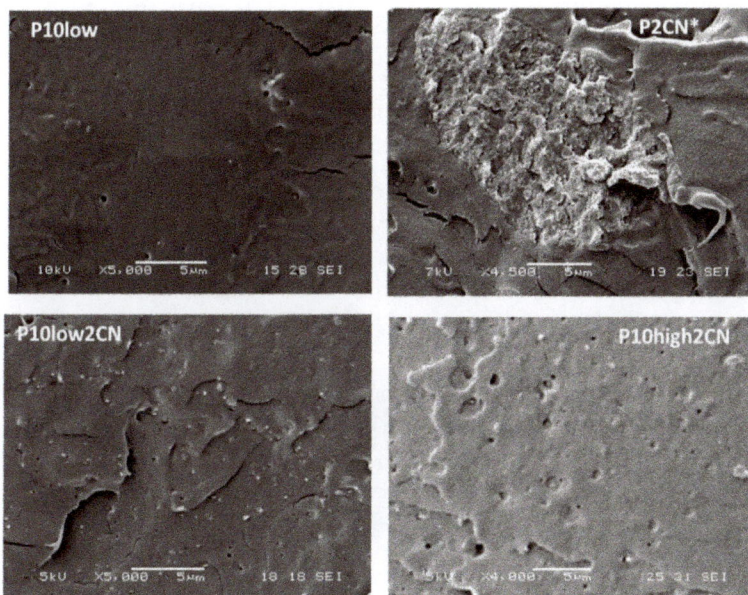

Figure 6. SEM micrographs obtained on cryofractured surfaces obtained from tensile specimens of different PLA/Chitin nanofibrils composites.

Interestingly, it was found that the dimensions of the PEG domains seemed to decrease by increasing the content of CN (Figure 7), reasonably indicating not negligible interactions between chitin nanofibrils and PEG 400, leading to a better dispersion of PEG. The interactions may be responsible for this decrease, considering that chitin nanofibrils are present both in the plasticized PLA matrix and in the PEG domains, thus acting as interfacial stabilizers. Moreover, micrometric PEG-based domains consist, during the melt extrusion of PLA nanocomposites, of liquid pools with a viscosity lower than that of the matrix, and these pools can contain CNs. The viscosity of such a liquid is increased because of the presence of chitin nanofibrils. In fact, in general, the viscosity of a suspension increases as a function of nanofibrils content [52,53], because of the tendency of gelling of the nanofibrils. The dimension of domains in immiscible polymer blends is reported to be dependent on the viscosity ratio, and in general it decreases when the ratio of the viscosity of the matrix and the dispersed phase is close to 1 [54,55]. The decrease of the PEG domain dimensions can be thus additionally ascribed to the viscosity variations in the two phases due to the presence in both of PEG and chitin nanofibrils.

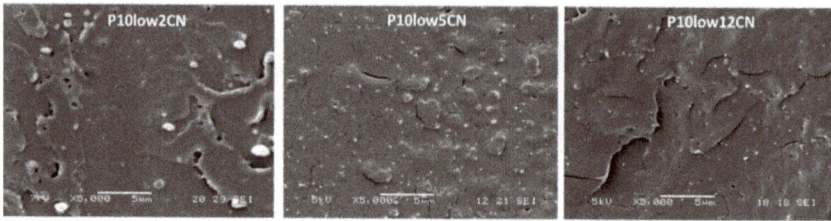

Figure 7. SEM micrographs obtained on cryofractured surfaces obtained from tensile specimens of different PLA/Chitin nanofibrils composites. From left to right the content of CN increases.

The composites P10low2CN was characterized by STEM microscopy (Figure 8) that revealed the presence of CNs as single fibers and bundles, dispersed at the nanometric scale. A good correspondence with the dimensions observed in FESEM micrographs of CNs (Figure 1) is evident.

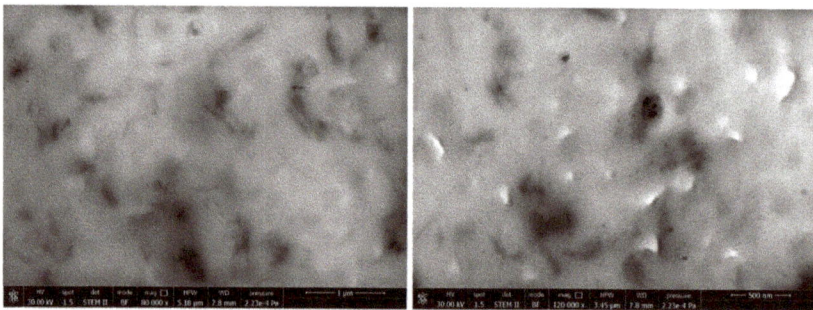

Figure 8. TEM micrographs obtained on the P10low2CN sample at different magnifications.

Infrared ATR spectroscopy was applied (Figure 9) to investigate the distribution of chitin nanofibrils on the surface of injection molded specimens. The ATR technique allowed us to record the vibrational spectrum of a material present on a surface at up to a few microns. In particular, the Smart itX ATR diamond plate allows a depth of penetration 2.03 micrometers at 1000 cm^{-1}. By overlaying the spectra of P10low, as a reference, with the spectra of the samples containing increasing amounts of CNs (2%, 5% and 12% by weight), it was found that some bands attributable to CNs are revealed. These bands are extremely weak in P10low2CN sample, whereas they are well evident in P10low5CN and P10low12CN. It is thus evident that a significant number of CNs are present on the specimen surface when the content of CNs is above 5% by weight. If the content is lower, the presence of CNs can not be revealed by this technique.

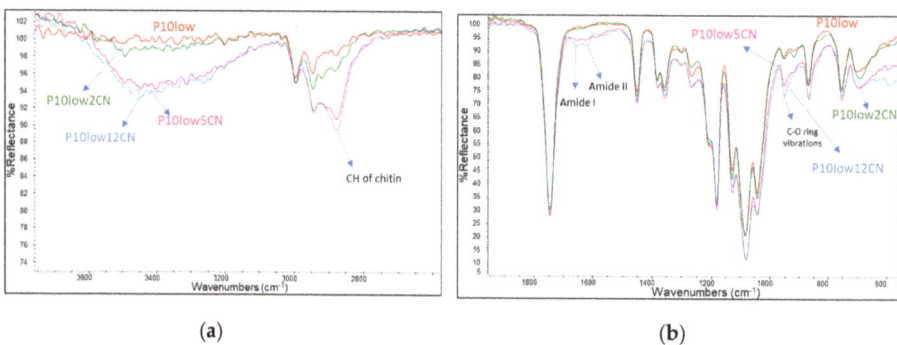

(a)

(b)

Figure 9. (a,b) Infrared ATR spectra of P10low, P10low2CN, P10low5CN and P10low12CN.

2.2. Tensile Properties

Tensile tests were performed on specimens produced by injection molding and conditioned for two weeks at 50% as relative humidity. The pure PLA is brittle and shows a high value of Young's Modulus (3.5 GPa), a high value of stress at break, but a low value of elongation at break (Table 2). When the PLA is plasticized using PEG 400 (trial P10low), a strong decrease in Young's Modulus and stress at break and an increase in elongation at break (up to 180%) were observed, in agreement with an increased ductility of the material. Interestingly, the addition of CN to plasticized PLA resulted in a decrease in Young's Modulus and stress at break, whereas the elongation at break was not significantly affected by the presence of CNs, also when their content was increased in the composites. Interestingly, the addition of PEG 400 and PEG 8000 resulted in similar properties, with the latter composite showing a higher Young's Modulus and stress at break than the former.

Table 2. Tensile properties of the PLA nanocomposites: E is the Young's Modulus, σ_y is the stress at yield, σ_b is the stress at break and ε_b is the elongation at break.

Samples	E (GPa)	σ_y (MPa)	σ_b (MPa)	ε_b (%)
P	3.5 ± 0.1	60.4 ± 0.3	57 ± 1	4.1 ± 0.5
P2CN* [a]	2.9 ± 0.1	–	58 ± 2	2.3 ± 0.4
P10low	2.3 ± 0.3	26 ± 0.3	33 ± 2	180 ± 10
P10low2CN	1.8 ± 0.3	23 ± 5	32 ± 2	160 ± 10
P10high2CN	2.5 ± 0.1	45 ± 5	34 ± 2	160 ± 10
P1low2CN	3.2 ± 0.8	52 ± 6	40 ± 7	10 ± 2
P5low2CN	2.8 ± 0.8	47 ± 3	32 ± 2	11.4 ± 0.9
P10low5CN	1.8 ± 0.3	34 ± 2	23 ± 2	160 ± 10
P10low12CN	1.7 ± 0.3	23 ± 5	22 ± 6	181 ± 6

[a] CN* is a sample of Chitin nanofibrils powder obtained by simply drying the CN suspension, without using PEG.

2.3. Thermal Properties

Thermal properties were recorded on nanocomposite injection molded specimens after a few months from preparation following a methodology consisting of a first heating step, cooling, and a second heating. Regarding the first heating (Table 3), it showed a glass transition T_g with an evident peak of enthalpy relaxation due to the ordering of PLA chains during the specimen storing period. The presence of PEG determined the decrease in the glass transition temperature and the decrease of the cold crystallization temperature. The cold crystallization temperature T_{cc} did not significantly change as a function of CN content, whereas the crystallinity significantly changed (Table 3).

Table 3. DSC results related to chitin nanofibrils PLA based nanocomposites (first heating).

	T_g (°C)	T_{cc} (°C)	ΔH_{cc} (J/g)	T_m (°C)	ΔH_m (J/g)	X_c (%)
P	57.7	106.9	21.24	149.2/157.4	21.4	0.2
P2CN*	57.5	107.7	23.21	148.8/155.9	25.5	2.4
P10low	42.1	77.6	17.8	154.7	27.1	9.9
P10low2CN	40.0	74.4	18.3	152.1	26.6	7.8
P10high2CN	43.8	76.6	6.7	154.7	33.3	28.4
P1low2CN	54.9	100.9	18.6	147.9/157.9	23.7	5.5
P5low2CN	47.7	88.6	15.6	(142.3)/155.9	19.2	3.8
P10low5CN	42.8	74.1	15.5	154.2	29.0	14.5
P10low12CN	40.8	75.5	11.4	152.3	29.1	19.0

T_g = glass transition temperature; T_{cc} = crystallization during heating peak temperature; ΔH_{cc} = enthalpy of crystallization during heating; T_m = melting peak temperature; ΔH_m = integral of the melting peak.

Regarding the second heating (Table 4), it is evident that the cold crystallization temperature significantly decreased as a function of CN content, in agreement with a slight nucleating action of the

CNs. This effect is evident only in the second heating thanks to the lower content of crystallinity X_c developed during the controlled cooling step.

Table 4. DSC results related to chitin nanofibrils PLA based nanocomposites (second heating).

	T_g (°C)	T_{cc} (°C)	ΔH_{cc} (J/g)	T_m (°C)	ΔH_m (J/g)	X_c (%)
P	59.1	109.4	23.8	148.3/157.4	26.3	2.8
P2CN*	55.2	108.8	23.3	147.9/155.8	24.5	1.3
P10low	36.1	83.8	21.3	153.7	32.6	12.1
P10low2CN	32.7	79.8	17.1	152.1	29.2	13.0
P10high2CN	31.4	80.5	17.5	154.2	29.3	12.7
P1low2CN	55.2	104.6	22.2	147.2/156.7	30.2	8.3
P5low2CN	45.3	92.4	24.6	141.4/157.4	27.2	2.7
P10low5CN	36.3	78.8	19.1	153.6	31.1	12.8
P10low12CN	29.8	73.3	20.4	151.6	28.8	9.0

T_g = glass transition temperature; T_{cc} = crystallization during heating peak temperature; ΔH_{cc} = enthalpy of crystallization during heating; T_m = melting peak temperature; ΔH_m = integral of the melting peak.

3. Discussion

The obtained results regarding phase morphology characterization, as well as the analysis of properties, agree with the achievement of a nanoscaled dispersion of chitin nanofibrils in plasticized PLA. The SEM characterization (Figures 6 and 7) and the observation of the optical properties of the injection molded specimens (Figure 5) evidenced this achievement and showed that it is dependent on the CN content and PEG content in the composites.

Thanks to infrared ATR characterization analysis, it was possible to show the clear presence of the CN bands on the injection molded specimens surface in samples containing more than 5% by weight of CN. Conditions for dispersing CNs in the bulk that resulted in the effective presence of CNs on the surface of injection molded specimens, for potentially exploiting their functional anti-microbial properties, were thus evidenced. It is interesting to note that, in the case where injection molded products are in a hot and humid environment (e.g., in applications related to human body, like implants or surgical suture wires), the surface of the PLA can degrade by hydrolysis, leaving the chitin nanofibrils to emerge on the surface. This behavior allows the preservation over time of the functional characteristics of CNs on the surface, even in the case of slow degradation of the PLA.

The thermal properties of the nanocomposites were analyzed as a function of CN and PEG 400 content (Figure 10). The trend of the glass transition as a function of the CN content is reported in Figure 10a. The values are almost constant but higher for the first heating than for the second. This difference can be ascribed to the formation of ordered structure in the sample injection molded and stored for some months (first heating), and is also responsible of the presence of the enthalpy relaxation peak in the glass transition range. Interestingly the glass transition temperature (T_g) showed an almost linear trend both in the first and second heating as a function of the PEG content (Figure 10b). The highest slope for the second heating trend can be explained by considering that in the first heating, the enthalpy relation made the samples less sensitive to plasticizer content. The linear fitting of the two trends allowed us to determine the intercept value of the line, corresponding to the glass transition temperature of nanocomposites without PEG. The values of extrapolated T_gs for pure PLA containing 2% of CN of 56.3 and 57.7 °C for first and second heating respectively. Interestingly the values recorded for the P2CN* sample were 57,5 and 55,2 °C for the first and second heatings, respectively (Tables 3 and 4). In the first heating, due to the uncontrolled storing conditions, the difference of about 1 °C seems to be irrelevant. In the second heating, recorded after a controlled cooling step, the increase of 2.5 °C is significant, and can be ascribed to the better dispersion achieved in the sample obtained by dispersing CN by using PEG, resulting in a better interaction between CNs and PLA matrix, than in the sample P2CN*, where CNs were present in agglomerated micrometric particles (Figure 6).

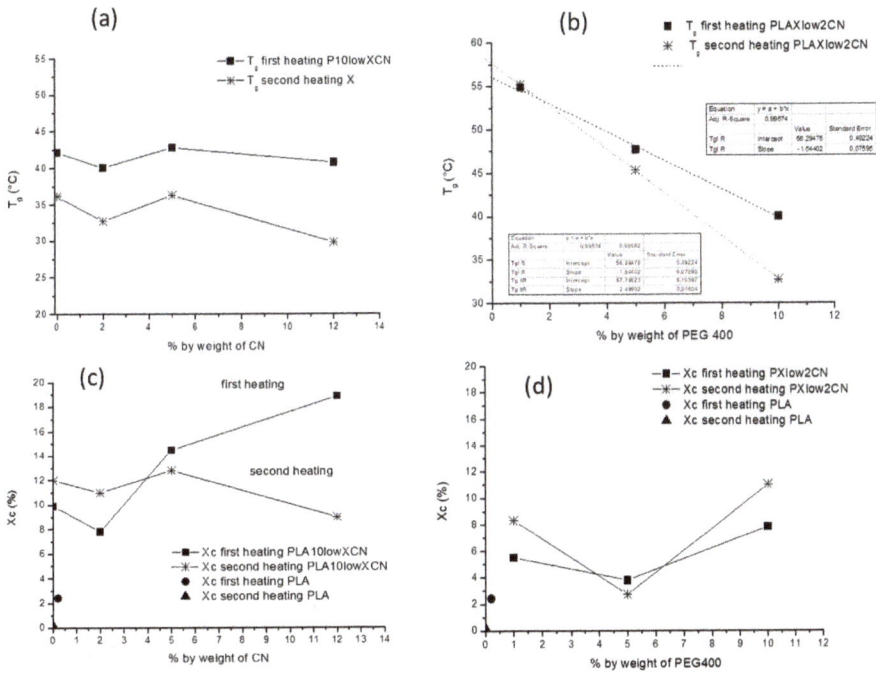

Figure 10. Analysis of thermal properties from first and second heating steps in PLA/Cn nanocomposites: (**a**) trend of Tg as a function of % by weight of CN; (**b**) trend of Tg as a function of PEG 400 content and dashed lines to extrapolate Tg at 0% by weight of PEG; (**c**) trend of crystallinity as function of % by weight of CN; (**d**) trend of crystallinity as a function of % by weight of PEG 400.

The crystallinity X_c was almost constant as a function of CN content in the second heating, showing an insignificant effect of CNs on controlled crystallization. In contrast, the crystallinity was significantly increased when the content of CN was 5% and 12% by weight. As the cooling of the injection molding process was very fast, this difference can be ascribed to the crystallinity developed during the storing of specimens. Hence, the presence of CNs in amounts higher than 2% resulted in an increase of crystallinity in the injection molding specimens during the storage that resulted in shrinkage or slow plasticizer expulsion. The surface of P10low5CN and P10low12CN became oily after some months from their preparation, whereas the P10low2CN specimens did not show any surface oiliness. This evidence may be relevant in view of the application of these nanocomposites to the injection molding sector.

The crystallinity as a function of % by weight of PEG400 for the nanocomposites at 2% by weight of CN showed a minimum value for both first and second heating at 5% by weight of PEG. The trend is like the one observed for plasticized and nucleated PLA by Fehri et al. [35]. When the concentration of the plasticizer is low (up to 5%), it hinders the crystal growth with respect to pure PLA. In contrast, when the concentration is high, the main effect of the plasticizer is to provide a higher free volume for segments motions, allowing the chain fragments to assemble more easily in crystals.

Based on the slight changes in properties observed in cases of relatively fast cooling conditions, the effect of thermal properties on tensile properties, performed a few days after the preparation of specimens by injection molding, can be considered almost negligible, in agreement with the studies carried out by Herrera et al. [31].

The stress strain curves recorded for the nanocomposites showed a trend like the one reported in Figure 10a as an example, where the stress at break resulted higher than the stress at yield. In Figure 11, a comparison between PEG 400 and PEG8000 regarding the tensile properties of the samples is proposed. The Young's Modulus (Figure 11b) decreased by adding the plasticizer alone, but the addition of CN resulted in a further decrease. Hence, the CNs, in the presence of 10% by weight of plasticizer, did not show a clear reinforcing action; this was also the case using PEG400 and PEG8000. The elongation at break (Figure 11e) increased by up to 180% by adding PEG 400 to PLA, and the presence of 2% by weight of nano-dispersed CNs did not result in a decrease in elongation at break both for PEG400 and PEG8000. The stress at break resulted in similar behavior for nanocomposites obtained with PEG400 and PEG8000, whereas the stress at yield resulted higher for PEG8000 than for PEG400. This can be ascribed to the higher mobility allowed in the system by the PEG400, with the lower molecular weight, allowing for easier sliding of macromolecules in correspondence with the beginning of the yield.

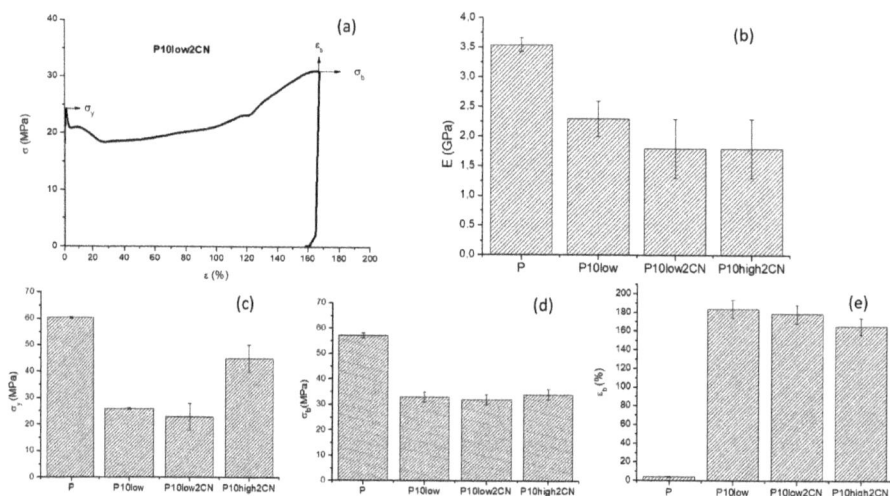

Figure 11. Tensile results to compare the use of PEG with high and low molecular weight: (**a**) example of stress vs. strain curve of the ductile composites; (**b**) Young's Modulus E; (**c**) Stress at yield σ_y; (**d**) stress at break σ_b; (**e**) elongation at break ε_b. Standard deviation is reported as error bars.

The tensile properties were also investigated by considering the trends of the different measurements as a function of chitin nanofibrils content and PEG content.

Regarding the chitin nanofibrils content investigated in the composite at 10% by weight of PEG400, the elastic Modulus slightly decreased by adding 2% by weight of CN, and remained almost constant at up to 12% by weight (Figure 12a). Stress at yield showed a maximum for the composite at 5% by weight of CN and the stress at break decreased as a function of CN content, showing the highest decrease, i.e., between 2% and 5% by weight, of CN. The nanocomposite at 5% seemed to be the most rigid; this was confirmed by the elongation at break data, showing a minimum for this composite. The variations of elongation at break are much limited in any case, which is in agreement with the good dispersion resulting from morphologic analysis.

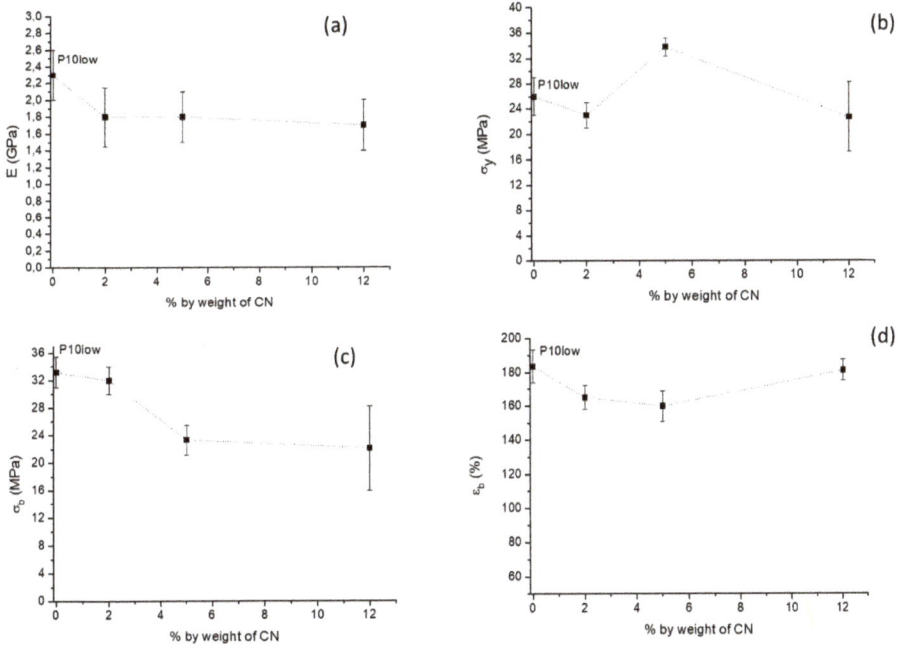

Figure 12. Tensile results as a function of CN content at constant PEG 400 concentration (10% by weight): (**a**) Young's Modulus E; (**b**) Stress at yield σ_y; (**c**) stress at break σ_b; (**d**) elongation at break ε_b.

The stress at yield significantly decreased as a function of PEG content (Figure 13b) because of the increased mobility due to plasticization contributing to decrease the energy required for the sliding of macromolecules with respect to the other. The stress at break slightly decreased as a function of PEG content because of a plasticization effect, decreasing the energy necessary for sample deformation. The elongation at break largely increased in between 5% and 10% by weight of PEG (Figure 13c). Only the nanocomposites obtained with a PEG content of 10% could reach elongation at break values above 150%. The Young's Modulus as a function of PEG content showed a decreasing linear trend (Figure 13a). The linear fitting allowed us to determine the value of the Modulus for the composite consisting of PLA and CN (Figure 13a) as the intercept of the obtained line. The obtained value by this methodology was 3.45 GPa, representing the Modulus of the PLA/CN 2% nanocomposite. It should be noted that the value obtained for the P2CN* sample presenting micrometric agglomerates of CN (Figure 6) was 2.9 GPa. The increase of 0,55 GPa can be reasonably attributed to the improved dispersion in the PLA matrix achieved thanks to the use of PEG 400. The increased interfacial surface between CN and PLA, favoring matrix-filler interactions, accounts for this improvement. It is important to note that the obtained value results were lower than the value determined for pure PLA (3.5 GPa). Hence the CNs could not reinforce the PLA matrix, either when micrometrically dispersed (P2CN*) or when nanometrically dispersed (extrapolated intercept value). These results are different from those obtained by Herrera et al. [31], who noticed an increase of Young's Modulus when only 1% of CNs was added. This effect on the mechanical properties can be attributed to the low affinity of the dispersing agent for the PLA matrix, TEC, used by these authors. This molecule, as well as other citrates, due to the presence of ester groups, showed a very high affinity for the PLA matrix [35–37], thus strongly favoring the nanodispersion of CN in the matrix and the formation of an extended CN/PLA interface. In contrast, PEG, having a high affinity for the polar groups of chitin, can support its nanodispersion, but can be more difficultly removed from CNs, remaining in between the CN surface and the PLA matrix. The PEG can thus coat the CN surface, lowering the reinforcing effect of nanofibers.

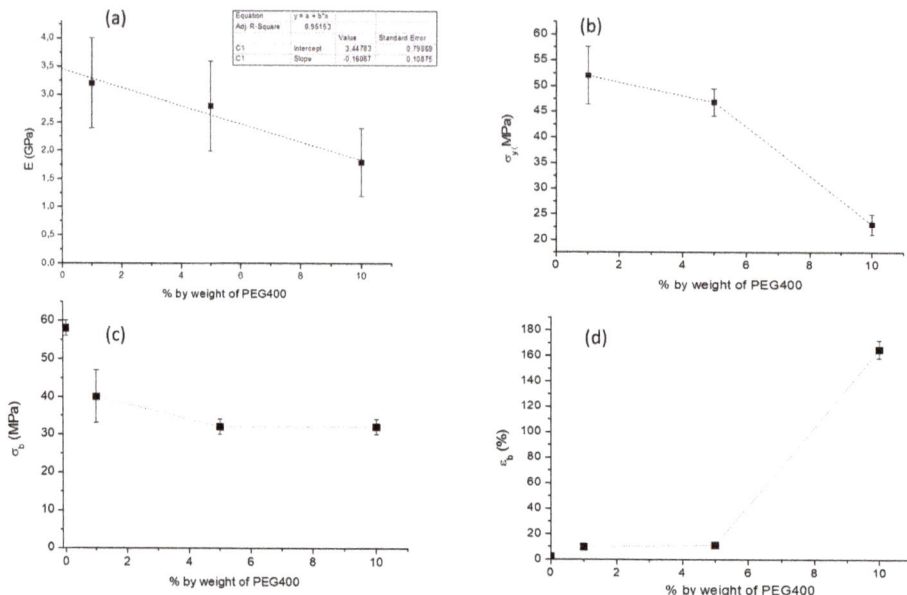

Figure 13. Tensile results as a function of PEG400 content at constant CN concentration (2% by weight): (a) Young's Modulus E; (b) Stress at yield σ_y; (c) stress at break σ_b; (d) elongation at break ε_b.

In literature, some samples prepared by the PEG method were studied, and it was noticed that some short fiber clusters are entangled with PLA matrix with a not evident phase separation between CNs and PLA. This observation revealed that PEG is a good interfacial compatibilizer for CNs and PLA. However, this reinforcing effect may be weakened due to a decrement of the CN aspect ratio. These aspects would require further research to better understand the effect of structural and morphologic parameters on the final properties.

4. Materials and Methods

4.1. Materials

Chitin nanofibrils (CN) water suspension at a concentration of 2% wt. was produced by MAVI SUD through its patented process [56], starting from chitin coming from seafood waste. For the preparation of the pre-composites, it was concentrated at 20% by weight.

Poly(ethylene glycol) (PEG), a liquid having a molecular weight of 400 (low), and PEG, a solid having a molecular weight of 1500, 4000, 6000 and 8000 (high), were purchased from Aldrich and used without any further purification.

PLA Ingeo™ 2003D, Extrusion Grade with density of 1.24 g/cm³, a melt index of 6 g/10 min at 210 °C and 2.16 Kg, produced by NatureWorks LLC. It has a molecular weight of 170,000 g/mol and contains up to 4.1% isomeric D units. It was dried in ventilated oven at 60 °C for 16 h before extrusion trials.

4.2. Materials Preparation

PEG400 (or PEG800) were added to concentrated chitin nanofibrils suspension and stirred for two hours at room temperature. The amount was calculated considering that in the final pre-composites, the weight ratio of CN and PEG was 1:1. The obtained semiliquid emulsion was dried in a ventilated oven at 50 °C up to constant weight to obtain a solid when PEG8000, PEG6000, PEG4000 and PEG1500 were used, and a sample with liquid highly viscous consistency when PEG 400 was used.

The extrusion of the PLA 2003D (Ingeo™ Nature Works, Minnetonka, MN, USA) in the presence of PEG8000 or PEG400 was carried out after drying the material for 16 h at 60 °C in a ventilated oven, using a TwinLab II Haake™ Rheomex CTW 5 laboratory screw extruder (Vreden, Germany). After homogenizing using mortar and pestle, the materials were fed into the extruder from the hopper at the beginning of the twin screws, and were mixed into the recirculating channel of the extruder. The extrusion was carried out at 180 °C and 90 rpm for one minute. After extrusion, the molten material was transferred through a preheated cylinder into the Haake™ MiniJet II mini injection molding machine to obtain Haake type III test specimens for tensile testing. The injection molding was carried out at 180 °C, 650 bars, holding time of 15 s, mold temperature of 35 °C.

4.3. Characterizations

The analysis of the average length and width of chitin nanofibrils was carried out using ImageJ software applied on micrographs obtained by using a FEI Quanta 450 ESEM FEG field emission instrument.

The morphology of master batches and composites was studied by scanning electron microscopy (SEM) using a JEOL JSM-5600LV instrument and analyzing cryo-fractured surfaces, previously subjected to sputtering with gold.

Infrared spectra were recorded in the 550–4000 cm^{-1} range with a Nicolet 380 Thermo Corporation Fourier Transform Infrared (FTIR) Spectrometer equipped with smart Itx ATR accessory, collecting 256 scans at 4 cm^{-1} resolution

Thermogravimetric analysis was performed on 10–20 mg of sample using a Mettler-Toledo TGA/SDTA 851 instrument (Columbus, OH, USA) operating with nitrogen as the purge gas (60 mL/min) at 10 °C·min^{-1} heating rate in the 25–800 °C temperature range.

The nanocomposites samples were cut with a Reichert Ultracut E ultramicrotome into ultrafine sheet (<1 micron thickness) and collected onto double folding copper grids (50/100 mesh). The grids were closed and coated with a thin layer of carbon by a EMITECH K950 Evaporator Coater Sputter (Laughton, UK) to make them electrically conductive. The micrographs were obtained with a FEI Quanta 450 ESEM FEG in STEM mode.

Tensile tests (UNI EN ISO527) were carried out using a universal INSTRON 5500R test machine with a 1 kN load cell at a speed of 10 mm/min onto specimens conditioned for 2 weeks at 25 °C and relative humidity of 50%.

Differential scanning calorimetry analyzes (DSC) were performed on material sampled from injection molded specimens using a TA Q200 instrument with nitrogen as carrier gas and indium as a calibration standard. The samples were heated from −100 °C to 250 °C at 10 °C/min and cooled from 250 °C to −100 °C at 20 °C/min. The second heating was carried out by heating analogously from −100 °C to 200 °C at 10 °C/min. The crystallinity was calculated by the formula $Xc = [(\Delta H_m - \Delta H_{cc})/(\Delta H_0 \cdot w)] \times 100$, ΔH_m is the melting enthalpy, ΔH_{cc} is the cold crystallization enthalpy, ΔH_0 is the melting enthalpy of PLA fully crystalline (the value of 93J/g was considered [52]) and w is the weight fraction of PLA in the composite.

5. Conclusions

In the present paper, the preparation by extrusion of nanocomposites consisting of a PLA matrix and a dispersed chitin nanofibrils (CN) phase was obtained using a methodology based on the preliminary preparation of pre-composites based on CN and poly(ethylene glycol) (PEG) having a molecular weight of 8000 or 400. The presence of PEG, making it possible to keep the CNs separated, avoiding the problem of their agglomeration due to inter-macromolecular interactions.

The pre-composites were dispersed in molten PLA to obtain nanocomposites with different content of PEG and CN in a laboratory extruder. The morphologic analysis by SEM demonstrated the absence of agglomerated CNs in the sample, in agreement with the presence of a nanostructured material.

The content of PEG and CN strongly influenced the colorlessness and the transparency of the injection molded specimens. In particular, a higher PEG content and a lower CN content resulted in more transparent and colorless specimens. The presence of CNs on the surface of the injection molded specimens was detected in the nanocomposites with a content of CN higher than 5%.

The thermal properties showed few relevant changes due to the addition of CNs to the samples, being only responsible of a slight nucleating action. Interestingly, it was found that in injection molded specimens, the polymeric chains slowly reorganize into ordered structures during storage at room temperature, thus showing both enthalpy relaxation peak and increased crystallinity. This reorganization is evident above 5% by weight of CNs.

The tensile properties did not show a reinforcing effect of CNs, but the achieved good dispersion allowed us to maintain high values of elongation at break (>150%), typical of plasticized PLA, up to 12% by weight of CN. The absence of a reinforcing effect, not in agreement with literature data, was tentatively explained by considering that PEG, having a high affinity for the polar groups of chitin, can support its nano-dispersion, but can be more difficult to remove from CNs, remaining in between CNs surface and PLA matrix. The PEG can thus coat the CN surface, lowering the reinforcing effect of nanofibers. This aspect, useful for selecting the correct CN/dispersing agent system, would necessitate further elucidation.

The methodology investigated and developed here, in comparison with the other methodologies for dispersing CNs in PLA, can offer the advantage of being easily applicable also at an industrial scale, and does not modify the thermo-mechanical properties typical of plasticized PLA. Moreover, this methodology can be advantageous in the case of the study of ductile materials for injection molding or flexible materials, to be applied to plastic films, both with low contents of CNs, in order to exploit their functional properties.

Author Contributions: Conceptualization, M.-B.C. and A.L.; methodology, M.-B.C., V.G., L.P. and L.A.; formal analysis, M.-B.C. and L.P.; resources, P.C. and P.M.; data curation, V.G., L.P. and L.A.; writing—original draft preparation, M.-B.C.; writing—review and editing, L.A., P.C. and A.L.; supervision, P.M. and A.L.; project administration, M.B.C. and P.C.; funding acquisition, P.C. and A.L.

Funding: This work was supported by European Union that financed the projects NANO-CHITOPACK (Sustainable technologies for the production of biodegradable materials based on natural chitin-nanofibrils derived by waste of fish industry, to produce food grade packaging, G.A. n°. 315233) and POLYBIOSKIN (High performance functional bio-based polymers for skin-contact products in biomedical, cosmetic and sanitary industry, G.A. n°. 745839). The latter is still ongoing.

Acknowledgments: Valter Castelvetro and Sabrina Bianchi, of the Department of Chemistry and Industrial Chemistry of the University of Pisa, are thanked for supporting in TGA characterization. Irene Anguillesi is thanked for technical support. Randa Ishak is thanked for the kind assistance in electron microscopy characterization.

Conflicts of Interest: The authors declare no conflict of interest.

References

1. Kim, S.K. (Ed.) *Chitin, Chitosan, Oligosaccharides and Their Derivatives*, 1st ed.; CRC Press: Boca Raton, FL, USA, 2011; p. 241.
2. Strelcova, Z.; Kulhanek, P.; Friak, M.; Fabritius, H.O.; Petrov, M.; Neugebauer, J.; Koca, J. The structure and dynamics of chitin nanofibrils in an aqueous environment revealed by molecular dynamics simulations. *RSC Adv.* **2016**, *6*, 30710–30721. [CrossRef]
3. Fazeni, K.; Lindorfer, J. The Energy Institute at the Johannes Kepler University, 2011. Available online: http://www.energyefficiency.at/web/projekte/chibio-1-53-1-53-1-53-1-53.html (accessed on 28 February 2015).
4. Nisticò, R. Aquatic-Derived Biomaterials for a Sustainable Future: A European Opportunity. *Resources* **2017**, *6*, 65. [CrossRef]
5. Younes, I.; Rinaudo, M. Chitin and Chitosan Preparation from Marine Sources. Structure, Properties and Applications. *Mar. Drugs* **2015**, *13*, 1133–1174. [CrossRef] [PubMed]
6. Morganti, P.; Muzzarelli, C. Spray Dried Chitin Nanofibrils, Method for Production and Uses Thereof. U.S. Patent 2009/0203642, 13 August 2009.

7. Gadgey, K.K.; Bahekar, A. Studies on extraction methods of chitin from crab shell and investigation of its mechanical properties. *Int. J. Mech. Eng. Technol.* **2017**, *8*, 220–231.

8. Wu, J.; Zhang, K.; Girouard, N.; Carson Meredith, J. Facile Route to Produce Chitin Nanofibers as Precursors for Flexible and Transparent Gas Barrier Materials. *Biomacromolecules* **2014**, *15*, 4614–4620. [CrossRef] [PubMed]

9. Gao, X.; Chen, X.; Zhang, J.; Guo, W.; Jin, F.; Yan, N. Transformation of Chitin and Waste Shrimp Shells into Acetic Acid and Pyrrole. *ACS Sustain. Chem. Eng.* **2016**, *4*, 3912–3920. [CrossRef]

10. De Azeredo, H.M.C. Antimicrobial nanostructures in food packaging. *Trends Food Sci. Technol.* **2013**, *30*, 56–69. [CrossRef]

11. Tsai, G.-J.; Su, W.-H.; Chen, H.-C.; Pan, C.-L. Antimicrobial activity of shrimp chitin and chitosan from different treatments and applications of fish preservation. *Fish. Sci.* **2002**, *68*, 170–177. [CrossRef]

12. Jayakumara, R.; Prabaharan, M.; Sudheesh Kumar, P.T.; Naira, S.V.; Tamura, H. Biomaterials based on chitin and chitosan in wound dressing applications. *Biotechnol. Adv.* **2011**, *29*, 322–337. [CrossRef]

13. Garrison, T.F.; Murawski, A.; Quirino, R.L. Bio-Based Polymers with Potential for Biodegradability. *Polymers* **2016**, *8*, 262. [CrossRef]

14. Pandey, A.; Chauhan, N.P.S.; Shabafrooz, V.; Ameta, R.; Mozafari, M. Polylactic acid and polyethylene glycol as antimicrobial agents. In *Biocidal Polymers*, 1st ed.; Chauhan, N.P.S., Ed.; Smither Rapra: Shawbury, UK, 2016; pp. 131–144.

15. Rabnawaz, M.; Wyman, I.; Auras, R.; Cheng, S. A roadmap towards green packaging: The current status and future outlook for polyesters in the packaging industry. *Green Chem.* **2017**, *19*, 4737–4753. [CrossRef]

16. Chen, Y.; Geever, L.M.; Killion, J.A.; Lyons, J.G.; Higginbotham, C.L.; Devine, D.M. Review of Multifarious Applications of Poly (Lactic Acid). *Polym.-Plast. Technol. Eng.* **2016**, *55*, 1057–1075. [CrossRef]

17. Saini, P.; Arora, M.; Kumar, M.N.V.R. Poly(lactic acid) blends in biomedical applications. *Adv. Drug Deliv. Rev.* **2016**, *107*, 47–59. [CrossRef] [PubMed]

18. Yu, H.Y.; Zhang, H.; Song, M.L.; Zhou, Y.; Yao, J.; Ni, Q.Q. From Cellulose Nanospheres, Nanorods to Nanofibers: Various Aspect Ratio Induced Nucleation/Reinforcing Effects on Polylactic Acid for Robust-Barrier Food Packaging. *ACS Appl. Mater. Interfaces* **2017**, *9*, 43920–43938. [CrossRef] [PubMed]

19. Basu, A.; Nazarkovsky, M.; Ghadi, R.; Khan, W.; Domb, A.J. Poly(lactic acid)-based nanocomposites. *Polym. Adv. Technol.* **2017**, *28*, 919–930. [CrossRef]

20. Scaffaro, R.; Botta, L.; Lopresti, F.; Maio, A.; Sutera, F. Polysaccharide nanocrystals as fillers for PLA based nanocomposites. *Cellulose* **2017**, *24*, 447–478. [CrossRef]

21. Jonoobi, M.; Harun, J.; Mathew, A.P.; Oksman, K. Mechanical properties of cellulose nanofiber (CNF) reinforced polylactic acid (PLA) prepared by twin screw extrusion. *Compos. Sci. Technol.* **2010**, *70*, 1742–1747. [CrossRef]

22. Trifol Guzman, J.; Plackett, D.; Sillard, C.; Szabo, P.; Bras, J.; Daugaard, A.E. Hybrid poly(lactic acid)/nanocellulose/nanoclay composites with synergistically enhanced barrier properties and improved thermomechanical resistance. *Polym. Int.* **2016**, *65*, 988–995. [CrossRef]

23. Nagarajan, V.; Mohanty, A.K.; Misra, M. Perspective on Polylactic Acid (PLA) based Sustainable Materials for Durable Applications: Focus on Toughness and Heat Resistance. *ACS Sustain. Chem. Eng.* **2016**, *4*, 2899–2916. [CrossRef]

24. Fan, Y.M.; Saito, T.; Isogai, A. Preparation of chitin nanofibers from squid pen beta-chitin by simple mechanical treatment under acid conditions. *Biomacromolecules* **2008**, *9*, 1919–1923. [CrossRef]

25. Nakagaito, A.N.; Kanzawa, S.; Takagi, H. Polylactic Acid Reinforced with Mixed Cellulose and Chitin Nanofibers—Effect of Mixture Ratio on the Mechanical Properties of Composites. *J. Compos. Sci.* **2018**, *2*, 36. [CrossRef]

26. Araki, J. Preparation of Sterically Stabilized Chitin Nanowhisker Dispersions by Grafting of Poly(ethylene glycol) and Evaluation of Their Dispersion Stability. *Biomacromolecules* **2015**, *16*, 379–388. [CrossRef] [PubMed]

27. Araki, J.; Wada, M.; Kuga, S. Steric Stabilization of a Cellulose Microcrystal Suspension by Poly(ethylene glycol) Grafting. *Langmuir* **2001**, *17*, 21–27. [CrossRef]

28. Araki, J.; Kuga, S.; Magoshi, J. Influence of reagent addition on carbodiimide-mediated amidation for poly(ethylene glycol) grafting. *J. Appl. Polym. Sci.* **2002**, *85*, 1349–1352. [CrossRef]

29. Zhang, Q.; Wei, S.; Huang, J.; Feng, J.; Chang, P.R. Effect of Surface Acetylated-Chitin Nanocrystals on Structure and Mechanical Properties of Poly (lactic acid). *J. Appl. Polym. Sci.* **2014**, *131*, 39809. [CrossRef]

30. Rizvi, R.; Cochrane, B.; Naguib, H.; Lee, P.C. Fabrication and characterization of melt-blended polylactide-chitin composites and their foams. *J. Cell. Plast.* **2011**, *47*, 283–300. [CrossRef]

31. Herrera, N.; Singh, A.A.; Salaberria, A.M.; Labidi, J.; Mathew, A.P.; Oksman, K. Triethyl Citrate (TEC) as a Dispersing Aid in Polylactic Acid/Chitin Nanocomposites Prepared via Liquid-Assisted Extrusion. *Polymers* **2017**, *9*, 406. [CrossRef]

32. Guan, Q.; Naguib, H.E. Fabrication and Characterization of PLA/PHBV-Chitin Nanocomposites and Their Foams. *J. Polym. Environ.* **2014**, *22*, 119–130. [CrossRef]

33. Herrera, N.; Salaberria, A.M.; Mathew, A.P.; Oksman, K. Plasticized polylactic acid nanocomposite films with cellulose and chitin nanocrystals prepared using extrusion and compression molding with two cooling rates: Effects on mechanical, thermal and optical properties. *Compos. Part A* **2016**, *83*, 89–97. [CrossRef]

34. Herrera, N.; Roch, H.; Salaberria, A.M.; Pino-Orellana, M.A.; Labidi, J.; Fernandes, S.C.M.; Radic, D.; Leiva, A.; Oksman, K. Functionalized blown films of plasticized polylactic acid/chitin nanocomposite: Preparation and characterization. *Mater. Des.* **2016**, *92*, 846–852. [CrossRef]

35. Fehri, M.K.; Mugoni, C.; Cinelli, P.; Anguillesi, I.; Coltelli, M.B.; Fiori, S.; Montorsi, M.; Lazzeri, A. Composition dependence of the synergistic effect of nucleating agent and plasticizer in poly(lactic acid): A Mixture Design study. *eXPRESS Polym. Lett.* **2016**, *10*, 274–288. [CrossRef]

36. Coltelli, M.B.; Della Maggiore, I.; Bertoldo, M.; Signori, F.; Bronco, S.; Ciardelli, F. Poly(lactic acid) Properties as a Consequence of Poly(butylene adipate-co-terephthalate) Blending and Acetyl Tributyl Citrate Plasticization. *J. Appl. Polym. Sci.* **2008**, *110*, 1250–1262. [CrossRef]

37. Quero, E.; Müller, A.J.; Signori, F.; Coltelli, M.B.; Bronco, S. Isothermal Cold-Crystallization of PLA/PBAT Blends with and without the Addition of Acetyl Tributyl Citrate. *Macromol. Chem. Phys.* **2012**, *213*, 36–48. [CrossRef]

38. Scatto, M.; Salmini, E.; Castiello, S.; Coltelli, M.B.; Conzatti, L.; Stagnaro, P.; Andreotti, L.; Bronco, S. Plasticized and nanofilled poly(lactic acid)-based cast films: Effect of plasticizer and organoclay on processability and final properties. *J. Appl. Polym. Sci.* **2013**, *127*, 4947–4956. [CrossRef]

39. Coltelli, M.B.; Cinelli, P.; Anguillesi, I.; Salvadori, S.; Lazzeri, A. Structure and properties of extruded composites based on bio-polyesters and nano-chitin, Session M: Functional Textiles-from research and development to innovations and industrial uptake. In Proceedings of the Symposium E-MRS Fall Meeting, Warsaw University of Technology, Warsaw, Poland, 15–18 September 2014; p. 192.

40. Cinelli, P.; Coltelli, M.B.; Mallegni, N.; Morganti, P.; Lazzeri, A. Degradability and sustainability of nanocomposites based on polylactic acid and chitin nano fibrils. *Chem. Eng. Trans.* **2017**, *60*, 1–6.

41. Coltelli, M.B.; Gigante, V.; Panariello, L.; Aliotta, L.; Morganti, P.; Danti, S.; Cinelli, P.; Lazzeri, A. Chitin nanofibrils in renewable materials for packaging and personal care applications. *Adv. Mater. Lett.* **2019**, in press.

42. Li, J.; Gao, Y.; Zhao, J.; Sun, J.; Li, D. Homogeneous dispersion of chitin nanofibers in polylactic acid with different pretreatment methods. *Cellulose* **2017**, *24*, 1705–1715. [CrossRef]

43. Nakagaito, A.N.; Yamada, K.; Ifuku, S.; Morimoto, M.; Saimoto, H. Fabrication of Chitin Nanofiber-Reinforced Polylactic Acid Nanocomposites by an Environmentally Friendly Process. *J. Biobased Mater. Bioenergy* **2013**, *7*, 152–156. [CrossRef]

44. Shamshina, J.L.; Zavgorodnya, O.; Berton, P.; Chhotaray, P.K.; Choudhary, H.; Rogers, R.D. An Ionic Liquid Platform for Spinning Composite Chitin-Poly(lactic acid) Fibers. *ACS Sustain. Chem. Eng.* **2018**, *6*, 10241–10251. [CrossRef]

45. Jolanta Kumirska, J.; Czerwicka, M.; Kaczyński, Z.; Bychowska, A.; Brzozowski, K.; Thöming, J.; Stepnowski, P. Application of Spectroscopic Methods for Structural Analysis of Chitin and Chitosan. *Mar. Drugs* **2010**, *8*, 1567–1636. [CrossRef]

46. Pearson, F.G.; Marchessault, R.H.; Liang, C.Y. Infrared spectra of crystalline polysaccharides. V. Chitin. *J. Polym. Sci.* **1960**, *13*, 101–116. [CrossRef]

47. Brunner, E.; Ehrlich, H.; Schupp, P.; Hedrich, R.; Hunoldt, S.; Kammer, M.; Machill, S.; Paasch, S.; Bazhenov, V.V.; Kurek, D.V.; et al. Chitin-based scaffolds are an integral part of the skeleton of the marine demosponge Ianthella basta. *J. Struct. Biol.* **2009**, *168*, 539–547. [CrossRef] [PubMed]

48. Focher, B.; Naggi, A.; Torri, G.; Cosani, A.; Terbojevich, M. Structural differences between chitin polymorphs and their precipitates from solutions-evidence from CP-MAS ^{13}CNMR, FT-IR and FT-Raman spectroscopy. *Carbohydr. Polym.* **1992**, *17*, 97–102. [CrossRef]

49. Cheng, D.; Wen, Y.; Wang, L.; An, X.; Zhua, X.; Ni, Y. Adsorption of polyethylene glycol (PEG) onto cellulose nano-crystals to improve its dispersity. *Carbohydr. Polym.* **2015**, *123*, 157–163. [CrossRef] [PubMed]

50. Paul, D.R.; Robeson, L.M. Polymer nanotechnology: Nanocomposites. *Polymer* **2008**, *49*, 3187–3204. [CrossRef]

51. Biswas, S.K.; Shams, M.I.; Das, A.K.; Islam, M.N.; Nazhad, M.M. Flexible and Transparent Chitin/Acrylic Nanocomposite Films with High Mechanical Strength. *Fibers Polym.* **2015**, *16*, 774–781. [CrossRef]

52. Mikesová, J.; Hasek, J.; Tishchenko, G.; Morganti, P. Rheological study of chitosan acetate solutions containing chitin nanofibrils. *Carbohydr. Polym.* **2014**, *112*, 753–757. [CrossRef] [PubMed]

53. Xu, X.; Wang, H.; Jiang, L.; Wang, X.; Payne, S.A.; Zhu, J.Y.; Li, R. Comparison between Cellulose Nanocrystal and Cellulose Nanofibril Reinforced Poly(ethylene oxide) Nanofibers and Their Novel Shish-Kebab-Like Crystalline Structures. *Macromolecules* **2014**, *47*, 3409–3416. [CrossRef]

54. Cardinaels, R.; Moldenaers, P. Morphology development in immiscible polymer blends. In *Polymer Morphology. Principles, Characterization and Properties*; Guo, Q., Ed.; John Wiley and Sons: Hoboken, NJ, USA, 2016; Chapter 19; pp. 348–373.

55. Potschke, P.; Paul, D.R. Formation of co-continuous structures in met-mixed immiscible polymer blends. *J. Macromol. Sci. C Polym. Rev.* **2003**, *43*, 87–141. [CrossRef]

56. Muzzarelli, C.; Morganti, P. Preparation of Chitin and Derivatives Thereof for Cosmetic and Therapeutic Use. U.S. Patent 8383157 B2, 26 February 2013.

© 2019 by the authors. Licensee MDPI, Basel, Switzerland. This article is an open access article distributed under the terms and conditions of the Creative Commons Attribution (CC BY) license (http://creativecommons.org/licenses/by/4.0/).

International Journal of
Molecular Sciences

MDPI

Article

PHB is Produced from Glycogen Turn-over during Nitrogen Starvation in *Synechocystis* sp. PCC 6803

Moritz Koch [1], Sofía Doello [1], Kirstin Gutekunst [2] and Karl Forchhammer [1,*]

[1] Interfaculty Institute of Microbiology and Infection Medicine Tübingen, Eberhard-Karls-Universität Tübingen, 72076 Tübingen, Germany; moritz.koch@uni-tuebingen.de (M.K.); sofia.doello@gmail.com (S.D.)
[2] Department of Biology, Botanical Institute, Christian-Albrechts-University, 24118 Kiel, Germany; kgutekunst@bot.uni-kiel.de
* Correspondence: karl.forchhammer@uni-tuebingen.de; Tel.: +49-7071-29-72096

Received: 4 April 2019; Accepted: 18 April 2019; Published: 20 April 2019

Abstract: Polyhydroxybutyrate (PHB) is a polymer of great interest as a substitute for conventional plastics, which are becoming an enormous environmental problem. PHB can be produced directly from CO_2 in photoautotrophic cyanobacteria. The model cyanobacterium *Synechocystis* sp. PCC 6803 produces PHB under conditions of nitrogen starvation. However, it is so far unclear which metabolic pathways provide the precursor molecules for PHB synthesis during nitrogen starvation. In this study, we investigated if PHB could be derived from the main intracellular carbon pool, glycogen. A mutant of the major glycogen phosphorylase, GlgP2 (*slr1367* product), was almost completely impaired in PHB synthesis. Conversely, in the absence of glycogen synthase GlgA1 (*sll0945* product), cells not only produced less PHB, but were also impaired in acclimation to nitrogen depletion. To analyze the role of the various carbon catabolic pathways (EMP, ED and OPP pathways) for PHB production, mutants of key enzymes of these pathways were analyzed, showing different impact on PHB synthesis. Together, this study clearly indicates that PHB in glycogen-producing *Synechocystis* sp. PCC 6803 cells is produced from this carbon-pool during nitrogen starvation periods. This knowledge can be used for metabolic engineering to get closer to the overall goal of a sustainable, carbon-neutral bioplastic production.

Keywords: cyanobacteria; bioplastic; PHB; sustainable; glycogen; metabolic engineering; Synechocystis

1. Introduction

Cyanobacteria are among the most widespread organisms on our planet. Their ability to perform oxygenic photosynthesis allows them to grow autotrophically with CO_2 as the sole carbon source [1]. Additionally, many cyanobacteria acquired the ability to fix nitrogen, one of the most limiting nutrients [2]. However, many others are not able to fix nitrogen, one of them being the well-studied model organism *Synechocystis* sp. PCC 6803 (hereafter: *Synechocystis*) [3]. Nitrogen starvation starts a well-orchestrated survival process in *Synechocystis*, called chlorosis [4]. During chlorosis, *Synechocystis* degrades not only its photosynthetic machinery, but also accumulates large quantities of biopolymers, namely glycogen and poly-hydroxy-butyrate (PHB) [5]. Glycogen synthesis following the onset of nitrogen starvation serves transiently as a major sink for newly fixed CO_2 [6] before CO_2 fixation is tuned down during prolonged nitrogen starvation. During resuscitation from chlorosis, a specific glycogen catabolic metabolism supports the re-greening of chlorotic cells [7]. By contrast to the pivotal role of glycogen, the function of the polymer PHB remains puzzling, since mutants impaired in PHB synthesis survived and recovered from chlorosis as awild-type [8,9]. Nevertheless, many different cyanobacterial species produce PHB, implying a hitherto unrecognized functional importance [10]. In other microorganisms PHB fulfills various functions during conditions of unbalanced nutrient availability and can also protect cells against low temperatures or redox stress [11–13]. Understanding the intracellular mechanisms that lead to PHB production could help to elucidate the physiological role of this polymer. Regardless of the physiological significance of PHB, this polymer has been recognized as a promising alternative for current plastics, which contaminate terrestrial and aquatic ecosystems [14]. PHB can serve as a basis for completely biodegradable plastics, with properties comparable to petroleum-derived plastics [15,16]. Since *Synechocystis* produces PHB only under nutrient limiting conditions, this phenomenon can be exploited to temporally separate the initial biomass production from PHB production induced by shifting cells to nitrogen limiting conditions [10].

One of the biggest obstacles preventing economic PHB production in cyanobacteria remains the low level of intracellular PHB accumulation [17]. While chemotrophic bacteria are capable of producing more than 80% PHB of their cell dry mass, (e.g., *Cupriavidus necator*), most cyanobacteria naturally produce less than 20% of their cell dry mass [15]. Additionally, their growth rate is too slow to compete with the PHB production in chemotrophic bacteria. There have been many attempts in the past to further improve the intracellular PHB production, often with limited [1] success [18–20]. One of the most successful approaches has been achieved by random mutagenesis, leading to up to 37% PHB of the cell dry mass [21]. However, more directed approaches involving genetic engineering are often limited by a lack of knowledge about how the cells' metabolism works in detail. For example, until today, it was still unknown from which carbon metabolites PHB was derived. There have been several different studies analyzing the intracellular fluxes in cyanobacteria [22]. However, most of them did not analyze the carbon flow during prolonged nitrogen starvation. One of these studies showed that in nitrogen-starved photosynthetically grown cyanobacteria up to 87% of the carbon in PHB is derived from intracellular carbon sources rather than from newly fixed CO_2 [23]. However, until now, it was not clearly resolved which metabolic routes provide the precursors for PHB synthesis. This knowledge would lay the foundation for future metabolic engineering approaches to create overproduction strains. Hence, the goal of this study was to find out where the carbon for the PHB production is coming from and which pathways it is taking until it reaches PHB.

It has been shown that disruption of PHB synthesis results in an increased production of glycogen; however, an overproduction of glycogen did not lead to higher amounts of PHB [24]. Another study that also investigated the accumulation of glycogen in a PHB-free mutant Δ*phaC*, could not detect any differences in growth or glycogen accumulation [8].

An important aspect in the issue concerning the relation between glycogen and PHB metabolism deals with the contribution of various carbon metabolic pathways for the production of precursors for PHB under conditions of nitrogen limitation. *Synechocystis* is able to catabolize glucose via three parallel operating glycolytic pathways [25] (Figure 1): the Embden-Meyerhof-Parnas (EMP) pathway, the oxidative pentose phosphate (OPP) pathway [26], and the Entner Doudoroff (ED) pathway [25]. When nitrogen-starved cells recover from chlorosis, they require the parallel operating OPP and ED pathways, whereas the EMP pathway seems dispensable [7]. Metabolic analysis of mutants overexpressing the transcriptional regulator *rre37* showed a correlated upregulation of PHB synthesis and EMP pathway genes (*phaAB* and *pfkA*, respectively) [27]. However, so far is has not been investigated, how important these pathways for the production of PHB during nitrogen starvation.

This work started with the initial aim to define whether PHB synthesis depends on the metabolism of glycogen. Since the initial results implied that PHB is strongly affected by glycogen catabolism, we further investigated the importance of the different carbon pathways EMP, ED and OPP for the production of PHB. These findings shall help to further understand the intracellular PHB metabolism in cyanobacteria, which can be used to create more efficient PHB overproduction strains, making the production of PHB as a bioplastic more cost efficient.

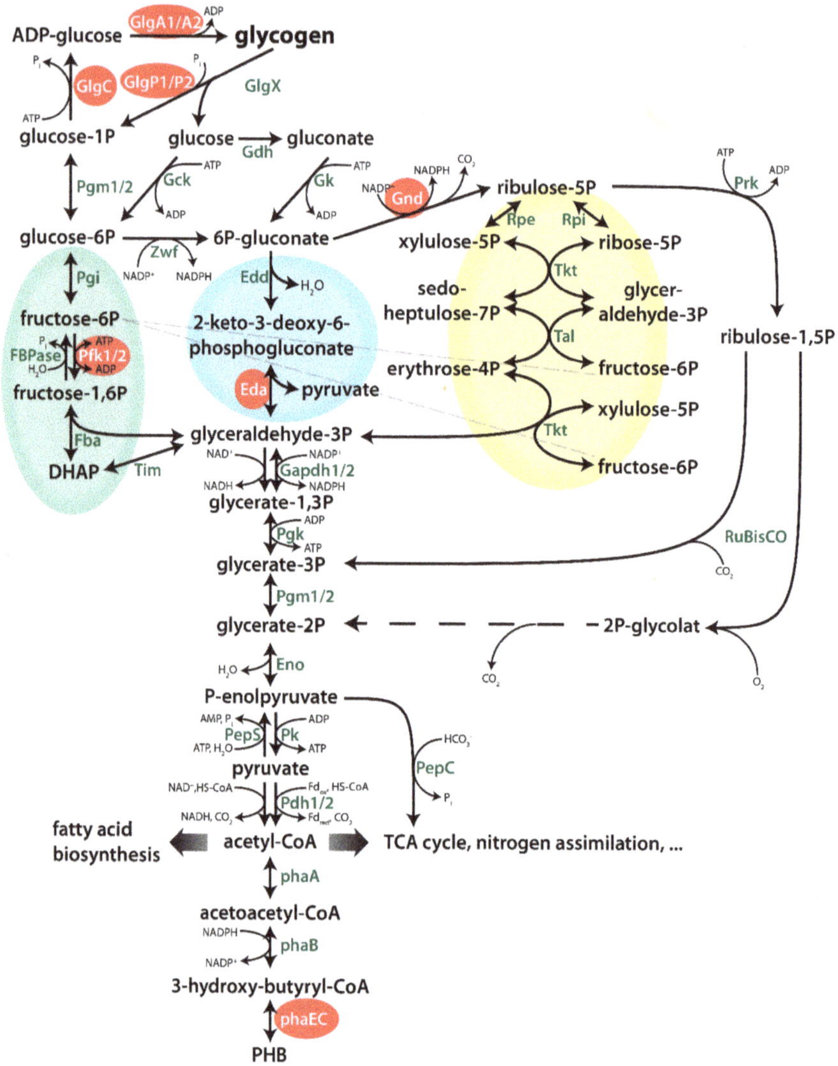

Figure 1. Overview of central metabolism of Synechocystis. Genes which were deleted in this study are highlighted in with a red background. Dotted lines represent several enzymatic reactions. The EMP, ED and OPP (Embden-Meyerhof-Parnas, Entner-Doudoroff, Oxidative Pentose Phosphate) pathways are highlighted in green, blue and yellow, respectively.

2. Results

Following the onset of nitrogen-starvation, large quantities of fixed carbon are stored in *Synechocystis* cells as glycogen granules. Long-term starvation experiments of *Synechocystis* cultures have shown that, while cells are chlorotic, glycogen is slowly degraded, following its initial rapid accumulation but PHB is slowly and steadily accumulating [9]. Considering that chlorotic cells are photosynthetically inactive, these data could indicate a potential correlation between the turn-over of glycogen and the synthesis of PHB. An overview of the metabolic pathways connecting the glycogen pool with PHB is shown in Figure 1. To substantiate the hypothesis that PHB might be derived from glycogen turn-over, we investigated PHB accumulation in various mutant strains, in which key steps in different pathways are interrupted. The respective mutations are shown in Figure 1. All strains used in this work were characterized previously, with their phenotypes, including growth behaviours, described in the respective publications (see Table A1). Furthermore, all mutants used in these studies were fully segregated to ensure clear phenotypes.

2.1. Impact of Glycogen Synthesis on PHB Production

To analyze the role of glycogen synthesis on the production of PHB, we first analyzed the accumulation of these biopolymers during nitrogen starvation in mutants with defects in glycogen synthesis. The double mutant of the two glycogen synthase genes *glgA1* (*sll0945*) and *glgA2* (*sll1393*) is unable to acclimate to nitrogen deprivation and rapidly dies upon shifting cells to nitrogen free BG11^0 medium [8] and, therefore, could not be analyzed. Instead, we used a knockout mutant of the glucose-1-phosphate adenylyltransferase (*glgC*, *slr1176*) and two knockout strains of each of the isoforms of the glycogen synthase, *glgA1* (*sll0945*) and *glgA2* (*sll1393*). Yoo et al. [28] reported that the single *glgA1* and *glgA2* mutants were still able to produce similar amounts of glycogen as the wild-type (WT), since one glycogen synthase is still present, and this seems to be sufficient to reach the wild-type levels of glycogen. However, the structure of the glycogen produced by the two isoforms seemed to slightly differ in chain-length distribution [28]. In that study, no distinguishing phenotype of the two mutant strains had been reported. In the present study, the cultures were shifted to nitrogen free medium BG11^0 and further incubated under constant illumination of 40 μmol photons m^{-2} s^{-1}. Under these experimental conditions, the Δ*glgA1* mutant showed an impaired chlorosis reaction, whereas the Δ*glgA2* mutant performed chlorosis as the wild-type strain (Figure 2A). To further determine the viability of two weeks nitrogen-starved cells, serial dilutions were dropped on nitrate-supplemented BG11 plates. As shown in Figure 2B, the Δ*glgA1* mutant was severely impaired in recovering from nitrogen starvation, whereas Δ*glgA2* could recover from chlorosis with the same efficiency as the wild-type (Figure 2B).

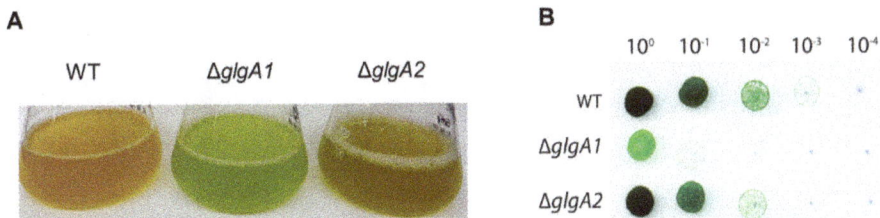

Figure 2. Characterization of the glycogen synthase mutants, Δ*glgA1* and Δ*glgA2*. (**A**) Cultures after five days of nitrogen starvation. (**B**) Recovery assay of chlorotic wild-type (WT) and mutants Δ*glgA1* and Δ*glgA2*, using the drop agar method. Cultures that were nitrogen-starved for 14 days were serially diluted from 1 to 1:10,000 and from each dilution, a drop of 5 μL was plated on BG11 agar and grown for seven days.

During the course of three weeks of nitrogen starvation, the quantities of PHB and glycogen that accumulate in the cells were determined (Figure 3A,B).

In the wild-type, the amount of glycogen already peaked after the first week and slowly decreased in the following two weeks (Figure 3B). As previously reported by Yoo et al. [28], the single Δ*glgA1* and Δ*glgA2* mutants initially accumulated similar amounts of glycogen to the wild-type, but in contrast to the wild-type, the level of glycogen remained high. The PHB content in the *glgA2* mutant was similar to the wild-type for the first seven days of nitrogen starvation, but PHB accumulation slowed down afterwards (Figure 3B). By contrast, the *glgA1* mutant was strongly impaired in PHB production. Together, the phenotype of the *glgA1* and *glgA2* mutants indicates that glycogen synthase GlgA1 plays a much more important role in nitrogen starvation acclimation than GlgA2, although the amount of glycogen produced by these two strains is almost the same. One explanation could be that the subtle differences in the glycogen produced form the two isoenzymes might result in different functions, with GlgA1-produced glycogen being much more relevant for the maintenance metabolism in chlorotic cells and for the resuscitation from chlorosis than glycogen produced by GlgA2. In clear correlation with the redundant role of GlgA2, the *glgA2* mutant was not impaired in PHB synthesis, whereas mutation of the functionally important *glgA1* gene resulted in strongly impaired PHB synthesis.

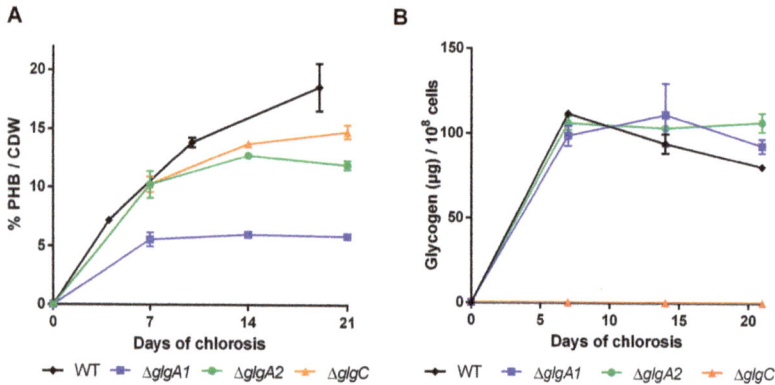

Figure 3. Polyhydroxybutyrate (PHB) content in percentage of cell dry weight (CDW) (**A**) and cellular glycogen content (**B**) of mutants impaired in the glycogen synthesis. Cultures were shifted to nitrogen free medium at day 0 and were subsequently grown for 21 days. Each point represents a mean of three independent biological replicates.

The *glgC* mutant was previously characterized by Grundel et al. [6]. They showed that ΔglgC is not able to perform a proper nitrogen-starvation acclimation response: it maintains its pigments while it loses viability, which was also observed in our experiments. Namakoshi et al. [29] showed that this mutant is unable to synthesize glycogen, which is also in line with our results. Under our conditions, unlike previously described by Damrow et al. [8], the *glgC* mutant did not show an increased amount of PHB compared to the WT, but seemed to accumulate less PHB instead (Figure 3A). It has to be noted though that Damrow et al. [8] investigated only one single timepoint, after seven days of nitrogen starvation. In addition, these results should be treated with care, since PHB content is normalized to cell dry weight, which shrinks in the *glgC* mutant due to progressive cell lysis. Consequently, the cell density was severely diminished at the end of the experiment (OD$_{750}$ of 0.51, compared to ~1.15 of other mutants and the wild-type). The differences between our study and that of Damrow et al. [8] thus may result from differences in cell lysis rather than from differences in PHB synthesis. When the relatively low cell density of the *glgC* mutant is considered, it produces much less PHB per volume compared to the wild-type.

2.2. Impact of Glycogen Degradation on PHB Production

If glycogen turn-over would result in PHB accumulation during chlorosis, synthesis of PHB should be abrogated when glycogen degradation is impaired. To test this assumption, mutants in catabolic glycogen phosphorylase genes (*glgP*) were investigated with the same methods as described above. Glycogen can be degraded by the two glycogen phosphorylase isoenzymes, encoded by *glgP1* (slr1356) and *glgP2* (slr1367) [7]. A detailed study by Doello et al. [7] showed that GlgP2 is the main enzyme responsible for glycogen degradation during resuscitation from nitrogen chlorosis. Knocking out GlgP1 (Δ*glgP1*) does not affect the efficiency of recovery, whereas knocking out GlgP2 (Δ*glgP2*) or both phosphorylases (Δ*glgP1/2*) completely impairs the ability to degrade glycogen [7]. Here, we investigated glycogen and PHB accumulation during three weeks of nitrogen starvation in these glycogen phosphorylases mutants.

Although the initial amount of glycogen was higher in the *glgP1* mutant compared to the WT (Figure 4B), the amount decreased during the course of the experiment. By contrast, no glycogen degradation occurred in the $\Delta glgP2$ and $\Delta glgP1/2$ double mutant. This correlates with the specific requirement of chlorotic cells for GlgP2 for resuscitation from nitrogen starvation, as it has been previously described [7].

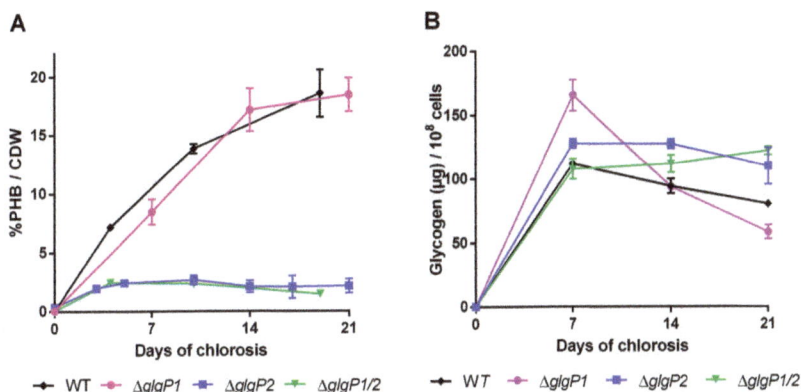

Figure 4. PHB content in percentage of cell dry weight (CDW) (**A**) and glycogen content (**B**) of mutants impaired in the glycogen degradation. Cultures were shifted to nitrogen free medium at day 0 and were subsequently grown for 21 days. Each point represents a mean of three independent biological replicates.

Intriguingly, the different mutants showed a drastic difference in the amounts of PHB being produced (Figure 4A): While the $\Delta glgP1$ strain produced similar amounts of PHB as the wild-type, a strong decrease was observed for the $\Delta glgP2$ strain. The same phenotype was observed for the double knockout mutant, indicating that origin of the effect is based on the absence of *glgP2*. The PHB synthesis phenotypes were further confirmed by fluorescence microscopy after staining PHB with Nile red (Figure 5). PHB granules appear as bright red fluorescing intracellular granular structures. In agreement with the results from PHB quantification by HPLC analysis, the $\Delta glgP1$ strain showed similar amounts and distribution of PHB granules than the wild-type. By contrast, only very small granules, if at all visible, could be detected in the strains $\Delta glgP2$ and $\Delta glgP1/2$.

Altogether, the inability of the mutants $\Delta glgP2$ and $\Delta glgP1/2$ to accumulate PHB demonstrates unequivocally that glycogen catabolism through GlgP2 is required for the ongoing PHB synthesis during prolonged nitrogen starvation.

2.3. Impact of Mutations in Carbon Catabolic Pathway on PHB Production

The experiments outlined above revealed that specific glycogen synthesizing or degrading enzymes have a strong effect on the amounts of PHB being produced and that glycogen turn-over via GlgP2 provides the carbon skeletons for PHB synthesis. To investigate how the released glucose phosphate molecules are metabolized downstream of glycogen, knockouts of the three most important glycolytic routes [25] were checked for their PHB and glycogen production during chlorosis. While the strain Δeda (*sll0107*) lacks the ability to metabolize molecules via the ED pathway, Δgnd is not able to use the OPP pathway. Additionally, the strain $\Delta pfk1/2$ lacks both phosphofructokinases, which causes an interruption of the EMP pathway. Also, the individual knockouts of both isoforms, $\Delta pfk1$ and $\Delta pfk2$, were investigated.

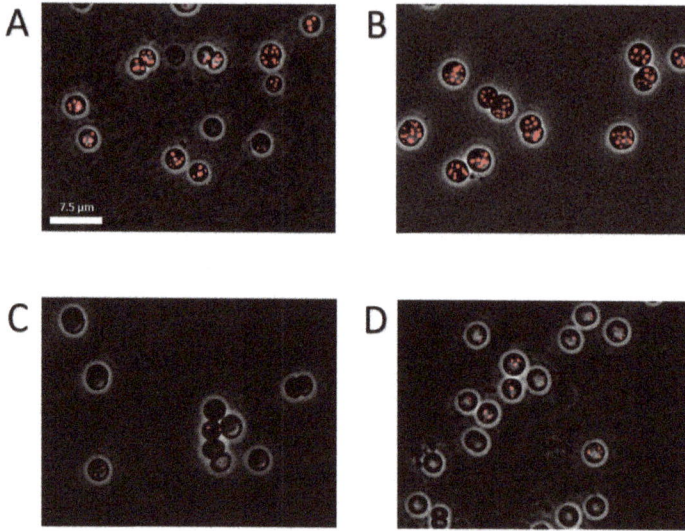

Figure 5. Fluorescence microscopic picture of Nile-red stained PHB granules in chlorotic cells. Cultures where grown for 14 days in nitrogen depleted medium BG11^0. Shown is an overlay of phase contrast with a CY3 channel of the WT (**A**), Δ*glgP1* (**B**), Δ*glgP2* (**C**) and Δ*glgP1/2* (**D**). Scale bar corresponds to 7.5 μm.

Again, the different mutant strains and a WT control were grown for three weeks under nitrogen deprived conditions and PHB and glycogen content was quantified (Figure 6).

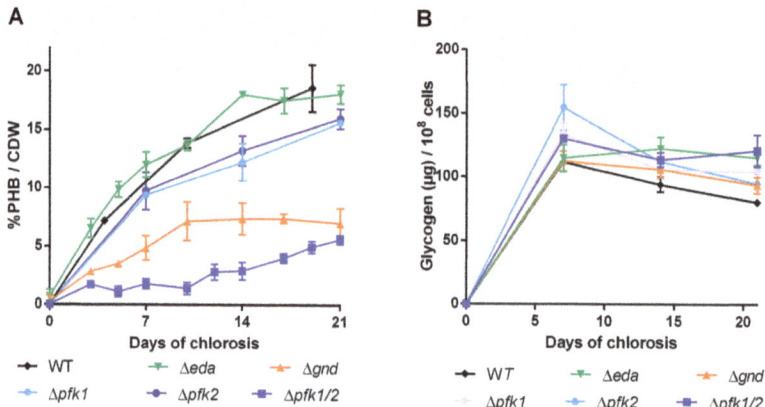

Figure 6. PHB content in percentage of cell dry weight (CDW) (**A**) and glycogen content (**B**) of mutants with disrupted carbon pathway. Cultures were shifted to nitrogen free medium at day 0 and were subsequently grown for 21 days. Each point represents a mean of three independent biological replicates.

Distortion of the ED pathway (Δ*eda*) did not result in any PHB phenotype different from the WT and the glycogen content remained high during the course of the experiment. In the Δ*gnd* mutant, a slower increase of PHB than in the WT was observed within the first ten days (Figure 6A) and PHB accumulation subsequently ceased. When the EMP pathway was blocked (Δ*pfk1/2*), only very little PHB was produced in the first two weeks of the chlorosis. Thereafter though, PHB production slightly increased and finally reached similar levels as in the Δ*gnd* mutant. The total amount of glycogen over the time of chlorosis did not decrease in this mutant. The single Δ*pfk* mutants showed PHB contents similar to the WT, indicating that the two isoenzymes are able to replace each other's function. Taken together, it appears that EMP and OPP pathways contribute to PHB production, whereas the ED pathway does not play a role.

2.4. Impact of PHB Formation on Glycogen Synthesis

In order to check how the PHB production affects the accumulation of glycogen, a PHB-free mutant, namely Δ*phaEC*, was checked for its production of carbon polymers (Figure 7).

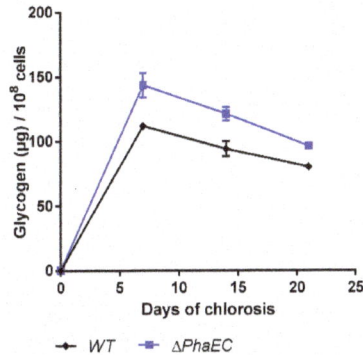

Figure 7. Glycogen content of wild-type and mutant lacking the PHB synthase genes (*PhaEC*). Cultures were shifted to nitrogen free medium at day 0 and were subsequently grown for 21 days. Each point represents a mean of three independent biological replicates.

As expected, the mutant was unable to synthesize PHB (data not shown [8]). Compared to the WT, the mutant produced moderately higher amounts of glycogen and degrades it slightly faster, so that at the end of the experiments, the glycogen levels were quite similar.

3. Discussion

As recently shown by isotope labeling experiments [23], the majority of the carbon from PHB is coming from intracellular metabolites, which contribute around 74% to the carbon within PHB. Additionally, a random mutagenized strain, which is an overproducer of PHB, shows also a strongly accelerated decay of glycogen [30]. Here, we provide clear evidence that the intracellular glycogen pool and its products provide the carbon metabolites for PHB synthesis during nitrogen starvation. In the absence of glycogen degradation, as it is the case in the Δ*glgP2* and the Δ*glgP1/2* double mutant, PHB synthesis is almost completely abrogated. In the Δ*glgP2* mutant, the remaining GlgP1 enzyme is apparently not efficiently catabolizing glycogen, which agrees with its lack of function for the resuscitation from chlorosis. By contrast, GlgP2 is required for glycogen catabolism and resuscitation from chlorosis [7]. From these data, it is reasonable to hypothesize that during chlorosis, GlgP2 slowly degrades glycogen, and the degradation products end up in the PHB pool.

Like the two glycogen phosphorylase isoenzymes, the two glycogen synthase isoenzymes GlgA1 and GlgA2 appear to have specialized functions. Even though both Δ*glgA1* and Δ*glgA2* synthesized similar amounts of glycogen, deletion of *glgA1* resulted in a mutant with reduced bleaching, viability and PHB content whereas Δ*glgA2* was less affected. This indicates that glycogen produced by GlgA1 is important for resuscitation and growth. On the other hand, GlgA2 produced glycogen appeared less important or its function is yet unknown. Previous publications did not see such a difference, which may be explained by a much shorter time of nitrogen starvation used in their study [6]. Taken together, those mutants (*glgP2* and Δ*glgA1*) that were less viable under nitrogen starvation, did also synthesize less PHB. It remains to be demonstrated, if the different impact that GlgA1 and GlgA2 exert on viability and PHB synthesis originates from the slightly different branching patterns [28] of the glycogen, that they synthesize.

In the Δ*glgC* mutant, PHB is formed, although no glycogen is produced (Figure 3), which seems to contradict the hypothesis of glycogen-derived PHB [8,28]. Taking into account the above results, it appears likely that the carbon metabolites used for PHB can under certain conditions bypass the glycogen pool. When *glgC* is knocked out, glucose-1P (and its precursor, glucose-6P) cannot be further converted into ADP-glucose and may accumulate. The glucose-phosphates could then be downstream metabolized to glyceraldehyde-3P and further converted into PHB. By contrast, when *glgA1* is mutated, the newly fixed carbon can be converted by GlgC to ADP-glucose and subsequently enters the GlgA2-synthesized inactive glycogen pool, where it cannot be further metabolized into PHB. A similar connection between PHB and glycogen has already been described in other organisms (*Sinorhizobium meliloti*), where PHB levels were lower in a mutant lacking *glgA1* [31]. Since the GlgA1/GlgA2 double mutant rapidly dies upon nitrogen starvation [6], the impact of the complete absence of glycogen due to glycogen synthase deficiency on PHB synthesis cannot be experimentally tested.

Furthermore, we observed that a slower degradation of glycogen often correlates with a low-PHB-phenotype, as seen in the case of Δ*glgA1*, Δ*glgP2*, Δ*glgP1/2* and Δ*pfk1/2* (Figures 3, 4 and 6, respectively). This further supports the hypothesis, that only when glycogen gets degraded during the process of chlorosis, PHB is formed. In two cases, Δ*glgP1/2* and Δ*pfk1/2*, the amount of glycogen was even increasing during the later course of nitrogen starvation. This hints towards an ongoing glycogen formation during nitrogen chlorosis, which gets only visible once glycogen degradation is disturbed. Apparently, glycogen metabolism is much more dynamic than presumed from the relative static pool size observed in the wild-type. A steady glycogen synthesis may be counterbalanced by ongoing degradation, together resulting in only a slow net change of its pool size.

The residual metabolism in nitrogen starved chlorotic cells [9,32] is probably required to ensure long-term survival through repair of essential biomolecules such as proteins, DNA and RNA, osmoregulation, regulated shifts in metabolic pathways and the preparation for a quick response as soon as nutrients are available again [33,34]. According to these needs, non-growing starved cells still require a constant supply of ATP, reduction equivalents and the ability to produce cellular building blocks for survival. In line with this, we observed that PHB production was mainly achieved via the EMP pathway, which has the highest ATP yield among the three main carbon catalytic pathways (EMP, OPP, ED). In addition, we found that the OPP pathway is involved in PHB synthesis as well. This pathway provides metabolites for biosynthetic purposes as the repair of biomolecules for maintenance. By contrast, deletion of the ED pathway, which has a lower ATP yield in comparison to the EMP pathway and is physiologically probably most important in connection with photosynthesis and the Calvin–Benson cycle [7,25], did not impair PHB production under nitrogen starvation. Nevertheless, the glycogen levels did not decrease in the Δ*gnd* mutant, implying that mutation of the ED pathway affects the dynamics of glycogen turn-over discussed above.

Finally, we observed that the various carbon catabolic pathways have different functional importance for PHB production, in a time-dependent manner: While the mutant Δ*gnd* (blocking the OPP pathway) produced PHB in the first phase of the experiment but later stopped its synthesis, the Δ*pfk1/2* mutant (impaired in the EMP pathway) was initially blocked in PHB accumulation but later started to produce it (Figure 4). Interestingly, the time point, at which the Δ*gnd* mutant stopped PHB production matched its start in the Δ*pfk1/2* mutant. This could indicate a consecutive role of EMP and OPP pathway during nitrogen chlorosis. Although the exact function of the EMP pathway remains unknown, we show here for the first time a phenotype of a cyanobacterial mutant lacking this pathway. This suggested to also investigate the deletion of the individual knockouts. Deletion of only one of the Pfk isoenzymes (*pfk1*: sll1196 and *pfk2*: sll0745) resulted in mutants that produced about 80% of the PHB of WT cells, whereas the PHB production of the double mutant Δ*pfk1/2* was severely reduced. Pfk1 and Pfk2 can thus obviously compensate for the loss of the respective other, even though PHB production is highest if both enzymes are present. This observation is well in line with transcriptomic studies, which detected an increase in expression of both Pfk isoenzymes during nitrogen starvation [5]. The observation that both EMP and OPP pathway are of importance during arrested growth under nitrogen starvation is in agreement with earlier investigations that reported the upregulation of the sugar catabolic genes *pfk1*, *pfk2*, *zwf*, *gnd* and *gap1* concomitantly with glycogen accumulation [5,35]. EMP and OPP pathway thus support PHB production in non-growing, nitrogen-starved cells, whereas ED and OPP pathway are most important during resuscitation form nitrogen chlorosis after feeding the cells with nitrate [7].

The mutant Δ*phaEC* did not produce any PHB and degraded glycogen similarly to the WT (Figure 7). This indicates that there is no direct feedback between these two polymer pools. In the absence of PHB synthesis, metabolites form glycogen degradation could be leaked by overflow reactions. Under unbalanced metabolic situations, it has been shown that cyanobacteria can excrete metabolites into the medium to control their intracellular energy status [6,36,37]. In any case, this result demonstrates that PHB and glycogen do not compete for CO_2 fixation products, but glycogen is epistatic over PHB synthesis.

Previous studies showed that, under nitrogen starvation, certain genes are upregulated, which are under the regulation of SigE, a group 2 σ factor [38]. Among these genes are glycogen degrading enzymes like *glgP1* and *glgP2* and *glgX*, but also the key enzymes for the pathways further downstream, namely *pfk* and *gnd*. The fact that all these genes are expressed simultaneously with the genes of the PHB synthesis [9], demonstrate that all relevant transcripts of the key enzymes required for the conversion from glycogen to PHB are present during nitrogen starvation. Our finding that PHB is mainly synthesized from glycogen degradation during nitrogen chlorosis is supported by a recent study, where a *Synechocystis* sp. PCC 6714 strain with enhanced PHB accumulation was created by random mutagenesis. Transcriptome analysis revealed that this strain exhibits an increased expression of glycogen phosphorylase [21]. This indicates that manipulation of glycogen metabolism may be a key for improved PHB synthesis.

Gaining further insights into the intracellular carbon fluxes could provide more information on how PHB production is regulated. Once the regulation is understood, this knowledge could be used to redirect the large quantities of glycogen towards PHB. This knowledge could be used in metabolic engineering approaches to either completely reroute the carbon from glycogen (making up more than 60% of the CDW) to PHB, for example by overexpression of glycogen degrading enzymes, or even from inorganic carbon to PHB directly. Therefore, the new insights from this work can be exploited for biotechnological applications to further increase the amounts of PHB being produced in cyanobacteria.

4. Materials and Methods

4.1. Cyanobacterial Cultivation Conditions

For standard cultivation, *Synechocystis* sp. PCC 6803 cells were grown in 200 mL BG_{11} medium, supplemented with 5 mM $NaHCO_3$ [39]. A list of the used strains of this study is provided in Table A1. Two different wild-type strains, a Glc sensitive and a Glc tolerant one, were used. Both strains showed the same behavior during normal growth as well as during chlorosis. Appropriate antibiotics were added to the different mutants to ensure the continuity of the mutation. The cells were cultivated at 28 °C, shaking at 120 rpm and constant illumination of 40–50 μmol photons m^{-2} s^{-1}. Nitrogen starvation was induced as described previously [40]. In short, exponentially growing cells (OD 0.4–0.8) were centrifuged for 10 min at 4000× *g*. The cells were washed in 100 mL of BG_0 (BG_{11} medium without $NaNO_3$) before they were centrifuged again. The resulting pellet was resuspended in BG_0 until it reached an OD of 0.4.

4.2. Microscopy and Staining Procedures

To observe PHB granules within the cells, 100 µL of cyanobacterial cells were centrifuged (1 min at 10,000× *g*) and 80 µL of the supernatant discarded. Nile Red (10 µL) was added and used to resuspend the pellet in the remaining 20 µL of the supernatant. From these mixtures, 10 µL were taken and applied on an agarose coated microscope slide to immobilize the cells. The Leica DM5500B microscope (Leica, Wetzlar, Germany) was used with a 100×/1.3 oil objective for fluorescence microscopy. To detect Nile red stained PHB granules, an excitation filter BP 535/50 was used, together with a suppression filter BP 610/75. A Leica DFC360FX (Leica, Wetzlar, Germany)) was used for image acquisition.

4.3. PHB Quantification

PHB content within the cells was determined as described previously [41]. Roughly 15 mL of cells were harvested and centrifuged at 4000× *g* for 10 min at 25 °C. The resulting pellet was dried for 3 h at 60 °C in a speed-vac (Christ, Osterode, Germany), before 1 mL of concentrated H_2SO_4 was added and boiled for 1 h at 100 °C to break up the cells and to convert PHB to crotonic acid. From this, 100 µL were taken and diluted in 900 µL 0.014 M H_2SO_4. To remove cell debris, the samples were centrifuged for 10 min at 10,000× *g*, before 500 µL of the supernatant were transferred to 500 µL 0.014 M H_2SO_4. After an additional centrifugation step with the same conditions as above, the supernatant was used for HPLC analysis on a Nucleosil 100 C 18 column (Agilent, Santa Clara, CA, USA) (125 by 3 mm). As a liquid phase, 20 mM phosphate buffer (pH 2.5) was used. Commercially available crotonic acids was used as a standard with a conversion ratio of 0.893. The amount of crotonic acid was detected at 250 nm.

4.4. Glycogen Quantification

Intracellular glycogen content was measured by harvesting 2 mL of cyanobacterial culture. The cells were washed twice with 1 mL of ddH_2O. Afterwards the pellet was resuspended in 400 µL KOH (30% *w/v*) and incubated for 2 h at 95 °C. For the subsequent glycogen precipitation, 1200 µL ice cold ethanol (final concentration of 70%) were added. The mixture was incubated at –20 °C for 2–24 h. Next, the solution was centrifuged at 4 °C for 10 min at 10,000× *g*. The pellet was washed twice with 70% and 98% ethanol and dried in a speed-vac for 20 min at 60 °C. Next, the pellet was resuspended in 1 mL of 100 mM sodium acetate (pH 4.5) and 8 µL of an amyloglucosidase solution (4.4 U/µL) were added. For the enzymatic digest, the cells were incubated at 60 °C for 2 h. For the spectrometical glycogen determination, 200 µL of the digested mixture was used and added to 1 mL of O-toluidine-reagent (6% O-toluidine in 100% acetic acid). The tubes were incubated for 10 min at 100 °C. The samples were cooled down on ice for 3 min, before the OD_{635} was measured. The final result was normalized to the cell density at OD_{750}, where $OD_{750} = 1$ represents 10^8 cells. A glucose standard curve was used to calculate the glucose contents in the sample from their OD540.

4.5. Drop Agar Method

Serial dilutions of chlorotic cultures were prepared (10^0, 10^{-1}, 10^{-2}, 10^{-3}, 10^{-4} and 10^{-5}) starting with an OD_{750} of 1. Five microliters of these dilutions were dropped on solid BG_{11} agar plates and cultivated at 50 µmol photons m^{-2} s^{-1} and 27 °C for 7 days.

Author Contributions: Conceptualization, M.K. and K.F.; Methodology, M.K., S.D. and K.F.; Investigation, M.K. and S.D.; Writing-Original Draft Preparation, M.K., S.D. and K.F.; Writing-Review & Editing, M.K., S.D., K.G. and K.F.; Supervision, K.F.; Project Administration, M.K. and K.F.

Funding: This research was funded by the Studienstiftung des Deutschen Volkes and the RTG 1708 "Molecular principles of bacterial survival strategies". We acknowledge support by Deutsche Forschungsgemeinschaft and Open Access Publishing Fund of University of Tübingen.

Acknowledgments: We would like to thank Yvonne Zilliges for providing the mutants ΔglgA1, ΔglgA2 and ΔglgC. Furthermore, we thank Eva Nussbaum for maintaining the strain collection and Andreas Kulick for assistance with HPLC analysis.

Conflicts of Interest: The authors declare no conflict of interest.

Appendix A

Table A1. List of used strains.

Strain	Background	Relevant Marker of Genotype	Reference
Synechocystis sp. PCC 6803 GS	GS	-	Pasteur culture collection
Synechocystis sp. PCC 6803 GT	GT	-	Chen et al. 2016
Δ*glgA1*	GT	*sll0945::kmR*	Gründel et al. 2012
Δ*glgA2*	GT	*sll1393::cmR*	Gründel et al. 2012
Δ*glgC*	GT	*slr1176::cmR*	Damrow et al. 2012
Δ*glgP1*	GS	*sll1356::kmR*	Doello et al. 2018
Δ*glgP2*	GS	*slr1367::spR*	Doello et al. 2018
Δ*glgP1/2*	GS	*sll1356::kmR, slr1367::spR*	Doello et al. 2018
Δ*eda*	GT	*sll0107::gmR*	Chen et al. 2016
Δ*gnd*	GT	*sll0329::gmR*	Chen et al. 2016
Δ*pfkB1/2*	GT	*sll1196::kmR, sll0745::spR*	Chen et al. 2016
Δ*phaEC*	GS	*slr1829, slr1830::kmR*	Klotz et al. 2016

References

1. Soo, R.M.; Hemp, J.; Parks, D.H.; Fischer, W.W.; Hugenholtz, P. On the origins of oxygenic photosynthesis and aerobic respiration in Cyanobacteria. *Science* **2017**, *355*, 1436–1440. [CrossRef] [PubMed]
2. Vitousek, P.M.; Howarth, R.W. Nitrogen Limitation on Land and in the Sea—How Can. It Occur. *Biogeochemistry* **1991**, *13*, 87–115. [CrossRef]
3. Kaneko, T.; Tanaka, A.; Sato, S.; Kotani, H.; Sazuka, T.; Miyajima, N.; Sugiura, M.; Tabata, S. Sequence analysis of the genome of the unicellular cyanobacterium *Synechocystis* sp. strain PCC6803. I. Sequence features in the 1 Mb region from map positions 64% to 92% of the genome. *DNA Res.* **1995**, *2*, 153–166. [CrossRef]
4. Forchhammer, K.; Schwarz, R. Nitrogen chlorosis in unicellular cyanobacteria—a developmental program for surviving nitrogen deprivation. *Environ. Microbiol.* **2018**. [CrossRef] [PubMed]
5. Osanai, T.; Oikawa, A.; Shirai, T.; Kuwahara, A.; Iijima, H.; Tanaka, K.; Ikeuchi, M.; Kondo, A.; Saito, K.; Hirai, M.Y. Capillary electrophoresis-mass spectrometry reveals the distribution of carbon metabolites during nitrogen starvation in *Synechocystis* sp. PCC 6803. *Environ. Microbiol.* **2014**, *16*, 512–524. [CrossRef] [PubMed]
6. Grundel, M.; Scheunemann, R.; Lockau, W.; Zilliges, Y. Impaired glycogen synthesis causes metabolic overflow reactions and affects stress responses in the cyanobacterium *Synechocystis* sp. PCC 6803. *Microbiology* **2012**, *158*, 3032–3043. [CrossRef] [PubMed]
7. Doello, S.; Klotz, A.; Makowka, A.; Gutekunst, K.; Forchhammer, K. A Specific Glycogen Mobilization Strategy Enables Rapid Awakening of Dormant Cyanobacteria from Chlorosis. *Plant Physiol.* **2018**, *177*, 594–603. [CrossRef]
8. Damrow, R.; Maldener, I.; Zilliges, Y. The Multiple Functions of Common Microbial Carbon Polymers, Glycogen and PHB, during Stress Responses in the Non-Diazotrophic Cyanobacterium *Synechocystis* sp. PCC 6803. *Front. Microbiol.* **2016**, *7*, 966. [CrossRef]
9. Klotz, A.; Georg, J.; Bučinská, L.; Watanabe, S.; Reimann, V.; Januszewski, W.; Sobotka, R.; Jendrossek, D.; Hess, W.R.; Forchhammer, K. Awakening of a Dormant Cyanobacterium from Nitrogen Chlorosis Reveals a Genetically Determined Program. *Curr. Biol.* **2016**, *26*, 2862–2872. [CrossRef]
10. Ansari, S.; Fatma, T. Cyanobacterial Polyhydroxybutyrate (PHB): Screening, Optimization and Characterization. *PLoS ONE* **2016**, *11*, e0158168. [CrossRef]
11. Nowroth, V.; Marquart, L.; Jendrossek, D. Low temperature-induced viable but not culturable state of Ralstonia eutropha and its relationship to accumulated polyhydroxybutyrate. *FEMS Microbiol. Lett.* **2016**, *363*, fnw249. [CrossRef]

12. Batista, M.B.; Teixeira, C.S.; Sfeir, M.Z.T.; Alves, L.P.S.; Valdameri, G.; Pedrosa, F.O.; Sassaki, G.L.; Steffens, M.B.R.; de Souza, E.M.; Dixon, R.; et al. PHB Biosynthesis Counteracts Redox Stress in Herbaspirillum seropedicae. *Front. Microbiol.* **2018**, *9*, 472. [CrossRef] [PubMed]

13. Urtuvia, V.; Villegas, P.; González, M.; Seeger, M. Bacterial production of the biodegradable plastics polyhydroxyalkanoates. *Int. J. Biol. Macromol.* **2014**, *70*, 208–213. [CrossRef] [PubMed]

14. Li, W.C.; Tse, H.F.; Fok, L. Plastic waste in the marine environment: A review of sources, occurrence and effects. *Sci. Total Environ.* **2016**, *566*, 333–349. [CrossRef] [PubMed]

15. Drosg, B.; Gattermayr, F.; Silvestrini, L. Photo-autotrophic Production of Poly(hydroxyalkanoates) in Cyanobacteria. *Chem. Biochem. Eng. Q.* **2015**, *29*, 145–156. [CrossRef]

16. Martin, K.; Lukas, M. Cyanobacterial Polyhydroxyalkanoate Production: Status Quo and Quo Vadis? *Curr. Biotechnol.* **2015**, *4*, 464–480.

17. Singh, A.; Sharma, L.; Mallick, N.; Mala, J. Progress and challenges in producing polyhydroxyalkanoate biopolymers from cyanobacteria. *J. Appl. Phycol.* **2017**, *29*, 1213–1232.

18. Lau, N.S.; Foong, C.P.; Kurihara, Y.; Sudesh, K.; Matsui, M. RNA-Seq Analysis Provides Insights for Understanding Photoautotrophic Polyhydroxyalkanoate Production in Recombinant *Synechocystis* sp. *PLoS ONE* **2014**, *9*, e86368. [CrossRef]

19. Khetkorn, W.; Incharoensakdi, A.; Lindblad, P.; Jantaro, S. Enhancement of poly-3-hydroxybutyrate production in *Synechocystis* sp. PCC 6803 by overexpression of its native biosynthetic genes. *Bioresour. Technol.* **2016**, *214*, 761–768. [CrossRef]

20. Carpine, R. Genetic engineering of *Synechocystis* sp. PCC6803 for poly-β-hydroxybutyrate overproduction. *Algal Res. Biomass Biofuels Bioprod.* **2017**, *25*, 117–127. [CrossRef]

21. Kamravamanesh, D.; Kovacs, T.; Pflügl, S.; Druzhinina, I.; Kroll, P.; Lackner, M.; Herwig, C. Increased poly-beta-hydroxybutyrate production from carbon dioxide in randomly mutated cells of cyanobacterial strain *Synechocystis* sp. PCC 6714: Mutant generation and characterization. *Bioresour. Technol.* **2018**, *266*, 34–44. [CrossRef]

22. Steuer, R.; Knoop, H.; Machné, R. Modelling cyanobacteria: From metabolism to integrative models of phototrophic growth. *J. Exp. Bot.* **2012**, *63*, 2259–2274. [CrossRef]

23. Dutt, V.; Srivastava, S. Novel quantitative insights into carbon sources for synthesis of poly hydroxybutyrate in Synechocystis PCC 6803. *Photosynth. Res.* **2018**, *136*, 303–314. [CrossRef]

24. Rajendran, V.; Incharoensakdi, A. Disruption of polyhydroxybutyrate synthesis redirects carbon flow towards glycogen synthesis in *Synechocystis* sp. PCC 6803 overexpressing glgC/glgA. *Plant Cell Physiol.* **2018**, *59*, 2020–2029.

25. Chen, X.; Schreiber, S.; Appel, J.; Makowka, A.; Fähnrich, B.; Roettger, M.; Hajirezaei, M.R.; Sönnichsen, F.D. The Entner-Doudoroff pathway is an overlooked glycolytic route in cyanobacteria and plants. *Proc. Natl. Acad. Sci. USA* **2016**, *113*, 5441–5446. [CrossRef]

26. Yu, J.; Liberton, M.; Cliften, P.F.; Head, R.D.; Jacobs, J.M.; Smith, R.D.; Koppenaal, D.W.; Brand, J.J.; Pakrasi, H.B. Synechococcus elongatus UTEX 2973, a fast growing cyanobacterial chassis for biosynthesis using light and CO2. *Sci. Rep.* **2015**, *5*, 8132. [CrossRef] [PubMed]

27. Osanai, T.; Oikawa, A.; Numata, K.; Kuwahara, A.; Iijima, H.; Doi, Y.; Saito, K.; Hirai, M.Y. Pathway-level acceleration of glycogen catabolism by a response regulator in the cyanobacterium Synechocystis species PCC 6803. *Plant Physiol.* **2014**, *164*, 1831–1841. [CrossRef] [PubMed]

28. Yoo, S.H.; Lee, B.H.; Moon, Y.; Spalding, M.H.; Jane, J.L. Glycogen Synthase Isoforms in *Synechocystis* sp. PCC6803: Identification of Different Roles to Produce Glycogen by Targeted Mutagenesis. *PLoS ONE* **2014**, *9*, e91524. [CrossRef]

29. Namakoshi, K.; Nakajima, T.; Yoshikawa, K.; Toya, Y.; Shimizu, H. Combinatorial deletions of glgC and phaCE enhance ethanol production in *Synechocystis* sp. PCC 6803. *J. Biotechnol.* **2016**, *239*, 13–19. [CrossRef]

30. Kamravamanesh, D.; Slouka, C.; Limbeck, A.; Lackner, M.; Herwig, C. Increased carbohydrate production from carbon dioxide in randomly mutated cells of cyanobacterial strain *Synechocystis* sp. PCC 6714: Bioprocess. understanding and evaluation of productivities. *Bioresour. Technol.* **2019**, *273*, 277–287. [CrossRef] [PubMed]

31. Wang, C.X.; Saldanha, M.; Sheng, X.; Shelswell, K.J.; Walsh, K.T.; Sobral, B.W.; Charles, T.C. Roles of poly-3-hydroxybutyrate (PHB) and glycogen in symbiosis of Sinorhizobium meliloti with *Medicago* sp. *Microbiology* **2007**, *153*, 388–398. [CrossRef] [PubMed]

32. Sauer, J.; Schreiber, U.; Schmid, R.; Völker, U.; Forchhammer, K. Nitrogen starvation-induced chlorosis in Synechococcus PCC 7942. Low-level photosynthesis as a mechanism of long-term survival. *Plant Physiol.* **2001**, *126*, 233–243. [CrossRef] [PubMed]

33. Lever, M.A.; Rogers, K.L.; Lloyd, K.G.; Overmann, J.; Schink, B.; Thauer, R.K.; Hoehler, T.M.; Jørgensen, B.B. Life under extreme energy limitation: A synthesis of laboratory- and field-based investigations. *FEMS Microbiol. Rev.* **2015**, *39*, 688–728. [CrossRef]

34. Kempes, C.P.; van Bodegom, P.M.; Wolpert, D.; Libby, E.; Amend, J.; Hoehler, T. Drivers of Bacterial Maintenance and Minimal Energy Requirements. *Front. Microbiol.* **2017**, *8*, 31. [CrossRef]

35. Osanai, T.; Azuma, M.; Tanaka, K. Sugar catabolism regulated by light- and nitrogen-status in the cyanobacterium *Synechocystis* sp. PCC 6803. *Photochem. Photobiol. Sci.* **2007**, *6*, 508–514. [CrossRef] [PubMed]

36. Cano, M.; Holland, S.C.; Artier, J.; Burnap, R.L.; Ghirardi, M.; Morgan, J.A.; Yu, J. Glycogen Synthesis and Metabolite Overflow Contribute to Energy Balancing in Cyanobacteria. *Cell Rep.* **2018**, *23*, 667–672. [CrossRef]

37. Benson, P.J.; Purcell-Meyerink, D.; Hocart, C.H.; Truong, T.T.; James, G.O.; Rourke, L.; Djordjevic, M.A.; Blackburn, S.I.; Price, G.D. Factors Altering Pyruvate Excretion in a Glycogen Storage Mutant of the Cyanobacterium, Synechococcus PCC7942. *Front. Microbiol.* **2016**, *7*, 475. [CrossRef]

38. Osanai, T.; Kanesaki, Y.; Nakano, T.; Takahashi, H.; Asayama, M.; Shirai, M.; Kanehisa, M.; Suzuki, I.; Murata, N.; Tanaka, K. Positive regulation of sugar catabolic pathways in the cyanobacterium *Synechocystis* sp. PCC 6803 by the group 2 sigma factor sigE. *J. Biol. Chem.* **2005**, *280*, 30653–30659. [CrossRef]

39. Rippka, R.; Deruelles, D.; Waterbury, J.B.; Herdman, M.; Stanier, R.Y. Generic Assignments, Strain Histories and Properties of Pure Cultures of Cyanobacteria. *J. Gen. Microbiol.* **1979**, *111*, 1–61. [CrossRef]

40. Schlebusch, M.; Forchhammer, K. Requirement of the Nitrogen Starvation-Induced Protein Sll0783 for Polyhydroxybutyrate Accumulation in *Synechocystis* sp. Strain PCC 6803. *Appl. Environ. Microbiol.* **2010**, *76*, 6101–6107. [CrossRef]

41. Taroncher-Oldenburg, G.; Nishina, K.; Stephanopoulos, G. Identification and analysis of the polyhydroxyalkanoate-specific beta-ketothiolase and acetoacetyl coenzyme A reductase genes in the cyanobacterium *Synechocystis* sp. strain PCC6803. *Appl. Environ. Microbiol.* **2000**, *66*, 4440–4448. [CrossRef] [PubMed]

© 2019 by the authors. Licensee MDPI, Basel, Switzerland. This article is an open access article distributed under the terms and conditions of the Creative Commons Attribution (CC BY) license (http://creativecommons.org/licenses/by/4.0/).

International Journal of
Molecular Sciences

MDPI

Article

Evaluation of Mechanical and Interfacial Properties of Bio-Composites Based on Poly(Lactic Acid) with Natural Cellulose Fibers

Laura Aliotta [1,2], Vito Gigante [1], Maria Beatrice Coltelli [1], Patrizia Cinelli [1,2,*] and Andrea Lazzeri [1,2,*]

[1] Department of Civil and Industrial Engineering, University of Pisa, Via Diotisalvi, 2, 56122 Pisa, Italy; laura.aliotta@dici.unipi.it (L.A.); vito.gigante@dici.unipi.it (V.G.); maria.beatrice.coltelli@unipi.it (M.B.C.)
[2] Interuniversity National Consortium of Materials Science and Technology (INSTM), Via Giusti 9, 50121 Florence, Italy
* Correspondence: patrizia.cinelli@unipi.it (P.C.); andrea.lazzeri@unipi.it (A.L.); Tel.: +39-050-2217869 (P.C.); +39-050-2217807 (A.L.)

Received: 31 December 2018; Accepted: 17 February 2019; Published: 22 February 2019

Abstract: The circular economy policy and the interest for sustainable material are inducing a constant expansion of the bio-composites market. The opportunity of using natural fibers in bio-based and biodegradable polymeric matrices, derived from industrial and/or agricultural waste, represents a stimulating challenge in the replacement of traditional composites based on fossil sources. The coupling of bioplastics with natural fibers in order to lower costs and promote degradability is one of the primary objectives of research, above all in the packaging and agricultural sectors where large amounts of non-recyclable plastics are generated, inducing a serious problem for plastic disposal and potential accumulation in the environment. Among biopolymers, poly(lactic acid) (PLA) is one of the most used compostable, bio-based polymeric matrices, since it exhibits process ability and mechanical properties compatible with a wide range of applications. In this study, two types of cellulosic fibers were processed with PLA in order to obtain bio-composites with different percentages of microfibers (5%, 10%, 20%). The mechanical properties were evaluated (tensile and impact test), and analytical models were applied in order to estimate the adhesion between matrix and fibers and to predict the material's stiffness. Understanding these properties is of particular importance in order to be able to tune and project the final characteristics of bio-composites.

Keywords: bio-composites; mechanical properties; poly(lactic acid); cellulose fibers

1. Introduction

The increasing environmental awareness coupled with the circular economy policy, supported by new regulations, are driving plastic industries as well as consumers toward the selection of ecologically friendly raw materials for their plastic products. Several products developed for large application fields are based on natural fibers in composites with a polypropylene matrix [1]. These materials are not compostable and are hardly recyclable; thus, work is in progress to investigate new composites with biopolymers as polymeric matrices (bio-composites), offering the advantage of bio-recycling options at the end of their service life through composting or anaerobic digestion. In this contest, bio-based polymers reinforced with natural fibers are beneficial to prepare biodegradable composite materials [2]. The most used biopolymers for this application are poly(lactic acid) (PLA), cellulose esters, polyhydroxyalkanoates (PHAs), and starch-based plastics [3,4].

In applications where biodegradability offers clear advantages for customers and the environment, such as single-use applications (packaging and agriculture), it is expected that the demand for these biopolymers will increase [5–7].

In this context, poly(lactic acid) (PLA) is certainly one of the best candidates, being compostable and produced from renewable resources such as sugar beets or corn starch [8,9]. In addition to its biodegradability and renewability, PLA exhibits at room temperature a Young's modulus of about 3 GPa, a tensile strength between 50 and 70 MPa with an elongation at break of about 4%, and an impact strength close to 2.5 kJ/m^2 [10].

Although PLA is considered a sustainable alternative to traditional petroleum-based plastics, many drawbacks must be overcome in order to enlarge its application field. In particular, PLA has a relatively higher cost (2.5–3.0 Euro/Kg) compared to commodity petro-derived polymers (1 Euro/Kg), it has low flexibility, bad impact resistance, low thermal stability (due to its high glass transition temperature, $T_g \approx 60$ °C), and low crystallization rates that could limit its applications [11].

Generally, composite materials show enhanced mechanical and physical properties when compared to their individual composite components [12,13]. However, especially when the fibers are very short and randomly oriented, the resulting composite does not necessarily provide enhanced properties. In this case, the benefit of composite production is envisaged in cost savings, lighter weight, and promoted degradability.

The study of the interaction between the fiber and the polymeric matrix in a composite plays an important role because it influences both physical and mechanical properties of the final materials. In particular, the adhesion—that is the ability to transfer stresses across the interface—is often related to a combination of different factors such as the interface thickness, the interphase layer, the adhesion strength, and the surface energy of the fibers [14–16].

Natural fibers have many advantages compared to synthetic ones. They are recyclable, biodegradable, renewable, have relatively high strength and stiffness, and do not cause skin irritation [17]. On the other hand, there are also some disadvantages such as moisture uptake, the presence of color, the presence of odor when heated or burned during processing, quality variations, and low thermal stability. Many investigations have been carried out on the potential of natural fibers as reinforcement for composites, and in several cases the results have shown that natural fiber composites reached a good stiffness, but their final strength was not improved [18,19].

Composite manufacturing industries are looking for plant-based natural fiber reinforcements, such as flax, hemp, jute, sisal, kenaf, and banana as alternative materials to replace synthetic fibers. Lignocellulose fibers have also been considered for replacing glass fibers [20] as lignocellulose fibers are cheaper, lighter than glass fibers, and safer to be handled by workers [21].

Due to their advantages of low cost, biodegradability, large availability, and valuable mechanical and physical properties [22], a wide variety of lignocellulose fibers and natural fillers—coming from agricultural and industrial crops such as corn, wheat, bagasse, orange and apple peel algae, and sea grasses-derived fibers—have been used in the production of composites in various industrial sectors, such as packaging, automotive industry, and building [23–26].

For these reasons, several bio-composites were produced with a polymeric biodegradable matrix such as PLA and natural fibers. Tserki et al. [27] investigated the usefulness of lignocellulose waste flours derived from spruce, olive husks, and paper flours as potential reinforcements for the preparation of cost-effective bio-composites using PLA as the matrix. Petinakis et al. [28] studied the effect of wood flour content on the mechanical properties and fracture behavior of PLA/wood flour composites. In several natural fiber bio-composites with PLA as the matrix, the interfacial adhesion between the polymeric matrix and the fibers was poor [29]. Thus, the incorporation of lignocellulose materials into biodegradable polymer matrices, such as PLA, generally has the effect of improving the mechanical properties, such as tensile modulus, but sometimes the strength and toughness of these bio-composites are not improved.

Although several reviews [30–33] deal with lignocellulose-based composites including preparation methods and properties, most of them do not consider a deep analysis of interfacial adhesion between fiber and matrix or the application of mathematical models to explain them, which are very useful for predicting and tuning the properties of bio-composites. Thus, work remains to be done on the collective analysis of various applications of cellulose-based material. Natural fibers contain large amount of cellulose, hemicelluloses, lignin, and pectin, tending to be polar and hydrophilic, while polymeric materials are generally not polar and exhibit significant hydrophobicity [34]. The weak interfacial bonding between highly polar natural fibers and a non-polar organophilic matrix can lead to the worsening of the final properties of the bio-composites, ultimately hindering their industrial usage. Different strategies have been applied to eliminate this deficiency in compatibility and interfacial bond strength, including the use of surface modification techniques [35].

The hydrophilic nature of natural fibers decreases their adhesion to a hydrophobic matrix and, as a result, it may cause a loss of strength. To prevent this, the fiber surface may be modified in order to promote adhesion. Several methods have been proposed to modify natural fibers' surface, such as graft copolymerization of monomers onto the fiber surface and the use of maleic anhydride copolymers, alkyl succinic anhydride, stearic acid, etc. [36].

In this work, different amounts of two types of short cellulosic fibers (with different aspect ratios) added in a PLA polymeric matrix were investigated to evaluate the final effect on the mechanical properties. Furthermore, in order to have an estimation of the matrix/fiber adhesion, the B parameter calculated from the Pukanszky's model [37] was determined. The increase in stiffness of the final composite was also investigated using different analytical models existing in the literature with the aim to find the best fit with experimental data.

2. Results and Discussion

Results of the thermal gravimetric analysis are reported in Figure 1. From the weight loss peaks in the weight-to-temperature graph, it is evident that the fibers will not degrade during extrusion and injection molding, since the maximum temperature reached during processing is similar to the extrusion temperature, that was equal to 190 °C. This is an advantage of cellulose fibers versus other natural fibers which very often present thermal degradation during processing with negative effects on color and odor of the produced bio-composites.

In the graph, a small weight loss at temperatures lower than 100 °C can be observed and attributed to the loss of the residual moisture trapped in the fibers. The degradation of the fibers occurs at relatively quite high temperatures, beyond 300 °C, well above those reached in the processing of composites. Consequently, we can expect that the fibers inside the composites are stable and are not degraded, as confirmed by the nice white color of the composites and the absence of odor.

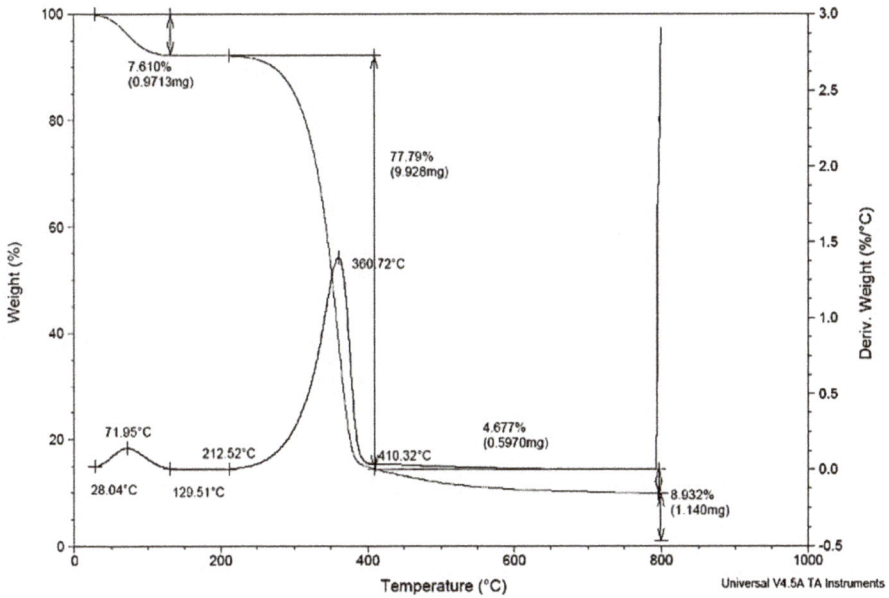

(a)

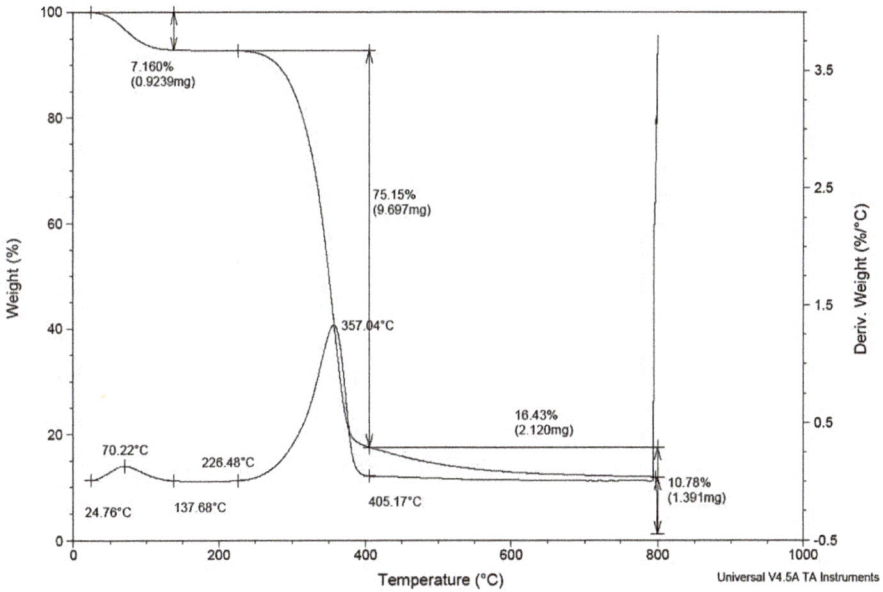

(b)

Figure 1. Thermogravimetric analysis (TGA) graphs of: (**a**) Arbocel® 600BE/PU and (**b**) Arbocel® BWW40.

From the results of the mechanical tests, it can be observed that, as expected, increasing the fiber content increases the elastic modulus of the composites (Figure 2a), in agreement with the trend normally observed in other studies in which cellulosic fibers were used [38] in polymeric matrices.

This behavior is very common, and the stiffness increment is generally related to the higher rigidity of the reinforcement versus the polymeric matrix However, BWW40 fibers show less marked increments in the Young's modulus compared to 600BE/PU fibers. This is likely due to the fibers' orientation and their higher aspect ratio with respect to the former. The higher aspect ratio for BWW40 can in fact cause twisting phenomena (that in general are encountered for natural fibers [39,40]) that can influence not only the elastic modulus but also the fibers' adhesion.

On the other hand, no significant increase in the final strength and strain at break of the composite are registered (Figure 2b,c). In particular for the composite with Arbocel® BWW 40, the stress at break decreases with the filler content. Even the Charpy impact resistance does not show significant improvements (Figure 2d). From these results, we can suppose that a very little or entirely null stress transfer takes place between the fiber and the matrix, due to a lack of fibers–matrix adhesion.

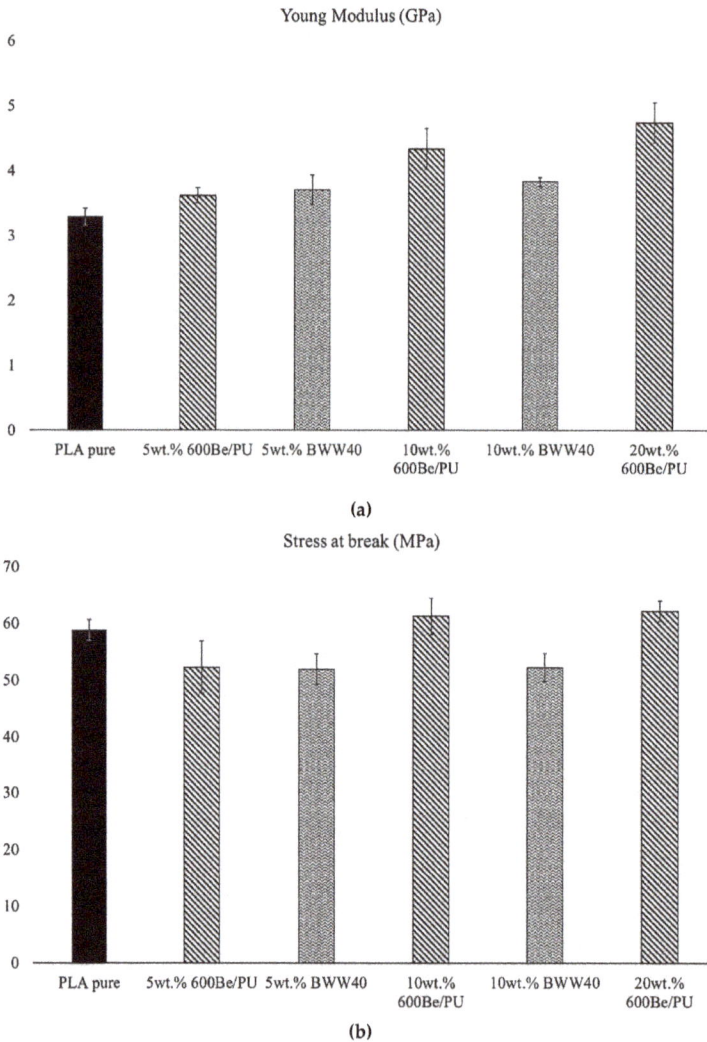

(a)

(b)

Figure 2. *Cont.*

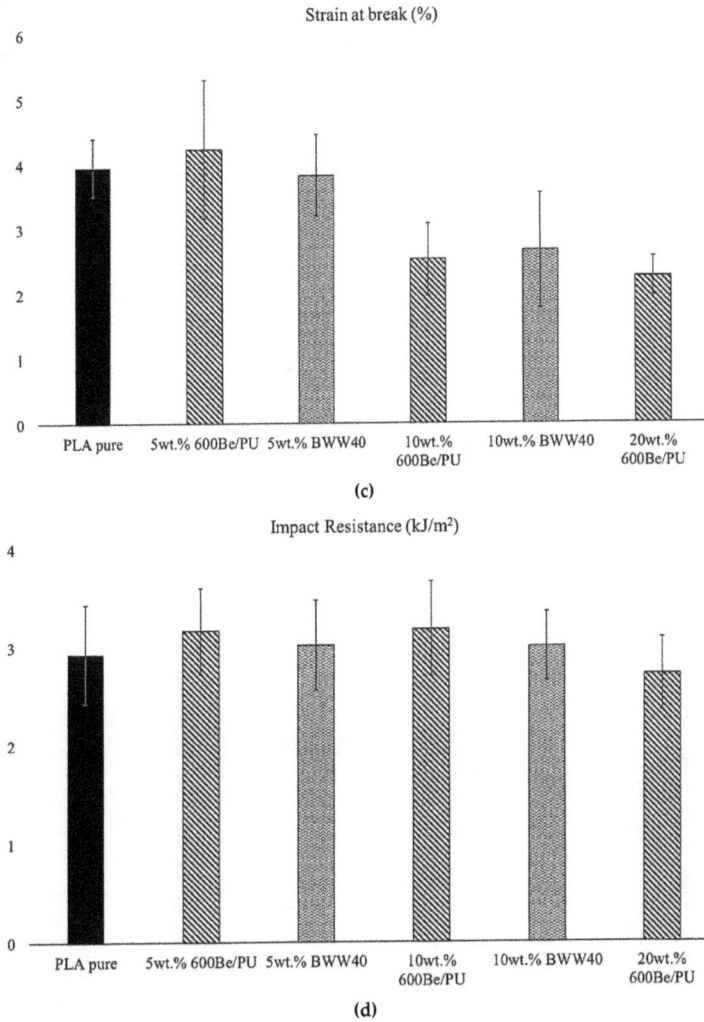

Figure 2. Mechanical properties of poly(lactic acid) (PLA)–Arbocel® composites: (**a**) Young's modulus, (**b**) stress at break, (**c**) strain at break, and (**d**) impact resistance.

In Figure 3, the Pukánszky's plot for the two different types of Arbocel® used is reported. An approximately linear trend of $ln\ \sigma_{red}$ can be extrapolated to calculate the B parameter.

The values obtained for the parameter B (reported in Table 1) confirmed that the adhesion between these cellulosic fibers and the PLA matrix is very low, explaining the results of the mechanical tests in which no significant improvements in strength were observed.

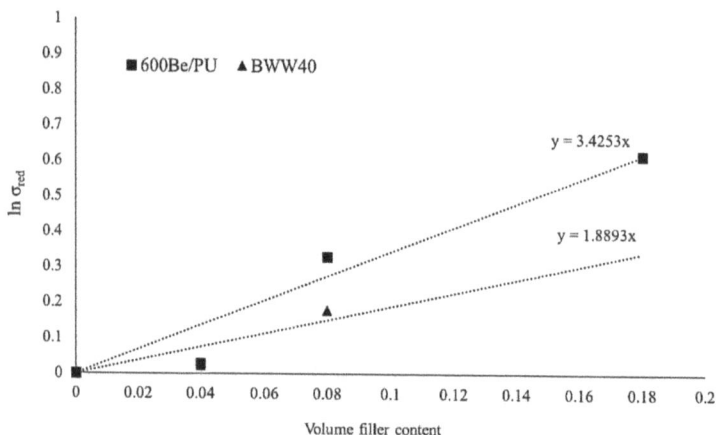

Figure 3. Pukánszky's plot for PLA–Arbocel® composites.

Table 1. B values.

Arbocel® Type	B
600BE/PU	3.42
BWW40	1.89

In particular, for BWW40 fibers, the B parameter is lower than that of 600BE/PU fibers. This means worse adhesion that can also be related to the higher aspect ratio and possible twisting phenomena. It is therefore explained why a moderate loss in tensile strength was observed for this type of composite.

In Figure 4, two SEM images are displayed reporting details of both Arbocel® fibers within the PLA matrix. We can observe a detachment and pull-out of the fibers due to the poor matrix–fibers adhesion.

Figure 4. SEM micrographs of composites: (**a**) PLA + 10 wt % Be600/PU, (**b**) PLA + 10 wt % BWW40.

The results obtained are consistent with those in the literature, in which a lack of adhesion was encountered in similar composite materials. The study of interfacial adhesion is, in fact, a well-known problem when natural fibers and synthetic polymers are used [4,41]. A compatibilization is necessary if we want to obtain a composite with tailored mechanical properties and good efficiency in the transferring of the stress from the matrix to the fibers.

Furthermore, in this work, multiple analytical models were applied to investigate which better predicts the experimental data. The results of these analyses are reported in Figure 5.

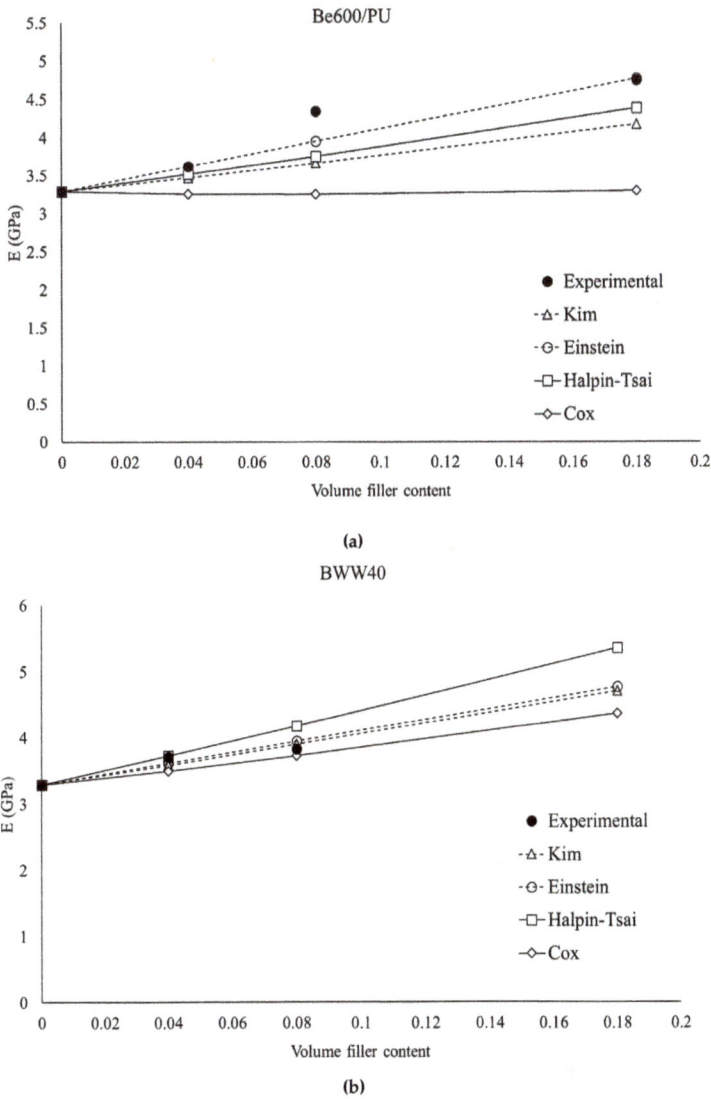

(a)

(b)

Figure 5. Comparison between experimental elastic modulus of the blends and mathematical models for (a) Be600/PU–PLA composites and (b) BWW40–PLA composites.

It can be observed that the Cox model provides underestimated stiffness predictions. This is due to the fact that the Cox model is referred to as the shear lag theory in which long, straight, and discontinuous fibers completely embedded in a continuous matrix were considered [42]. Consequently, this model is not very accurate when the fibers' aspect ratios are very small [43] and the adhesion is not good. This explains why the prediction is completely inaccurate for Be600/PU fibers, which have a very short aspect ratio (equal to 3), and improves with BWW40 fibers (aspect ratio equal to 10).

The Kim's model derives from the shear lag theory (like the Cox model) but was extended to resolve discrepancies of the Cox model in the case of short fiber-reinforced composites. In this case, the predicted values of the elastic modulus were similar to the experimental ones [43]. Effectively, in our case, this model gives good results for both types of fiber and consequently may be efficiently applied to these systems.

The Einstein's method is very simple and in general is applied for spheres, hence for fillers having a low aspect ratio. This model does not contain information about the geometry of the reinforcement, but the stiffness of the composites, depends only on the filler volume content [44]. Despite its simplicity, this model is able to efficiently estimate the composites' stiffness, probably because of the fibers' low aspect ratio.

The Halpin-Tsai equation also fits the experimental data with good accuracy. This model is often used to predict the elastic modulus of fibers that are randomly oriented on a plane [30]. In our case, the fibers are very short and not aligned. As a consequence, this model that contains an expression for the evaluation of both longitudinal and transversal moduli provides a good, but not perfect fitting.

3. Material and Methods

3.1. Materials and Characterization

The PLA used was 2003D, derived from natural resources and purchased from NatureWorks (Minnetonka, MN, USA) (grade for thermoforming and extrusion processes) [melt flow index (MFI): 6 g/10 min (210 °C, 2.16 kg), nominal average molar mass: 200,000 g/mol]. This type of PLA contains about 3–6% of D-lactic acid units in order to lower the melting point and the crystallization tendency, improving the processing ability. Two different types of commercial cellulosic short fibers, kindly provided by J Rettenmaier Sohne® (Rosenberg, Germany), were used. The trade names and the main properties of these two types of fibers are:

- ARBOCEL® 600BE/PU (mean diameter 20 μm, mean fiber length 60 μm, and consequently, aspect ratio 3, bulk density: 200–260 g/L, fiber density 1.44 g/cm^3)
- ARBOCEL® BWW40 (mean diameter 20 μm, mean fiber length 200 μm, and consequently, aspect ratio 10, bulk density: 110–145 g/L, fiber density 1.44 g/cm^3)

The morphology of these Arbocel® fibers before processing is shown in Figure 6.

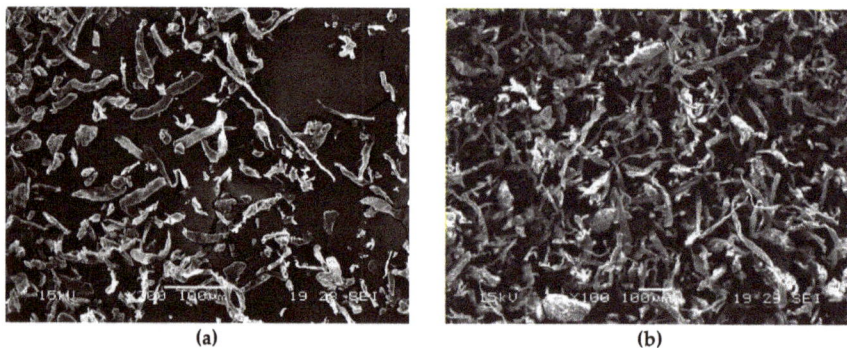

| (a) | (b) |

Figure 6. Scanning Electron Microscope (SEM) micrograph showing micro-cellulose fibers before processing: (**a**) Arbocel® 600BE7PU and (**b**) Arbocel® BWW40.

An increasing amount of cellulose fibers (at 5, 10, and 20 wt % corresponding to 4, 8, and 18 vol %, respectively) were added to the PLA matrix in order to produce bio-composites.

The materials, dried for at least 24 h in an air-circulated oven, were mixed in the correct quantities and then processed on a Thermo Scientific MiniLab Haake (Vreden. Germany)twin-screw extruder at a screw rate of 110 rpm/min and a cycle time of 60 s. After extrusion, the molten materials were transferred through a preheated cylinder to a Thermo Scientific Haake MiniJet II mini injection molder, for the preparation of the specimens for the Charpy and tensile tests. The Haake MiniJet II was equipped with an internal microprocessor capable of monitoring all the working parameters such as time, temperature, and injection pressure. The operative conditions of extrusion and injection molding are reported in Table 2.

Table 2. Processing conditions of Minilab and Minijet.

Minilab	
Extrusion temperature (°C)	190
Cycle time (s)	60
Screw rate (rpm)	110
Minijet	
Cylinder temperature (°C)	190
Mould temperature (°C)	60
Pressure (bar)	680
Residence time (s)	15

It is important to observe that for the blends containing ARBOCEL® BWW40, it was impossible to produce specimens containing 20 wt % of fibers because the molten material was too viscous, and the specimens were not consistent.

The tensile and impact properties of pure PLA and its composites containing different percentages of Arbocel® fibers were determined.

The tensile tests were carried out at room temperature, at a crosshead speed of 10 mm/min on an Instron universal testing machine 5500R equipped with a 10 kN load cell and interfaced with a computer running MERLIN software (INSTRON version 4.42 S/N–014733H) 24 h after specimen production. At least five specimens (gauge dimensions: 25 × 5 × 1.5 mm) were tested for each sample, and the average values reported.

The impact tests were performed on V-notched specimens (width: 10 mm, length: 80 mm, thickness: 4 mm, V-notch 2 mm) using a 15 J Charpy pendulum of an Instron CEAST 9050. The standard ISO179:2000 was followed. At least 10 specimens for each blend were tested at room temperature.

The fibers and their composites were investigated by Scanning Electron Microscope (SEM) (FEI Quanta 450 FEG).

The thermogravimetric analysis (TGA) of Arbocel® fibers was also performed on a TGA Rheometric Scientific at a scanning velocity of 10 °C/min from room temperature up to 1000 °C, using nitrogen as purge gas.

3.2. Theoretical Analysis

In this work, different analytical models were applied in order to estimate the fiber/matrix adhesion and to predict the elastic modulus of PLA–cellulose composites containing different amount of fibers.

It is well known that the strength of a composite varies on the basis of its fiber content. Adhesion between fibers and polymeric matrix has a very large effect on this property, and in particular, it was demonstrated that the reinforcement characteristics seem to have a larger effect on strength than on stiffness [45]. For rigid fillers and for fibers with a low aspect ratio (as in this study), the reinforcing effect of a filler or a fiber can be expressed quantitatively by the following equation, proposed by Pukánszky [14]:

$$\sigma_c = \sigma_m \frac{1 - \varphi_f}{1 + 2.5\varphi_f} exp\left(B\varphi_f\right) \tag{1}$$

where the terms σ_c and σ_m, in this case, are the tensile stress of the composite and of the matrix, respectively, φ_f is the volumetric filler fraction, while the term $(1 - \varphi_f)/(1 + 2.5\varphi_f)$ indicates the decreasing of the effective load-bearing cross section due to reinforcement introduction. Finally, the term $\exp{(B\varphi_f)}$ takes into account the filler–matrix interactions, by means of the interaction parameter B [41]. We can write Equation (1) in linear form:

$$ln(\sigma_{red}) = log\frac{\sigma_c \left(1 + 2.5\varphi_f\right)}{\sigma_m \left(1 - \varphi_f\right)} = B\varphi_f \tag{2}$$

Plotting the natural logarithm of Pukánszky's reduced tensile strength (that is adimensional) against volume fraction (in the following graph this will be named Pukánszky's plot) results in a linear correlation in which the linear slope is proportional to the interaction parameter B [37]. In this way, by applying Equation (2), it is possible to calculate the B parameter for the two different types of fibers and consequently obtain a simple estimation of their adhesion to the PLA matrix.

For the prediction of the elastic modulus, the present system, based on a thermoplastic matrix in which random short fibers are dispersed, is not easy to evaluate. In fact, in this case, a great number of geometric, topological, and mechanical parameters are necessary [46]. Theoretical approaches usually attempt to exploit as much readily available information (which in most cases consists of the mechanical properties of matrix and fibers and the reinforcement volume fraction) as possible, while suitable assumptions cover missing data. Referring in particular to the elastic modulus, the existing expressions can be obtained from the elasticity theory—from a sort of mixture rule—or they are simply an attempt to match theoretical curves with experimental data [42–44,47,48]. Some of these analytical models (reported in Table 3) consider in particular the aspect ratio, the packing factor, and the Poisson ratio, in order to better predict the elastic modulus of composites containing increasing amounts of reinforcement.

Table 3. List of the analytical expressions used in this work for the prediction of the composites' Young's modulus.

Model	$E_{composite}$
Einstein	$E_c = E_m \left(1 + 2.5 \; \varphi_f\right)$
Kim	$E_c = \varphi_m E_m + \varphi_f E_f \cdot \left\{1 + \left(\sqrt{\frac{E_m}{E_f}} - 1\right) \cdot \frac{\tanh{(n \cdot a_r)}}{(n \cdot a_r)}\right\}$
Cox	$E_c = \varphi_m E_m + \varphi_f E_f \cdot \left(1 - \frac{\tanh{(n \cdot a_r)}}{(n \cdot a_r)}\right)$
Halpin-Tsai	$E_c = \frac{3}{8}E_l + \frac{5}{8}E_t$

In Table 2, E_f and E_m are the elastic modulus of the fibers and matrix, respectively, φ_f is the fibre volume fraction, a_r is the fibers' aspect ratio. The adimensional parameter n is defined as:

$$\frac{2E_m}{E_f(1 + v)ln\left(\frac{P}{\varphi_f}\right)} \tag{3}$$

where v is the Poisson ratio of the matrix (≈ 0.4), and P is the fibers' packing factor with the value $2\pi/\sqrt{3}$.

Furthermore, in the Halpin-Tsai model, the two terms E_l and E_t are, respectively, the longitudinal and the tangential modulus, quantified by the following expressions:

$$E_l = E_m \cdot \frac{1 + 2 \cdot a_r \cdot \left(\frac{\frac{E_f}{E_m} - 1}{\frac{E_f}{E_m} + 2a_r}\right) \cdot \varphi_f}{1 - \left(\frac{\frac{E_f}{E_m} - 1}{\frac{E_f}{E_m} + 2a_r}\right) \cdot \varphi_f} \tag{4}$$

$$E_t = E_m \cdot \frac{1 + 2 \cdot a_r \cdot \left(\frac{\frac{E_f}{E_m} - 1}{\frac{E_f}{E_m} + 2} \right) \cdot \varphi_f}{1 - \left(\frac{\frac{E_f}{E_m} - 1}{\frac{E_f}{E_m} + 2} \right) \cdot \varphi_f} \tag{5}$$

4. Conclusions

In this study PLA–cellulose Arbocel® fiber composites were produced and studied. Two different types of cellulose fibers having different aspect ratios were used for producing cohesive, white, nice-looking, and odorless bio-based composite materials, whose mechanical and thermal properties still meet the requirements for practical applications, such as in the packaging and agricultural sectors.

The addition of 600BE/PU cellulose fibers up to 20 wt % does not worsen the starting PLA properties (unlike with BWW 40 fibers). Consequently, 600BE/PU fibers can be used without any compatibilization in order to lower the final product cost and at the same time increase the stiffness and promote the biodegradability of the materials [20]. A compatibilization between polymeric matrix and fibers would be necessary (as verified by Pukánszky's B parameter) if there is an interest in obtaining composites with improved tensile and Charpy impact properties with respect to those of the raw PLA-based materials, as these might be required in more demanding sectors such as automotive or electronics.

The stiffness of the composites was predicted by applying and comparing different analytical models. It was observed that a very simple model such as the Einstein's model gives positive results. At the same time, it is not adequate to apply the Cox's model and it is necessary to use an adjustment of it (Kim's model). However, because of the random orientation of the fibers, the Halpin-Tsai model gives a good estimation and is preferable because it is able to provide information not only on the final composite stiffness, but also on the transversal and longitudinal composite stiffness.

Author Contributions: A.L. supervised the study results, and discussion. P.C. co-supervised the study and reviewed the paper. L.A. performed the experimental research and wrote a draft of the paper. V.G. performed the mechanical analysis and collaborated to write the paper, M.B.C. analyzed the data and contributed to the results and discussion.

Funding: Financial support was provided by the European Commission's Horizon 2020, Bio-Based Industry Consortium, under Grant Agreement n. 720719, AGRIMAX, "Agri and food waste valorisation co-ops based on flexible multi-feedstocks biorefinery processing technologies for new high added value applications" is acknowledged.

Acknowledgments: Authors acknowledge J Rettenmaier Sohne, Germany, for cooperation in providing the cellulose fibers.

Conflicts of Interest: The authors declare no conflict of interest.

References

1. Hornsby, P.; Hinrichsen, E.; Tarverdi, K. Preparation and properties of polypropylene composites reinforced with wheat and flax straw fibres—Part 2: Analysis of composite microstructure and mechanical properties. *J. Mater. Sci.* **1997**, *1*, 1009–1015. [CrossRef]
2. Liang, J.Z.; Ma, W.Y. Young's modulus and prediction of plastics/elastomer blends. *J. Polym. Eng.* **2012**, *32*, 343–348. [CrossRef]
3. Bledzki, A.K.; Reihmane, S.; Gassan, J. Properties and modification methods for vegetable fibers for natural fiber composites. *J. Appl. Polym. Sci.* **1996**, *59*, 1329–1336. [CrossRef]
4. Oksman, K.; Skrifvars, M.; Selin, J.F. Natural fibres as reinforcement in polylactic acid (PLA) composites. *Compos. Sci. Technol.* **2003**, *63*, 1317–1324. [CrossRef]
5. Râpă, M.; Popa, M.; Cinelli, P.; Lazzeri, A.; Burnichi, R.; Mitelut, A.; Grosu, E. Biodegradable Alternative to Plastics for Agriculture Application. *AcRomanian Biotechnol. Lett.* **2011**, *16*, 59–64.

6. Garrison, T.F.; Murawski, A.; Quirino, R.L. Bio-Based Polymers with Potential for Biodegradability. *Polymers* **2016**, *8*, 262. [CrossRef]

7. Künkel, A.; Becker, J.; Borger, L.; Hamprecht, J.; Koltzenburg, S.; Loos, R.; Schick, M.B.; Schlegel, K.; Sinkel, C.; Skupin, G.; Yamamoto, M. Polymers, Biodegradable. *Ullmann's Encycl. Ind. Chem.* **2016**, 1–29. [CrossRef]

8. Murariu, M.; Dechief, A.-L.; Ramy-Ratiarison, R.; Paint, Y.; Raquez, J.-M.; Dubois, P. Recent advances in production of poly(lactic acid) (PLA) nanocomposites: A versatile method to tune crystallization properties of PLA. *Nanocomposites* **2015**, *1*, 71–82. [CrossRef]

9. Aliotta, L.; Cinelli, P.; Coltelli, M.B.; Righetti, M.C.; Gazzano, M.; Lazzeri, A. Effect of nucleating agents on crystallinity and properties of poly (lactic acid) (PLA). *Eur. Polym. J.* **2017**, *93*, 822–832. [CrossRef]

10. Raquez, J.M.; Habibi, Y.; Murariu, M.; Dubois, P. Polylactide (PLA)-based nanocomposites. *Prog. Polym. Sci.* **2013**, *38*, 1504–1542. [CrossRef]

11. Liu, H.; Zhang, J. Research progress in toughening modification of poly(lactic acid). *J. Polym. Sci. Part B Polym. Phys.* **2011**, *49*, 1051–1083. [CrossRef]

12. Yu, L.; Dean, K.; Li, L. Polymer blends and composites from renewable resources. *Prog. Polym. Sci.* **2006**, *31*, 576–602. [CrossRef]

13. Bajpai, P.K.; Singh, I.; Madaan, J. Development and characterization of PLA-based green composites: A review. *J. Thermoplast. Compos. Mater.* **2012**. [CrossRef]

14. Lazzeri, A.; Phuong, T.V. Dependence of the Pukánszky's Interaction Parameter B on the Interface Shear Strength (IFSS) of Nanofiller- and Short Fiber-Reinforced Polymer Composites. *Composites Science and Technology* **2014**, *93*, 106–113. [CrossRef]

15. Lauke, B. Determination of adhesion strength between a coated particle and polymer matrix. *Compos. Sci. Technol.* **2006**, *66*, 3153–3160. [CrossRef]

16. Zappalorto, M.; Salviato, M.; Quaresimin, M. Influence of the interphase zone on the nanoparticle debonding stress. *Compos. Sci. Technol.* **2011**, *72*, 49–55. [CrossRef]

17. Peijs, T.; Garkhail, S.; Heijenrath, R.; van Den Oever, M.; Bos, H. Thermoplastic composites based on flax fibres and polypropylene: Influence of fibre length and fibre volume fraction on mechanical properties. *Macromol. Symp.* **2011**, *127*, 193–203. [CrossRef]

18. Oksman, K.; Wallström, L.; Berglund, L.A.; Filho, R.D.T. Morphology and mechanical properties of unidirectional sisal- epoxy composites. *J. Appl. Polym. Sci.* **2002**, *84*, 2358–2365. [CrossRef]

19. Mathew, A.P.; Oksman, K.; Sain, M. Mechanical properties of biodegradable composites from poly lactic acid (PLA) and microcrystalline cellulose (MCC). *J. Appl. Polym. Sci.* **2005**, *97*, 2014–2025. [CrossRef]

20. Lawrence, T.; Drzal, A.K.; Mohanty, M.M. Bio-composite materials as alternatives to petroleum-based composites for automotive applications. In Proceedings of the Automotive Composites Conference, Troy, MI, USA, 19–20 September 2001.

21. Jawaid, M.; Abdul Khalil, H.P.S. Cellulosic/synthetic fibre reinforced polymer hybrid composites: A review. *Carbohydr. Polym.* **2011**, *86*, 1–18. [CrossRef]

22. Ferrero, B.; Boronat, T.; Moriana, R.; Fenollar, O.; Balart, R. Green composites based on wheat gluten matrix and posidonia oceanica waste fibers as reinforcements. *Polym. Compos.* **2013**, *34*, 1663–1669. [CrossRef]

23. Seggiani, M.; Cinelli, P.; Verstichel, S.; Puccini, M.; Anguillesi, I.; Lazzeri, A. Development of Fibres-Reinforced Biodegradable Composites. *Chem. Eng. Trans.* **2015**, *43*, 1813–1818. [CrossRef]

24. Chiellini, E.; Cinelli, P.; Imam, S.H.; Mao, L. Composite Films Based on Biorelated Agro-Industrial Waste and Poly(vinyl alcohol). Preparation and Mechanical Properties Characterization. *Biomacromolecules* **2001**, *2*, 1029–1037. [CrossRef] [PubMed]

25. Seggiani, M.; Cinelli, P.; Geicu, M.; Elen, P.M.; Monica, P.; Lazzeri, A. Microbiological valorisation of bio-composites based on polylactic acid and wood fibres. *Chem. Eng. Trans.* **2016**, *49*, 127–132. [CrossRef]

26. Seggiani, M.; Cinelli, P.; Balestri, E.; Mallegni, N.; Stefanelli, E.; Rossi, A.; Id, C.L.; Id, A.L. Novel Sustainable Composites Based on Degradability in Marine Environments. *Materials* **2018**, *11*, 772. [CrossRef] [PubMed]

27. Tserki, V.; Matzinos, P.; Panayiotou, C. Effect of compatibilization on the performance of biodegradable composites using cotton fiber waste as filler. *J. Appl. Polym. Sci.* **2003**, *88*, 1825–1835. [CrossRef]

28. Petinakis, E.; Yu, L.; Edward, G.; Dean, K.; Liu, H.; Scully, A.D. Effect of Matrix–Particle Interfacial Adhesion on the Mechanical Properties of Poly(lactic acid)/Wood-Flour Micro-Composites. *J. Polym. Environ.* **2009**, *17*, 83. [CrossRef]

29. Tao, Y.; Wang, H.; Li, Z.; Li, P.; Shi, S.Q. Development and Application of Wood Flour-Filled Polylactic Acid Composite Filament for 3D Printing. *Materials* **2017**, *10*, 339. [CrossRef]

30. Petinakis, E.; Yu, L.; Simon, G. Natural Fibre Bio-Composites Incorporating Poly (Lactic Acid). In *Fiber Reinforced Polymers—The Technology Applied for Concrete Repair*; Masuelli, M., Ed.; Intechopen: London, UK, 2013; pp. 41–59, ISBN 978-953-51-0938-9. [CrossRef]

31. Gurunathan, T.; Mohanty, S.; Nayak, S.K. A review of the recent developments in biocomposites based on natural fibres and their application perspectives. *Compos. Part A Appl. Sci. Manuf.* **2015**, *77*, 1–25. [CrossRef]

32. Muthuraj, R.; Misra, M.; Mohanty, A.K. *Biocomposites*; Elsevier: Amsterdam, The Netherlands, 2015; ISBN 9781782423737.

33. La Mantia, F.P.; Morreale, M. Green composites: A brief review. *Compos. Part A Appl. Sci. Manuf.* **2011**, *42*, 579–588. [CrossRef]

34. Satyanarayana, K.G.; Arizaga, G.G.C.; Wypych, F. Biodegradable composites based on lignocellulosic fibers—An overview. *Prog. Polym. Sci.* **2009**, *34*, 982–1021. [CrossRef]

35. Huda, M.S.; Drzal, L.T.; Misra, M.; Mohanty, A.K. Wood-fiber-reinforced poly(lactic acid) composites: Evaluation of the physicomechanical and morphological properties. *J. Appl. Polym. Sci.* **2006**, *102*, 4856–4869. [CrossRef]

36. Herrera-Franco, P.J.; Valadez-Gonzalez, A. A study of the mechanical properties of short natural-fiber reinforced composites. *Compos. Part B Eng.* **2005**, *36*, 597–608. [CrossRef]

37. Pukánszky, B. Influence of interface interaction on the ultimate tensile properties of polymer composites. *Composites* **1990**, *21*, 255–262. [CrossRef]

38. Phuong, V.T.; Gigante, V.; Aliotta, L.; Coltelli, M.B.; Cinelli, P.; Lazzeri, A. Reactively extruded ecocomposites based on poly(lactic acid)/bisphenol A polycarbonate blends reinforced with regenerated cellulose microfibers. *Compos. Sci. Technol.* **2017**, *139*, 127–137. [CrossRef]

39. Tucker III, C.L.; Liang, E. Stiffness predictions for unidirectional short-fiber composites: Review and evaluation. *Compos. Sci. Technol.* **1999**, *59*, 655–671. [CrossRef]

40. Gigante, V.; Aliotta, L.; Phuong, T.V.; Coltelli, M.B.; Cinelli, P.; Lazzeri, A. Effects of waviness on fiber-length distribution and interfacial shear strength of natural fibers reinforced composites. *Compos. Sci. Technol* **2017**, *152*, 129–138. [CrossRef]

41. Huber, T.; Mussig, J. Fibre matrix adhesion of natural fibres cotton, flax and hemp in polymeric matrices analyzed with the single fibre fragmentation test. *Compos. Interfaces* **2008**, *15*, 335–349. [CrossRef]

42. Cox, H.L. The elasticity and strength of paper and other fibrous materials. *Br. J. Appl. Phys.* **1952**, *3*, 72–79. [CrossRef]

43. Kim, H.G.; Kwac, L.K. Evaluation of elastic modulus for unidirectionally aligned short fiber composites. *J. Mech. Sci. Technol.* **2009**, *23*, 54–63. [CrossRef]

44. Ahmed, S.; Jones, F.R. A review of particulate reinforcement theories for polymer composites. *J. Mater. Sci.* **1990**, *25*, 4933–4942. [CrossRef]

45. Renner, K.; Kenyó, C.; Móczó, J.; Pukánszky, B. Micromechanical deformation processes in PP/wood composites: Particle characteristics, adhesion, mechanisms. *Compos. Part A Appl. Sci. Manuf.* **2010**, *41*, 1653–1661. [CrossRef]

46. Bourkas, G.; Prassianakis, I.; Kytopoulos, V.; Sideridis, E.; Younis, C. Estimation of Elastic Moduli of Particulate Composites by New Models and Comparison with Moduli Measured by Tension, Dynamic, and Ultrasonic Tests. *Adv. Mater. Sci. Eng.* **2010**, *2010*, 891824. [CrossRef]

47. Halpin, J.C.; Kardos, J.L. The Halpin-Tsai equations: A review. *Polym. Eng. Sci.* **1976**, *16*, 344–352. [CrossRef]

48. Fu, S.Y.; Feng, X.Q.; Lauke, B.; Mai, Y.W. Effects of particle size, particle/matrix interface adhesion and particle loading on mechanical properties of particulate-polymer composites. *Compos. Part B Eng.* **2008**, *39*, 933–961. [CrossRef]

© 2019 by the authors. Licensee MDPI, Basel, Switzerland. This article is an open access article distributed under the terms and conditions of the Creative Commons Attribution (CC BY) license (http://creativecommons.org/licenses/by/4.0/).

International Journal of
Molecular Sciences

MDPI

Article

Thermal, Mechanical, and Rheological Properties of Biocomposites Made of Poly(lactic acid) and Potato Pulp Powder

Maria Cristina Righetti [1,*], Patrizia Cinelli [1,2,*], Norma Mallegni [1], Carlo Andrea Massa [1], Simona Bronco [1], Andreas Stäbler [3] and Andrea Lazzeri [1,2]

[1] CNR-IPCF, National Research Council—Institute for Chemical and Physical Processes, Via Moruzzi 1, 56124 Pisa, Italy; norma.mallegni@pi.ipcf.cnr.it (N.M.); carlo.andrea.massa@pi.ipcf.cnr.it (C.A.M.); simona.bronco@pi.ipcf.cnr.it (S.B.); andrea.lazzeri@unipi.it (A.L.)
[2] Department of Civil and Industrial Engineering, University of Pisa, Largo Lucio Lazzarino 1, 56122 Pisa, Italy
[3] Fraunhofer Institute for Process Engineering and Packaging IVV, Giggenhauser Straße, 35, 85354 Freising, Germany; andreas.staebler@ivv.fraunhofer.de
*   Correspondence: cristina.righetti@pi.ipcf.cnr.it (M.C.R.); patrizia.cinelli@unipi.it (P.C.)

Received: 31 December 2018; Accepted: 1 February 2019; Published: 5 February 2019

Abstract: The thermal, mechanical, and rheological properties of biocomposites of poly(lactic acid) (PLA) with potato pulp powder were investigated in order to (1) quantify how the addition of this filler modifies the structure of the polymeric material and (2) to obtain information on the possible miscibility and compatibility between PLA and the potato pulp. The potato pulp powder utilized is a residue of the processing for the production and extraction of starch. The study was conducted by analyzing the effect of the potato pulp concentration on the thermal, mechanical, and rheological properties of the biocomposites. The results showed that the potato pulp powder does not act as reinforcement but as filler for the PLA polymeric matrix. A progressive decrease in elastic modulus, tensile strength, and elongation at break was observed with increasing the potato pulp percentage. This moderate loss of mechanical properties, however, still meets the technical requirements indicated for the production of rigid packaging items. The incorporation of potato pulp powder to PLA offers the possibility to reduce the cost of the final products and promotes a circular economy approach for the valorization of agro-food waste biomass.

Keywords: bio-based polymers; natural fibers; biomass; biocomposites; fiber/matrix adhesion

1. Introduction

Biodegradable bio-based polymers obtained from renewable resources represent an important alternative to petroleum-derived non-degradable polymers. For this reason, bio-based polymers—for example, poly(lactic acid) (PLA), polyhydroxylalkanoates (PHA), in particular polyhydroxylbutyrate (PHB) and its copolymers poly(hydroxylbutyrate-co-valerate) (PHBV), cellulose, and starch derived plastics, which are produced from renewable agricultural and biomass feedstock—have become a subject of crucial interest for academia and industry. Among them, PLA is the biodegradable polymer most present on the market and widely used for packaging purposes. Many properties of PLA, such as strength and stiffness, are comparable to those of traditional petroleum-based polymers. On the other hand, the drawbacks for its applications are the low toughness and the relatively low glass transition, which limits its use at a temperature above about 50 °C, because, due to its slow crystallization rate, processing of PLA generally results in amorphous products [1]. In addition, bio-based polymers are generally more expensive than conventional petroleum-derived polymers and often exhibit worse thermal and mechanical properties, which make them unacceptable for several applications.

In order to balance the cost and achieve suitable properties for different applications, a possible solution is offered by biocomposites. Biocomposites are a special class of composite materials obtained by blending natural fibers with bio-based polymers. Biocomposites represent an ecological and low-cost alternative to conventional petroleum-derived materials and, for this reason, are becoming progressively more utilized for a wide variety of uses. Natural fiber reinforced biocomposites have been reviewed in several articles [2–7].

Besides the benefit of being biodegradable and produced from renewable resources, natural fibers are also less abrasive to processing equipment than synthetic fibers like ceramic or glass fibers and traditional fillers like mica and glass, and they exhibit a lower density, which makes biocomposites lightweight and economical materials [8].

The chemical composition of the natural fibers, which varies depending on the plant from which they are derived, consists mainly of cellulose (50–70 wt%), hemicellulose (10–20 wt%), lignin (10–30 wt%), and pectin and waxes in smaller amounts [7]. The physico-mechanical properties of the natural fibers also depend on the original plants. The density, elastic modulus, tensile strength, and elongation at break are approximately in the ranges 0.8–1.5 g/cm^3, 5–20 GPa, 200–900 MPa, and 1.5–20%, respectively [7].

The mechanical properties of a biocomposite results from both the matrix and fiber properties [9,10] and strongly depend on the matrix/fiber interface [11]. The tensile strength is more sensitive to the matrix/fiber adhesion, whereas the modulus depends, in general, on both the matrix and the fiber properties. The mass or volume fraction of the fibers, the fiber aspect ratio (length to width ratio), the fiber–matrix adhesion, and the fiber orientation are crucial factors responsible for the final properties of natural fiber reinforced composites [12]. In a composite, the matrix plays the role of transferring the applied stress to the fibers. This occurs at the interface, and therefore a good matrix/fiber adhesion is necessary. Poor adhesion often characterizes biocomposites made of hydrophobic polymers and hydrophilic natural fibers, which can lead to scarce mechanical properties due to the tendency of the fibers to form aggregates during processing. In addition, the hydrophilic nature of the lignocellulosic fibers induces absorption of moisture in an amount that can vary up to about 10%, with the result that if the fibers are not efficiently dried previously, the process ability is worsened, and the formation of porous products is thus unluckily favored. The fiber aspect ratio strongly influences the tensile modulus and the fracture properties, and a critical fiber length is necessary to develop composites with high stiffness and strength. Fibers with low aspect ratio and irregular shape behave as fillers and not as reinforcement. Thus, multiple factors influence the mechanical properties of a composite, not only the intrinsic properties of the fibers but also their shape and dimensions and their dispersion in the polymeric matrix together with the interactions that are established at the matrix/filler interface.

In regard to PLA, several types of natural fibers have been used to produce biocomposites [13–18]. The effect of the processing conditions and, in some cases, surface chemical modification of the fibers on mechanical, rheological, and thermal properties of PLA/fibers biocomposites has been widely investigated [13–18].

Organic wastes have sometimes been utilized as reinforcements or additives for various polymers, examples being charcoal and lignin [19]. By considering that significant amounts of organic wastes from industry and agriculture are unutilized, it follows that the use of organic residue materials in biocomposites can represent an ecologically friendly method to produce materials for different applications. This can also allow the reduction of the cost of the products.

Potato wastes are biomasses rich in starch and lignocellulosic constituents. After extraction of the starch, the potato pulp accumulates in high amounts—approximately 0.75 tons of pulp arises per ton of purified starch. Within the European Union, about 140,000 tons of dried potato pulp are produced annually in the starch industry [20]. The original potato pulp contains water up to about

90%, but de-watering processes generally results in an increase in the dry matter up to about 90 wt%. Dried potato pulp, which consists mainly of lignocellulosic fibers, starch, and proteins, can be used as filler or for a reinforcement action on bio-based polymers. The cost of this raw material is low, which makes potato pulp even more interesting for industrial utilization [21].

In the present work, biocomposites made of poly(lactic acid) and potato pulp powder (PPP) have been produced by extrusion followed by injection molding and characterized in terms of process ability and thermal, mechanical, and rheological properties. It is worth pointing out that the composition of the potato pulp powder is expected to depend on the geographical origin of the agricultural product and on the climatic and harvesting conditions as well as on the industrial processing methods. For this reason, the present study provides only a general description of the properties of these PLA-based biocomposites, which could slightly change with varying the original materials. In order to make the processing of PLA with the potato pulp powder easier, a plasticizer, acetyl tributyl citrate (ATBC), was used. ATBC is an efficient plasticizer of PLA [22]. It is derived from naturally occurring citric acid, non-toxic, and accepted for food-contact [23]. In addition, calcium carbonate ($CaCO_3$) was used in low percentages as an inert filler to facilitate the removal of the injection-molded specimens from the mold. The thermal, mechanical, and rheological properties of the PLA-based matrix were investigated as preliminary step in order to better quantify the influence of the reinforcing fibers on the material properties.

The PLA-based samples investigated in the present study, with the relative composition, are listed in Table 1. All the samples were processed as described in Section 3.

Table 1. Composition of the poly(lactic acid) (PLA)-based matrix and biocomposites.

	Potato Pulp
PLA (100%)	-
[PLA (90 wt%) + ATBC (10 wt%)]	-
[PLA (85 wt%) + ATBC (10 wt%) + $CaCO_3$ (5 wt%)]	-
[PLA (85 wt%) + ATBC (10 wt%) + $CaCO_3$ (5 wt%)] (95 wt%)	PPP (5 wt%)
[PLA (85 wt%) + ATBC (10 wt%) + $CaCO_3$ (5 wt%)] (90 wt%)	PPP (10 wt%)
[PLA (85 wt%) + ATBC (10 wt%) + $CaCO_3$ (5 wt%)] (80 wt%)	PPP (20 wt%)

2. Results and Discussion

2.1. Thermogravimetric Analysis of the Potato Pulp Powder, the PLA-Based Matrix, and the Biocomposite [PLA (85 wt%) + ATBC (10 wt%) + $CaCO_3$ (5 wt%)] + PPP (20 wt%)]

The main drawback of the use of natural fibers as reinforcement or fillers in biocomposites is the relative low processing temperature required due to the thermal degradation that occurs at high temperatures which can irreversibly damage them. As all the components of the potato pulp can be subjected to thermal degradation during composite processing, it is of practical significance to investigate the thermal decomposition process of the potato pulp powder in order to estimate the temperature range in which the biocomposites can be processed.

The thermal stability of the potato pulp powder was determined by means of the thermogravimetric analysis under nitrogen flow, because the contact of the material with air is reduced in the extruder and molder. Figure 1 shows the thermogravimetric curve of the potato pulp powder, which reports the change in weight according to a fixed temperature program. The initial weight loss detected at temperatures lower than 150 °C is due to water vaporization. The water content in PPP is approximately 3 wt%. The weight residue that is observed at high temperature is due to the carbon deposit that remains in the presence of an inert atmosphere. The thermal degradation of natural fibers generally has multiple processes due to the different components [24,25]. Degradation of hemicellulose occurs mainly in the temperature range of 200–350 °C with the maximum mass loss rate at around 300 °C, whereas cellulose pyrolysis takes places at higher temperatures, between 250 and 400 °C. At high temperatures, the solid residue of pure hemicellulose and cellulose is about 20%. Lignin is the most difficult component to decompose; its degradation takes place slowly and continuously from about 250 °C, with a high solid residue that remains at high temperatures (around 50%). Starch degradation extends from approximately 300 to 350 °C, and its solid residue at 500 °C is approximately 20% [26]. Degradation of proteins also occurs in a similar temperature range, between 200 and 400 °C, with a maximum loss rate at around 300 °C and 20% residue at high temperatures [27]. From Figure 1, the potato pulp powder appears stable up to about 180 °C. This thermal stability assures that the potato pulp powder does not undergo substantial degradation during the processing of the PLA biocomposites at 180 °C, being that the residence time at this temperature is no longer than 1.5 min. It can therefore be concluded that the potato pulp can be confidently used for the production of PLA-based biocomposites.

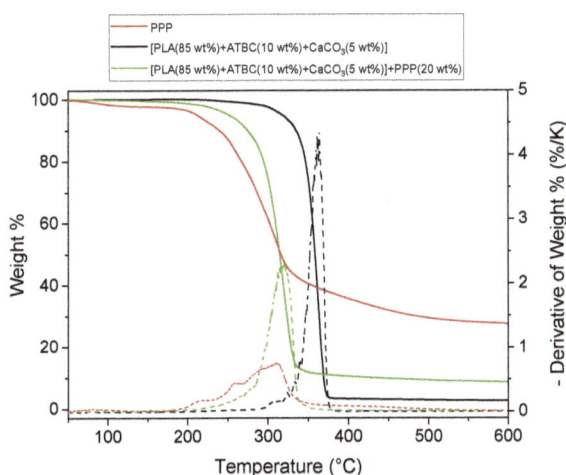

Figure 1. Thermogravimetric curves of the potato pulp powder (PPP), the poly(lactic acid) (PLA)-based matrix andbiocomposite [PLA (85 wt%) + acetyl tributyl citrate (ATBC; 10 wt%) + CaCO$_3$ (5 wt%)] + potato pulp powder (PPP; 20 wt%) at 10 K/min under nitrogen flow. The dotted lines are the derivatives of the weight % curves.

Figure 1 shows also the thermogravimetric curves of the PLA-based matrix and biocomposite with 20 wt% of PPP. Both these thermal degradations occur in a single step in a narrow temperature range. The initial degradation temperature of the PLA-based matrix is located at about 300 °C, whereas the maximum degradation rate is centered at about 365 °C, which is in excellent agreement with previous studies [28–30] in which the independence of the PLA thermal stability on the molar mass was also demonstrated [28]. The biocomposite with 20 wt% of potato pulp powder starts to degrade at about 180 °C, whereas the maximum degradation rate shifts to around 320 °C, thus approaching the thermal degradation of plain potato pulp powder. As expected, the residue of the biocomposite at 600 °C is higher compared to that of the PLA based matrix due to the presence of the lignocellulosic residue. In conclusion, Figure 1 reveals that the processing of the biocomposites at 180 °C does not substantially affect the potato pulp powder structure because the degradation of the biocomposite starts at the same temperature of the original unprocessed potato pulp powder.

2.2. Scanning Electron Microscopy of the Potato Pulp Powder

The morphologies of the potato pulp powder were investigated with scanning electron microscopy (SEM), and SEM images are shown in Figure 2. A homogeneous distribution of pulp fragments, which appear as small aggregates, can be observed. The aggregates are relatively large (200 µm and more).

The round-shaped particles detected at 1200× magnification in the potato pulp powder are either starch or pectin because they disappear after treatment with amylase and pectinase.

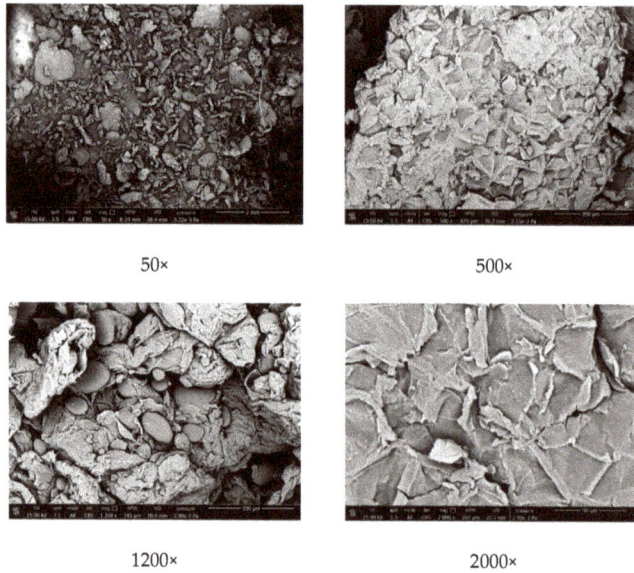

50×   500×

1200×   2000×

Figure 2. Scanning electron microscopy (SEM) images of the potato pulp powder (PPP) at the magnifications indicated.

2.3. Thermal, Mechanical, and Rheological Properties of the PLA-Based Matrix

The thermal, mechanical, and rheological properties of the PLA-based matrix were investigated as a preliminary step in order to better quantify the influence of the reinforcing fibers on the material properties.

The specific heat capacity (c_p) curves of PLA and PLA mixed with (1) the plasticizer ATBC and (2) with ATBC and the mineral filler CaCO$_3$, measured at 10 °C/min, are shown in Figure 3. As described in Section 3, the samples were injection-molded for 1 min at 90 °C. The glass transition, which occurs in the proximity of 60 °C, which is in agreement with literature data [31], is overlapped by an enthalpy recovery peak due to the permanence of the samples at room temperatures for one day. Before the melting endotherm, all the curves displayed an intense cold crystallization peak located approximately in the interval between 90 and 130 °C. In the temperature range between 100 and 120 °C, both the crystalline α'- and α-forms grew [32,33]. Thus the melting behavior that extends from approximately 130 to 160 °C results from the fusion of both the α- and the α'-crystals. At the heating rate of 10 K/min, the α'-crystals transform into the more ordered α-phase via melting and almost simultaneous recrystallization, with the result that a multiple melting behavior, independent of the molar mass, is commonly observed [34,35]. Reorganization and recrystallization events overlap the entire fusion process, which generally takes place in semi-crystalline polymers at a relatively low heating rate [36,37].

Table 2 lists the T_g values together with the enthalpy of cold crystallization (Δh_c) and the enthalpy of fusion (Δh_m) of the samples investigated in this section, calculated from the c_p curves shown in Figure 3. The Δh_c and Δh_m values collected in Table 2 are normalized to the PLA content. From these experimental values, an estimation of the crystalline weight fraction growing during the cold crystallization process (w_{Cc}) and disappearing during the melting process (w_{Cm}) was obtained by dividing Δh_c and Δh_m by the enthalpy of fusion of 100% crystalline PLA phase ($\Delta h_m°$) at the crystallization and melting peak temperatures, respectively. As both the α'- and α-forms grow during cold crystallization and disappear during the melting process, the average values between the enthalpy of fusion of the α'- and α-forms were utilized [38], i.e., $\Delta h_m° = 101$ J/g for the cold crystallization centered at about 110 °C, $\Delta h_m° = 96$ J/g for the cold crystallization centered at about 100 °C, and $\Delta h_m° = 119$ J/g for the melting process centered approximately at 150 °C. The w_{Cc} and w_{Cm} values listed in Table 2 reveal that the PLA and the PLA samples with the addition of (1) only the plasticizer ATBC and (2) the plasticizer ATBC and the mineral filler $CaCO_3$, after processing for 1 min at 90 °C, are completely amorphous. This can be explained by considering the short times that characterize a typical injection molding and the slow crystallization kinetics of PLA [39]. In the presence of ATBC, the crystallinity that grows during the heating run is higher with respect to PLA, which can be ascribed to the higher mobility achieved by the PLA chains. The plasticizing effect of ATBC on PLA is evidenced by the decrease in the T_g value, the reduced peak temperature of the cold crystallization, and the higher crystallinity that develops upon heating, whereas the inertia of $CaCO_3$ towards the PLA phase transitions is proven by the unchanged thermal parameters.

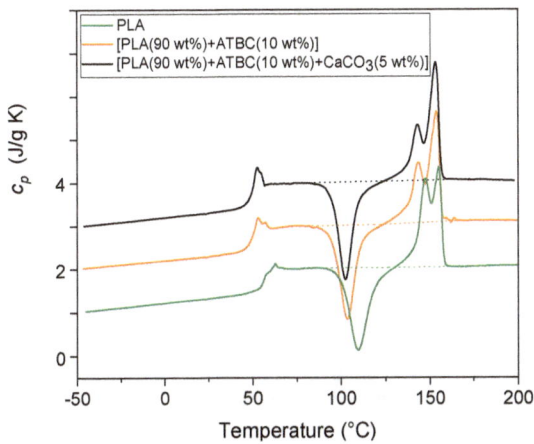

Figure 3. Specific heat capacity (c_p) of PLA and PLA mixed with (1) ATBC and (2) ATBC and the mineral filler $CaCO_3$ at the concentrations indicated. The curves were obtained upon heating at 10 K/min after previous fast cooling to −50 °C. The dotted lines are the baselines used for the calculation of the Δh_c and Δh_m values. The ordinate values refer only to the bottom curve. All the other curves are shifted vertically for the sake of clearness.

Table 2. Glass transition temperatures (T_g), enthalpy of cold crystallization (Δh_c), enthalpy of fusion (Δh_m), and crystalline weight fraction growing during cold crystallization (w_{Cc}) and disappearing during fusion (w_{Cm}) for PLA and PLA mixed with (1) ATBC and (2) ATBC and the mineral filler $CaCO_3$ (estimated errors: ± 0.5 °C for T_g; ± 1 J/g for Δh_c and Δh_m, and ± 0.02 for w_{Cc} and w_{Cm}).

	T_g (°C)	Δh_c (J/g)	w_{Cc}	Δh_m (J/g)	w_{Cm}
PLA	58	24	0.24	28	0.24
[PLA (90 wt%) + ATBC (10 wt%)]	50	27	0.28	33	0.28
[PLA (85 wt%) + ATBC (10 wt%) + $CaCO_3$ (5 wt%)]	50	27	0.28	33	0.28

The mechanical properties of PLA and of PLA after processing in the presence of (1) the plasticizer ATBC and (2) the plasticizer ATBC with the mineral filler $CaCO_3$ are summarized in Figure 4. PLA is a quite brittle polymer with a high elastic modulus and a high tensile strength. The addition of the plasticizer ATBC, as expected, modifies the mechanical properties: The elastic modulus does not substantially change, the tensile strength decreases, and conversely, the elongation at break greatly increases. In the presence of the plasticizer ATBC, the intermolecular forces between the PLA chains decrease, the mobility of the polymeric chains enhances, and an increase in flexibility and ductility is produced. A similar behavior has been reported for other PLA plasticized systems [40,41].

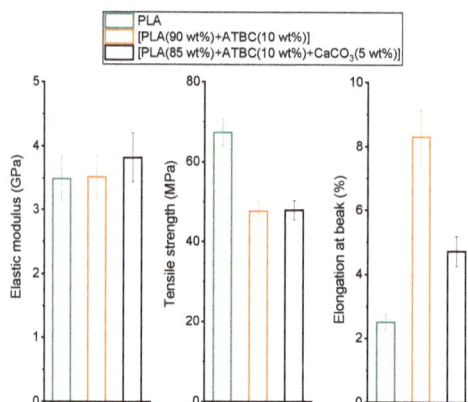

Figure 4. Elastic modulus, tensile strength and elongation at break of PLA and PLA mixed with (1) ATBC and (2) ATBC and the mineral filler CaCO$_3$ at the concentrations indicated.

The addition of the mineral filler CaCO$_3$ produces a slight increase in the elastic modulus and a strong decrease in the elongation at break, whereas it has negligible effect on the tensile strength. This behavior is in agreement with literature data. The modulus increase is the result of the reinforcement effect of CaCO$_3$ particles. However, if the mineral filler is present in small percentages, the influence on the elastic modulus is small because it can be assumed that the perturbed polymer fraction around each mineral particle is low compared with the unperturbed one [42,43]. The tensile strength remains unaffected by the presence of the mineral filler, which means that the mobility of the polymer chains is not influenced by the small percentage of CaCO$_3$ as also proven by the constant T_g value. Conversely, the elongation at break strongly decreases because the presence of micro-sized mineral particles modifies the fracture mechanism and hinders the elongation of the material. Mineral particles generally act as stress concentrators capable of initiating cracking and favoring specific fracture mechanisms.

The study of rheological properties of PLA is crucial for gaining a fundamental understanding of the processability of PLA materials, especially for injection molding and extrusion processes. PLA behaves like a pseudo-plastic, a non-Newtonian fluid, and a typical shear thinning fluid in which, at high shear rates, the macromolecules orient and decrease the entanglement number [44,45].

Figure 5 shows the dependence of the modulus of the complex viscosity on the deformation frequency for PLA and PLA with the addition of (1) the plasticizer ATBC, and (2) the plasticizer ATBC and the mineral filler CaCO$_3$. The rheological measurements were performed from low to high frequencies. All the curves indicate a decrease in complex viscosity with increasing rotational frequency. In the terminal zone, which defines the zero-shear viscosity η_0, a small decrease in viscosity is observed, probably due an original entangled structure not completely destroyed before the beginning of the rheological test. The η_0 value for PLA at 175 °C is about 1.8×10^6 mPa·s. This value is in the range typical of commercial polymers (M_w = 140–160 kg/mol), i.e., 10^5–10^7 mPa·s [46]. The incorporation of ATBC reduces the viscosity due to its plasticizing effect, which leads to a decrease in the intermolecular forces and an increase in the mobility of the polymeric chains [47,48]. The addition of CaCO$_3$ causes a further small reduction of viscosity, which is in excellent agreement with literature data [49]. A low CaCO$_3$ content is insufficient to form a filler network. As a consequence, the incorporation of the mineral filler weakens the interaction between PLA chains segments, thus producing a viscosity reduction.

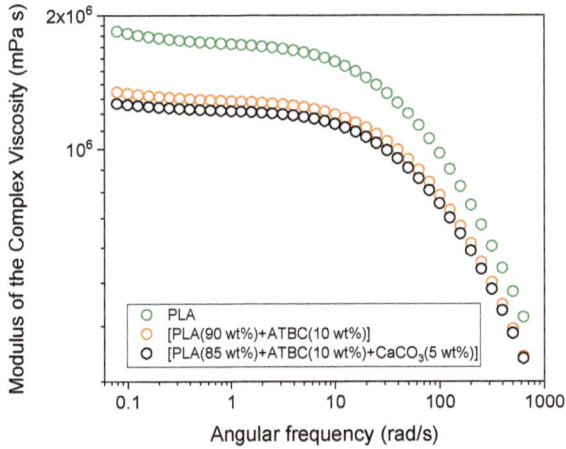

Figure 5. Modulus of the complex viscosity vs. angular frequency at 175 °C for PLA and PLA with the addition of (1) the plasticizer ATBC, and (2) the plasticizer ATBC and the mineral filler CaCO₃.

2.4. Thermal, Mechanical and Rheological Properties of the PLA-Based Biocomposites

The thermal, mechanical, and rheological properties of the biocomposites of PLA with potato pulp powder were investigated in order to quantify how the addition of these fibers modifies the structure of the polymeric material. These data together provide information on the possible miscibility and compatibility between PLA and potato pulp. The PLA-based biocomposites with potato pulp were studied by analyzing the effect of potato pulp concentration.

The specific heat capacity (c_p) curves of the [PLA (85 wt%) + ATBC (10 wt%) + CaCO₃ (5 wt%)] matrix with the addition of an increasing percentage of PPP (5 wt%, 10 wt%, and 20 wt%), measured at 10 K/min, are shown in Figure 6.

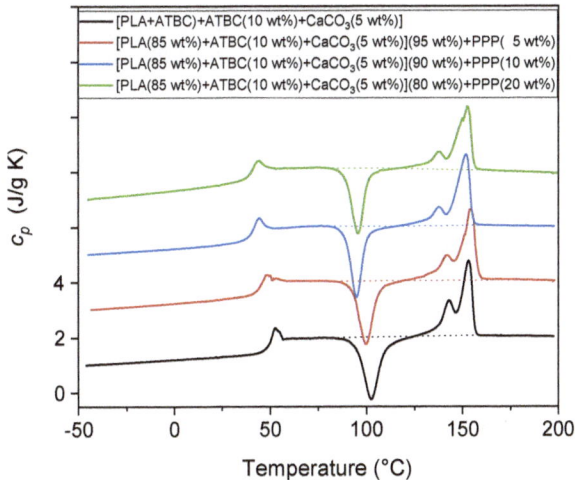

Figure 6. Specific heat capacity (c_p) of the PLA-based matrix and biocomposites indicated in the legend. The curves were obtained upon heating at 10 K/min after previously fast cooling to -50 °C. The dotted lines are the baselines used for the calculation of the Δh_c and Δh_m values. The ordinate values refer only to the bottom curve. All the other curves are shifted vertically for the sake of clarity.

Table 3 lists the T_g values together with the enthalpy of cold crystallization (Δh_c) and the enthalpy of fusion (Δh_m) of the samples investigated in this section as calculated from the c_p curves shown in Figure 6. The addition of PPP affects the glass transition temperature, which decreases with increasing the PPP percentage. This reveals that the polymer chain mobility increases after addition of the potato pulp powder. This can be ascribed to the development of weak intermolecular interactions between PLA and the potato pulp particles, which could cause the formation of outspread free volume at the matrix/fiber interface.

Table 3. Glass transition temperatures (T_g), enthalpy of cold crystallization (Δh_c), enthalpy of fusion (Δh_m), and crystalline weight fraction growing during cold crystallization (w_{Cc}) and disappearing during fusion (w_{Cm}) for the PLA-based matrix and biocomposites (estimated errors: $\pm$ 0.5 °C for T_g; $\pm$ 1 J/g for Δh_c and Δh_m, and $\pm$ 0.02 for w_{Cc} and w_{Cm}).

	T_g (°C)	Δh_c (J/g)	w_{Cc}	Δh_m (J/g)	w_{Cm}
[PLA (85 wt%) + ATBC (10 wt%) + CaCO$_3$ (5 wt%)]	50	27	0.28	33	0.28
[PLA (85 wt%) + ATBC (10 wt%) + CaCO$_3$ (5 wt%)] (95 wt%) + PPP (5 wt%)	45	26	0.27	34	0.28
[PLA (85 wt%) + ATBC (10 wt%) + CaCO$_3$ (5 wt%)] (90 wt%)+PPP (10 wt%)	40	26	0.32	35	0.32
[PLA (85 wt%) + ATBC (10 wt%) + CaCO$_3$ (5 wt%)] (80 wt%)+PPP (20 wt%)	40	28	0.34	37	0.34

The peak temperature of the cold crystallization process decreases in the biocomposites, proving that PPP plays a nucleating role during the heating run, which accelerates the PLA cold crystallization process. This a general behavior of the natural fibers, as also found in other PLA/natural fibers composites [15,50], which has been explained as due to the formation of a transcrystalline layer around the fibers. The occurrence of transcrystallinity was reported for a large combination of semi-crystalline thermoplastic matrices and fibers [51,52]. For biocomposites containing 10 and 20 wt% of PPP, cold crystallization occurs almost completely at temperatures lower than 100 °C, which means that only the α'-form develops [32,33]. At the heating rate of 10 K/min, however, these α'-crystals are expected to transform into the more ordered α-phase [34]. For biocomposites with 10 and 20 wt% of PPP, w_{Cc} was calculated by dividing the measured Δh_c by the enthalpy of fusion of the α'-form at about 95 °C ($\Delta h_m° = 81$ J/g), whereas w_{Cm} was obtained from the ratio between Δh_m and the enthalpy of fusion of the α'-form at about 150 °C ($\Delta h_m° = 108$ J/g) [38], as the enthalpy of crystallization and the enthalpy of fusion of the α-crystals cancel out. For the biocomposite with 5 wt% of PPP, for which cold crystallization extends up to above 110 °C, the average values between the enthalpy of fusion of the α'- and α-forms were utilized, i.e., $\Delta h_m° = 96$ J/g for the cold crystallization centered at about 100 °C, and $\Delta h_m° = 119$ J/g for the melting process centered approximately at 150 °C [38].

The w_{Cc} and w_{Cm} values listed in Table 3, as derived from the measured Δh_c and Δh_m values, respectively, reveal that the PLA-based biocomposites containing PPP, processed for 1 min at 90 °C, are completely amorphous. Despite the nucleating action of PPP, which leads to a progressively and slightly higher crystallinity to develop upon heating, the residence time of the materials in the mold is too short to allow the crystallization of PLA.

The mechanical properties of the [PLA (85 wt%) + ATBC (10 wt%) + CaCO3 (5 wt%)] matrix and the biocomposites with potato pulp powder are summarized in Figure 7.

As a general rule, the modulus of a composite depends on the modulus of both the matrix and the reinforcement and filler, and the modulus of the natural fibers is generally higher than the modulus of the polymeric matrices [12]. Figure 7 reveals that the elastic modulus decreases by increasing the PPP percentage, which could mean that PPP acts for PLA matrix as filler and not as reinforcement, or it could be ascribed to PLA degradation during the processing step induced by fiber moisture [53]; although, both PLA and the PPP powder were previously dried for a long time, and the water content of the potato pulp powder was about 3 wt% (Figure 1). The number-average molar mass

(M_n) and the mass-average molar mass (M_w) of PLA of both the matrix and the biocomposite with 20 wt% of potato pulp powder processed under identical conditions were measured by size exclusion chromatography (SEC). For the PLA-based matrix, M_n = 92,000 g/mol and M_w = 170,000 g/mol, whereas for the biocomposite [PLA (85 wt%) + ATBC (10 wt%) + CaCO$_3$ (5 wt%)] (80 wt%) + PPP (20 wt%), M_n = 70,000 g/mol and M_w = 130,000 g/mol were obtained. The observed decrease of about 20% in the PLA molar mass can thus be ascribed to the PLA degradation occurring during processing in the presence of the moisture introduced by the potato pulp powder. This could explain the loss in the mechanical properties. However, for M_w higher than about 100,000 g/mol, the mechanical properties of PLA were not strongly influenced by the molar mass [54]. Thus, besides the possible effect of PLA degradation induced by the fiber moisture, the reduction in the elastic modulus of the PLA-based biocomposites with potato pulp powder has to be ascribed also to poor adhesion between the matrix and the filler. Potato pulp powder is composed mainly of lignocellulosic fibers and starch, which are highly hydrophilic, so poor interactions are expected at the interface with the less hydrophilic PLA. Poor adhesion between the potato pulp powder and the polymeric matrix is also evidenced by the tensile strength, which decreases with PPP percentage. The elongation at break is reduced drastically because the dispersed filler particles act as stress concentrators at the polymer/filler interface by inhibiting the deformation, which leads to reduced ductility of the material.

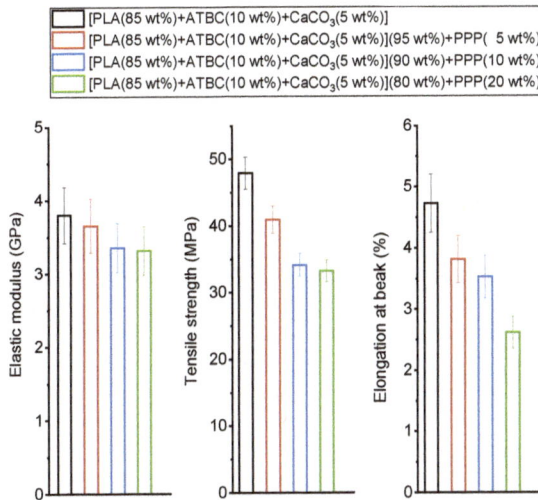

Figure 7. Elastic modulus, tensile strength, and elongation at break of the PLA-based matrix and biocomposites indicated in the legend.

Figure 8 shows the dependence of the modulus of the complex viscosity on the deformation frequency at 175 °C for the PLA-based biocomposites containing potato pulp powder. The experiments were performed from high to low frequencies in a narrow deformation frequency range outside but not far from the Newtonian plateau in order to considerably reduce the measurement times, and consequently, fiber degradation. This also allows the comparison between the viscoelastic behaviors of the different biocomposites. Figure 8 indicates that the viscosity of the biocomposites is smaller than that of the polymeric matrix and decreases with an increase in PPP percentage. For M_w > 100,000 g/mol, the zero shear viscosity η_o, referring to the Newtonian plateau, can be described by the relationship: $\log \eta_o = \log K + 3.4 \log M_w$ [55,56]. Assuming for the PLA-based matrix that $\log \eta_o \approx 6.1$ (Figure 5), the molar mass reduction detected for the [PLA (85 wt%) + ATBC (10 wt%) + CaCO$_3$ (5 wt%)] (80 wt%) + PPP (20 wt%) biocomposite should produce a decrease in $\log \eta_o$ of about 7%. The decrease in the modulus of the complex viscosity for the biocomposite with 20 wt% of PPP is actually about 16% in the investigated deformation frequency range, which is close to the Newtonian plateau. This finding further confirms that the presence of free volume at the polymer/fiber interfaces due to poor adhesion between potato pulp powder and PLA reduces the viscosity in the biocomposites and favors the flow of the PLA chains.

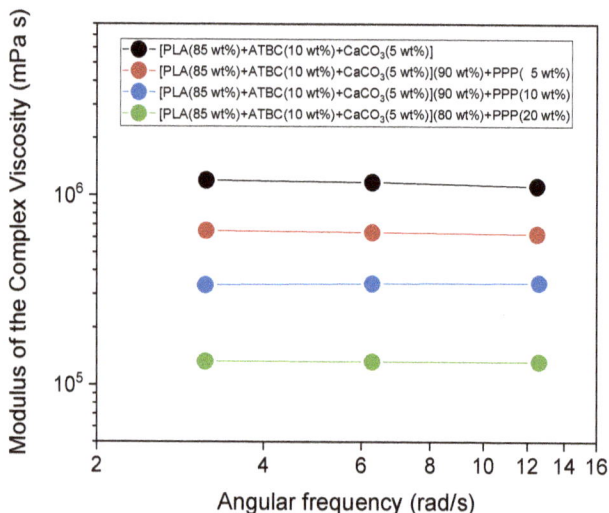

Figure 8. Modulus of the complex viscosity vs. angular frequency at 175 °C for PLA-based matrix and biocomposites indicated in the legend.

2.5. Morphological Properties of the PLA-Based Biocomposites

The morphology of fracture surfaces of the PLA-based biocomposites with PPP from dog-bone specimens was studied by scanning electron microscopy in order to investigate the dispersion of the potato pulp particles and the compatibility between the matrices and the fibers.

Figure 9 shows that the potato pulp particles are well dispersed within the matrix and their distribution is uniform, which means that they were satisfactorily separated during the extrusion process. The micrographs clearly show that the interfacial adhesion between the PLA-based matrix and the potato pulp powder is quite poor because a gap is well evident between the polymeric matrix and the potato pulp particles as indicated by the black arrows. Poor adhesion leads to brittle materials as also proven by the fiber pullout indicated by the red arrow, which illustrates how, in these biocomposites, failure occurs at the matrix/fiber interface. The finding is in agreement with the decrease in the tensile strength discussed above (Figure 7).

300× 2400×

[PLA (85 wt%) + ATBC (10 wt%) + CaCO₃ (5 wt%)] (90 wt%) + PPP (10 wt%)

300× 2400×

[PLA (85 wt%) + ATBC (10 wt%) + CaCO₃ (5 wt%)] (80 wt%) + PPP (20 wt%)

Figure 9. SEM images of the [PLA (85 wt%) + ATBC (10 wt%) + CaCO$_3$ (5 wt%)] (90 wt%) + PPP (10 wt%) and [PLA (85 wt%) + ATBC (10 wt%) + CaCO$_3$ (5 wt%)] (80 wt%) + PPP (20 wt%) biocomposite at the indicated magnification.

3. Materials and Methods

3.1. Materials

Poly(lactic acid) (PLA), derived from natural resources, was obtained from 2003D NatureWorks (Minnetonka, MN, USA). The product grade was suitable for thermoforming and extrusion processes and contained 3% D-lactic acid units (melt flow index (MFI): 6 g/10 min (210 °C, 2.16 kg); nominal average molar mass: 200,000 g/mol).

The plasticizer acetyl tributyl citrate (ATBC) was purchased from Sigma Aldrich S.R.L (Milan, Italy).

The calcium carbonate (CaCO$_3$) OMYACARB® was an inert filler supplied by the company OMYA (Oftringen, Switzerland). The powder had fine granulometry with particle size distribution centered at 12 μm.

Dried potato pulp powder (PPP) was produced by the company SüdStärke (Schrobenhausen, Germany). The moisture content was about 3 wt%, and the composition of the dry matter was as follows: proteins 7 wt%, starch 25 wt%, cellulose 16 wt%, hemicellulose 7 wt%, lignin 20 wt%, pectin 17 wt%, and ash 5 wt%.

3.2. Composite Preparation

Biocomposites of PLA with potato pulp powder were prepared by adding the filler PPP in different percentages to the polymeric matrix constituted by the bio-based PLA (with a concentration of 85 wt%), the plasticizer ATBC (with a concentration of 10 wt%), and CaCO$_3$ (with a concentration of 5 wt%). For comparison, pure PLA and PLA mixed only with the plasticizer ATBC were also processed in the same way.

Before processing, the PLA and potato pulp powder were dried at a temperature of 60 °C for at least 24 h. The PLA-based matrix and biocomposites were prepared by using a MiniLab II HAAKE Rheomex CTW 5—a co-rotating conical twin-screw extruder. The molten materials were transferred from the mini extruder through a preheated cylinder to a mini injection molder (Thermo Scientific HAAKE MiniJet II), which allows the preparation of dog-bone tensile bar specimens to be used for thermal, mechanical, and rheological characterization. The dimensions of the dog-bone tensile bars were as follows: Width in the larger section—10 mm, width in the narrow section—4.8 mm, thickness—1.35 mm, and length—90 mm. The extruder operating conditions adopted for all the formulations are reported in Table 4.

Table 4. Operating condition use for the extrusion and injection molding process.

Extrusion Temperature (°C)	Screw Speed (rpm)	Cycle Time (s)	Injection Temperature (°C)	Injection Pressure (bar)	Molding Time (s)	Mold Temperature (°C)
180	100	90	180	500	60	90

After preparation, all the samples were stored in a desiccator and analyzed the day after in order to avoid physical ageing effects on the physical properties investigated.

3.3. Composite Characterization

The thermal stability of the potato pulp powder and selected samples was investigated by thermogravimetric analysis (TGA) carried out on about 10 mg of sample by using a SII TG/DTA 7200 EXSTAR Seiko (Minato, Tokyo, Japan) under nitrogen flow (200 mL/min) and at a heating speed of 10 K/min from 50 to 600 °C.

Number-average molar mass (M_n) and weight-average molar mass (M_w) were determined using size exclusion chromatography (SEC) with the Agilent Technologies 1200 Series (Santa Clara, CA, USA) and calculated with the Agilent ChemStation Software. The instrument was equipped with an Agilent degasser, an isocratic HPLC pump, two PLgel 5 µm MiniMIX-D columns conditioned at 35 °C, and an Agilent refractive index (RI) detector. The mobile phase was chloroform ($CHCl_3$) at a flow rate of 0.3 mL/min. The system was calibrated with polystyrene standards in a range from 500 to 3×10^5 g/mol. Samples were dissolved in $CHCl_3$ (2 mg/mL) and filtered through a 0.20 µm syringe filter before analysis.

The morphology of the PPP powder was investigated by scanning electron microscopy (SEM) with an FEI-Quanta 450 ESEM instrument (ThermoFisher, Waltham, MA, USA). The micrographs of samples fractured with liquid nitrogen and etched with gold were collected. Backscattered electrons generated the images, the resolution for which was provided by beam deceleration with a landing energy of 2 kV.

Differential scanning calorimetry (DSC) measurements were performed with a Calorimeter DSC 8500 Perkin Elmer (Waltham, MA, USA) equipped with an IntraCooler III as a refrigerating system. The instrument was calibrated to temperature with high purity standards (indium, naphthalene, cyclohexane) according to the procedure for standard DSC [57]. Energy calibration was performed with indium. Dry nitrogen was used as a purge gas at a rate of 30 mL/min. The as prepared samples were analyzed from −50 to 200 °C at the heating rate of 10 K/min and with a single thermal scan.

Tensile tests on the samples prepared with the injection molder were performed at room temperature at a crosshead speed of 10 mm/min and by means of an INSTRON 5500 R universal testing machine (Instron Corp., Canton, MA, USA) equipped with a 10 kN load cell and interfaced with a computer running the Testworks 4.0 software (MTS Systems Corporation, Eden Prairie, MN, USA). At least five specimens were tested for each sample in accordance to ASTM D 638, and the average values were reported.

Oscillatory shear measurements were performed by means of a Rheometer Anton Paar MCR 302 (Graz, Austria) equipped with parallel plates of 25 mm diameter under nitrogen flow to minimize oxidation and to maintain a dry environment. Frequency sweep experiments were performed at 175 °C at a fixed strain (3%) with a gap of 1 mm, in the linear regime, in order to measure the modulus of the complex viscosity. For the analysis of the PLA-based matrix, the angular frequencies were swept from 0.068 to 628 rad/s with five points per decade. For the analysis of the biocomposites, a narrower angular frequency was covered to reduce the time measurement and avoid degradation of the potato pulp.

4. Conclusions

Potato pulp powder, which is an organic waste from the production and extraction of starch, can be suitable for processing in the melt with the bio-based and biodegradable PLA, presenting no main problems in processing. Due to the low aspect ratio of the potato pulp powder, there is no reinforcing effect on the polymeric matrix, but there is a moderate loss in mechanical properties. The potato pulp power utilized to produce PLA-based biocomposites acts as a filler and not as a reinforcement for the polymeric matrix. The loss in mechanical properties is, however, not dramatically high, because the elastic modulus decreases by only 13% when the PPP content is 20 wt%. The adhesion between the potato pulp powder and the polymeric matrices is poor, as attested by the tensile strength, which appears progressively smaller for the biocomposites with respect to the PLA matrix. The ductility of the biocomposites is also smaller with respect to that of the PLA matrix because the dispersed potato pulp particles act as stress concentrators and promote microcrack formation. Conversely, the lower viscosity of the biocomposites containing PPP is a factor that favors the processing and the molding of these materials.

Notwithstanding, the present PLA-based biocomposites still present properties valuable for practical applications and still meet the technical requirements indicated for rigid packaging production. In addition, the presence of natural fibers promotes and accelerates the biodegradation of the biocomposite material [58].

Finally, the incorporation of about 20 wt% of potato pulp powder, no-food competition biomass, to PLA offers the possibility to markedly reduce the cost of the final products when considering the relatively high cost of PLA. This opens an innovative option for the valorization of an abundant agro-food biomass such as potato pulp and thus meeting circular economy expectations.

Author Contributions: M.C.R. contributed to funding acquisition, supervised the study, and wrote and revised the paper. P.C. contributed to funding acquisition, and co-supervised the study and revised the paper. N.M. prepared the biocomposites, and performed the mechanical and morphological characterization. C.A.M. performed the rheological measurements. S.B. performed the thermal characterization and the SEC measurements. A.S. provided the potato pulp powder and performed the composition analysis. A.L. contributed to funding acquisition and revised the paper.

Funding: This research was funded by the European Commission's Horizon 2020 Research and Innovation Programme (2014–2020)—Sustainable techno-economic solutions for the agricultural value chain (AgroCycle), under Grant Agreement n. 690142.

Conflicts of Interest: The authors declare no conflict of interest.

References

1. Hamad, K.; Kaseem, M.; Ayyoob, M.; Joo, J.; Deri, F. Polylactic acid blends: The future of green, light and tough. *Prog. Polym. Sci.* **2018**, *85*, 83–127. [CrossRef]
2. Bledzki, A.K.; Gassan, J. Composites reinforced with cellulose based fibres. *Prog. Polym. Sci.* **1999**, *24*, 221–274. [CrossRef]
3. Mohanty, A.K.; Misra, M.; Hinrichsen, G. Biofibers, biodegradable polymers and biocomposites: An overview. *Macromol. Mater. Eng.* **2000**, *276–277*, 1–24. [CrossRef]
4. Yu, L.; Dean, K.; Lin, L. Polymer blends and composites from renewable resources. *Prog. Polym. Sci.* **2006**, *31*, 576–602. [CrossRef]

5. Maya Jacob, J.; Sabu, T. Review—Biofibres and biocomposites. *Carbohydr. Polym.* **2008**, *71*, 343–364. [CrossRef]

6. Satyanarayana, K.G.; Arizaga, G.G.C.; Wypych, F. Biodegradable composites based on lignocellulosic fibers—An overview. *Prog. Polym. Sci.* **2009**, *34*, 982–1021. [CrossRef]

7. Faruk, O.; Bledzki, A.K.; Fink, H.P.; Sain, M. Biocomposites reinforced with natural fibers: 2000–2010. *Prog. Polym. Sci.* **2012**, *37*, 1552–1596. [CrossRef]

8. Pandey, J.K.; Ahn, S.H.; Lee, C.S.; Mohanty, A.K.; Misra, M. Recent Advances in the Application of Natural Fiber Based Composites. *Macromol. Mater. Eng.* **2010**, *295*, 975–989. [CrossRef]

9. Nielsen, L.E.; Landel, R.F. *Mechanical Properties of Polymers and Composites*, 2nd ed.; Faulkner, L.L., Ed.; CRC Press: New York, NY, USA, 1994; ISBN 0824761837.

10. Ku, H.; Wang, H.; Pattarachaiyakoop, N.; Trada, M. A review on the tensile properties of natural fiber reinforced polymer composites. *Compos. Part B* **2011**, *42*, 856–873. [CrossRef]

11. Signori, F.; Pelagaggi, M.; Bronco, S.; Righetti, M.C. Amorphous/Crystal and Polymer/Filler Interphases in Biocomposites from Poly(butylene succinate). *Thermochim. Acta* **2012**, *543*, 74–81. [CrossRef]

12. Saheb, D.N.; Jog, J.P. Natural Fiber polymer composites: A Review. *Adv. Polym. Technol.* **1999**, *18*, 351–363. [CrossRef]

13. Li, J.; He, Y.; Inoue, Y. Thermal and mechanical properties of biodegradable blends of poly(L-lactic acid) and lignin. *Polym. Int.* **2003**, *52*, 949–955. [CrossRef]

14. Porras, A.; Maranon, A. Development and characterization of a laminate composite material from polylactic acid (PLA) and woven bamboo fabric. *Compos. Part B* **2012**, *43*, 2782–2788. [CrossRef]

15. Plackett, D.; Andersen, T.L.; Pedersen, W.B.; Nielsen, L. Biodegradable composites based on L-polylactide and jute fibres. *Compos. Part A* **2003**, *63*, 1287–1296. [CrossRef]

16. Masirek, R.; Kulinski, Z.; Chionna, D.; Pirokowska, E.; Pracella, M. Composites of poly(L-lactide) with hemp fibers: Morphology and thermal and mechanical properties. *J. Appl. Polym. Sci.* **2007**, *105*, 255–268. [CrossRef]

17. Johari, A.P.; Mohanty, S.; Kurmvanshi, S.K.; Nayak, S.K. Influence of different treated cellulose fibers on the mechanical and thermal properties of poly(lactic acid). *ACS Sustain. Chem. Eng.* **2016**, *4*, 1619–1629. [CrossRef]

18. Dong, Y.; Ghataura, A.; Takagi, H.; Haroosh, H.J.; Nakagaito, A.N.; Lau, K.T. Polylacticacid (PLA) biocomposites reinforced with coir fibres: Evaluation of mechanical performance and multifunctional properties. *Compos. Part A* **2014**, *63*, 76–84. [CrossRef]

19. Väisänen, T.; Haapala, A.; Lappalainen, R.; Tomppo, L. Utilization of agricultural and forest industry waste and residues in natural fiber-polymer composites: A review. *Waste Manag.* **2016**, *54*, 62–73. [CrossRef]

20. Fritsch, C.; Staebler, A.; Happel, A.; Cubero Márquez, M.A.; Aguiló-Aguayo, I.; Abadias, M.; Gallur, M.; Cigognini, I.M.; Montanari, A.; Jose López, M.; et al. Processing, Valorization and Application of Bio-Waste Derived Compounds from Potato, Tomato, Olive and Cereals: A Review. *Sustainability* **2017**, *9*, 1492. [CrossRef]

21. Schmid, M.; Herbst, C.; Mueller, K.; Staebler, A.; Schlemmer, D.; Coltelli, M.B.; Lazzeri, A. Effect of potato pulp filler on the mechanical properties and water vapor transmission rate of thermoplastic WPI/PBS blends. *Polym. Plast. Technol. Eng.* **2016**, *55*, 510–517. [CrossRef]

22. Mekonnen, T.; Mussone, P.; Khalil, H.; Bressler, D. Progress in bio-based plastics and plasticizing modifications. *J. Mater. Chem. A* **2013**, *1*, 13379–13398. [CrossRef]

23. European Food Safety Authority. Scientific Opinion on Flavouring Group Evaluation 10, Revision 3 (FGE.10Rev3): Aliphatic primary and secondary saturated and unsaturated alcohols, aldehydes, acetals, carboxylic acids and esters containing an additional oxygenated functional group and lactones from chemical groups 9, 13 and 30. *EFSA J.* **2012**, *10*, 25–63. [CrossRef]

24. Yang, H.; Yan, R.; Chen, H.; Lee, D.H.; Zheng, C. Characteristic of hemicellulose, cellulose and lignin pyrolysis. *Fuel* **2007**, *86*, 1781–1788. [CrossRef]

25. Garcia-Perez, M.; Chaala, A.; Yang, J.; Roy, C. Co-pyrolysis of sugarcane bagasse with petroleum residue. Part I: Thermogravimetric analysis. *Fuel* **2001**, *80*, 1245–1258. [CrossRef]

26. Aggarwal, P.; Dollimore, D.; Heon, K. Comparative thermal analysis study of two biopolymers, starch and cellulose. *J. Therm. Anal.* **1997**, *50*, 7–17. [CrossRef]

27. Kamur, P.; Sandeep, K.P.; Alavi, S.; Truong, V.D.; Gorga, R.E. Preparation and characterization of bio-nanocomposite films based on soy protein isolate and montmorillonite using melt extrusion. *J. Food Eng.* **2010**, *100*, 480–489. [CrossRef]

28. McNeill, I.C.; Leiper, H.A. Degradation Studies of Some Polyesters and Polycarbonates-1. Polylactide: General Features of the Degradation Under Programmed Heating Conditions. *Polym. Degrad. Stab.* **1985**, *1*, 264–285. [CrossRef]

29. Awal, A.; Rana, M.; Sain, M. Thermorheological and mechanical properties of cellulose reinforced PLA bio-composites. *Mech. Mater.* **2015**, *80*, 87–95. [CrossRef]

30. Zhao, Y.-Q.; Cheung, H.-Y.; Lau, K.-T.; Xu, C.-L.; Zhao, D.-D.; Li, H.-L. Silkworm silk/poly(lactic acid) biocomposites: Dynamic mechanical, thermal and biodegradable properties. *Polym. Degrad. Stab.* **2010**, *95*, 1978–1987. [CrossRef]

31. Pyda, M.; Bopp, R.C.; Wunderlich, B. Heat capacity of poly(lactic acid). *J. Chem. Thermodyn.* **2004**, *36*, 731–742. [CrossRef]

32. Pan, P.; Inoue, Y. Polymorphism and isomorphism in biodegradable polyesters. *Prog. Polym. Sci.* **2009**, *34*, 605–640. [CrossRef]

33. Di Lorenzo, M.L.; Rubino, P.; Luijkx, R.; Helou, M. Influence of chain structure on crystal polymorphism of poly(lactic acid). Part 1: Effect of optical purity of the monomer. *Colloid Polym. Sci.* **2014**, *292*, 399–409. [CrossRef]

34. Androsch, R.; Schick, C.; Di Lorenzo, M.L. Melting of conformationally disordered crystals α'-phase of poly(L-lactic acid). *Macromol. Chem. Phys.* **2014**, *215*, 1134–1139. [CrossRef]

35. Pan, P.; Kai, W.; Zhu, B.; Dong, T.; Inoue, Y. Polymorphs Crystallization and Multiple Melting Behavior of Poly(L-lactide): Molecular Weight Dependence. *Macromolecules* **2007**, *40*, 6898–6905. [CrossRef]

36. Righetti, M.C.; Laus, M.; Di Lorenzo, M.L. Temperature dependence of the rigid amorphous fraction in poly(ethylene terephthalate). *Eur. Polym. J.* **2014**, *58*, 60–68. [CrossRef]

37. Righetti, M.C.; Di Lorenzo, M.L. Rigid amorphous fraction and multiple melting behavior in poly(butylene terephthalate) and isotactic polystyrene. *J. Therm. Anal Calorim.* **2016**, *126*, 521–530. [CrossRef]

38. Righetti, M.C.; Gazzano, M.; Di Lorenzo, M.L.; Androsch, R. Enthalpy of melting of α' and α-crystals of poly(L-lactic acid). *Eur. Polym. J.* **2015**, *70*, 215–220. [CrossRef]

39. Saeidlou, S.; Huneault, M.A.; Li, H.; Park, C.B. Poly(lactic acid) crystallization. *Prog. Polym Sci.* **2012**, *37*, 1657–1677. [CrossRef]

40. Baiardo, M.; Frisoni, G.; Scandola, M.; Rimelen, M.; Lips, D.; Ruffieux, K.; Wintermantel, E. Thermal and mechanical properties of plasticized poly(l-lactic acid). *J. Appl. Polym. Sci.* **2003**, *90*, 1731–1738. [CrossRef]

41. Jacobsen, S.; Fritz, H.G. Plasticizing polylactide—The effect of different plasticizers on the mechanical properties. *Polym. Eng. Sci.* **1999**, *39*, 1303–1310. [CrossRef]

42. Kim, H.S.; Park, B.H.; Choi, J.H.; Yoon, J.S. Mechanical Properties and Thermal Stability of Poly(L-lactide)/Calcium Carbonate Composites. *J. Appl. Polym. Sci.* **2008**, *109*, 3087–3092. [CrossRef]

43. Molnár, K.; Móczól, J.; Murariu, M.; Dubois, P.; Pukánszky, B. Factors affecting the properties of PLA/CaSO4 composites: Homogeneity and interactions. *eXPRESS Polym. Lett.* **2009**, *3*, 49–61. [CrossRef]

44. Lim, L.T.; Auras, R.; Rubino, M. Processing technologies for poly(lactic acid). *Prog. Polym. Sci.* **2008**, *33*, 820–852. [CrossRef]

45. Auras, R.A.; Lim, L.T.; Selke, S.E.M.; Tsuji, H. *Poly(Lactid Acid) Synthesis, Structures, Properties, Processing, and Applications*; John Wiley & Sons: Hoboken, NJ, USA, 2010; ISBN 978-0-470-29366-9.

46. Domenek, S.; Fernandes-Nassar, S.; Ducruet, V. Rheology, Mechanical Properties, and Barrier Properties of Poly(lactic acid). *Adv. Polym. Sci.* **2017**, *279*, 303–342. [CrossRef]

47. Wang, N.; Zhang, X.; Ma, X.; Fang, J. Influence of carbon black on the properties of plasticized poly(lactic acid) composites. *Polym. Degrad. Stab.* **2008**, *93*, 1044–1052. [CrossRef]

48. Xu, Y.Q.; Qu, J.P. Mechanical and Rheological Properties of Epoxidized Soybean Oil Plasticized Poly(lactic acid). *J. Appl. Polym. Sci.* **2009**, *112*, 3185–3191. [CrossRef]

49. Gu, S.Y.; Zou, C.Y.; Zhou, K.; Ren, J. Structure-Rheology Responses of Polylactide/Calcium, Carbonate Composites. *J. Appl. Polym. Sci.* **2009**, *114*, 1648–1655. [CrossRef]

50. Shibata, M. Poly(lactic acid)/Cellulosic Fiber composites. In *Biodegradable Polymer Blends and Composites from Renewable Resources*; John Wiley & Sons, Inc.: Hoboken, NJ, USA, 2009. [CrossRef]

51. Zafeiropoulos, N.E.; Baillie, C.A.; Matthews, F.L. A study of transcrystallinity and its effect on the interface in flax fibre reinforced composite materials. *Composits Part A* **2001**, *32*, 525–543. [CrossRef]
52. Quan, H.; Li, Z.M.; Yang, M.B.; Huang, R. On transcrystallinity in semi-crystalline polymer composites. *Compos. Sci. Technol.* **2005**, *65*, 999–1102. [CrossRef]
53. Middleton, J.C.; Tipton, A.J. Synthetic biodegradable polymers as orthopedic devices. *Biomaterials* **2000**, *21*, 2335–2346. [CrossRef]
54. Maharana, T.; Mohanty, B.; Negi, Y.S. Melt-solid polycondensation of lactic acid and its biodegradability. *Prog. Polym. Sci.* **2009**, *34*, 99–124. [CrossRef]
55. Palade, L.-I.; Lehermeier, H.J.; Dorgan, J.R. Melt Rheology of High L-Content Poly(lactic acid). *Macromolecules* **2001**, *34*, 1384–1390. [CrossRef]
56. Othman, N.; Acosta-Ramirez, A.; Mehrkhodavandi, P.; Dorgan, J.R.; Hatzikiriakos, S.G. Solution and melt viscoelastic properties of controlled microstructure poly(lactide). *J. Rheol.* **2011**, *55*, 987–1004. [CrossRef]
57. Sarge, S.M.; Hemminger, W.; Gmelin, E.; Höhne, G.W.H.; Cammenga, H.K.; Eysel, W. Metrologically based procedures for the temperature, heat and heat flow rate calibration of DSC. *J. Therm. Anal.* **1997**, *49*, 1125–1134. [CrossRef]
58. John, M.J. Environmental degradation in biocomposites. In *Biocomposites for High-Performance Applications*; Woodhead Publishing: Cambridge, UK, 2017. [CrossRef]

© 2019 by the authors. Licensee MDPI, Basel, Switzerland. This article is an open access article distributed under the terms and conditions of the Creative Commons Attribution (CC BY) license (http://creativecommons.org/licenses/by/4.0/).

International Journal of
Molecular Sciences

MDPI

Article

The Degradation Properties of MgO Whiskers/PLLA Composite In Vitro

Yun Zhao [1,2], Bei Liu [1], Hongwei Bi [3], Jinjun Yang [4], Wei Li [1,2], Hui Liang [1], Yue Liang [1], Zhibin Jia [1], Shuxin Shi [1] and Minfang Chen [1,2,*]

1   School of Materials Science and Engineering, Tianjin University of Technology, Tianjin 300384, China;
    yun_zhaotju@163.com (Y.Z.); liubei8679@163.com (B.L.); earlybird.184@126.com (W.L.);
    huiliang2014@126.com (H.L.); 15102260655@163.com (Y.L.); zhibin225225@163.com (Z.J.);
    13752048706@163.com (S.S.)
2   Key Laboratory of Display Materials and Photoelectric Device (Ministry of Education), Tianjin University of
    Technology, Tianjin 300384, China
3   Tianjin Sannie Bioengineering Technology Co., Ltd., Tianjin 300384, China; bluewanli@126.com
4   School of Environmental Science and Safety Engineering, Tianjin University of Technology, Tianjin 300384,
    China; tjyjj_2014@tjut.edu.cn
*   Correspondence: mfchentj@126.com

Received: 14 August 2018; Accepted: 30 August 2018; Published: 13 September 2018

Abstract: In this study, composite films of stearic acid–modified magnesium oxide whiskers (Sa–w–MgO)/poly-L-lactic acid (PLLA) were prepared through solution casting, and the in vitro degradation properties and cytocompatibility of the composites with different whisker contents were investigated. The results showed that the degradation behavior of the composite samples depended significantly on the whisker content, and the degradation rate increased with the addition of MgO content. Furthermore, the degradation of the composites with higher contents of whiskers was influenced more severely by the hydrophilicity and pH value, leading to more final weight loss, but the decomposition rate decreased gradually. Furthermore, the pH value of the phosphate buffer solution (PBS) was obviously regulated by the dissolution of MgO whiskers through neutralization of the acidic product of PLLA degradation. The cytocompatibility of the composites also increased remarkably, as determined from the cell viability results, and was higher than that of PLLA at the chosen whisker content. This was beneficial for the cell affinity of the material, as it notably led to an enhanced biocompatibility of the PLLA, in favor of promoting cell proliferation, which significantly improved its bioactivity, as well.

Keywords: biopolymers composites; MgO whiskers; PLLA; in vitro degradation

---

1. Introduction

Over the last few decades, poly(L-lactide) (PLLA) has been given significant attention, owing to its advantages of good biocompatibility and processability, biodegradability, and bioresorbability, which are desirable in the biomedical field, especially for bone repair [1–4]. However, some drawbacks limit its wider application [5–10], particularly its degradation properties, which are an extremely important factor in bone repair, as accumulation of the lactic acid degradation product from PLLA can cause issues such as aseptic inflammation in vivo, hydrophobicity and lower biological activity that are unfavorable to the viability of the bone cell, and biodegradation that cannot be well controlled according to the requirement of bone formation.

To solve these problems, many researchers have attempted to improve PLLA properties through various methods [11–15]. Generally, a composite prepared by adding inorganic fillers is very promising, as it has the potential to enhance the mechanical and thermal properties, to enhance the hydrophilicity and bioactivity, and to reduce aseptic inflammation, because the fillers used have good bioactivity and similar constituents to the inorganic compounds in bone, such as calcium phosphate. However, these fillers, such as hydroxyapatite (HA) [4,7,11,13], β-tricalcium phosphate (β-TCP) [5,6,15], and bioactive glass [16,17], are usually slightly soluble [18], which tends to limit their impact on neutralizing the acidic environment induced by PLLA hydrolysis.

Magnesium oxide (MgO), another inorganic filler with good biocompatibility, and bioactive capacity that is not toxic, has attracted much attention [19–22] for polymer modification. In addition to the aforementioned advantages, the Mg^{2+} released from MgO dissolving in water is beneficial to the protein synthesized via the activation of many enzymes, which contribute to the excellent bioactivity of MgO. It has been reported that the dissolution of MgO in vitro prominently influences the solution pH through alkaline degradation [19]. Recently, the researchers utilized the different shapes of MgO [9,19–24] as fillers and introduced them to PLLA for preparing composites. The results mostly showed that the mechanical properties and thermal behavior of PLLA improved significantly with the presence of MgO fillers, which was also proved in our works [24]. However, few reports have focused on the effect of MgO contents on the degradation behavior of the composite during the decomposition process. Ma et al. [19] obtained the n-MgO/PLLA composite by adding n-MgO modified by poly(ε-caprolactone) into a PLLA matrix, and the reduction of pH was suppressed to a certain extent, which was also demonstrated by us [22]. Luo et al. [9] studied the in vitro degradation properties of the MgO whiskers (w-MgO)/PLLA composite; their work mainly focused on the effect of whiskers on PLLA degradation for a long decomposition period. It is worth mentioning that because such composites have the potential for application in bone repair, it is crucial to investigate their degradability and the cytocompatibility performance of the initial degradation stage, especially in terms of how it is affected by the filler contents, which is greatly important for the materials used, as implant inescapably contacted with bone cell and tissue [18]. Meanwhile, it was also noted that the change in the hydrophilicity of the PLLA matrix improved with the addition of fillers such as organic montmorillonite (OMMT), which prominently accelerated the hydrolytic degradation of PLLA, playing another key role in controlling the degradation process of PLLA [25,26]. As for MgO, they greatly impacted not only the hydrolysis behavior of PLLA, but also its degradation conditions, by controlling the pH. Therefore, the complex process in which the hydrolysis of PLLA could be facilitated can be controlled by varying the whisker content as well as be retarded by changes in the pH value, simultaneously [19,27]. Furthermore, it was reported that the hydrolysis process of PLLA was also favorable for promoting the growth of an ordered crystalline structure, due to the easier movement of the chains and segment under the degrading state [28,29]. Additionally, studying the degradation behavior in vitro can provide fundamental experimental results for the degrading properties of PLLA used in bone repair in vivo.

In this work, the MgO whiskers were modified using stearic acid, and the chemical bondings were obtained through the interaction between them, as demonstrated by our previous work [24]. Because stearic acid is effective in modifying inorganic fillers like hydroxyapatite [30] and TiO_2 [31] through changing the polarities, it was greatly helpful in improving the distribution of MgO whiskers and enhancing the interface of the PLLA composites. Then, this work undertook the detailed characterization of w-MgO/PLLA soaked in phosphate buffer solution (PBS) for different degrading periods and analyzed the influence of the whisker content on the composite degrading behaviors. The pH stability and cell affinity of the composite were also studied and evaluated by relative measurements and cell cultures to relate the changes in the composite degrading procedure to the material's biocompatibility and bioactivity.

2. Results

2.1. XRD Measurement

In order to investigate the effect of MgO whiskers on the properties of PLLA, such as its crystalline structure and crystallization behavior, the samples of PLLA and composites were tested after being soaked using XRD and DSC, and the results of XRD testing are shown in Figure 1 and Table 1. Figure 1 shows that the intensity of diffraction characteristic peaks of PLLA, for example with 2θ values of 17.76° and 18.68°, corresponding to the (110/200) and (203) diffraction planes, respectively, clearly strengthened gradually, probably due to the surface decomposition and increment of PLLA crystallinity during degrading periods. Meanwhile, a slight shift of these peaks was noted with an increase in the immersion time, indicating the generation and transformation of different types of crystalline PLLA in the degrading process [32], especially between the forms of α' and α. Specifically, the strength of 2θ at 24.82° for the (206) diffraction plane generated by the α' form became more intense after 28 days' immersion. Furthermore, these similar variations were also observed for the composite specimens, and the peak shifts were more prominent. This phenomenon can account for the accelerated degradation of the PLLA matrix, owing to the presence of MgO whiskers because of increased hydrophilicity [29]. Thus, a strong shift of the peaks was noted after 14 days. The little decrease in the 2θ value after 28 days of degradation seen for all composite samples, in comparison to that noted after 14 days, was attributed to the change in the relative content of the α' and α forms caused by PLLA decomposition. Additionally, the characteristic peaks of MgO located at 43.5°, 62.5°, and 78.6°, corresponding to the (200), (220), and (222) diffraction planes, respectively, can also be found in the graphs of the composites, and PLLA3, with the highest content of 3 wt% MgO, showing the most significant peaks. However, the intensity of these peaks of MgO weakened gradually with progress in the experimental period: This change suggested a loss in the amount of MgO on the testing surface of composite films in the degrading process, showing no obvious peaks of the whiskers in the patterns of PLLA1, PLLA2, and PLLA3 for the 7 days in Figure 1.

Table 1. The 2θ values of samples on Day 0.

Sample	Diffraction Plane								
	(010)	(110/200)	(203)	(204)	(015)	(016)	(206)	(207)	(018)
	$2\theta(°)$								
PLLA	14.69	17.76	18.68	21.36	22.45	23.78	24.82	26.62	28.71
PLLA1	14.63	16.48	18.96	21.36	22.28	23.72	24.87	26.60	28.78
PLLA2	14.63	16.67	19.02	21.39	22.43	23.83	24.09	26.64	28.93
PLLA3	14.74	16.67	19.06	21.43	22.34	23.79	24.94	26.64	28.82

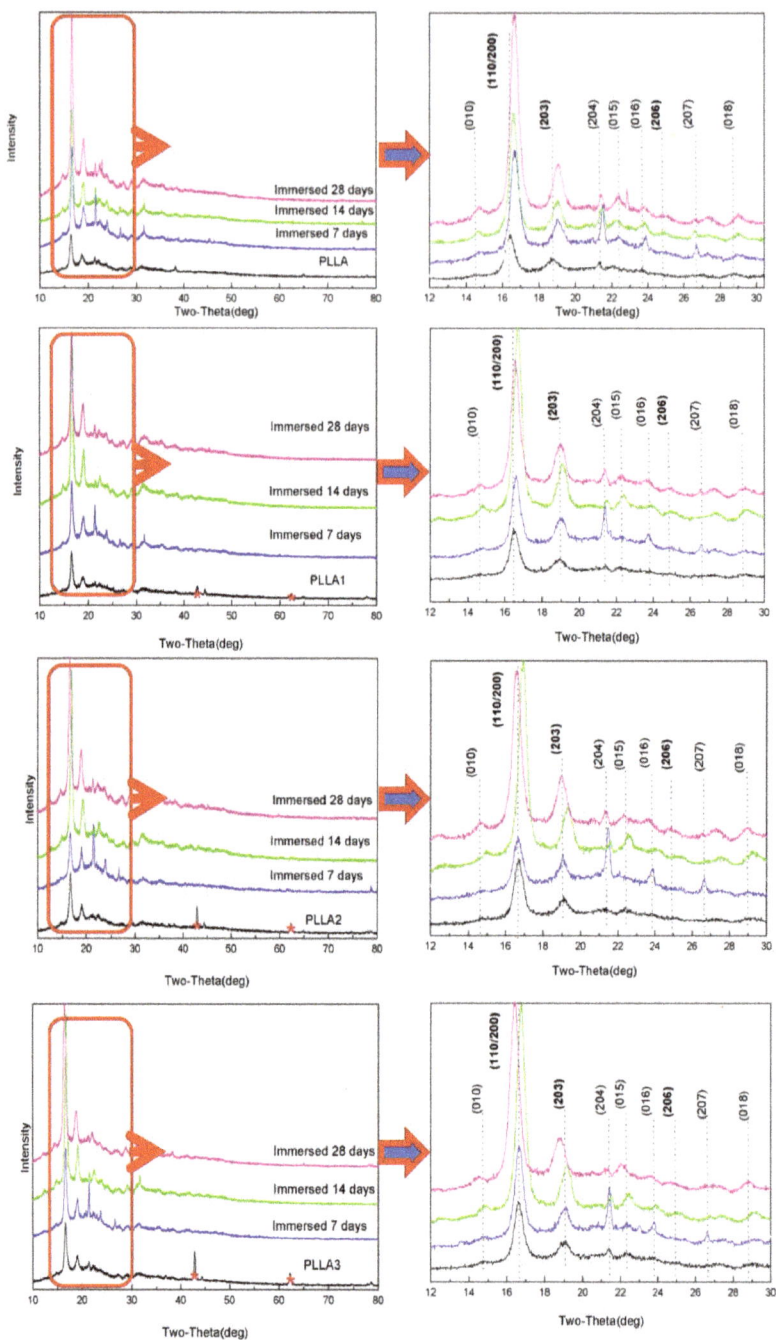

Figure 1. X-ray powder diffraction (XRD) patterns of the poly(L-lactide) (PLLA) and composite films at different degrading times and under partial enlargement.

2.2. DSC Analysis

The DSC data in the PLLA show the endothermic peaks of T_g and exothermic peaks of T_{cc} for the neat and composite materials for Day 0 in Figure 2. During the degradation process, the exothermic peaks observed around T_g and T_{cc} for the neat PLLA samples disappeared gradually, and meanwhile one weak exothermic peak was seen at the left side of the main endothermic peak, which was due to the $\alpha' \rightarrow \alpha$ transition [28] in the substrate from 7 days' to 28 days' immersion. Furthermore, it can be seen in Table 2 that the T_{mc} of the PLLA sample increased when it was immersed initially, and a small variation was observed in the value over the course of the experiment. This suggested an increased crystallinity of the specimen at the beginning of the procedure, caused by enhancement of the chain mobility, and particularly an increase in the amount of α form crystals, which was consistent with the XRD observation where a larger value of 2θ corresponding to the (110)/(200) crystal plane was obtained for PLLA after 28 days' immersion. Compared to the control, the composites exhibited different crystalline behaviors. Because the whiskers effectively promoted PLLA bulk crystallization, there were obvious exothermic peaks of $\alpha' \rightarrow \alpha$ transitions observed for the composite samples on Day 0. Double fusion peaks and a prominent reduction in the final T_{mc} value were observed after 28 days, indicating that the composites decomposed faster than the control sample under the same degrading procedure. Moreover, the PLLA3 sample, after 28 days' immersion, showed double fusion peaks similar to the ones shown by the PLLA1 and PLLA2 samples after 14 days, suggesting that an increased whisker amount was beneficial to inhibiting the degrading extent of the composites. This was also demonstrated by the T_{mc} value of PLLA1, 164.6 °C, which was the lowest among those exhibited by all specimens.

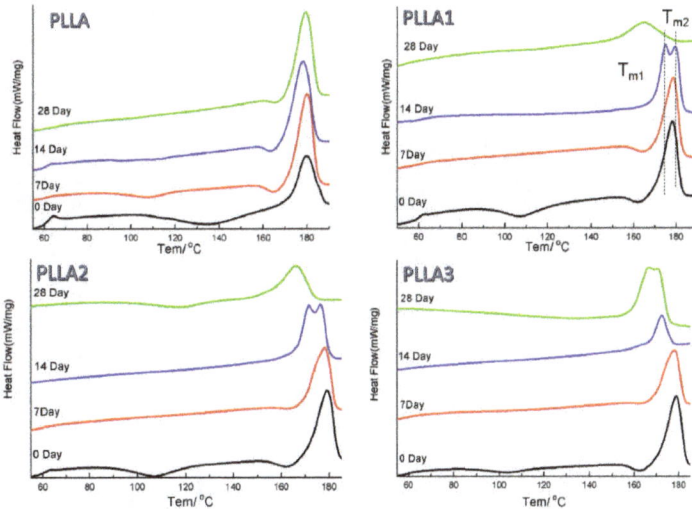

Figure 2. Secondary heating curves obtained by differential scanning calorimetry (DSC) for the samples after soaking in PBS for different degrading times.

Table 2. Melting points of the samples used in DSC measurements.

Degradation Time	T_m/°C			
	PLLA	PLLA1	PLLA2	PLLA3
0	179.8	178.2	179.2	179.1
7	180.0	178.5	178.1	177.9
14	178.6	174.9/179.4	171.4/176.4	172.4
28	179.6	164.6	166.3	166.8/170.4

2.3. Surface and Fracture Morphologies of Sample Films after Degradation

The surface and fracture morphologies (treated by liquid nitrogen for brittle fracture) of the sample films immersed in PBS for various degradation periods were obtained and are shown in Figure 3. It is obvious that the surfaces of all specimens degraded for 7 days became a little rougher in comparison to their initial smooth state, and it was also observed that the control film had a larger rough region than the composite ones, indicating an increase in the crystallinity of the composite on addition of MgO whiskers. This roughness observed after 7 days was probably due to the generation of the α′ form under the degradation, as shown in Figure 3a, of PLLA, PLLA1, and PLLA2, which had the bunch shape [33] because hydrolysis can facilitate matrix crystallinity through the movement of molecular chains [34,35] or reorganization and recrystallization in amorphous areas of the bulk [28]. The roughness was not obvious and not even observed after 14 days and 28 days, and a few small holes were found to be dispersed on the surface over the course of the sequential degrading process. This observation may be ascribed to the degradation of crystalline areas mentioned above with the progress of time [9], during which the rough surfaces disappeared.

To evaluate the effect of degradation on the inner structure of the films, the fracture morphology of the specimens after being treated for various degrading times are shown in Figure 3b. It shows that the samples had no obvious differences and still kept the dense microstructure for up to 28 days, and the fracture morphology of the composites still exhibited better toughness than neat PLLA. However, some micropores can be seen on the composites and can probably be attributed to the loss of the non-crystalline region during the PLLA hydrolysis degrading procedure, although more residual parts of the crystalline component were retained on them, which is consistent with Luo's results [9].

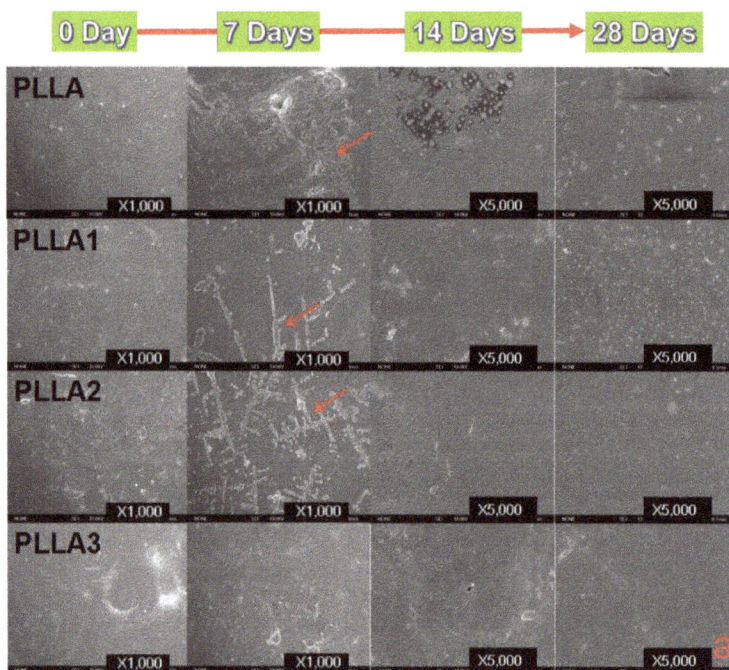

Figure 3. *Cont.*

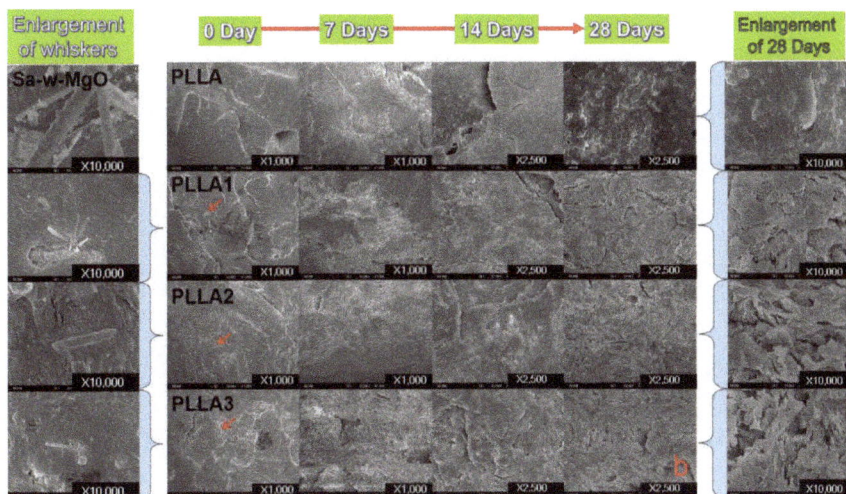

Figure 3. SEM micrographs of PLLA and composite films at different degradation times: (**a**) surface morphology (red arrows: crystalline region from the degradation); (**b**) fracture morphology (red arrows: MgO whiskers).

2.4. The pH Values of the Immersing Solution, the Weight Loss and Mechanical Properties of Sample Films after Degradation

The pH values of the PBS after the degradation of neat PLLA and composite films are shown in Figure 4. It shows that the pH changes of the film samples in the testing period were not obvious, as pH varied in the range of 7.0 to 7.5. With an increase in the degradation time, the pH value of the soaking solution in the presence of neat PLLA film decreased from 7.4 (Day 0) to 7.25 (Day 28), and this was ascribed to the acidic product after the degradation of neat PLLA. Meanwhile, a slight buffering effect can be seen, attributed to the OH- released after dissolution of MgO in the PBS, owing to the added MgO whiskers in PLLA1. Obviously, increasing the incorporated content of whiskers in PLLA2 and PLLA3 resulted in an effective buffering on the pH value of the PBS, as the pH values observed were higher than those of the former. This suggested that the higher the amount of MgO whiskers added, the higher the pH value obtained would be and the more notable the buffering effect on the degrading solution, caused by the release of OH- because of MgO dissolving. However, it should also be noted that the higher value of pH, like alkalinity condition, was also able to accelerate PLLA degradation, and the degradation was even faster than its acidic autocatalytic degradation [36,37]. Generally, the degradation rate of PLLA in ambient pH value follows the order alkaline > acid > neutral. Therefore, it is illustrated that the much higher content of whiskers was not favorable for further enhancing the stability of the pH value of the PBS during the initial degrading stage of PLLA-based composite films.

For further analysis of the whiskers affecting the degradation of the PLLA bulk, the decomposition rate of the samples over time was investigated through their weight losses, and the results are shown in Figure 5. The weight of the sample films was reduced gradually during the degrading procedure. The weight losses of the composites PLLA2 and PLLA3 were comparatively more obvious, suggesting that a composite with a higher content of MgO whiskers exhibited a faster degradation rate, due to the increase in the pH value of the PBS, owing to MgO dissolution. Then, the weights of the films changed slightly after 28 days, because the non-crystalline region was degraded by the hydrolysis process and the residual part that was highly crystalline could not be hydrolyzed in the PBS, owing to the tight and regular arrangement of its molecular chains. Furthermore, PLLA1 degraded more slowly compared with PLLA in the later stage, which was consistent with the pH results and could be accounted for by the buffering effect of MgO neutralizing the acid from PLLA hydrolysis.

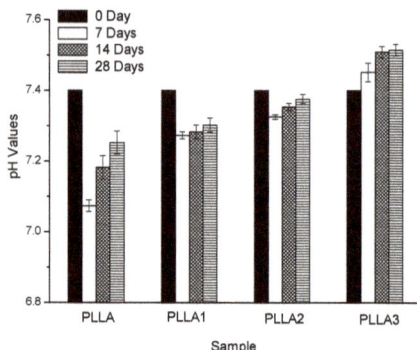

Figure 4. The pH value as a function of degradation time of the neat PLLA and composite films immersed in the phosphate buffer solution (PBS).

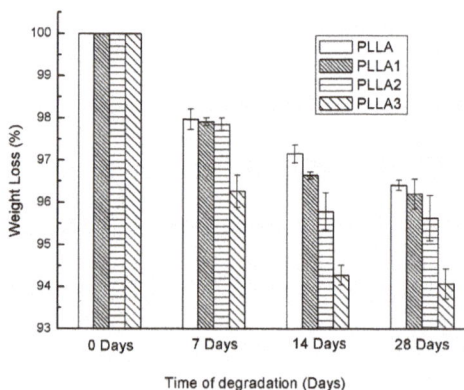

Figure 5. Weight losses as a function of degradation time of neat PLLA and composite films immersed in PBS.

The variation of mechanical properties of the samples during the degrading process is shown in Figure 6. Basically, the mechanical properties of PLLA and composite, including tensile strength, elastic modulus, and elongation at break, decreased as immersing time went. With increasing whisker content, the mechanical properties of the composite decreased notably; especially, the PLLA 3 with 3% whisker content exhibited a more obvious reduction in the tensile strength and Young modulus in comparison to pure PLLA. This deteriorated performance was probably due to the higher pH value of PLLA3 presented during the degradation. Meanwhile, it can be seen that the mechanical properties of composite were changed more greatly within the initially degrading stage, and this can probably be ascribed to the increasing hydrophilicity of composite induced by whiskers added [24].

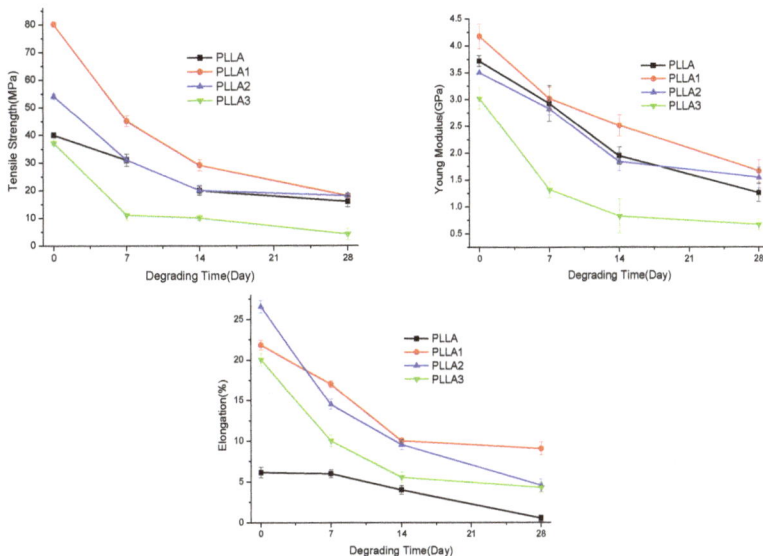

Figure 6. Mechanical properties of PLLA and composites at different degrading time.

2.5. Cell Experiments Results of Sample Films with Different Whisker Contents

The results of cell morphology observation and viability evaluations of the sample films are displayed in Figures 7 and 8. It can be seen that small cells exhibited a good shuttle shape with an adherent and spreading morphology, whereas floating cells with spherical or round shapes observed in Figure 7 indicate a shape change and activity loss of the cells. Compared to that of the negative control group, the amount of cells showing loss in activity in PLLA was comparatively higher, but contrarily a decrease in the comparative number of floating cells was also observed in the composite samples in Figure 7c–e, with the cell shape being maintained similar to that of the negative control group. The variations of the cell states suggested that cells can show good growth and proliferation in the degrading solution of the composite films, because of the beneficial bioactivity and biocompatibility of MgO whiskers.

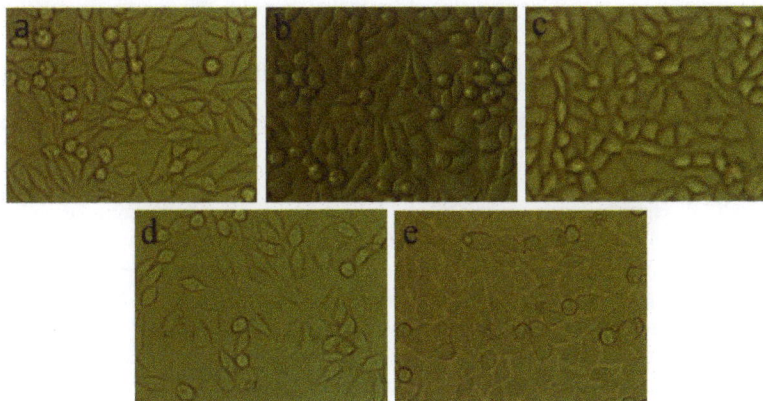

Figure 7. Morphology of the cell cultivated in leach liquor for 2 days (×100): (**a**) Negative control, (**b**) PLLA, (**c**) PLLA1, (**d**) PLLA2, (**e**) PLLA3.

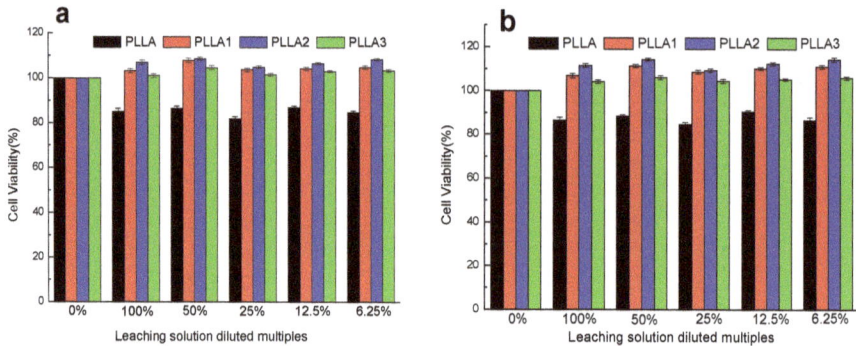

Figure 8. Cell viability of on PLLA, PLLA1, PLLA2, PLLA3 diluted 0, 2, 4, 8, and 16 times in leach liquor and culture for (**a**) 4 days and (**b**) 7 days, respectively.

With the addition of MgO whiskers, cell viability in the composite films increased significantly compared to that of neat PLLA, as shown in Figure 8. Consistent with the cell morphology results shown in Figure 7, it exhibited a relatively higher cell viability on the composite than on neat PLLA, especially in the presence of PLLA2, with a 2 wt% content of whiskers. On extending the culture time, the composite had a higher cell number, indicating prominent cell proliferation. The improved cell viability in the presence of MgO whiskers was probably due to the buffering effect of MgO whiskers on the PBS and significant Mg^{2+} bioactivity as a cofactor in the cell metabolism acting on the cells. The former is helpful in providing a relatively stable pH state, and the latter is beneficial for enhancing cell growth rate.

3. Discussion

PLLA, as a widely used biodegradable polymer for biomedical applications, usually shows different degradation properties, depending on the degrading environmental conditions. In addition to the impact of enzymes in the body on its decomposition, the behavior of bulk hydrolysis mainly accounts for its degradation in vivo, and the performance of the implant in the early degrading process is extremely important, in particular, because new bone is usually intensively active during the early recovery stage [29–32].

Generally, the features of this hydrolysis include two sections with the decomposition of the non-crystalline region and that of the crystalline part, in that order [9]. The results of the XRD patterns and DSC curves for the pristine PLLA in Figures 1 and 2 showed that the relative crystallinity of PLLA increased after degradation in PBS, as observed in Figure 3, demonstrating that the non-crystalline region in matrix was hydrolyzed first and, simultaneously, the crystalline structures of PLLA as α'-form were fabricated on inducing of hydrolysis through a chain or fragment movement in this process. On adding MgO whiskers, the crystallization degree, mechanical properties, and hydrophilicity of PLLA improved significantly, and the content of the whisker was undoubtedly also a key factor that affected the PLLA degrading properties [28,35]. Regarding the composite degradation behaviors, the initial change was similar to the degrading behaviors of the control PLLA, and the amorphous regions in the PLLA matrix were gradually destroyed, as shown by the corresponding changes in the patterns and curves of XRD and DSC. However, the hydrolysis rate of the PLLA bulk increased remarkably in the presence of whiskers, and the decomposing rate of the composites was prominently faster than that of the crystalline structure forming, which could be attributed to the hydrophilicity of the PLLA enhanced by the MgO whiskers. This hydrophilicity resulted in the notable acceleration and increment of water molecular seeping into the bulk (as shown in Figure 9), as reported previously [28]. As noted in Figures 1 and 2, this increase in the degradation was limited in the specimen of PLLA3, which showed a lower rate of hydrolysis in comparison to PLLA2 in 28 days, and PLLA3 degradation showed a tendency of accelerating and then retarding, depending on the variation of T_m, as shown in Table 2.

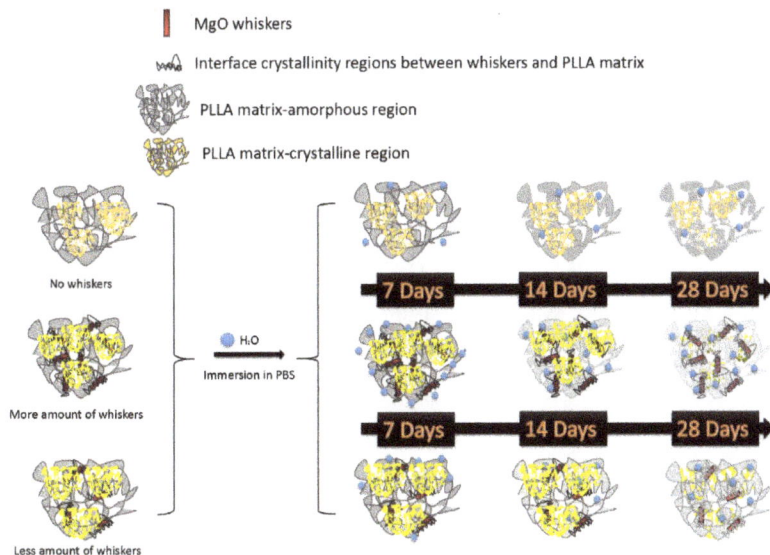

Figure 9. Diagram of in vitro degradation process for PLLA/magnesium oxide (MgO) whiskers composite films with different whisker contents.

This was different from previously reported studies [25,28] mentioning an increase in the degradation rate with an increase in the amount of inorganic fillers, such as carbon nanotube–organic montmorillonite. This was probably due to two aspects: one was that the effect of the whiskers' heterogeneous nucleation promoted the crystallization of PLLA, and an interface crystallinity [38] was also inclined to be introduced between the matrix and the whiskers, especially around the whisker regions, owing to the tight and regular packing of molecular chains. The numerous crystalline structures favorably prevented the water molecules permeating, in spite of the MgO whiskers increasing the hydrophilicity of the matrix, thereby inhibiting the water uptake of the materials. The other factor was the neutralizing effect of MgO whiskers that can effectively control the pH value of the PBS through their alkaline characteristics in solution. The MgO whiskers can impede the effect of acidic conditions and the autocatalysis caused by the PLLA hydrolysis, as a result of slowing down the degrading procedure of the PLLA. This hypothesis agrees with the SEM results of the surface and fracture morphologies of the samples, as shown in Figure 3. It should be noted that the influence of the dissolution of MgO on the pH of the PBS was more prominent for higher whisker contents. PLLA3 exhibited a higher degradation at the beginning of the immersion over 7 days, with a relatively higher weight loss, whereas the weight change was not obvious from 14 days to 28 days, which was similar to the mechanical properties in Figure 6. This was ascribed to the combined effects of the hydrophilicity and pH increase of the composites, which resulted in the fast loss of the amorphous region of the composites occurring initially. Then, the higher amount of OH- could control the pH value favorably, and this was beneficial in retarding the degrading course of the composites. Meanwhile, a small negative effect on cell viability of the PLLA3, as seen in Figures 7 and 8, was also probably due to the pH increase of the sample at the earlier stage, and this can also be detrimental to the films' biocompatibility and biological activity in terms of cell proliferation and viability. Simultaneously, the samples with 1 wt% and 2 wt% whisker contents were relatively more conductive for cell performance, and this was different from the results in previously reported studies, which selected a weight content of 5 wt% for the whiskers for investigating the effect of MgO whiskers on cell behaviors. Compared to the results above, the composites of our study showed faster degradation and beneficial pH regulation during the PLLA hydrolysis. The difference in the optimal whisker content in the studies performed was probably due to the type of surface modifiers of MgO whiskers. Therefore, the surface modifiers of whiskers can be an effective approach to controlling the degradation behaviors of the PLLA, and specific studies will be conducted on the modifier as a factor on composite material performance in following studies.

4. Materials and Methods

4.1. Materials

Powdered PLLA (M_w = 417,000) was purchased from Daigang Biological Engineering Co., Ltd. (Jinan, China), and MgO whiskers were prepared in the laboratory according to the procedure described in the literature [24]. All other agents used were of analytical grade and did not require further treatment.

4.2. Preparation of MgO Whiskers and Their Modification

Using the procedure described in Reference [24], MgO whiskers were prepared according to a sequence of steps based on the chemical synthesis of a whisker precursor (magnesium carbonate hydrate) and its calcination. Na_2CO_3 (0.6 M) was added dropwise into an equal volume of $MgCl_2$ (0.6 M) and stirred for 20 min. The mixture was aged at room temperature for 10 h and then filtered, washed, and dried at 80 °C for 3–4 h. The precursor was calcined at 750 °C for 4 h (the heating rate was 5 °C/min). The final product of this process was MgO whiskers (w-MgO).

Mixtures of 0.5 g of whiskers and 50 mL of ethanol were placed in three-necked flasks, which were subjected to ultrasonication in a bath to achieve full dispersion of the mixtures. The mixtures were then heated to 45 °C under reflux condensation. Stearic acid (Sa, 0.005 g) was dissolved into 20 mL of ethanol, added dropwise into each MgO suspension, and allowed to react for 1 h. The mixtures were then centrifuged, washed, and dried. The resultant product was denoted as Sa–w-MgO.

4.3. Preparation of Sa–w-MgO/PLLA Composite Films

PLLA/Sa–w-MgO composites with three different Sa–w-MgO contents (1, 2, and 3 wt%) were prepared through the solution casting method. The detailed procedure has been described in our previous reports [24]. Detailed information on the samples prepared is given in Table 3.

Table 3. Detailed information of the samples.

Abbreviation	Sample	Whisker Content (*w/w*)
PLLA	PLLA	0
PLLA1	1 wt% Sa–w-MgO/PLLA	1/100
PLLA2	2 wt% Sa–w-MgO/PLLA	2/100
PLLA3	3 wt% Sa–w-MgO/PLLA	3/100

4.4. Degradation of the Composite Material in PBS

We carried out in vitro degradation of pure PLLA and composite samples of the dimensions 1 cm × 1 cm with a thickness of 1 mm by immersing them in 10 mL of a phosphate buffer solution in a rocking water bath at 37 °C for 7, 14, and 28 days. The specimens of each material were removed from PBS at the end of these periods. After rinsing with deionized water and removing the surface water using filter paper, the samples were weighed and then dried in vacuum at 40 °C to a constant weight. The weight losses were calculated using the following equation:

$$W_{Loss}(\%) = \frac{m_0 - m_d}{m_0} \times 100\% \tag{1}$$

where W_{Loss} is the average degrading rate, and m_0 and m_d are the initial and final weights (after drying to constant weight), respectively. When the degradation was complete, the pH value of the PBS was determined by averaging the results of 3 independent measurements, obtained from three identical samples for each content. The process of film preparation and degradation is displayed in Figure 10.

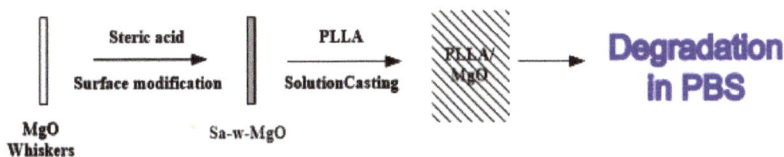

Figure 10. Schematic illustration of the preparation and degradation of the composite.

4.5. Characterization

X-ray powder diffraction (XRD, Rigaku D/max/2500PC, Tokyo, Japan) was performed to obtain XRD patterns of the samples, using Cu Kα radiation with λ = 1.5418 Å, operating at 40 kV/100 mA, with a scanning speed of 8°/min.

The morphology of the samples was characterized by field-emission scanning electron microscopy (FESEM, JOEL 6700F, Osaka, Japan, operating at 10 kV).

Differential scanning calorimetry (DSC) analyses were performed with a SETRAM SETSYS EVOLUTION16/18 (DSC 200F3, Netzsch Co., Selb, Germany). The samples, weighing approximately 5–8 mg each, were sealed in an aluminum pan, heated under nitrogen flow from room temperature to 220 °C at a heating rate of 20 °C/min, isothermally conditioned at 220 °C for 2 min, cooled to 0 °C, and then reheated to 220 °C at a heating rate of 10 °C/min.

The Figure 11 shows the detailed information of the samples' shape used in mechanical testing. The measurements were taken in tension using a universal testing machine (Instron 5845, Boston, MA, USA) at an extension rate of 0.5 mm/min. There were three replicates of each sample tested, and the average and standard deviation of each set of replicates were calculated.

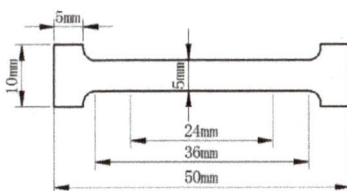

Figure 11. Detailed information of the samples used in mechanical testing.

4.6. Cytotoxicity Testing

4.6.1. Preparation of Material Extracts

With the ratio of film samples to culture medium set at 1 cm^2/mL, the disinfected pristine PLLA and composite samples (1 cm × 1 cm) were immersed in an RPMI-1640 solution (Gibco, Grand Island, NY, USA) under a humidified atmosphere with 5% CO$_2$ at 37 °C for 3 days to get the extraction medium. Prior to cell seeding, extracts were sterile-filtered through 0.2 mL syringe filters and then diluted by 2, 4, 8, and 16 times. The control groups involved the use of 10% fetal calf serum and 90% 1640 medium as negative controls.

4.6.2. Culture of Murine Fibroblast Cells (L929)

Murine fibroblast cells (L929) were used to evaluate the in vitro cytotoxicity of the sample films. L-929 cells (Dingguo, Tianjin, China) were cultured in RPMI-1640 medium, supplemented with 10% fetal bovine serum (FBS) (Hyclone, Logan, UT, USA), 100 U/mL penicillin, and 100 U/mL streptomycin in a humidified incubator at 95% relative humidity and 5% CO_2 at 37 °C. Then, the cells were cultured through the enzyme (Tryosin-EDTA, Sorabio, Beijing, China) digesting method.

4.6.3. Cell Behavior and Viability

(1) After well sterilized, the pure PLLA and composite films were fixed in the 24-well flat-bottomed cell culture plates. Then, 10^4 cells/500 µL medium were added in each well with the plates incubated for 2 days at 37 °C/5% CO_2, and cell morphology was consequently observed by optical microscopy (Nikon ECLIPSE Ti inverted microscope, Tokyo, Japan).

(2) We incubated 24-well flat-bottomed cell culture plates with 10^4 cells/500 µL medium in each well for 24 h to allow attachment. After the growth states were confirmed, the medium in the experimental group was removed and 500 µL extracts with various concentrations (dilutions of 0, 2, 4, 8, and 16 times) were placed in cell culture plates. The medium in the negative control group was replaced with 500 µL RPMI-1640 and 10% fetal calf serum, and at least 3 samples from each group were tested to confirm reproducibility. After incubating the cells in a humidified atmosphere for 4 and 7 days at 37 °C/5% CO_2, 500 µL of 3-(4,5-dimethyl-2-thiazolyl)-2,5-diphenyl-2-H-tetrazolium bromide (MTT) (sigma, Burlington, MA, USA) (5 mg/mL) was added to each well and the samples were incubated with MTT in the incubator at 37 °C for 3 h, and 350 µL DMSO (Sorabio, China) was added into each well. Finally, the absorbance was recorded by a multimode detector on a microplate reader (Epoch, BioTek, Winooski, VT, USA) at a wavelength of 570 nm, and the OD value of each group was statistically analyzed to calculate cell viability. The cell viability were calculated using the following equation:

$$\text{Cell viability}(\%) = \frac{OD_{\text{Experiment}}}{OD_{\text{Negative control}}} \times 100\% \tag{2}$$

where $OD_{\text{Experiment}}$ is the OD value from the sample films and $OD_{\text{Negative control}}$ is the OD value from the negative control group, respectively.

5. Conclusions

In this study, the composites of PLLA/MgO whiskers with different whisker contents were soaked in PBS for relatively short treating periods to investigate their initial in vitro degrading properties while used as implants in the context of bone repair. It was demonstrated that an increase in the whisker content can accelerate the degradation rate of the PLLA matrix. The degrading procedure of composites with a higher content of whiskers was influenced more significantly by the hydrophilicity and pH value. As the results showed, the degradation of the composite began with loss of the non-crystalline region prior to destruction of the crystalline part, and the content of MgO whiskers present was a significant factor that influenced the degrading procedure. The cell morphology observation and cell viability measurements suggested that, in the presence of MgO whiskers, the composite showed higher cell viability and led to cells with flattened and narrow rhombus shapes, which enhanced the bioactivity of the PLLA matrix significantly. This PLLA/MgO biocomposite can be potentially utilized for bone repair with better performance.

Author Contributions: Conceptualization: M.C., J.Y. and H.B.; Data curation: B.L., Y.Z. and H.L.; Investigation: Y.L.; Resources: B.L. and S.S.; Validation: Z.J.; Writing: Y.Z., W.L., M.C.

Acknowledgments: The authors acknowledge the financial support for this work from Key projects of the Joint Foundation of the National Natural Science Foundation of China (U1764254), Tianjin Natural Science Foundation (No. 17JCQNJC03100), and the National Natural Science Foundation of China (No. 51501129).

Conflicts of Interest: The authors declare no conflicts of interest.

References

1. Alsberg, E.; Kong, H.J.; Hirano, Y.; Smith, M.K.; Albeiruti, A.; Mooney, D.J. Regulating bone formation via controlled scaffold degradation. *J. Dent. Res.* **2003**, *82*, 903–908. [CrossRef] [PubMed]

2. Kakinoki, S.; Uchida, S.; Ehashi, T.; Murakami, A.; Yamaoka, T. Surface Modification of Poly(L-lactic acid) Nanofiber with Oligo(D-lactic acid) Bioactive-Peptide Conjugates for Peripheral Nerve Regeneration. *Polymers* **2011**, *3*, 820–832. [CrossRef]

3. Rogina, A.; Pribolšan, L.; Hanžek, A.; Gómez-Estrada, L.; Ferrer, G.G.; Marijanović, I.; Ivanković, M.; Ivanković, H. Macroporous poly(lactic acid) construct supporting the osteoinductive porous chitosan-based hydrogel for bone tissue engineering. *Polymer* **2016**, *98*, 172–181. [CrossRef]

4. Wang, X.J.; Song, G.J.; Lou, T. Fabrication and characterization of nano composite scaffold of poly(L-lactic acid)/hydroxyapatite. *J. Mater. Sci.* **2010**, *21*, 183–188. [CrossRef] [PubMed]

5. Yang, Y.F.; Zhao, Y.H.; Tang, G.W.; Li, H.; Yuan, X.Y.; Fan, Y.B. In vitro degradation of porous poly(L-lactide-co-glycolide)/β-tricalcium phosphate (PLGA/β-TCP) scaffolds under dynamic and static conditions. *Polym. Degrad. Stab.* **2008**, *93*, 1838–1845. [CrossRef]

6. Liao, L.; Chen, L.; Chen, A.Z.; Pu, X.M.; Kang, Y.Q.; Yao, Y.D.; Liao, X.M.; Huang, Z.B.; Yin, G.F. Preparation and characteristics of novel poly-L-lactide/β-calcium metaphosphate fracture fixation composite rods. *J. Mater. Res.* **2007**, *22*, 3324–3329. [CrossRef]

7. Shikinami, Y.; Matsusue, Y.; Nakamura, T. The complete process of bioresorption and bone replacement using devices made of forged composites of raw hydroxyapatite particles/poly L-lactide (F-u-HA/PLLA). *Biomaterials* **2005**, *26*, 5542–5551. [CrossRef] [PubMed]

8. Li, X.; Chu, C.L.; Liu, L.; Liu, X.K.; Bai, J.; Guo, C.; Xue, F.; Lin, P.H.; Chu, P.K. Biodegradable poly-lactic acid based-composite reinforced unidirectionally with high-strength magnesium alloy wires. *Biomaterials* **2015**, *49*, 135–144. [CrossRef] [PubMed]

9. Wen, W.; Zou, Z.P.; Luo, B.H.; Zhou, C.R. In vitro degradation and cytocompatibility of g-MgO whiskers/PLLA composites. *J. Mater. Sci.* **2017**, *52*, 2329–2344. [CrossRef]

10. Nampoothiri, K.M.; Nair, N.R.; John, R.P. An overview of the recent developments in polylactide (PLA) research. *Bioresour. Technol.* **2010**, *101*, 8493–8501. [CrossRef] [PubMed]

11. Jiang, L.Y.; Xiong, C.D.; Jiang, L.X.; Xu, L.J. Effect of hydroxyapatite with different morphology on the crystallization behavior, mechanical property and in vitro degradation of hydroxyapatite/poly(lactic-co-glycolic) composite. *Compos. Sci. Technol.* **2014**, *93*, 61–67.

12. Cifuentes, S.C.; Gavilán, R.; Lieblich, M.; Benavente, R.; González-Carrasco, J.L. In vitro degradation of biodegradable polylactic acid/magnesium composites: Relevance of Mg particle shape. *Acta Biomater.* **2016**, *32*, 348–357. [CrossRef] [PubMed]

13. Shalumon, K.T.; Sheu, C.; Fong, Y.T.; Liao, H.T.; Chen, J.P. Microsphere-Based Hierarchically Juxtapositioned Biphasic Scaffolds Prepared from Poly(Lactic-co-Glycolic Acid) and Nanohydroxyapatite for Osteochondral Tissue Engineering. *Polymers* **2016**, *8*, 429. [CrossRef]

14. Kothapalli, C.R.; Shaw, M.T.; Wei, M. Biodegradable HA-PLA 3-D porous scaffolds: Effect of nano-sized filler content on scaffold properties. *Acta Biomater.* **2005**, *1*, 653–662. [CrossRef] [PubMed]

15. Loher, S.; Reboul, V.; Brunner, T.J.; Simonet, M.; Dora, C.; Neuenschwander, P.; Stark, W.J. Improved degradation and bioactivity of amorphous aerosol derived tricalcium phosphate nanoparticles in poly(lactide-co-glycolide). *Nanotechnology* **2006**, *17*, 2054–2061. [CrossRef]

16. Yun, H.; Kim, S.; Park, E.K. Bioactive glass-poly(ε-caprolactone) composite scaffolds with 3 dimensionally hierarchical pore networks. *Mater. Sci. Eng. C* **2016**, *31*, 198–205. [CrossRef]

17. Fernandes, J.S.; Gentile, P.; Martins, M.; Neves, N.M.; Miller, C.; Crawford, A.; Pires, R.A.; Hatton, P.; Reis, R.L. Reinforcement of poly-L-lactic acid electrospun membranes with strontium borosilicate bioactive glasses for bone tissue engineering. *Acta Biomater.* **2016**, *44*, 168–177. [CrossRef] [PubMed]

18. Johnson, A.J.W.; Herschler, B.A. A review of the mechanical behavior of CaP and CaP/polymer composites for applications in bone replacement and repair. *Acta Biomater.* **2011**, *7*, 16–30. [CrossRef] [PubMed]

19. Ma, F.Q.; Lu, X.L.; Wang, Z.M.; Sun, Z.J.; Zhang, F.F.; Zheng, Y.F. Nanocomposites of poly(L-lactide) and surface modified magnesia nanoparticles: Fabrication, mechanical property and biodegradability. *J. Phys. Chem. Solids* **2011**, *72*, 111–116. [CrossRef]

20. Kum, C.H.; Cho, Y.; Seo, S.H.; Joung, Y.K.; Ahn, D.J.; Han, D.K. A poly(lactide) stereocomplex structure with modified magnesium oxide and its effects in enhancing the mechanical properties and suppressing inflammation. *Small* **2014**, *10*, 3783–3794. [CrossRef] [PubMed]

21. Kum, C.H.; Cho, Y.; Joung, Y.K.; Choi, J.; Park, K.; Seo, S.H.; Park, Y.S.; Ahn, D.J.; Han, D.K. Biodegradable poly(L-lactide) composites by oligolactide-grafted magnesium hydroxide for mechanical reinforcement and reduced inflammation. *J. Mater. Chem. B* **2013**, *1*, 2764–2772. [CrossRef]

22. Yang, J.J.; Cao, X.X.; Zhao, Y.; Wang, L.; Liu, B.; Jia, J.P.; Liang, H.; Chen, M.F. Enhanced pH stability, cell viability and reduced degradation rate of poly(L-lactide)-based composite in vitro: Effect of modified magnesium oxide nanoparticles. *J. Biomater. Sci. Polym. Ed.* **2017**, *28*, 486–503. [CrossRef] [PubMed]

23. Wen, W.; Luo, B.; Qin, X.; Li, C.; Liu, M.; Ding, S.; Zhou, C. Strengthening and toughening of poly(L-lactide) composites by surface modified MgO whiskers. *Appl. Surf. Sci.* **2015**, *332*, 215–223. [CrossRef]

24. Zhao, Y.; Liu, B.; You, C.; Chen, M.F. Effects of MgO whiskers on mechanical properties and crystallization behavior of PLLA/MgO composites. *Mater. Des.* **2016**, *89*, 573–581. [CrossRef]

25. Chen, H.M.; Chen, J.W.; Chen, J.; Yang, J.H.; Huang, T.; Zhang, N.; Wan, Y. Effect of organic montmorillonite on cold crystallization and hydrolytic degradation of poly(L-lactide). *Polym. Degrad. Stab.* **2012**, *97*, 2273–2283. [CrossRef]

26. Paul, M.A.; Delcourt, C.; Alexandre, M.; Degée, P.; Monteverde, F.; Dubois, P. Polylactide/montmorillonite nanocomposites: Study of the hydrolytic degradation. *Polym.Degrad. Stab.* **2005**, *87*, 535–542. [CrossRef]

27. Fukushima, K.; Tabuani, D.; Dottori, M.; Armentano, I.; Kenny, J.M.; Gamino, G. Effect of temperature and nanoparticle type on hydrolytic degradation of poly(lactic acid) nanocomposites. *Polym. Degrad. Stab.* **2011**, *96*, 2120–2129. [CrossRef]

28. Chen, H.M.; Feng, C.X.; Zhang, W.B.; Yang, J.H.; Huang, T.; Zhang, N.; Wang, Y. Hydrolytic degradation behavior of poly(L-lactide)/carbon nanotubes nanocomposites. *Polym. Degrad. Stab.* **2013**, *98*, 198–208. [CrossRef]

29. Bose, S.; Tarafder, S. Calcium phosphate ceramic systems in growth factor and drug delivery for bone tissue engineering: A review. *Acta Biomater.* **2012**, *8*, 1401–1421. [CrossRef] [PubMed]

30. Li, Y.B.; Weng, W.J. Surface modification of hydroxyapatite by stearic acid: Characterization and in vitro behaviors. *J. Mater. Sci. Mater. Med.* **2008**, *19*, 19–25. [CrossRef] [PubMed]

31. Zhang, L.; Chen, L.; Wan, H.Q.; Chen, J.M.; Zhou, H.D. Synthesis and Tribological Properties of Stearic Acid-Modified Anatase (TiO$_2$) Nanoparticles. *Tribol. Lett.* **2011**, *41*, 409–416. [CrossRef]

32. Barry, M.; Pearce, H.; Cross, L.; Tatullo, M.; Gaharwar, A.K. Advances in Nanotechnology for the Treatment of Osteoporosis. *Curr. Osteoporos. Rep.* **2016**, *14*, 87–94. [CrossRef] [PubMed]

33. Hutmacher, D.W. Scaffolds in tissue engineering bone and cartilage. *Biomaterials* **2000**, *21*, 2529–2543. [CrossRef]

34. Rezwan, K.; Chen, Q.Z.; Blaker, J.J.; Boccaccini, A.R. Biodegradable and bioactive porous polymer/inorganic composite scaffolds for bone tissue engineering. *Biomaterials* **2006**, *27*, 3413–3431. [CrossRef] [PubMed]

35. Chen, H.M.; Shen, Y.; Yang, J.H.; Huang, T.; Zhang, N.; Wang, Y.; Zhou, Z.W. Molecular ordering and α′-form formation of poly(L-lactide) during the hydrolytic degradation. *Polymer* **2013**, *54*, 6644–6653. [CrossRef]

36. Tsuji, H.; Nakahara, K. Poly(L-lactide). IX. Hydrolysis in Acid Media. *J. Appl. Polym. Sci.* **2002**, *86*, 186–194. [CrossRef]

37. Li, S.M. Hydrolytic degradation characteristics of aliphatic polyesters derived from lactic and glycolic acids. *J. Biomed. Mater. Res.* **1999**, *48*, 342–353. [CrossRef]

38. Ning, N.Y.; Fu, S.R.; Zhang, W.; Chen, F.; Wang, K.; Deng, H.; Zhang, Q.; Fu, Q. Realizing the enhancement of interfacial interaction in semicrystalline polymer/filler composites via interfacial crystallization. *Prog. Polym. Sci.* **2012**, *37*, 1425–1455. [CrossRef]

© 2018 by the authors. Licensee MDPI, Basel, Switzerland. This article is an open access article distributed under the terms and conditions of the Creative Commons Attribution (CC BY) license (http://creativecommons.org/licenses/by/4.0/).

International Journal of
Molecular Sciences

MDPI

Article

Processability and Degradability of PHA-Based Composites in Terrestrial Environments

Patrizia Cinelli *, Maurizia Seggiani *, Norma Mallegni, Vito Gigante and Andrea Lazzeri

Department of Civil and Industrial Engineering, University of Pisa, Largo Lucio Lazzarino 2, 56122 Pisa, Italy;
norma.mallegni@gmail.com (N.M.); vito.gigante@dici.unipi.it (V.G.); andrea.lazzeri@unipi.it (A.L.)
* Correspondence: patrizia.cinelli@unipi.it (P.C.); maurizia.seggiani@unipi.it (M.S.);
  Tel.: +39-050-221-7869 (ext. 881) (P.C. & M.S.)

Received: 21 December 2018; Accepted: 9 January 2019; Published: 12 January 2019

Abstract: In this work, composites based on poly(3-hydroxybutyrate-3-hydroxyvalerate) (PHB-HV) and waste wood sawdust (SD) fibers, a byproduct of the wood industry, were produced by melt extrusion and characterized in terms of processability, thermal stability, morphology, and mechanical properties in order to discriminate the formulations suitable for injection molding. Given their application in agriculture and/or plant nursery, the biodegradability of the optimized composites was investigated under controlled composting conditions in accordance with standard methods (ASTM D5338-98 and ISO 20200-2004). The optimized PHB-HV/SD composites were used for the production of pots by injection molding and their performance was qualitatively monitored in a plant nursery and underground for 14 months. This study presents a sustainable option of valuation of wood factory residues and lowering the production cost of PHB-HV-based compounds without affecting their mechanical properties, improving their impact resistance and biodegradability rates in terrestrial environments.

Keywords: biocomposites; natural fibers; poly(3-hydroxybutyrate-3-hydroxyvalerate); biodegradation; impact properties

---

1. Introduction

Petroleum-based plastics are light, strong, durable, and demonstrate good resistance to degradation [1]. They offer a wide range of applications in the domestic, medical, and industrial fields in the form of single-use gears, packaging, furniture, machine chassis, and accessories to improve life quality [2]. For these reasons, approximately 150 million tons of plastic are consumed worldwide each year and this consumption is expected to continue growing until 2020 [3].

Large-scale dependence on petroleum-derived plastics leads to serious pollution problems; the methodologies used for plastic waste disposal are challenging. In landfills, the degradation rates are tremendously slow [4,5]. Incineration generates harmful by-products if advanced combustion designs, optimized operating practices, and effective emission-control technologies are not adopted. Chemical and mechanical recycling is not always possible, requires an advanced collection system, and causes a deterioration of the properties of plastic materials compromising their reuse [6,7].

Given this scenario, with the aim of decreasing plastic environmental impact, the use of bio-based and/or biodegradable polymers, such as polylactide acid (PLA), aliphatic polyesters, and polyhydroxyalkanoates (PHAs), having similar physicochemical properties as conventional plastics represent a valuable solution to the current plastic pollution problem [8–11]. However, the replacement of nondegradable with biodegradable plastics requires the complete knowledge of the biodegradability of these plastics and their blends in controlled and uncontrolled environments to have a real positive environmental impact. It is necessary to know their biodegradability and biodegradation rates both in managed (industrial and home composting, anaerobic digestion) and unmanaged (soil, oceans and rivers) environments for postconsumer management. Only a few biopolymers (PHAs and thermoplastic starch, TPS) are degradable in a wide range of managed and unmanaged environmental conditions; whereas most of them (e.g., PLA, polycaprolactone (PCL), poly(butylene succinate) (PBS), and polyhydroxyoctanoate (PHO)) degrade only in a narrower set of environmental conditions [12]. Narancic et al. [12] identified synergies but also antagonisms between polymers in blends that affect their biodegradability and biodegradation rates in different environments. Consequently, biodegradable plastics such as PHAs that degrade in a wide variety of controlled and uncontrolled environments are the most promising candidates for terrestrial and marine applications where their release into the environment does not cause plastic pollution. PHAs are a family of polyesters of several *R*-hydroxyalkanoic acids, synthetized by several microorganisms in the presence of excess carbon and when essential nutrients such as oxygen, nitrogen, or phosphorus are limiting or after pH shift [13–18]. PHAs have thermoplastic properties similar to those of polypropylene, good mechanical properties, and excellent biodegradability in various ecosystems [19–25]. The most common PHAs are the poly([R]-3-hydroxybutyrate) (PHB) and its copolyester with [R]-3-hydroxyvalerate (PHB-HV), which is well suited for food packaging [21]. Despite their good properties and excellent biodegradability, their relatively high cost (€7–12/kg) [14], compared to other biopolymers such as poly-lactic acid (PLA) (€2.5–3/kg) [26], has limited their use in commodities such as packaging and service items, restricting their use to high-value applications in the medical and pharmaceutical sectors. Efforts have been made to incorporate low-value materials, such as starch, into PHAs in order to reduce the cost of the final products [27–29]. Waste lignocellulose fibers, highly-available and at low-cost, sourced from agricultural and industrial crops, have been investigated as fillers in PHA-based composites [30–34].

In some previous cases, the developed composites displayed promising mechanical and physical properties. Natural fiber-reinforced composites show higher degradation when subjected to outdoor applications compared with composites with synthetic fibers [35]. Biodegradation of a composite occurs with the degradation of its individual components as well as with the loss of interfacial strength between them [36]. The weak interfacial bonding between highly polar natural fiber and non-polar matrix can lead to a reduction in final properties of the composite, ultimately hindering their industrial usage. Different methods have been applied to improve the compatibility and interfacial bond strength, including the use of various surface modification techniques [37,38]. It is harder to obtain high degrees of alignment with natural fibers than for synthetic fibers, since during the extrusion process, the long natural fibers tend to twist randomly [39]. This behavior compromises the mechanical properties of the composite as most of the natural fibers are not aligned parallel to the direction of the applied load.

Composites containing sawdust fibers and petroleum-based plastics have been well-studied. For example, they were processed by Sombatsompop et al. [40,41] with polypropylene (PP) and polyvinylchloride (PVC), revealing a reduction in the overall strength and toughness of the composites with increasing the wood fiber content. In a previous work [42], composites based on PHB-HV and fibers of *Posidonia oceanica*, a dominant Mediterranean seagrass, were successfully produced by melt extrusion and their degradability was investigated in sea water. The results showed an increase in the impact resistance of the composites with increasing fiber content. The presence of fibers favored the physical disintegration of the composite, increasing its biodegradation rate under simulated and real marine conditions.

In the present work, composites based on PHB-HV and waste sawdust fibers (SD), derived from the wood industry, were produced by extrusion and characterized in terms of processability, rheological, and mechanical properties. Given their use in terrestrial applications, biodegradation tests were completed on the developed composites under simulated composting conditions in accordance with standard methods and we preliminarily evaluated the degradability in soil of PHB-HV-SD-based molded specimens (pots with thickness of 1 mm).

2. Results and Discussion

2.1. Composite Processing

The torque measurements of polymeric melts characterize their flow behavior and reflect the trend of viscosity, as demonstrated by Melik et al. [43]. The objective of these measurements was to quantify the processing behavior of the investigated composites and, in particular, to evaluate the effect of the fibers on the melt fluidity. The average torque-time curves obtained at 170 °C and a rotor speed of 100 rpm with a HAAKE MiniLab are reported in Figure 1 for the PCA and PHB-HV/SD composites.

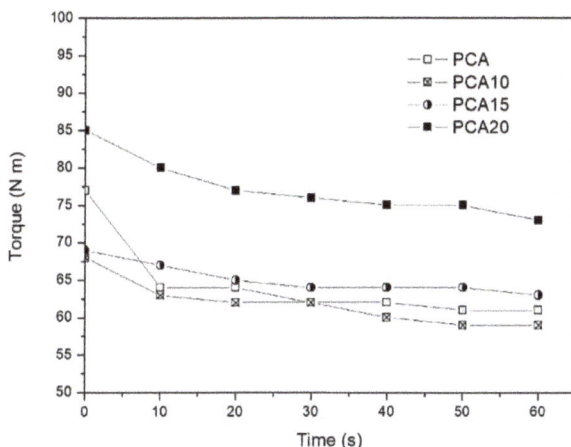

Figure 1. Torque-time curves obtained at 170 °C.

As shown, the incorporation of the SD fibers in the PHB-HV-based compound increased the torque and, consequently, the energy required for the melt mixing. With up to 15 wt % of fibers, there was a moderate increase in torque with respect to the pure matrix, whereas for PCA20, the torque values almost doubled, significantly affecting the processability of the biocomposite.

2.2. Composite Characterization

The thermogravimetric (TG) and their derivative (DTG) curves of SD fibers, PHB-HV, ATBC, and the developed composites are shown in Figure 2.

SD fibers showed an initial weight loss at around 100 °C, attributable to loss in the residual humidity. Then, the sharp decrease observed at 250–350 °C can be related to the degradation of hemicellulose, cellulose, and lignin. Hemicellulose decomposes easily with respect to the other components. Typically, the pyrolysis of hemicellulose occurs between 200 and 280 °C, resulting in formation of CO, CO_2, condensable vapors, and organic acids [44].

Therefore, the thermal degradation of the SD fibers occurs at temperatures higher than 200 °C, confirming their suitability of being processed with thermoplastic polymer matrices, such as PHB-HV, without thermal degradation occurring.

PHB-HV showed the onset of degradation at about 260 °C with maximum weight loss rate at 305 °C and no residue was recorded above 350 °C. For all of the produced composites, the thermal degradation started at temperatures above 200 °C, with the main degradation peaks occurring above 250 °C.

The composite mechanical properties are shown in Table 1.

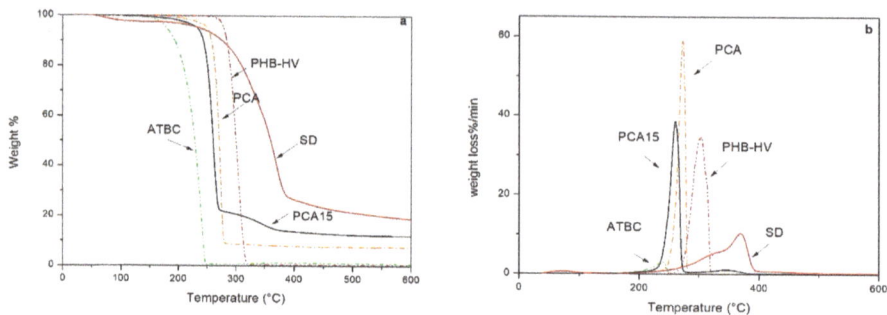

Figure 2. (**a**) TG and (**b**) DTG curves of the SD fibres, PHB-HV, ATBC, and developed composites.

Table 1. Mechanical properties of the composites with different SD fiber contents.

Sample	Young's Modulus (GPa)	Stress at Break (MPa)	Elongation (%)	Charpy Impact Energy (kJ/m^2)
PCA	2.64 ± 0.28	25.62 ± 2.11	2.14 ± 0.50	3.57 ± 0.36
PCA10	2.35 ± 0.24	21.02 ± 0.94	2.05 ± 0.28	6.17 ± 0.24
PCA15	2.52 ± 0.15	18.52 ± 0.84	1.35 ± 0.14	12.24 ± 0.50
PCA20	2.94 ± 0.35	20.93 ± 1.57	1.35 ± 0.13	5.91 ± 0.40

The values are the mean $\pm$ SD of at least five determinations.

The results of tensile tests showed that the elastic modulus was almost constant, except in the case of PCA20 where a higher fiber content led to a significant increase in stiffness, causing a decrease in break stress and elongation. This behavior is typical of composites with poor or no compatibility between the components, preventing the stress transfer phenomena from occurring and the filler particle becomes a stress concentrator, leading to brittle fracture [45].

To better understand these results, SEM analysis was performed on the cross-section of the specimens before the tensile test and on the fractured sections obtained after testing (Figure 3).

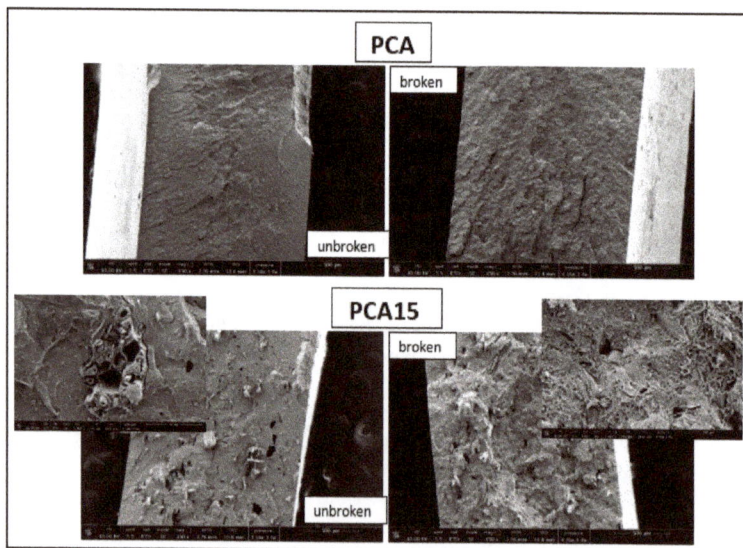

Figure 3. SEM images of the cross-sections of the PCA and PCA15 specimens before (unbroken samples) and after tensile tests (broken samples).

The unbroken PCA15 specimen showed a good dispersion of SD fibers that are homogeneously distributed within the thermoplastic matrix. In the broken specimen section, it was possible to observe a significant fiber pullout. This means that the interfacial interactions between the fibers and the matrix were not sufficiently strong to maintain their cohesion during the tensile test.

Natural fibers are rich in cellulose, hemicellulose, lignin, and pectin; consequently, they tend to be active polar hydrophilic materials, whereas polymeric materials are nonpolar and show considerable hydrophobicity. The hydrophilic nature of natural fibers reduces the adhesion to a hydrophobic matrix and, as a result, a loss in strength may be induced. To prevent this, the fiber surface can be modified by several methods proposed in the literature, such as graft copolymerization of monomers onto the fiber surface, the use of maleic anhydride copolymers, alkyl succinic anhydride, stearic acid, etc. [38,46,47]. Therefore, a compatibilization method may be necessary to produce composites with tailored tensile properties.

Even without compatibilizers, interesting results were obtained from the impact test, in which the PCA15 compound showed a higher impact resistance compared with the matrix without fibers. Factors such as fiber/matrix de-bonding, fiber and/or matrix fracture, and fiber pull out improve the impact performance. Fiber fracture dissipates less energy compared to fiber pull-out, which is common in composites with strong interfacial bonds. A high impact energy is a sign of weak fiber/matrix bond [48,49]. In this case, the absorbed energy value increased up to 15 wt. % SD fibers, and then decreased at higher loadings, which caused an increase in material brittleness [50,51].

2.3. Biodegradability in Lab-Scale Terrestrial Environments

2.3.1. Mineralization

Figure 4 shows the aerobic biodegradation curves obtained on the lab-scale. As shown, after six months, the PCA15 composite reached a mineralization percentage of about 78%, higher than those of the control sample (filter paper) and the composite without fibers (58%). This behavior can be explained by the presence of the fibers favoring the disintegration of the sample, increasing its susceptibility to microbial attack.

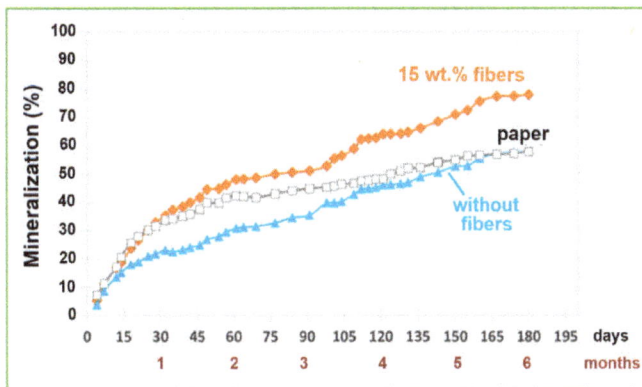

Figure 4. Mineralization curves under simulated terrestrial environmental conditions.

2.3.2. Disintegration

The average results of the disintegration test for each sample are reported in Table 2.

Table 2. Average percentage of disintegration after 90 days under simulated composting conditions.

Sample	Sample No.	Disintegration (%)	Average Disintegration (%)
PCA	1	87.2	
	2	94.3	92.6
	3	96.4	
PCA10	1	83.4	
	2	100.0	93.2
	3	96.2	
PCA15	1	96.3	
	2	88.6	94.2
	3	97.7	

After 90 days, all the samples examined showed greater than 90% disintegration. The composite containing the higher amounts of fibers presented the highest degrees of disintegration, showing that the presence of fibers inside the matrix facilitated the degradation of the material according to the results of the mineralization test.

Plasticizers, including ATBC, can accelerate the disintegration of polymeric matrices under composting conditions due to the greater mobility of the polymer chains, as observed by Arrieta et al. [52] for PHB-based blends. Narancic et al. [12] evidenced, in some cases, synergies between biodegradable polymers in blends that improve their biodegradation rates. A possible explanation of the synergy observed in some environments (e.g., home composting or anaerobic digestion) for PLA-PCL (80/20), PHB−PCL (60/40), and PCL−TPS (70/30) blends was that the addition of amorphous PCL, by lowering the crystallinity of the blends, improves their degradation rate.

Consequently, and also in this case, ATBC may have contributed, lowering the crystallinity of the PHB-HV together with the fibers to accelerate the disintegration of the specimens, thus increasing the biodegradation rate of the PHB-HV.

Degradation of Pots in Soil

The performance and degradation of 180 pots (produced by injection molding with thickness of 1 mm and weight of about 115 g) was evaluated in three different environments: 60 were placed in a greenhouse (15 pots for each composite: PCA, PCA10, and PCA15, and 15 pots of PP, as reference); 60 pots were located outside on a cloth and not in contact with the soil (15 pots for each material); and 60 pots were buried into the soil up to the upper profile, leaving only the plants outside (Figure 5).

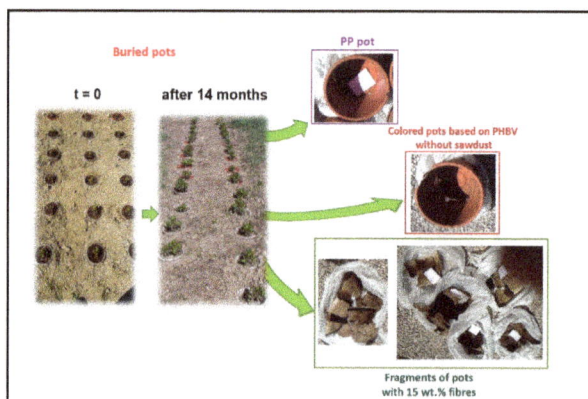

Figure 5. Buried PP and PHB-HV based pots without fibers and with 15 wt % sawdust fibers.

After 14 months of experimentation, the PP pots showed no evident signs of degradation, as expected, whereas the pots based on PHB-HV showed initial signs of degradation, in particular those containing 15 wt % sawdust fibers. The plant growth performance was the same in all pots (PHB-HV, PHB-HV/SD, PP).

Figure 5 shows the results obtained with the buried pots: the PP pots remained intact as expected. The pots based on PHB-HV without fibers were slightly damaged on the bottom, while the pots containing 15 wt % SD fibers were totally fragmented and markedly degraded, confirming the results of the lab-scale degradation tests.

3. Material and Methods

3.1. Materials

PHB-HV (PHI002) with 5 wt % valerate content was supplied in pellets by Naturplast® (Caen, France); it is characterized by a density of 1.25 g/cm^3 and melt flow index (190 °C, 2.16 kg) of 15–20 g/10 min. According to the supplier data sheet, PHB-HV is a semi crystalline polymer with a glass transition temperature of around 5 °C and a melting temperature of around 155 °C.

Sawdust fibers (SD) were obtained from a soft wood-processing company. SD was dried at 80° C in a vacuum oven for at least 24 h, then milled using a lab-scale grinder and screened with a 500-μm sieve.

To improve the flexibility and processability of the composites with high content of fiber, a bio-based plasticizer, acetylbutylcitrate (ATBC), was used. ATBC was purchased from Tecnosintesi® (Bergamo, Italy); it is a colorless liquid, soluble in organic solvents, produced by acetylation of tri-n-butylcitrate, having a density of 1.05 g/cm^3, and a molecular weight of 402.5 g/mol. ATBC is widely used as food packaging as it is non-toxic.

Micro-sized calcium carbonate particles, produced from a high purity white marble, with mean particle size (D_{50}) of 2.6 μm and top cut diameter (D_{98}) of 15 μm, cubic shape (aspect ratio = 1), and density of 2.7 g/cm^3, provided by OMYA® (Avenza Carrara, Italy) with trade name OMYACARB 2-AZ, were used as rigid inorganic filler. $CaCO_3$ is one of the most commonly used fillers in thermoplastics to reduce both their production cost and to improve their properties, such as stiffness, impact strength, processability, and thermal and dimensional stability [53].

3.2. Composite Preparation

The composites (whose compositions by weight are reported in Table 3) were prepared by mixing the various components at 170 °C for one minute in a Thermo Scientific Haake microcompounder (HAAKE™, Karlsruhe, Germany) with a twin-conic screw system at 100 rpm speed. The labels PCA, PCA10, PCA15 and PCA20 indicate the composite without SD fibers (PHB-HV/CaCO$_3$/ATBC 80/10/10 $w/w/w$%) and with 10, 15, and 20 wt % SD fibers, respectively.

Table 3. Composite formulations.

Composite	Weight Percentage			
	PHB-HV	CaCO$_3$	ATBC	Sawdust Fibers
PCA	80	10.0	10.0	0
PCA10	72	9.0	9.0	10
PCA15	68	8.5	8.5	15
PCA20	64	8.0	8.0	20

The miniLab compounder was equipped with a backflow channel (Figure 6) in which the recirculation of the melt occurs. During the melt mixing, the torque was recorded to evaluate the fiber effect on the melt fluidity. The extruded filaments were cut to obtain pellets used for the successive thermal characterizations, lab-scale biodegradation, and disintegration tests.

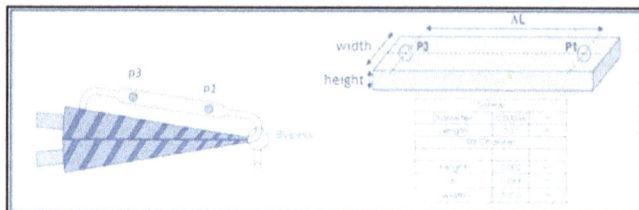

Figure 6. MiniLab Backflow channel and twin-screw conic system.

For each formulation, tensile test specimens (Haake type 3 dog-bone bar) and Charpy Impact test samples (80 × 10 × 4 mm parallelepiped) were produced by feeding the molten material from the minilab into a mini-injection press (Thermo Scientific Haake MiniJet II, HAAKE™, Karlsruhe, Germany) at 210 bar with a mold heated at 60 °C.

Industrial compounding of the selected formulations was carried out using a COMAC EBC 50HT extruder (COMAC, Milan, Italy) with a temperature profile from 135 °C in the feeding zone to 170 °C at the screw terminal zone. The resultant pellets were then used to produce items such as pots using an injection molding machine (Negri Bossi VS250, Negri Bossi, Milan, Italy) at Femto Engineering (Florence, Italy).

3.3. Composite Characterization

The thermal properties of the raw materials and composites in pellet form were evaluated by thermogravimetric analysis (TGA). These measurements were carried out, in duplicate, on 20 mg of sample using a Q500 TGA (TA Instruments; New Castle, DE, USA), under nitrogen flow (100 mL/min), at a heating speed of 10 °C/min from room temperature to 600 °C.

Tensile tests were performed on the injection molded Haake Type 3 (557-2290) dog-bone tensile bars in accordance with ASTM D 638. Stress-strain tests were carried out at room temperature, at a crosshead speed of 10 mm/min, by an Instron 5500R universal testing machine (Canton, MA, USA) equipped with a 10-kN load cell, and interfaced with a computer running MERLIN Instron software, version 4.42 S/N–014733H (Instron®, Canton, MA, USA).

The Charpy impact tests were completed at room temperature on 2-mm V-notched specimens using a 15 J Charpy pendulum (CEAST 9050, Instron®, Canton, MA, USA) following the standard method ISO 179:2000. For each mechanical test, at least five replicates were carried out.

The morphology of the composite tensile specimens, before and after the tensile test, was investigated by scanning electron microscopy (SEM) using a FEI Quanta 450 FEG SEM (Thermo Fisher Scientific, Waltham, MA, USA). The SEM analysis was performed under high-vacuum conditions because air can inhibit the electron beam. The undamaged samples were frozen under liquid nitrogen and then fractured. The surfaces were metalized with gold using a Sputter Coater Edward S150B to avoid charge build up.

3.4. Biodegradation in Terrestrial Environment

3.4.1. Mineralization Test in Compost

Given the application of composites in terrestrial environments, a mineralization test in compost was carried out following a laboratory method [10] based on ASTM D5338 [54]. The setup of the system is shown in Figure 7. The test was carried out in glass vessels (1 L capacity) containing a layer of a mixture of 3 g compost, 4 g perlite, and 0.5 g of sample in form of pellets, wetted with 15 mL distilled water, in order to operate with a weight ratio of sample/compost of 1:6 in accordance with ASTM D5338. Finally, the mixtures were sandwiched between two layers consisting of 5 g perlite wetted with 20 mL of water. Perlite (tradename Agrilit 2), a hygroscopic aluminum silicate, was supplied in granules of 1–2 mm by Perlite Italiana Srl (Milan, Italy). Perlite is able to hold 3 to 4 times its weight in water, largely used in horticultural applications where it provides aeration and optimal moisture conditions for plant growth. In the mineralization test, the perlite mixed with the compost maintained the hydration of the medium and increased the contact surface between the sample and the inoculum of the compost.

The used compost was derived from the organic fraction of solid municipal waste, kindly supplied by Cermec SpA, Consorzio Ecologia e Risorse (Massa Carrara, Italy). Compost had 23.8 wt % carbon content, 2.2 wt % nitrogen content, with a weight C/N ratio of 10.7.

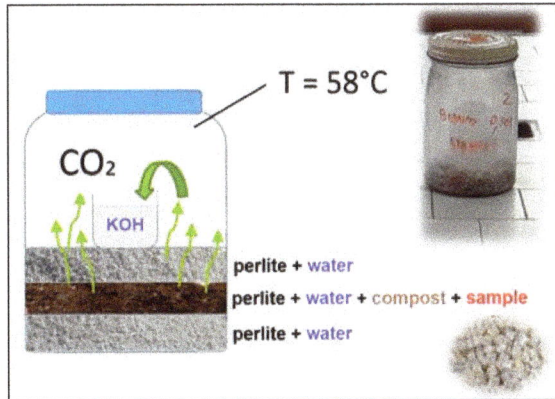

Figure 7. Scheme and photo of the apparatus used for the mineralization test.

A 100 mL beaker was inserted in each test vessel containing 50 mL of 0.1 M KOH. The basic solution trapped the carbon dioxide evolved from the sample (represented schematically by the green arrows in Figure 7) on the basis of the following reaction (Equation (1)):

$$2KOH + CO_2 \rightarrow K_2CO_3 + H_2O \tag{1}$$

Every 2–5 days, the absorbing solution was titrated with 0.1 M HCl after addition of 0.5 N BaCl$_2$ to avoid the presence of soluble carbonates.

The amount of evolved carbon dioxide was evaluated using the following equation:

$$mg\ CO_2 = (V_{KOH} \cdot C_{KOH} - V_{HCl} \cdot C_{HCl}) \cdot \frac{44}{2} \tag{2}$$

where mgCO$_2$ is the CO$_2$ evolved in mg from sample in a single test session; V_{KOH} is the volume in mL of KOH (50 mL) in flask at the beginning of test session; V_{HCl} is the volume in mL of HCl needed for titration of KOH at the end of test session; C_{KOH} and C_{HCl} are the concentrations of KOH and HCl in mol/L, respectively; 44 is molecular weight of CO$_2$; and 1/2 is the stoichiometric coefficient (mol CO$_2$/mol KOH) of the CO$_2$ absorption reaction, as shown in Equation (1).

The mineralization extent of each sample was calculated as neat percentage (corrected for the inoculum endogenous emission from the blank vessels) of the overall theoretical CO$_2$ (ThCO$_2$) amount, calculated on the basis of the initial organic carbon content of the testing sample:

$$Mineralization\% = \frac{\Sigma mgCO_2 - \Sigma mgCO_{2,blank}}{ThCO_2} \cdot 100 \tag{3}$$

The mineralization test was carried out in triplicate and the average values are reported.

3.4.2. Disintegration Test

According to ISO 20200 [55], the determination of the disintegration degree of polymeric material under simulated composting conditions [56] was carried out on PCA, PCA10, and PCA15 using a synthetic compost prepared by mixing the different components listed in Table 4. Then, tap water was added to the mixture to adjust its final water content to 50 wt % in total.

Samples in the form of hot-pressed films (about 2.5 g) with the dimensions 2.5 × 2.5 cm and thickness of 1 mm (Figure 8) were used. Before mixing with the synthetic compost, the samples were dried in an oven at $40 \pm 2\,^{\circ}$C under vacuum for the time needed to reach constant mass.

The test was carried out in flasks made of polypropylene, with dimensions of 30 × 20 × 10 cm (l, w, h), with two 5 mm holes (6.5 cm from the bottom) to provide gas exchange between the inner atmosphere and the outside environment. In each flask, 500 g of synthetic compost, mixed with three pieces of sample, were placed on the bottom, forming a homogeneous layer.

Table 4. Compost composition.

Material	Dry Mass %
Sawdust	40
Rabbit-feed	30
Ripe compost	10
Corn starch	10
Saccharose	5
Cornseed oil	4
Urea	1
Total	100

Figure 8. Sample used for the disintegration test.

The flasks were covered with a lid ensuring a tight seal to avoid excessive evaporation and placed in an air-circulation oven at a constant temperature of $58 \pm 2\,^{\circ}$C for 90 days. Following the procedure reported in ISO 20200, the gross mass of the flasks filled with the mixture was determined at the beginning of the composting process and they were weighed at determined times and, if needed, the initial mass was restored totally or in part by adding distilled water. At the end of the test, the lid of each flask was removed and the flasks were placed in the air-circulation oven at $58 \pm 2\,^{\circ}$C for 48 h to dry the content. Then, the compost of each reactor was sieved using three standard sieves (ISO 3310 [57]): 10 mm, 5 mm, and finally a 2 mm sieve. Any residual pieces of sample that did not pass through the sieves were collected, cleaned to remove the compost that covered them, and dried in an oven at $40 \pm 2\,^{\circ}$C under vacuum to constant mass. The total material recovered from the sieving procedure was considered to be non-disintegrated material. So, the degree of disintegration was calculated, in percent, using the Equation (4):

$$\text{Disintegration } (\%) = \frac{m_{i,\text{sample}} - m_{f,\text{sample}}}{m_{i,\text{sample}}} \cdot 100 \tag{4}$$

where $m_{i,sample}$ and $m_{f,sample}$ represent the initial sample mass and the final dry mass of the sample recovered after sieving, respectively.

The disintegration test was carried out in triplicate on each sample.

4. Conclusions

In this work, we explored the use of waste wood fibers in combination with PHB-HV to manufacture biodegradable wood plastic composites in accordance with the circular economy principles.

We produced composites based on PHB-HV and different amounts (10, 15, and 20 wt %) of soft wood sawdust, a byproduct of the wood industry, by extrusion in the presence of appropriate amounts of ATBC as a plasticizer and calcium carbonate as an inorganic filler.

Using appropriate amounts of ATBC, smooth processing was achieved for up to 20 wt % f fibers, despite the reduction in the melt fluidity observed with increasing fiber loading. The tensile modulus remained almost constant up to a 15 wt % fiber content, whereas the tensile strength and the elongation slightly decreased by increasing the fiber content up to 20 wt %. The impact resistance of the composites increased markedly with increasing SD amounts: the Charpy's impact energy increased from 3.6 (without fiber) to 12.2 kJ/m^2 for the composite with 15 wt % fiber. The results of the mineralization and disintegration tests in compost showed that the developed composites are compostable in accordance with EN 13427:2000, and the presence of fibers favored the physical disintegration of the composite, increasing the biodegradation rate of the polymeric matrix. After six months in compost, the composite with 15 wt % fiber showed a biodegradability of 78% compared with 58% for the composites without fibers and the paper (control sample). After 14 months, the pots based on PHB-HV/SD retained adequate mechanical performance and physical integrity inside the plant nursery and the plant growth performance was the same as that in traditional polypropylene pots, but the buried pots containing 15 wt % fibers were completely fragmented compared with those without fibers, confirming the accelerating effect of the fibers on the degradation of the polymeric matrix in soil.

In conclusion, the developed composites based on PHB-HV and waste wood sawdust are particularly suitable for production by the extrusion of relatively low-cost items such as pots, which are compostable and biodegradable in soil and usable in agriculture or plant nursery.

Author Contributions: P.C. and M.S. conceived and designed the experiments; N.M. performed the experiments; N.M. and V.G. analyzed the data; P.C., M.S. and V.G. wrote the paper; A.L. revised the paper.

Funding: This research was funded by Tuscany Region (Grant number: 3389.30072014.068000241).

Acknowledgments: The authors would like to thank the Tuscany Region for the financial support of the project PHA through the fund POR FESR 2014–2020 (Bando 2—RSI 2014) (Grant number: 3389.30072014.068000241). We would also like to thank Zefiro and Femto Engineering for the scale-up of the process, Rosa Galante for support in carrying out the biodegradation and disintegration tests, in terrestrial environments and at the lab-scale, respectively, and Randa Ishak for the SEM analysis.

Conflicts of Interest: The authors declare no conflict of interest.

References

1. Song, J.H.; Murphy, R.J.; Narayan, R.; Davies, G.B.H. Biodegradable and compostable alternatives to conventional plastics. *Philos. Trans. R. Soc. B Biol. Sci.* **2009**, *364*, 2127–2139. [CrossRef] [PubMed]
2. Muthuraj, R.; Misra, M.; Mohanty, A.K. Studies on mechanical, thermal, and morphological characteristics of biocomposites from biodegradable polymer blends and natural fibers. In *Biocomposites*; Elsevier: Amsterdam, The Netherlands, 2015; pp. 93–140, ISBN 9781782423737.
3. Koronis, G.; Silva, A.; Fontul, M. Green composites: A review of adequate materials for automotive applications. *Compos. Part B Eng.* **2013**, *44*, 120–127. [CrossRef]
4. Khanna, S.; Srivastava, A.K. On-line Characterization of Physiological State in Poly(β-Hydroxybutyrate) Production by Wautersia eutropha. *Appl. Biochem. Biotechnol.* **2009**, *157*, 237–243. [CrossRef] [PubMed]

5.   Castilho, L.R.; Mitchell, D.A.; Freire, D.M.G. Production of polyhydroxyalkanoates (PHAs) from waste materials and by-products by submerged and solid-state fermentation. *Bioresour. Technol.* **2009**, *100*, 5996–6009. [CrossRef] [PubMed]

6.   Hopewell, J.; Dvorak, R.; Kosior, E. Plastics recycling: Challenges and opportunities. *Philos. Trans. R. Soc. B Biol. Sci.* **2009**, *364*, 2115–2126. [CrossRef] [PubMed]

7.   Jose, J.; George, S.M.; Thomas, S. Recycling of polymer blends. In *Recent Developments in Polymer Recycling*; Kluwer Academic Publishers: Dordrecht, The Netherlands, 2011; Volume 37, pp. 187–214, ISBN 978-81-7895-524-7.

8.   Murariu, M.; Dubois, P. PLA composites: From production to properties. *Adv. Drug Deliv. Rev.* **2016**, *107*, 17–46. [CrossRef] [PubMed]

9.   Kourmentza, C.; Plácido, J.; Venetsaneas, N.; Burniol-Figols, A.; Varrone, C.; Gavala, H.N.; Reis, M.A.M. Recent Advances and Challenges towards Sustainable Polyhydroxyalkanoate (PHA) Production. *Bioengineering* **2017**, *4*, 55. [CrossRef]

10. Chiellini, E.; Cinelli, P.; Chiellini, F.; Imam, S.H. Environmentally Degradable Bio-Based Polymeric Blends and Composites. *Macromol. Biosci.* **2004**, *4*, 218–231. [CrossRef]

11. Yu, L.; Dean, K.; Li, L. Polymer blends and composites from renewable resources. *Prog. Polym. Sci.* **2006**, *31*, 576–602. [CrossRef]

12. Narancic, T.; Verstichel, S.; Reddy Chaganti, S.; Morales-Gamez, L.; Kenny, S.T.; De Wilde, B.; Padamati, R.B.; O'Connor, K.E. Biodegradable Plastic Blends Create New Possibilities for End-of-Life Management of Plastics but They Are Not a Panacea for Plastic Pollution. *Environ. Sci. Technol.* **2018**, *52*, 10441–10452. [CrossRef]

13. Law, K.H.; Cheng, Y.C.; Leung, Y.C.; Lo, W.H.; Chua, H.; Yu, H.F. Construction of recombinant Bacillus subtilis strains for polyhydroxyalkanoates synthesis. *Biochem. Eng. J.* **2003**, *16*, 203–208. [CrossRef]

14. Bugnicourt, E.; Cinelli, P.; Lazzeri, A.; Alvarez, V. Polyhydroxyalkanoate (PHA): Review of synthesis, characteristics, processing and potential applications in packaging. *Express Polym. Lett.* **2014**, *8*, 791–808. [CrossRef]

15. Steinbüchel, A.; Lütke-Eversloh, T. Metabolic engineering and pathway construction for biotechnological production of relevant polyhydroxyalkanoates in microorganisms. *Biochem. Eng. J.* **2003**, *16*, 81–96. [CrossRef]

16. Koller, M.; Marsalek, L.; de Sousa Dias, M.M.; Braunegg, G. Producing microbial polyhydroxyalkanoate (PHA) biopolyesters in a sustainable manner. *New Biotechnol.* **2017**, *37*, 24–38. [CrossRef] [PubMed]

17. Salehizadeh, H.; Van Loosdrecht, M.C.M. Production of polyhydroxyalkanoates by mixed culture: Recent trends and biotechnological importance. *Biotechnol. Adv.* **2004**, *22*, 261–279. [CrossRef] [PubMed]

18. Anjum, A.; Zuber, M.; Zia, K.M.; Noreen, A.; Anjum, M.N.; Tabasum, S. Microbial production of polyhydroxyalkanoates (PHAs) and its copolymers: A review of recent advancements. *Int. J. Biol. Macromol.* **2016**, *89*, 161–174. [CrossRef] [PubMed]

19. Numata, K.; Hideki, A.; Tadahisa, I. Biodegradability of Poly(hydroxyalkanoate) Materials. *Materials* **2009**, *2*, 1104–1126. [CrossRef]

20. Breulmann, M.; Künkel, A.; Philipp, S.; Reimer, V.; Siegenthaler, K.O.; Skupin, G.; Yamamoto, M. Polymers, Biodegradable. In *Ullmann's Encyclopedia of Industrial Chemistry*; VCH Publishers: New York, NY, USA, 2009.

21. Philip, S.; Keshavarz, T.; Roy, I. Polyhydroxyalkanoates: Biodegradable polymers with a range of applications. *J. Chem. Technol. Biotechnol.* **2007**, *82*, 233–247. [CrossRef]

22. Padovani, G.; Carlozzi, P.; Seggiani, M.; Cinelli, P. PHB-Rich Biomass and BioH 2 Production by Means of Photosynthetic Microorganisms. *Chem. Eng. Trans.* **2016**, *49*, 55–60. [CrossRef]

23. Deroiné, M.; Le Duigou, A.; Corre, Y.-M.; Le Gac, P.-Y.; Davies, P.; César, G.; Bruzaud, S. Seawater accelerated ageing of poly(3-hydroxybutyrate-co-3-hydroxyvalerate). *Polym. Degrad. Stab.* **2014**, *105*, 237–247. [CrossRef]

24. Volova, T.G.; Boyandin, A.N.; Vasiliev, A.D.; Karpov, V.A.; Prudnikova, S.V.; Mishukova, O.V.; Boyarskikh, U.A.; Filipenko, M.L.; Rudnev, V.P.; Bá Xuân, B.; et al. Biodegradation of polyhydroxyalkanoates (PHAs) in tropical coastal waters and identification of PHA-degrading bacteria. *Polym. Degrad. Stab.* **2010**, *95*, 2350–2359. [CrossRef]

25. Musioł, M.; Sikorska, W.; Janeczek, H.; Wałach, W.; Hercog, A.; Johnston, B.; Rydz, J. (Bio)degradable polymeric materials for a sustainable future—Part 1. Organic recycling of PLA/PBAT blends in the form of prototype packages with long shelf-life. *Waste Manag.* **2018**, 1–8. [CrossRef]

26. Aliotta, L.; Cinelli, P.; Coltelli, M.B.; Righetti, M.C.; Gazzano, M.; Lazzeri, A. Effect of nucleating agents on crystallinity and properties of poly (lactic acid) (PLA). *Eur. Polym. J.* **2017**, *93*, 822–832. [CrossRef]
27. Reis, K.C.; Pereira, J.; Smith, A.C.; Carvalho, C.W.P.; Wellner, N.; Yakimets, I. Characterization of polyhydroxybutyrate-hydroxyvalerate (PHB-HV)/maize starch blend films. *J. Food Eng.* **2008**, *89*, 361–369. [CrossRef]
28. Godbole, S.; Gote, S.; Latkar, M.; Chakrabarti, T. Preparation and characterization of biodegradable poly-3-hydroxybutyrate–starch blend films. *Bioresour. Technol.* **2003**, *86*, 33–37. [CrossRef]
29. Zhang, M.; Thomas, N.L. Preparation and properties of polyhydroxybutyrate blended with different types of starch. *J. Appl. Polym. Sci.* **2009**, *116*, 688–694. [CrossRef]
30. Seggiani, M.; Cinelli, P.; Mallegni, N.; Balestri, E.; Puccini, M.; Vitolo, S.; Lardicci, C.; Lazzeri, A. New Bio-Composites Based on Polyhydroxyalkanoates and Posidonia oceanica Fibres for Applications in a Marine Environment. *Materials* **2017**, *10*, 326. [CrossRef]
31. Seggiani, M.; Cinelli, P.; Verstichel, S.; Puccini, M.; Anguillesi, I.; Lazzeri, A. Development of Fibres-Reinforced Biodegradable Composites. *Chem. Eng. Trans.* **2015**, *43*, 1813–1818. [CrossRef]
32. Imam, S.H.; Cinelli, P.; Gordon, S.H.; Chiellini, E. Characterization of Biodegradable Composite Films Prepared from Blends of Poly(Vinyl Alcohol), Cornstarch, and Lignocellulosic Fiber. *J. Polym. Environ.* **2005**, *13*, 47–55. [CrossRef]
33. Seggiani, M.; Cinelli, P.; Geicu, M.; Elen, P.M.; Puccini, M.; Lazzeri, A. Microbiological valorisation of bio-composites based on polylactic acid and wood fibres. *Chem. Eng. Trans.* **2016**, *49*, 127–132. [CrossRef]
34. Chiellini, E.; Cinelli, P.; Imam, S.H.; Mao, L. Composite Films Based on Biorelated Agro-Industrial Waste and Poly(vinyl alcohol). Preparation and Mechanical Properties Characterization. *Biomacromolecules* **2001**, *2*, 1029–1037. [CrossRef] [PubMed]
35. Pandey, J.K.; Nagarajan, V.; Mohanty, A.K.; Misra, M. 1—Commercial potential and competitiveness of natural fiber composites. In *Woodhead Publishing Series in Composites Science and Engineering*; Misra, M., Pandey, J.K., Mohanty, A.K.B.T.-B., Eds.; Woodhead Publishing: Cambridge, UK, 2015; pp. 1–15, ISBN 978-1-78242-373-7.
36. Azwa, Z.N.; Yousif, B.F.; Manalo, A.C.; Karunasena, W. A review on the degradability of polymeric composites based on natural fibres. *Mater. Des.* **2013**, *47*, 424–442. [CrossRef]
37. Herrera-Franco, P.J.; Valadez-Gonzalez, A. A study of the mechanical properties of short natural-fiber reinforced composites. *Compos. Part B Eng.* **2005**, *36*, 597–608. [CrossRef]
38. Satyanarayana, K.G.; Arizaga, G.G.C.; Wypych, F. Biodegradable composites based on lignocellulosic fibers—An overview. *Prog. Polym. Sci.* **2009**, *34*, 982–1021. [CrossRef]
39. Gigante, V.; Aliotta, L.; Phuong, V.T.; Coltelli, M.B.; Cinelli, P.; Lazzeri, A. Effects of waviness on fiber-length distribution and interfacial shear strength of natural fibers reinforced composites. *Compos. Sci. Technol.* **2017**, *152*, 129–138. [CrossRef]
40. Sombatsompop, N.; Yotinwattanakumtorn, C.; Thongpin, C. Influence of type and concentration of maleic anhydride grafted polypropylene and impact modifiers on mechanical properties of PP/wood sawdust composites. *J. Appl. Polym. Sci.* **2005**, *97*, 475–484. [CrossRef]
41. Sombatsompop, N.; Chaochanchaikul, K. Average mixing torque, tensile and impact properties, and thermal stability of poly(vinyl chloride)/sawdust composites with different silane coupling agents. *J. Appl. Polym. Sci.* **2005**, *96*, 213–221. [CrossRef]
42. Seggiani, M.; Cinelli, P.; Balestri, E.; Mallegni, N.; Stefanelli, E.; Rossi, A.; Lardicci, C.; Lazzeri, A. Novel Sustainable Composites Based on Poly(hydroxybutyrate-co-hydroxyvalerate) and Seagrass Beach-CAST Fibers: Performance and Degradability in Marine Environments. *Materials* **2018**, *11*, 772. [CrossRef]
43. Melik, H.D.; Schechtman, A.L. Biopolyester melt behavior by torque rheometry. *Polym. Eng. Sci.* **1995**, *35*, 1795–1806. [CrossRef]
44. Baysal, E.; Deveci, I.; Turkoglu, T.; Toker, H. Thermal analysis of oriental beech sawdust treated with some commercial wood preservatives. *Maderas. Cienc. Tecnol.* **2017**, *19*, 329–338. [CrossRef]
45. Bledzki, A.K.; Reihmane, S.; Gassan, J. Properties and modification methods for vegetable fibers for natural fiber composites. *J. Appl. Polym. Sci.* **1996**, *59*, 1329–1336. [CrossRef]
46. Renner, K.; Kenyó, C.; Móczó, J.; Pukánszky, B. Micromechanical deformation processes in PP/wood composites: Particle characteristics, adhesion, mechanisms. *Compos. Part A Appl. Sci. Manuf.* **2010**, *41*, 1653–1661. [CrossRef]

47. Gurunathan, T.; Mohanty, S.; Nayak, S.K. A review of the recent developments in biocomposites based on natural fibres and their application perspectives. *Compos. Part A Appl. Sci. Manuf.* **2015**, *77*, 1–25. [CrossRef]
48. Graupner, N.; Müssig, J. A comparison of the mechanical characteristics of kenaf and lyocell fibre reinforced poly(lactic acid) (PLA) and poly(3-hydroxybutyrate) (PHB) composites. *Compos. Part A Appl. Sci. Manuf.* **2011**, *42*, 2010–2019. [CrossRef]
49. Ganster, J.; Fink, H.-P. Novel cellulose fibre reinforced thermoplastic materials. *Cellulose* **2006**, *13*, 271–280. [CrossRef]
50. Ku, H.; Wang, H.; Pattarachaiyakoop, N.; Trada, M. A review on the tensile properties of natural fiber reinforced polymer composites. *Compos. Part B Eng.* **2011**, *42*, 856–873. [CrossRef]
51. Imre, B.; Pukánszky, B. Recent advances in bio-based polymers and composites: Preface to the BiPoCo 2012 Special Section. *Eur. Polym. J.* **2013**, *49*, 1146–1150. [CrossRef]
52. Arrieta, M.P.; Samper, M.D.; López, J.; Jiménez, A. Combined Effect of Poly(hydroxybutyrate) and Plasticizers on Polylactic acid Properties for Film Intended for Food Packaging. *J. Polym. Environ.* **2014**, *22*, 460–470. [CrossRef]
53. Lin, Y.; Chan, C.-M. 3—*Calcium Carbonate Nanocomposites. Advances in Polymer Nanocomposites, Types and Applications*; Series in Composites Science and Engineering; Woodhead Publishing: Cambridge, UK, 2012; pp. 55–90, ISBN 978-1-84569-940-6.
54. American Society for Testing and Materials ASTM D5338. *Standard Test Method for Determining Aerobic Biodegradation of Plastic Materials under Controlled Composting Conditions*; American Society for Testing and Materials: West Conshohocken, PA, USA, 1998.
55. International Organization for Standardization ISO 20200. *Plastics Determination of the Degree of Disintegration of Plastic Materials under Simulated Composting Conditions in a Laboratory-Scale Test*; International Organization for Standardization: Geneva, Switzerland, 2015.
56. Vaverková, M.; Toman, F.; Adamcová, D.; Kotovicová, J. Study of the Biodegrability of Degradable/Biodegradable Plastic Material in a Controlled Composting Environment. *Ecol. Chem. Eng. S* **2012**, *19*, 347–358. [CrossRef]
57. International Organization for Standardization ISO 3310. *Test Sieves—Technical Requirements and Testing—Part 1: Test Sieves of Metal Wire Cloth*; International Organization for Standardization: Geneva, Switzerland, 2016; pp. 1–15.

© 2019 by the authors. Licensee MDPI, Basel, Switzerland. This article is an open access article distributed under the terms and conditions of the Creative Commons Attribution (CC BY) license (http://creativecommons.org/licenses/by/4.0/).

International Journal of
Molecular Sciences

MDPI

Article

Toughness Enhancement of PHBV/TPU/Cellulose Compounds with Reactive Additives for Compostable Injected Parts in Industrial Applications

Estefanía Lidón Sánchez-Safont [1], Alex Arrillaga [2], Jon Anakabe [2], Luis Cabedo [1] and Jose Gamez-Perez [1,*]

[1]   Polymers and Advanced Materials Group (PIMA), Universitat Jaume I, 12071 Castellón, Spain; esafont@uji.es (E.L.S.-S.); lcabedo@uji.es (L.C.)
[2]   Leartiker S. Coop., Xemein Etorbidea 12A, 48270 Markina-Xemein, Spain; aarrillaga@leartiker.com (A.A.); janakabe@leartiker.com (J.A.)
*   Correspondence: gamez@uji.es; Tel.: +34-964-728194

Received: 25 June 2018; Accepted: 17 July 2018; Published: 19 July 2018

Abstract: Poly(3-hydroxybutyrate-co-3-valerate), PHBV, is a bacterial thermoplastic biopolyester that possesses interesting thermal and mechanical properties. As it is fully biodegradable, it could be an alternative to the use of commodities in single-use applications or in those intended for composting at their end of life. Two big drawbacks of PHBV are its low impact toughness and its high cost, which limit its potential applications. In this work, we proposed the use of a PHBV-based compound with purified α-cellulose fibres and a thermoplastic polyurethane (TPU), with the purpose of improving the performance of PHBV in terms of balanced heat resistance, stiffness, and toughness. Three reactive agents with different functionalities have been tested in these compounds: hexametylene diisocianate (HMDI), a commercial multi-epoxy-functionalized styrene-co-glycidyl methacrylate oligomer (Joncryl® ADR-4368), and triglycidyl isocyanurate (TGIC). The results indicate that the reactive agents play a main role of compatibilizers among the phases of the PHBV/TPU/cellulose compounds. HMDI showed the highest ability to compatibilize the cellulose and the PHBV in the compounds, with the topmost values of deformation at break, static toughness, and impact strength. Joncryl® and TGIC, on the other hand, seemed to enhance the compatibility between the fibres and the polymer matrix as well as the TPU within the PHBV.

Keywords: biopolyester; compatibilizer; cellulose; elastomer; toughening; biodisintegration; heat deflection temperature

1. Introduction

Nowadays, the use of plastics is widely extended in almost all production fields, such as packaging, electronics, automotive, household, etc., and the market is dominated by the so-called commodities, traditional oil-based plastics. The growing concern over the environmental problems involved with petroleum-based polymers related to their non-renewable origin and poor biodegradability is leading the industry to replace current materials with biodegradable alternatives [1]. Therefore, researchers have been looking for alternatives that may be more environmentally sustainable, especially in short- and medium-term applications, such as packaging. Within this context, biopolyesters have received great attention, especially those that are bio-sourced and biodegradable, as a way to overcome some of the waste management issues [2].

Among the different commercially available biopolyesters, one of the most promising candidates to replace commodities is the poly(3-hydroxybutyrate-co-3-valerate) (PHBV) [3–6]. PHBV is a bacterial thermoplastic biopolyester from the polyhydroxyalcanoates family that possesses physical properties comparable to conventional polyolefins, high static mechanical performance [7], and relatively high thermal resistance [8], while being fully biodegradable. However, two important drawbacks of PHBV are its low impact toughness and its cost, which is still quite high [9,10]. These disadvantages are a serious handicap for its use in applications in rigid packaging parts, for instance, that could be obtained by injection moulding.

One of the most promising eco-friendly approaches to reduce the manufacturing costs of PHBV while maintaining its biodegradability and sustainability is the development of natural fibre-based polymer composites. Indeed, it also improves its mechanical performance in terms of stiffness as well as thermal resistance [11]. On the other hand, in order to enhance the toughness of PHBV, several attempts have been reported in the literature, some of them related to blending with other polymers such as poly(butylene adipate-co-terephthalate) (PBAT) [12,13], polybutylene succinate (PBS) [14,15], or polycaprolactone (PCL) [12] or by the addition of impact modifiers such as ethylene vinyl acetate [16], epoxidized natural rubber [13,17], or thermoplastic polyurethane (TPU) [12,18–20], showing in all cases great improvements in elongation at break.

In this work, we proposed the use of a purified α-cellulose fibres and a thermoplastic polyurethane (TPU) with the purpose of improving the performance of PHBV in terms of balanced heat resistance, stiffness, and toughness without compromising biodisintegrability in composting conditions.

However, previous works have shown that the interaction of these fillers with PHA matrices was not very strong, resulting in low toughness and tensile strength [21,22]. Nonetheless, some strategies to improve the chemical affinity between the cellulose and other polyesters have been used in order to increase the reinforcement effect of the cellulose, such as fibre treatments or use of compatibilizers (reviewed by [11,23–26]).

From an industrial point of view, reactive extrusion is a convenient, cost-effective approach to improve the interfacial adhesion of the different phases via an in situ reaction during melt processing [27]. Within this objective, three reactive agents have been tested: (a) hexametylene diisocianate (HMDI); (b) (Joncryl® ADR-4368), a commercial multi-epoxy-functionalized styrene-acrylic oligomer; and (c) triglycidyl isocyanurate (TGIC) (Figure 1). These reactive agents possess three different functional groups that could potentially react with the hydroxyl groups present at the cellulose surface and the ones from the alcohol and carboxylic acid groups at the polymer chain ends [28]. Some reports have been found in the literature about the use of diisocyanates as compatibilizers in biopolyester/fibre composites [29–31], PHBV/polylactic acid (PLA) blends [32], and PLA/TPU blends [33,34], showing good improvements in interfacial adhesion. Hao et al. showed improved interfacial adhesion in PLA/sisal fibre composites using Joncryl® [35] and Nanthananon et al. reported similar improvements in PLA/eucalyptus fibre systems [36]. Furthermore, the use of Joncryl® has also been proved efficient in the compatibilization of POM/TPU blends [37]. TGIC was successfully used to compatibilize polylactide/sisal fibre biocomposites [38].

In this work, the combined effect of TPU, cellulose fibres, and the use of three different reactive agents (HMDI, Joncryl®, and TGIC) is explored in order to improve the interfacial adhesion and compatibility of PHBV, TPU, and cellulose through reactive extrusion. This strategy is aimed at building a ternary system that will overcome the handicaps of PHBV that prevent its usage in injection-moulded applications in terms of cost, toughness, and thermal resistance.

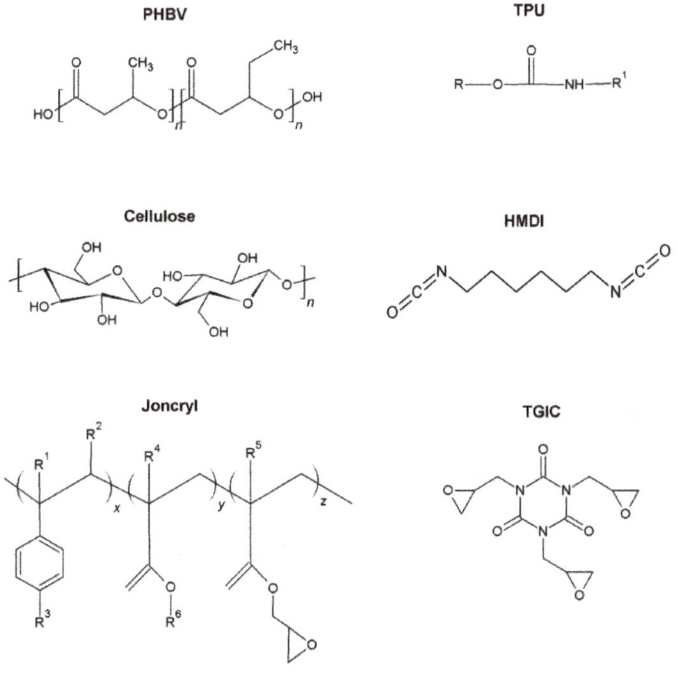

Figure 1. Chemical structures of materials used in this study.

2. Results and Discussion

2.1. Preparation of Compounds and Analysis of Their Processability

PHBV/TPU/Cellulose triple systems with different content of additives (TPU and cellulose) and reactive agents (HMDI, Joncryl®, and TGIC) were prepared by a co-rotating twin-screw extruder in the proportions described in Table 1.

The melt flow index is a useful tool to predict the processability of materials in industrial equipment such as injection moulding and gives an indirect measurement of melt viscosity, as it is indirectly proportional to viscosity. Figure 2 represents the melt flow index values of neat PHBV and the compounds (PHBV/30T/10C and PHBV/30T/30C) with 0, 0.3, 0.5, and 1 phr reactive agents content.

As seen in Figure 2, the addition of TPU and cellulose significantly decreases the melt fluidity of PHBV, especially for the highest cellulose content. This increment in melt viscosity is typical in fibre-based composites because of the increased shear produced by the restricted chain mobility induced by the fibres [39]. The addition of the different reactive agents in 0.3 phr leads to a further drastic reduction of the melt fluidity. As the reactive agent content increases, the MFI values decrease slightly, except in the case of TGIC, where the compounds have similar melt indexes. Among the three reactive agents, the highest reduction in MFI is found in the compositions with HMDI. This reduction in fluidity with the incorporation of the reactive agents is indicative of some reactivity with the components of the system and can be related to a compatibilization between the fibres and the polymers and/or between the PHBV and the TPU [27].

With respect to the processability, the reduced fluidity of the compositions with the reactive agents led to increased injection pressure values. However, despite the low MFI values of the compounds, the injected samples were successfully obtained without any change in the processing parameters with respect to neat PHBV.

Table 1. List of compounds and their composition.

Sample	TPU	Cellulose	HMDI	Joncryl®	TGIC
			(phr) *		
Neat PHBV	-	-	-	-	-
PHBV/30T/10C **	30	10	-	-	-
PHBV/30T/10C-0.3HMDI	30	10	0.3	-	-
PHBV/30T/10C-0.5HMDI	30	10	0.5	-	-
PHBV/30T/10C-1HMDI	30	10	1	-	-
PHBV/30T/10C-0.3Joncryl	30	10	-	0.3	-
PHBV/30T/10C-0.5Joncryl	30	10	-	0.5	-
PHBV/30T/10C-1Joncryl	30	10	-	1	-
PHBV/30T/10C-0.3TGIC	30	10	-	-	0.3
PHBV/30T/10C-0.5TGIC	30	10	-	-	0.5
PHBV/30T/10C-1TGIC	30	10	-	-	1
PHBV/30T/30C ***	30	30	-	-	-
PHBV/30T/30C-0.3HMDI	30	30	0.3	-	-
PHBV/30T/30C-0.5HMDI	30	30	0.5	-	-
PHBV/30T/30C-1HMDI	30	30	1	-	-
PHBV/30T/30C-0.3Joncryl	30	30	-	0.3	-
PHBV/30T/30C-0.5Joncryl	30	30	-	0.5	-
PHBV/30T/30C-1Joncryl	30	30	-	1	-
PHBV/30T/30C-0.3TGIC	30	30	-	-	0.3
PHBV/30T/30C-0.5TGIC	30	30	-	-	0.5
PHBV/30T/30C-1TGIC	30	30	-	-	1

* phr refers to the 100 unit weight PHBV matrix; ** PHBV/30T/10C corresponds to 71.4 wt % PHBV, 21.4 wt % TPU, and 7.2 wt % cellulose; *** PHBV/30T/30C corresponds to 62.5 wt % PHBV, 18.8 wt % TPU, and 18.8 wt % cellulose.

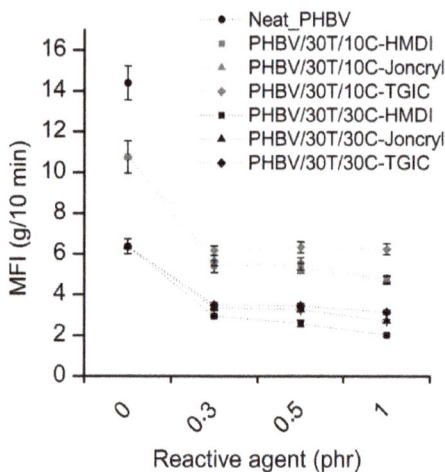

Figure 2. Melt flow index (MFI) of neat PHBV, PHBV/30T/10C, and PHBV/30T/30C with 0, 0.3, 0.5, and 1 phr reactive agent content.

2.2. Characterization

The morphology of the PHBV/TPU/cellulose triple systems was analysed by scanning electron microscopy (SEM). Micrographs of PHBV/30T/10C and PHBV/30T/30C without reactive agents and with 1 phr of HMDI, Joncryl®, and TGIC are depicted in Figures 3 and 4, respectively.

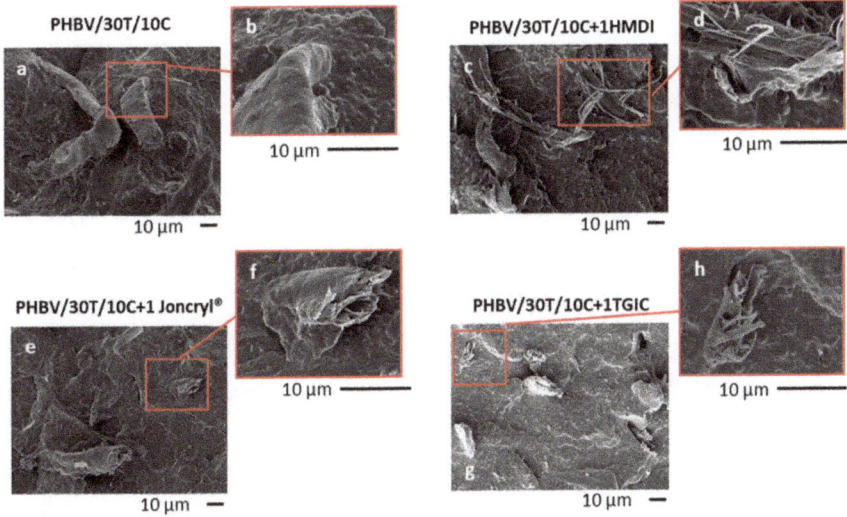

Figure 3. SEM micrographs of PHBV/30T/10C (**a,b**) and PHBV/30T/10C with 1 phr of HMDI (**c,d**), Joncryl® (**e,f**), and TGIC (**g,h**).

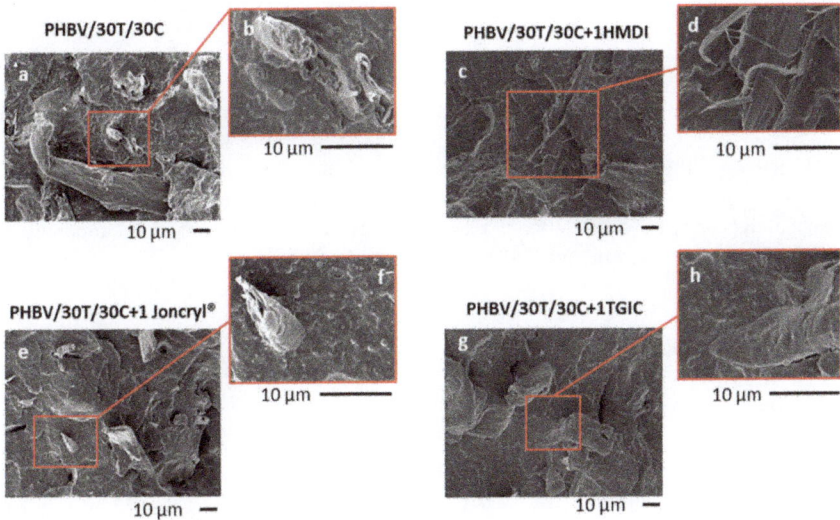

Figure 4. SEM micrographs of PHBV/30T/30C (**a,b**) and PHBV/30T/30C with 1 phr of HMDI (**c,d**), Joncryl® (**e,f**), and TGIC (**g,h**).

With respect to the fillers, in samples without reactive agent (Figures 3a and 4a), a good distribution of the fibres was observed, indicating good compounding and certain affinity between the fibres and the polymer matrix probably due to the formation of hydrogen bonds between the –OH groups of the fibre surface and PHBV [21,40,41]. However, some detachment of the fibres is also observed, indicating that the hydrogen bonding type is not enough to provide a strong adhesion between these phases. Regarding the fibre distribution, no remarkable differences were observed with the addition of the different reactive agents (Figures 3 and 4). Nevertheless, regarding the fibre–matrix interface, with the presence of the reactive agents the fibres seem to be well trapped by the polymer matrix, as broken fibres covered by the polymer were observed in all cases. In particular, in the case of HMDI (Figure 3c,d and Figure 4c,d), the fibres appear broken in the longitudinal direction, and defibrillation was observed, indicating a cohesive failure. These observations suggest the strongest adhesion between the cellulose fibres and the PHBV, indicating a compatibilization effect.

With respect to the polymeric matrix, all the compositions present a drop in matrix morphology, where the disperse phase is the TPU, as shown in Figures 5 and 6. These figures show the SEM images of the polymeric matrix for the PHBV/30T/10 and PHBV/30T/30C composites, with and without the reactive agents. The droplet size distributions of the dispersed phase are also included in the aforementioned figures. The average domain size (d), the estimated ligament distance, and the d10, d50, and d90 values are summarized in Table 2.

According to the measurements performed, the average domain size (d) of the TPU is 0.416 and 0.420 μm in PHBV/30T/10C and PHBV/30T/30C, respectively. Although some detachment of TPU is observed, the small size of the dispersed phase domains indicates a certain affinity between the phases. With the incorporation of the reactive agents the average TPU droplet size was reduced, as shown in Table 2. The highest droplet size reductions were obtained in the compounds containing the highest amount of cellulose with the reactive agents TGIC and Joncryl®. In these cases, indeed, a slight dependence on the average domain size (d) as the reactive agent content increases was observed in compounds with 10 phr of Celullose, but not in those with 30 phr Cellulose.

Regarding compounds with HMDI, the reduction in the average domain size (d) is lower with respect to the other reactive agents. In fact, the compound PHBV/30T/10C with 1 phr HMDI shows a similar value of d as the compound without reactive agents, with the domain size distribution being slightly displaced to bigger sizes, presenting the highest d90 value among all compounds. Nevertheless, the matrix ligament thickness of this system is in the same range as in the rest of the composites.

Despite the differences in the size distributions, the droplet size and the estimated matrix ligament thickness (T) are quite small for all compositions. As reported by Wu [42], the matrix ligament thickness plays an important role for rubber toughening in polymer blends. If the average matrix ligament thickness, defined as the surface-to-surface interparticle distance, is smaller than a critical value, the blend will be tough, whereas on the contrary the blend will be brittle. It can be concluded, from this analysis, that the use of TGIC and Joncryl® reactive agents produced an enhanced compatibilization effect on the TPU domains within the PHBV/cellulose matrix.

Table 2. Estimated d10, d50 and d90, average droplet size and ligament distance values of the TPU dispersed phase.

(phr)	PHBV/30T/10C					PHBV/30T/30C				
	d10 (μm)	d50 (μm)	d90 (μm)	d (μm)	T (μm)	d10 (μm)	d50 (μm)	d90 (μm)	d (μm)	T (μm)
0	0.17	0.34	0.79	0.42	0.13	0.17	0.35	0.77	0.42	0.13
0.3 HMDI	0.14	0.31	0.79	0.40	0.12	0.09	0.20	0.57	0.27	0.08
0.5 HMDI	0.11	0.21	0.51	0.26	0.08	0.09	0.18	0.76	0.31	0.10
1 HMDI	0.11	0.26	1.01	0.43	0.13	0.09	0.21	0.77	0.33	0.10
0.3 Joncryl	0.09	0.20	0.52	0.26	0.08	0.07	0.14	0.38	0.19	0.06
0.5 Joncryl	0.09	0.17	0.43	0.22	0.07	0.08	0.14	0.48	0.21	0.07
1 Joncryl	0.09	0.16	0.39	0.20	0.06	0.07	0.14	0.36	0.18	0.06
0.3 TGIC	0.10	0.21	0.57	0.28	0.09	0.07	0.14	0.43	0.20	0.06
0.5 TGIC	0.09	0.21	0.52	0.26	0.08	0.08	0.14	0.39	0.20	0.06
1 TGIC	0.08	0.17	0.54	0.25	0.08	0.09	0.17	0.42	0.21	0.07

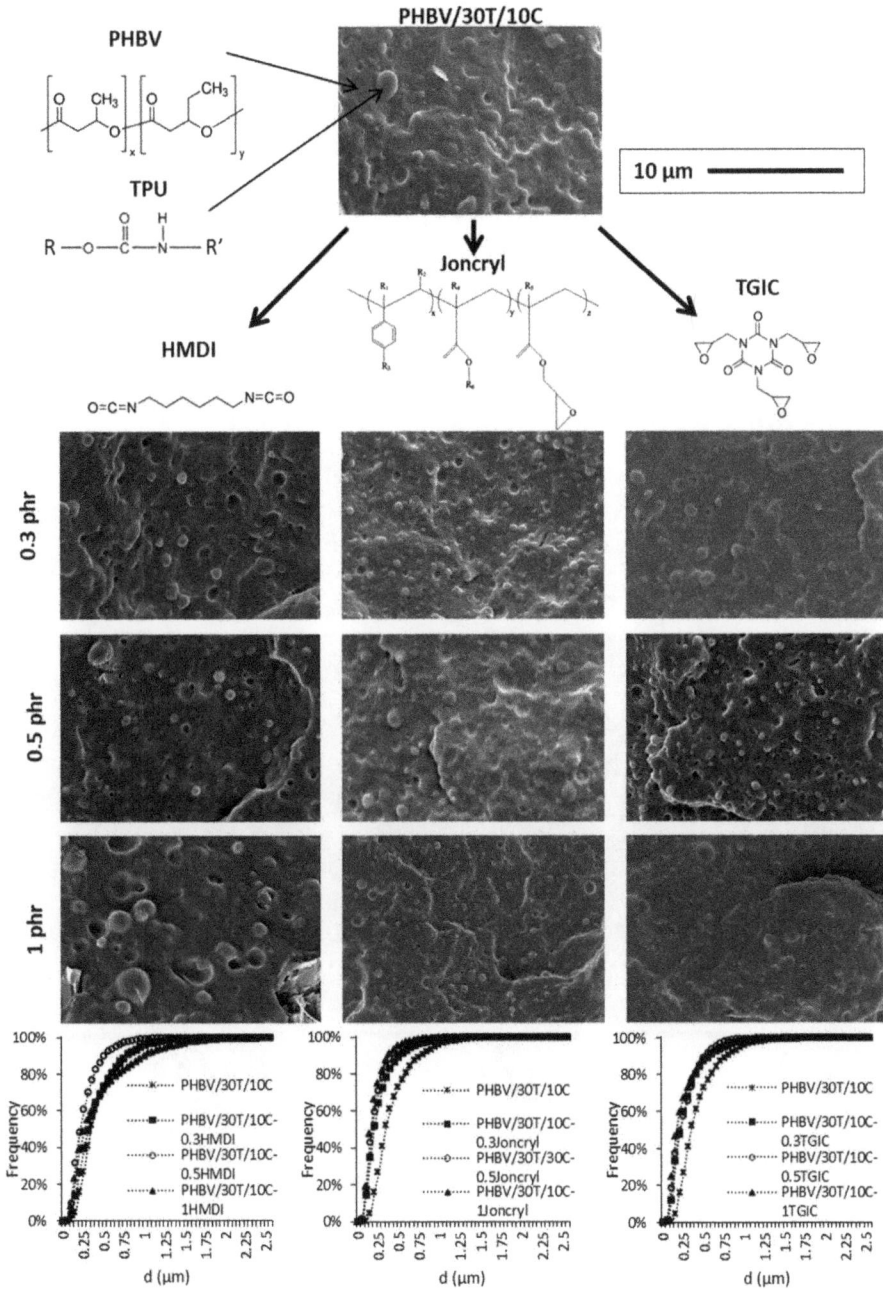

Figure 5. SEM images of the PHBV/30T/10C composites with the different reactive agents and cumulative frequency droplet size histograms of the dispersed phase.

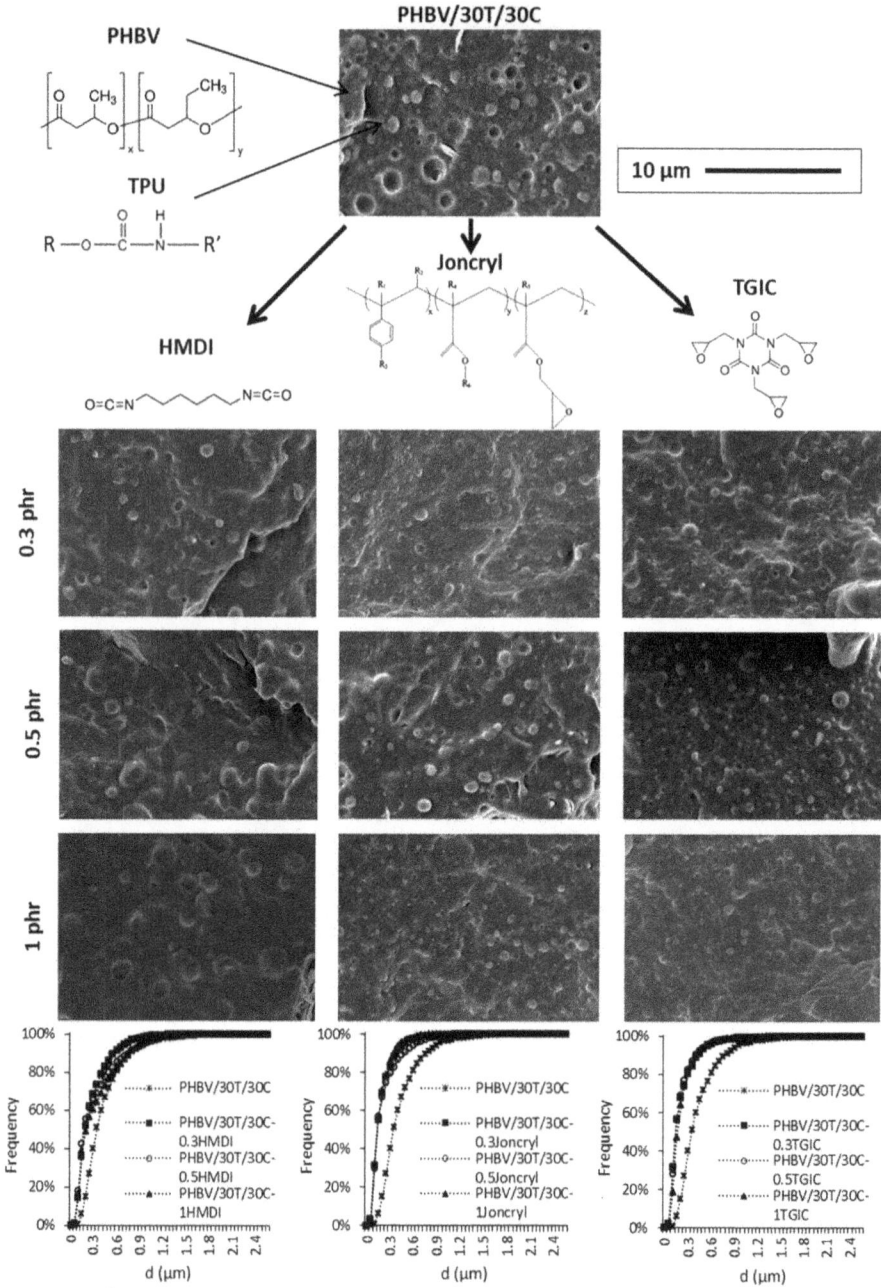

Figure 6. SEM images of the PHBV/30T/30C composites with the different reactive agents and cumulative frequency droplet size histograms of the dispersed phase.

2.2.1. Mechanical Properties

Tensile tests up to failure were conducted in order to study the mechanical properties of neat PHBV and the compounds with and without reactive agents (HMDI, Joncryl®, and TGIC). The Young's modulus, tensile yield strength, and elongation at break of the different compositions are shown in Figure 7. The representative strain–stress curves of neat PHBV and PHBV/TPU/cellulose composites with and without the highest level of reactive agents (1 phr) are also represented for the sake of clarity.

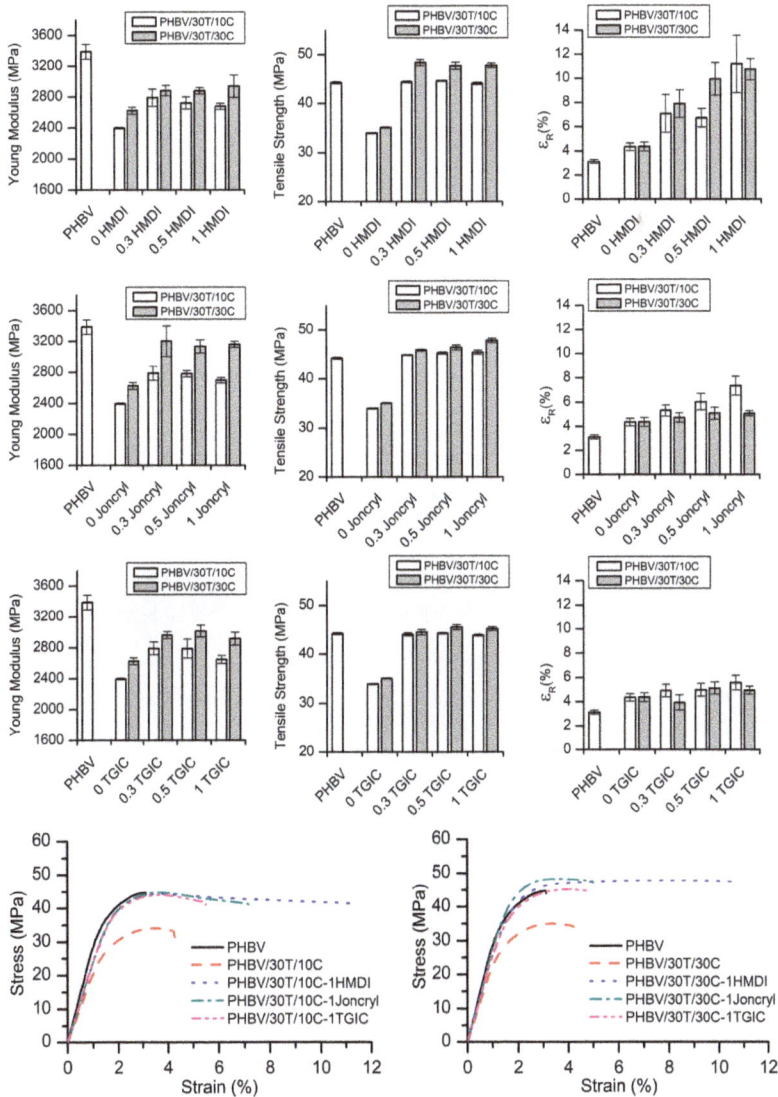

Figure 7. Mechanical properties of the neat PHBV and PHBV/TPU/Cellulose blends and representative stress–strain curves of neat PHBV, PHBV/30T/10C, and PHBV/30T/30C systems with and without 1 phr reactive agents.

PHBV presents a typical stiff and brittle mechanical performance, with high values of elastic modulus and tensile strength and low elongation at break (<5%). With respect to neat PHBV, the incorporation of TPU (30 phr) and cellulose (10 or 30 phr) leads to a reduction in the rigidity (about 25%) and the tensile strength (about 20%) and an enhancement in elongation at break (ca. 40%) and static toughness (25% and 32%, respectively) [43]. This increase in elongation at break is related with both the good distribution and small droplet size of the dispersed elastomeric phase (TPU) [19,20]. On the other hand, although there is a certain affinity among the phases, this limited interaction is not enough to ensure an efficient load transfer to the cellulose fibres. Without strong adhesion, the fibres detach at low deformation values, lowering the tensile strength of the compound and acting as stress concentrators for premature material failure [44]. With the addition of the reactive agents to the compounds, the elastic modulus and the tensile strength clearly increase with respect to the PHBV/TPU/cellulose without them. These parameters are strongly influenced by the matrix–fibre interaction and their improvement indicates a better load transfer due to an enhanced adhesion [30]. For the 10 phr cellulose compounds, the different reactive agents show a similar impact in these parameters, regardless of their content, supposing an improvement in the elastic modulus of about 15% and an increase in the tensile strength of around 30%. For the 30 phr cellulose compounds, the highest rise in the tensile modulus was obtained with the addition of Joncryl® (20% vs. ca. 10% for HMDI and TGIC). On the other hand, the tensile strength was improved by around 40%, 30%, and 35% with the HMDI, Joncryl®, and TGIC, respectively, reaching that of neat PHBV.

These results reveal that the three tested reactive agents are effective at improving the interfacial adhesion of the cellulose with the polymeric matrix and are in accordance with the SEM observations and the MFI values, pointing to an increased interaction among the phases [45]. This conclusion is in agreement with some other works that have been reported in the literature, on biopolyester–fibre composites compatibilized with diisocyanates [29–31], Joncryl® [35,36], or TGIC [38].

The biggest difference among the tested reactive agents with respect to their influence on the mechanical performance of the compounds is in the elongation at break. In all cases, this parameter was improved with respect to both neat PHBV and the compound without reactive agents. In Figure 6, looking at any PHBV/30T/10C compounds, as the reactive agent addition increases, the elongation at break rises too. On the other hand, in the case of the PHBV/30T/30C compounds, only the compounds with HMDI show an increase of elongation at break as the reactive agent content increases. This difference may point that the role of HMDI may not be the same as TGIC or Joncryl®.

In fact, the compounds with the highest TGIC level (1 phr) show an increase in elongation at break of ca. 28% with respect to the uncompatibilized PHBV/30T/10C system and 13% for PHBV/30T/30C. Similarly, the addition of 1 phr Joncryl® improved elongation at break by 70% in PHBV/30T/10C and 16% in PHBV/30T/30C. However, the compounds with 1 phr of HMDI showed an extraordinary enhancement of the elongation at break; the elongation at break was improved by 160% and 150% for compounds with 10 and 30 phr cellulose, respectively. Moreover, the static toughness (calculated from the area below the stress–strain curve) was enhanced by 320% and 340% with respect to the compound without reactive agents, and 420% and 450% with respect to neat PHBV.

2.2.2. Impact Resistance

Figure 8 summarizes the values obtained from unnotched and notched Charpy's impact tests, along with the static toughness from tensile tests of neat PHBV and the compounds with and without the reactive agents.

It is known that PHBV is very brittle and therefore it presents very low values of resilience in both unnotched and notched Charpy's impact tests and low static toughness, as shown in Figure 7. In this figure it can be observed that the compounds with TPU and cellulose clearly show an improvement in toughness resistance in the case of unnotched impact tests (Figure 8a,d,g) [43]. This improvement could be attributed to the positive role that the elastomeric TPU phase plays in absorbing impact energy [46]. However, concerning the notched tests (Figure 8b,e,h), there is no such increase in the impact energy absorbed, probably because there is a preferred crack propagation pathway through the matrix/fibre interfaces, where the adhesion is not very strong, as previously pointed out when discussing the variations in the elastic modulus of the compounds.

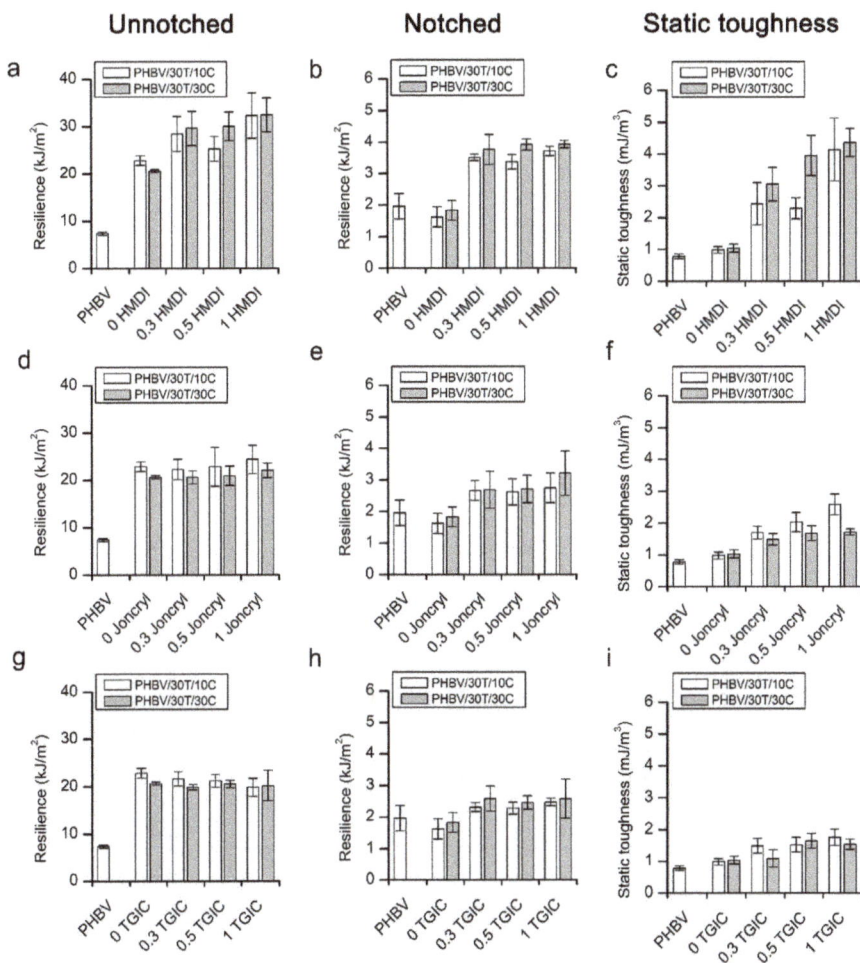

Figure 8. Charpy's impact results for unnotched specimens (**a,d,g**), notched specimens (**b,e,h**) and static toughness from the area below the strain–stress curve (**c,f,i**).

Nevertheless, the addition of reactive agents significantly improves the notched impact properties of the composites, which is in agreement with the static toughness determined from the area below the stress–strain curves (Figure 8c,f,i). This enhancement of the matrix–fibre interface adds to the effect of the small droplet size of the elastomeric phase that implies a low ligament thickness [42,47] and to by the enhanced interfacial interactions between TPU and PHBV [37], resulting in higher impact resistance in the presence of a notch.

When analysing the influence of the different reactive agents, the greatest increase in impact strength is obtained for composites with HMDI. For these composites, the absorbed impact energy was highly improved in both unnotched and notched impact tests, as well as in the static toughness (Figure 8a–c). Attending to the SEM micrographs of the impact fractured surfaces, with the HMDI addition (Figures 3 and 4) most of the fibres appear broken at their longitudinal direction, thus indicating a cohesive failure that confirms the presence of a very strong interface. This was not the case in compounds with the addition of TGIC or Joncryl®.

In polymer matrix composites, when there is a weak interface between the second phase and the polymeric matrix, the detachment of the particles during tensile deformations leads to the formation of flaws and voids at the interface of the fibre and the matrix. Those voids can coalesce and act as either crack initiators or provide a fast propagation crack pathway, which eventually leads to the premature failure of the material [13,48]. With a stronger particle–matrix adhesion, the possibility of growth and merge of those internal flaws is reduced, so there is an effective load transfer between the two phases, improving the fracture toughness [30]. It can be said that when the shear strength at the particle–matrix interface is higher than the shear yielding of any of the phases, plastic deformation of any of them can occur, thus increasing the energy absorbed. Thus, reactive agents can play different roles, increasing the adhesion between the PHBV/TPU, PHBV/cellulose and TPU/cellulose interfaces.

The impact performance of the compounds with HMDI stands out over the other ones. In this case, it seems that HMDI strongly increases the adhesion between PHBV and cellulose, which results in a synergetic effect with the addition of the TPU. The well-dispersed elastomeric phase decreases the yield strength of the polymer matrix and the strong interaction between the polymer and the fibres allows effective load transfer without producing flaws at the interfaces. Moreover, the exceptional mechanical performance of these compositions in terms of elongation at break also suggests that HMDI could play a positive role in enhancing the interfacial adhesion between PHBV and TPU. Indeed, diisocyanates have demonstrated effectiveness in improving the compatibility of biopolyester/TPU blends, as has been reported by Dogan et al. [33,34].

TGIC and Joncryl®, according to this reasoning, would not be so effective at enhancing the cellulose/PHBV interface, thus showing limited values of impact resistance, especially in the presence of a notch.

2.2.3. Heat Deflection Temperature HDT-A

The thermal resistance of neat PHBV and PHBV/TPU/cellulose composites was evaluated by means of heat deflection temperature (HDT-A) measurements. The results are grouped in Table 3.

Table 3. HDT-A values for neat PHBV and PHBV/TPU/cellulose systems with and without reactive agents.

Sample		HDT-A (°C)		
		0.3 phr	0.5phr	1 phr
PHBV	108 ± 1			
PHBV/30T/10C	94 ± 3			
PHBV/30T/10C + HMDI		93 ± 1	98 ± 3	95 ± 4
PHBV/30T/10C + Joncryl		95 ± 1	98 ± 3	97 ± 1
PHBV/30T/10C + TGIC		90 ± 3	90 ± 1	90 ± 3
PHBV/30T/30C	96 ± 1			
PHBV/30T/30C + HMDI		97 ± 2	99 ± 2	100 ± 1
PHBV/30T/30C + Joncryl		98 ± 4	97 ± 3	94 ± 1
PHBV/30T/30C + TGIC		104 ± 3	99 ± 3	99 ± 1

The PHBV presents a relatively high thermal resistance, showing an HDT-A value of 108 °C, in agreement with previously reported values [8,49,50]. The HDT values obtained for PHBV/30T/10C and PHBV/30T/30C are 94 and 96 °C, respectively. In spite of the relatively high content of the elastomeric additive (30 phr), the thermal resistance is not that much lower. This is due to the positive role played by the cellulose fibres in terms of reinforcement. As is widely reported in literature, in fibre-based polymer composites the restricted mobility of polymer chains in the presence of fibres leads to an increase in the dimensional stability and, thus, higher temperatures are required to deform them [50].

The use of reactive agents did have a significant influence on HDT values, but a trend was not seen with variation on their relative content. For the compounds with the lowest cellulose content (10 phr), HDT values ranged between 90 °C (TGIC) and 98 °C (Joncryl®), compared with a value of 94 °C for the compound without reactive agents. On the other hand, for the PHBV/30T/30C compounds, the thermal resistance was in almost all cases improved with HDT-A values around 100 °C (especially HMDI and TGIC), being the HDT value for the compound without reactive agents 96 °C.

These results are in agreement with the improved rigidity of the samples in the presence of reactive agents. Indeed, since there is no dependence of the HDT value on increasing the content of reactive agents, crosslinking reactions among the polymer chains can be discarded, thus indicating that the effect of the reactive agents is only a consequence of the compatibilization of the different phases. The increase in the HDT of fibre-based composites is therefore related to the reinforcement of the cellulose fibres [51] and, along with the use of the studied reactive agents, allows for enhancing the toughness and mechanical performance of PHBV without drastically decreasing its thermal resistance.

2.2.4. Biodisintegration in Composting Conditions

To explore the influence of cellulose content and the different reactive agents on the compostability of the PHBV/TPU/cellulose ternary systems, biodisintegration tests were conducted according to the ISO 20200 standard. The disintegration (weight loss) level over composting time is represented in Figure 9.

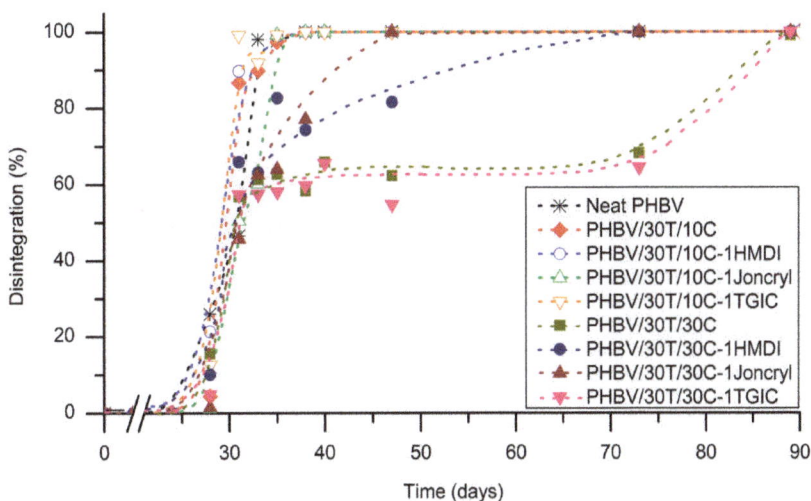

Figure 9. Disintegration in composting conditions of neat PHBV and PHB/TPU/cellulose systems with and without 1 phr reactive agents.

In general, all the compositions studied can be considered biodisintegrable in composting conditions according to ISO 20200. As shown in Figure 9, the PHBV disintegration process starts after an incubation period of 28 days. At this time the disintegration rate drastically increases to achieve total disintegration at 38 days of composting, in accordance with previous works [20,52,53]. No differences in the biodisintegration rate were detected for the PHBV/TPU/cellulose composites containing 10 phr cellulose, independent of the presence of reactive agents or the reactive agent type.

When the cellulose content was increased, the biodisintegration rate was, oddly, significantly reduced. To understand this occurrence, it must be taken into account how the samples were prepared and how biodisintegration takes place. For the composting tests, the specimens were obtained by hot pressing. Under the hot pressing conditions, the formation of a percolation mesh of interconnected cellulose fibres is favoured due to the high fibre–fibre affinity of the cellulose. This cellulose mesh is partially covered by TPU, which possess a low biodisintegration rate [20] with respect to PHBV and cellulose. Then, during the incubation time, a biofilm is formed at the surface of the testing specimen and the microbial advance occurs from the surface to the bulk, preferentially through the PHBV phase, as it is deduced by the stabilization of the disintegrated mass at around 60 wt % (approximately, the PHBV weight content) after 35 days of composting. We think that the TPU droplets, which take longer to biodisintegrate and are quite sticky at high temperature and moisture content, cover the fibres, limiting the access of the microbial advance to the cellulose.

This phenomenon causes a slowdown in the biodisintegration rate, but when the microorganisms have access to cellulose the weight loss rises rapidly and total disintegration is achieved within 90 days of composting. To validate this hypothesis, similar samples to those used for biodisintegration were placed for Soxhlet extraction of the PHBV phase with chloroform, and the resulting morphology analysed by SEM (Figure 10). TPU can be seen covering the fibres, partially confirming this reasoning.

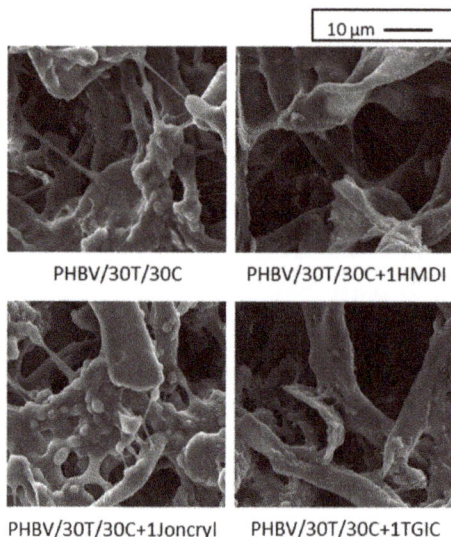

Figure 10. SEM micrographs of PHBV/30T/30C samples with and without 1 phr reactive agents after Soxhlet extraction.

Furthermore, PHBV/30T/30C composites showed different behaviour depending on the reactive agent added. The composition with TGIC presented a similar trend to the composition without reactive agent (that is, a slowdown at 60% weight loss); the ones with HMDI presented a fast biodisintegration of about 80% after 33 days of composting, reaching complete disintegration at day 73; and the composition with Joncryl® was totally degraded after 47 days. These differences in the biodisintegration rates among PHBV/30T/30C composites with different reactive agents could be influenced by the interactions of the reactive agents with the fibres, the PHBV, and the TPU. It is hypothesized that when there is a high interaction between the cellulose and the PHBV (promoted by the reactive agents), the microorganisms can access the cellulose more easily and the biodisintegration is completed earlier. On the contrary, when the PHBV–cellulose interaction is weak, the TPU can be easily located at the fibres surface, hindering the microbial advance. When looking at the SEM pictures of the compounds after Soxhlet extraction (Figure 10), there is more polymer covering the fibres in the case of no reactive agent or TGIC addition than in the case of the compounds with HMDI and Joncryl®, supporting the aforementioned reasoning.

3. Experimental

3.1. Materials

Poly(3-hydroxybutyrate-co-3-hydroxyvalerate) (PHBV) commercial grade with 3 wt % valerate content was purchased from Tianan Biologic Material Co. (Ningbo, China) in pellet form (ENMAT Y1000P). Thermoplastic polyurethane (TPU) Elastollan® 890 A 10FC was supplied by BASF (Ludwigshafen, Germany). Purified alpha-cellulose fibre grade (TC90) (alpha-cellulose content >99.5%) from CreaFill Fibers Corp. (Chestertown, MD, USA) was used. The reactive agents hexamethylene diisocyanate (HMDI) and triglycidyl isocyanurate (TGIC) were supplied by Sigma-Aldrich (Spain) and the Joncryl® 4368 was purchased from BASF (Ludwigshafen, Germany).

3.2. Sample Preparation

The PHBV and TPU used in this study were dried at 80 °C for at least 6 h in a DESTA DS06 HT dehumidifying dryer and the cellulose was dried in a lab oven (Memmert universal oven U, Schwabach, Germany) at 90 °C for a minimum of 16 h prior to the blending step, whilst the three reactive agents (HMDI, Joncryl®, and TGIC) were used as received.

PHBV/TPU/cellulose triple systems with different content of additives (TPU and cellulose) and reactive agents (HMDI, Joncryl®, and TGIC) (see Table 1) were prepared in a Labtech LTE (Samutprakarn, Thailand) (Ø = 26 mm, L/D ratio = 40) co-rotating twin-screw extruder. The temperature profile was set at 145/155/160/170 °C from hopper to nozzle, the rotation speed was 250 rpm, and the feeding speed was about 5 kg/h. All the components were manually dry-mixed before extrusion except HMDI, which was dispensed at the feeding zone by means of a peristaltic pump (Watson Marlow 120 S/R, Sondika, Spain). The extruded material was cooled in a water bath and pelletized (MAAG PRIMO S pelletizer, Stuttgart, Germany).

Material pellets were dried again at 80 °C for 8 h (DESTA DS06 HT) before the injection process. Standardized tensile specimens (ISO-527 Type 1A) were injection-moulded in a DEMAG IntElect 100 T injection moulding machine (Schwaig, Germany) with an injection temperature of 185 °C at the nozzle. A holding pressure of 600 bars was applied for 12 s, followed by 40 s of cooling time. For the sake of comparison, neat PHBV was also processed under identical conditions.

Prior to any characterization, all the samples were annealed at 80 °C for 48 h in order to obtain equivalent crystallinity and mechanical performance to aged samples.

3.3. Characterization

The melt flow index (MFI) of the different compounds was measured in a Tinius Olsen MP600 (Surrey, England) melt flow indexer according to the ISO 1133 standard. The tests were performed at 185 °C and 2.16 kg load.

The morphology of the PHBV/30T/10C and PHBV/30T/30C triple systems with and without reactive agents (HMDI, Joncryl® and TGIC) was examined by scanning electron microscopy (SEM) using a high-resolution field-emission JEOL 7001F microscope (Japan). The fracture surfaces from impact-fractured specimens were previously coated by sputtering with a thin layer of Pt. From selected representative SEM images (at 2500× magnification), the diameters of the droplets corresponding to the dispersed phase were measured using Fiji® software (ImageJ 1.51j8). The number of droplets measured in all cases was higher than 600. From the individual measures, the following parameters were determined: the average droplet size (*d*) and the droplet size distribution parameters d10, d50, and d90 (corresponding to the size where 10%, 50%, and 90% of the droplets are included, respectively). The matrix ligament thickness (T) was also estimated, according to Wu's equation [42]:

$$T = d\left[\left(\frac{\pi}{6\varphi_r}\right)^{\frac{1}{3}} - 1\right], \tag{1}$$

where d is the average domain size of the dispersed phase and φ_r is the volume fraction of the dispersed phase, determined as follows:

$$\varphi_r = \frac{\rho_m\, w_r}{(\rho_r w_m + \rho_m w_r)}, \tag{2}$$

where ρ_m and ρ_r are the densities of the matrix and dispersed phases, respectively, and w_m and w_r are their weight fractions.

Tensile tests were conducted on ISO-527 type 1A injection-moulded specimens in a Hounsfield H25K universal testing machine (Surrey, England) equipped with a 25 kN load cell according to the ISO-527-1:2012 standard.

Notched and unnotched Charpy impact tests were carried out by means of an ATS faar IMPats-15 (Segrate, Italy) impact pendulum with a 4 J hammer according to the ISO 179 standard. Samples were cut from injection-moulded bars.

Heat deflection temperature (HDT) analyses were performed using a Deflex 687-2 (Barcelona, Spain). A heating rate of 120 °C/h was used with an applied load of 1.8 MPa in accordance with Method A of ISO 75 standard. The temperature was recorded until the sample deflects 0.35 mm.

Biodisintegration tests were carried out with samples (15 × 15 × 0.2 mm³) obtained from hot-pressed plates (180 °C, 5 min, and ca. 40 bar). Tests were performed according to the ISO 20200 standard [54]. Solid synthetic waste was prepared by mixing 10% of activated mature compost (VIGORHUMUS H-00, purchased from Buras Profesional, S.A., Girona, Spain), 40% sawdust, 30% rabbit feed, 10% corn starch, 5% sugar, 4% corn seed oil, and 1% urea. The water content of the mixture was adjusted to 55%. The samples were placed inside mesh bags to simplify their extraction and allow the contact of the compost with the specimens, and then were buried in compost bioreactors at 4–6 cm depth. Bioreactors were incubated at 58 °C. The aerobic conditions were guaranteed by periodically mixing the synthetic waste and adding water according to the standard requirements. Three replicates of each sample were removed from the boxes at different composting times for analysis. Samples were washed with water and dried under a vacuum at 40 °C until a constant mass. The disintegration degree was calculated by normalizing the sample weight to the initial weight with Equation (3):

$$D = \frac{m_i - m_f}{m_i} \times 100 \tag{3}$$

where m_i is the initial dry mass of the test material and m_f is the dry mass of the test material recovered at different incubation stages. Moreover, the morphology of the films prepared for the composting tests was analysed by SEM after the Soxhlet extraction with chloroform of the PHBV phase.

4. Conclusions

In this study three reactive agents used in reactive extrusion (HMDI, Joncryl®, and TGIC) were tested in PHBV/TPU/cellulose for injection moulding applications that require biodisintegration in composting conditions. The influence of the cellulose content, the reactive agent type and the reactive agent content, were analysed. It was observed that the incorporation of TPU and cellulose in PHBV led to a reduction in the tensile elastic modulus and tensile strength, but an enhancement in elongation at break, with an overall increase in static toughness attributed to the toughening effect of the TPU. However, the addition of the reactive agents to the compounds resulted in a rise in the tensile strength and elastic modulus up to values close to or higher than neat PHBV and an increase in the value of strain at break with respect to the compounds without reactive agents.

In terms of impact resistance, the addition of the reactive agents improved the toughness of the compounds in notched and unnotched configurations. Furthermore, even though the TPU in the compounds causes a decrease in the thermal strength with respect to neat PHBV, the addition of cellulose up to 30 phr with the reactive agents was able to moderate this drop.

Those results indicate that the reactive agents play a main role as compatibilizers among the phases of the PHBV/TPU/cellulose compounds. HMDI showed the highest ability to compatibilize the cellulose and the PHBV in the compounds, with the topmost values of deformation at break and static toughness. Joncryl® and TGIC, on the other hand, seemed to enhance the compatibility between the fibres and the polymer matrix as well as the TPU within the PHBV.

The findings of this work point to a route to modify the properties of PHBV (and PHAs in general) through blending with reactive agents, which can help to overcome some of the difficulties that these materials encounter in standard applications.

Author Contributions: Formal analysis, E.L.S.-S.; Funding acquisition, L.C. and J.G.-P.; Investigation, E.L.S.-S.; Methodology, E.L.S.-S., A.A., J.A., L.C. and J.G.-P.; Resources, A.A. and J.A.; Supervision, J.G.-P.; Writing—original draft, E.L.S.-S.; Writing—review & editing, A.A., J.A., L.C. and J.G.-P.

Acknowledgments: The authors acknowledge financial support for this research from the Ministerio de Economia y Competitividad (AGL2015-63855-C2-2-R) and the Pla de Promoció de la Investigació de la Universitat Jaume I (UJI-B2016-35). The authors thank Servicios Centrales de Instrumentación (SCIC) of Universitat Jaume I for the use of SEM. We are also grateful to Raquel Oliver and Jose Ortega for experimental support.

Conflicts of Interest: The authors declare no conflict of interest.

References

1. Plastics Europe-Association of Plastics Manufacturers. Plastics-the Facts 2017: Analysis of European Plastics Production, Demand and Waste Data. Available online: http://www.plasticseurope.org/en/resources/publications/plastics-facts-2017 (accessed on 2 February 2018).

2. Rujnić-Sokele, M.; Pilipović, A. Challenges and opportunities of biodegradable plastics: A mini review. *Waste Manag. Res.* **2017**, *35*, 132–140. [CrossRef] [PubMed]

3. Cunha, M.; Fernandes, B.; Covas, J.A.; Vicente, A.A.; Hilliou, L. Film blowing of PHBV blends and PHBV-based multilayers for the production of biodegradable packages. *J. Appl. Polym. Sci.* **2016**, *133*. [CrossRef]

4. Laycock, B.; Halley, P.; Pratt, S.; Werker, A.; Lant, P. The chemomechanical properties of microbial polyhydroxyalkanoates. *Prog. Polym. Sci.* **2013**, *38*, 536–583. [CrossRef]

5. Albuquerque, P.B.S.; Malafaia, C.B. Perspectives on the production, structural characteristics and potential applications of bioplastics derived from polyhydroxyalkanoates. *Int. J. Biol. Macromol.* **2018**, *107*, 615–625. [CrossRef] [PubMed]

6. Wang, Y.; Yin, J.; Chen, G.Q. Polyhydroxyalkanoates, challenges and opportunities. *Curr. Opin. Biotechnol.* **2014**, *30*, 59–65. [CrossRef] [PubMed]

7. Pilla, S. *Handbook of Bioplastics and Biocomposites Engineering Applications*; John Wiley & Sons: Hoboken, NJ, USA, 2011; ISBN 9780470626078.

8. Peelman, N.; Ragaert, P.; Ragaert, K.; De Meulenaer, B.; Devlieghere, F.; Cardon, L. Heat resistance of new biobased polymeric materials, focusing on starch, cellulose, PLA, and PHA. *J. Appl. Polym. Sci.* **2015**, *132*. [CrossRef]

9. Keskin, G.; Kızıl, G.; Bechelany, M.; Pochat-Bohatier, C.; Öner, M. Potential of polyhydroxyalkanoate (PHA) polymers family as substitutes of petroleum based polymers for packaging applications and solutions brought by their composites to form barrier materials. *Pure Appl. Chem.* **2017**, *89*, 1841–1848. [CrossRef]

10. Bugnicourt, E.; Cinelli, P.; Lazzeri, A.; Alvarez, V. Polyhydroxyalkanoate (PHA): Review of synthesis, characteristics, processing and potential applications in packaging. *Express Polym. Lett.* **2014**, *8*, 791–808. [CrossRef]

11. Väisänen, T.; Haapala, A.; Lappalainen, R.; Tomppo, L. Utilization of agricultural and forest industry waste and residues in natural fiber-polymer composites: A review. *Waste Manag.* **2016**, *54*, 62–73. [CrossRef] [PubMed]

12. Jost, V.; Miesbauer, O. Effect of different biopolymers and polymers on the mechanical and permeation properties of extruded PHBV cast films. *J. Appl. Polym. Sci.* **2018**, *135*, 46153. [CrossRef]
13. Zhang, K.; Misra, M.; Mohanty, A.K. Toughened sustainable green composites from poly(3-hydroxybutyrate-co-3-hydroxyvalerate) based ternary blends and miscanthus biofiber. *ACS Sustain. Chem. Eng.* **2014**, *2*, 2345–2354. [CrossRef]
14. Chikh, A.; Benhamida, A.; Kaci, M.; Pillin, I.; Bruzaud, S. Synergistic effect of compatibilizer and sepiolite on the morphology of poly(3-hydroxybutyrate-co-3-hydroxyvalerate)/poly(butylene succinate) blends. *Polym. Test.* **2016**, *53*, 19–28. [CrossRef]
15. Ma, P.; Hristova-Bogaerds, D.G.; Lemstra, P.J.; Zhang, Y.; Wang, S. Toughening of PHBV/PBS and PHB/PBS Blends via In situ Compatibilization Using Dicumyl Peroxide as a Free-Radical Grafting Initiator. *Macromol. Mater. Eng.* **2012**, *297*, 402–410. [CrossRef]
16. El-Taweel, S.H.; Khater, M. Mechanical and Thermal Behavior of Blends of Poly(hydroxybutyrate-co-hydroxyvalerate) with Ethylene Vinyl Acetate Copolymer. *J. Macromol. Sci. Part. B* **2015**, *54*, 1225–1232. [CrossRef]
17. Adams, B.; Abdelwahab, M.; Misra, M.; Mohanty, A.K. Injection-Molded Bioblends from Lignin and Biodegradable Polymers: Processing and Performance Evaluation. *J. Polym. Environ.* **2017**, 1–14. [CrossRef]
18. Wang, S.; Chen, W.; Xiang, H.; Yang, J.; Zhou, Z.; Zhu, M. Modification and Potential Application of Short-Chain-Length Polyhydroxyalkanoate (SCL-PHA). *Polymers* **2016**, *8*, 273. [CrossRef]
19. González-Ausejo, J.; Sánchez-Safont, E.; Cabedo, L.; Gamez-Perez, J. Toughness Enhancement of Commercial Poly (Hydroxybutyrate-co-Valerate) (PHBV) by Blending with a Thermoplastic Polyurethane (TPU). *J. Multiscale Model.* **2016**, *7*, 1640008. [CrossRef]
20. Martínez-Abad, A.; González-Ausejo, J.; Lagarón, J.M.; Cabedo, L. Biodegradable poly(3-hydroxybutyrate-co-3-hydroxyvalerate)/thermoplastic polyurethane blends with improved mechanical and barrier performance. *Polym. Degrad. Stab.* **2016**, *132*, 52–61. [CrossRef]
21. Bhardwaj, R.; Mohanty, A.K.; Drzal, L.T.; Pourboghrat, F.; Misra, M. Renewable resource-based green composites from recycled cellulose fiber and poly(3-hydroxybutyrate-co-3-hydroxyvalerate) bioplastic. *Biomacromolecules* **2006**, *7*, 2044–2051. [CrossRef] [PubMed]
22. Sánchez-Safont, E.L.; Aldureid, A.; Lagarón, J.M.; Gámez-Pérez, J.; Cabedo, L. Biocomposites of different lignocellulosic wastes for sustainable food packaging applications. *Compos. Part. B Eng.* **2018**, *145*, 215–225. [CrossRef]
23. Satyanarayana, K.G.; Arizaga, G.G.C.; Wypych, F. Biodegradable composites based on lignocellulosic fibers—An overview. *Prog. Polym. Sci.* **2009**, *34*, 982–1021. [CrossRef]
24. Pereira, P.H.F.; Rosa, M.D.F.; Cioffi, M.O.H.; Benini, K.C.C.D.C.; Milanese, A.C.; Voorwald, H.J.C.; Mulinari, D.R. Vegetal fibers in polymeric composites: A review. *Polímeros* **2015**, *25*, 9–22. [CrossRef]
25. Wei, L.; McDonald, A. A Review on Grafting of Biofibers for Biocomposites. *Materials* **2016**, *9*, 303. [CrossRef] [PubMed]
26. Misra, M.; Pandey, J.K.; Mohanty, A.K. *Biocomposites: Design and Mechanical Performance*; Elsevier Inc.: Amsterdam, The Netherlands, 2015; ISBN 9781782423942.
27. Muthuraj, R.; Misra, M.; Mohanty, A.K. Biodegradable compatibilized polymer blends for packaging applications: A literature review. *J. Appl. Polym. Sci.* **2018**, *135*, 45726. [CrossRef]
28. Stenstad, P.; Andresen, M.; Tanem, B.S.; Stenius, P. Chemical surface modifications of microfibrillated cellulose. *Cellulose* **2008**, *15*, 35–45. [CrossRef]
29. Jiang, L.; Chen, F.; Qian, J.; Huang, J.; Wolcott, M.; Liu, L.; Zhang, J. Reinforcing and Toughening Effects of Bamboo Pulp Fiber on Poly(3-hydroxybutyrate-co-3-hydroxyvalerate) Fiber Composites. *Ind. Eng. Chem. Res.* **2010**, *49*, 572–577. [CrossRef]
30. Zarrinbakhsh, N.; Mohanty, A.K.; Misra, M. Improving the interfacial adhesion in a new renewable resource-based biocomposites from biofuel coproduct and biodegradable plastic. *J. Mater. Sci.* **2013**, *48*, 6025–6038. [CrossRef]
31. Anderson, S.; Zhang, J.; Wolcott, M.P. Effect of Interfacial Modifiers on Mechanical and Physical Properties of the PHB Composite with High Wood Flour Content. *J. Polym. Environ.* **2013**, *21*, 631–639. [CrossRef]
32. González-Ausejo, J.; Sánchez-Safont, E.; Lagarón, J.M.; Balart, R.; Cabedo, L.; Gámez-Pérez, J. Compatibilization of poly(3-hydroxybutyrate-co-3-hydroxyvalerate)–poly(lactic acid) blends with diisocyanates. *J. Appl. Polym. Sci.* **2017**, *134*, 1–11. [CrossRef]

33. Dogan, S.K.; Reyes, E.A.; Rastogi, S.; Ozkoc, G. Reactive compatibilization of PLA/TPU blends with a diisocyanate. *J. Appl. Polym. Sci.* **2014**, *131*. [CrossRef]

34. Dogan, S.K.; Boyacioglu, S.; Kodal, M.; Gokce, O.; Ozkoc, G. Thermally induced shape memory behavior, enzymatic degradation and biocompatibility of PLA/TPU blends: "Effects of compatibilization". *J. Mech. Behav. Biomed. Mater.* **2017**, *71*, 349–361. [CrossRef] [PubMed]

35. Hao, M.; Wu, H.; Qiu, F.; Wang, X. Interface Bond Improvement of Sisal Fibre Reinforced Polylactide Composites with Added Epoxy Oligomer. *Materials* **2018**, *11*, 398. [CrossRef] [PubMed]

36. Nanthananon, P.; Seadan, M.; Pivsa-Art, S.; Hiroyuki, H.; Suttiruengwong, S. Biodegradable polyesters reinforced with eucalyptus fiber: Effect of reactive agents. In *AIP Conference Proceedings*; AIP Publishing: Melville, NY, USA, 2017; p. 70012.

37. Tang, W.; Wang, H.; Tang, J.; Yuan, H. Polyoxymethylene/thermoplastic polyurethane blends compatibilized with multifunctional chain extender. *J. Appl. Polym. Sci.* **2013**, *127*, 3033–3039. [CrossRef]

38. Hao, M.; Wu, H. Effect of in situ reactive interfacial compatibilization on structure and properties of polylactide/sisal fiber biocomposites. *Polym. Compos.* **2018**, *39*, E174–E187. [CrossRef]

39. Ferrero, B.; Fombuena, V.; Fenollar, O.; Boronat, T.; Balart, R. Development of natural fiber-reinforced plastics (NFRP) based on biobased polyethylene and waste fibers from Posidonia oceanica seaweed. *Polym. Compos.* **2015**, *36*, 1378–1385. [CrossRef]

40. Hameed, N.; Guo, Q.; Tay, F.H.; Kazarian, S.G. Blends of cellulose and poly(3-hydroxybutyrate-co-3-hydroxyvalerate) prepared from the ionic liquid 1-butyl-3-methylimidazolium chloride. *Carbohydr. Polym.* **2011**, *86*, 94–104. [CrossRef]

41. Mechanical and biodegradation performance of short natural fibre polyhydroxybutyrate composites. *Polym. Test.* **2013**, *32*, 1603–1611.

42. Wu, S. Phase structure and adhesion in polymer blends: A criterion for rubber toughening. *Polymer* **1985**, *26*, 1855–1863. [CrossRef]

43. Sánchez-Safont, E.L.; Arrillaga, A.; Anakabe, J.; Gamez-Perez, J.; Cabedo, L. PHBV/TPU/Cellulose compounds for compostable injection molded parts with improved thermal and mechanical performance. *J. Appl. Polym. Sci.*. submitted.

44. Seggiani, M.; Cinelli, P.; Mallegni, N.; Balestri, E.; Puccini, M.; Vitolo, S.; Lardicci, C.; Lazzeri, A. New Bio-Composites Based on Polyhydroxyalkanoates and Posidonia oceanica Fibres for Applications in a Marine Environment. *Materials* **2017**, *10*, 326. [CrossRef] [PubMed]

45. Yatigala, N.S.; Bajwa, D.S.; Bajwa, S.G. Compatibilization improves physico-mechanical properties of biodegradable biobased polymer composites. *Compos. Part. A Appl. Sci. Manuf.* **2018**, *107*, 315–325. [CrossRef]

46. Wang, S.; Xiang, H.; Wang, R.; Peng, C.; Zhou, Z.; Zhu, M. Morphology and properties of renewable poly(3-hydroxybutyrate-co-3-hydroxyvalerate) blends with thermoplastic polyurethane. *Polym. Eng. Sci.* **2014**, *54*, 1113–1119. [CrossRef]

47. Margolina, A.; Wu, S. Percolation model for brittle-tough transition in nylon/rubber blends. *Polymer* **1988**, *29*, 2170–2173. [CrossRef]

48. Nagarajan, V.; Misra, M.; Mohanty, A.K. New engineered biocomposites from poly(3-hydroxybutyrate-co-3-hydroxyvalerate) (PHBV)/poly(butylene adipate-co-terephthalate) (PBAT) blends and switchgrass: Fabrication and performance evaluation. *Ind. Crops Prod.* **2013**, *42*, 461–468. [CrossRef]

49. Muthuraj, R.; Misra, M.; Mohanty, A.K. Reactive compatibilization and performance evaluation of miscanthus biofiber reinforced poly(hydroxybutyrate-co-hydroxyvalerate) biocomposites. *J. Appl. Polym. Sci.* **2017**, *134*. [CrossRef]

50. Rossa, L.V.; Scienza, L.C.; Zattera, A.J. Effect of curauá fiber content on the properties of poly(hydroxybutyrate-co-valerate) composites. *Polym. Compos.* **2013**, *34*, 450–456. [CrossRef]

51. Buchdahl, R. Mechanical properties of polymers and composites—Vols. I and II, Lawrence, E. Nielsen, Marcel Dekker, Inc.; New York, 1974, Vol. I 255 pp. Vol. II 301 pp. Vol. I $24.50, Vol. II $28.75. *J. Polym. Sci. Polym. Lett. Ed.* **1975**, *13*, 120–121. [CrossRef]

52. González-Ausejo, J.; Sanchez-Safont, E.; Lagaron, J.M.; Olsson, R.T.; Gamez-Perez, J.; Cabedo, L. Assessing the thermoformability of poly(3-hydroxybutyrate-co-3-hydroxyvalerate)/poly(acid lactic) blends compatibilized with diisocyanates. *Polym. Test.* **2017**, *62*, 235–245. [CrossRef]

53. Sánchez-Safont, E.L.; González-Ausejo, J.; Gámez-Pérez, J.; Lagarón, J.M.; Cabedo, L. Poly(3-Hydroxybutyrate-co-3-Hydroxyvalerate)/Purified Cellulose Fiber Composites by Melt Blending: Characterization and Degradation in Composting Conditions. *J. Renew. Mater.* **2016**, *4*, 123–132. [CrossRef]
54. ISO. *Determinación del Grado de Desintegración de Materiales Plásticos Bajo Condiciones de Compostaje Simuladas en un Laboratorio*; UNE-EN ISO UNE-EN ISO 20200; ISO: Geneva, Switzerland, 2006.

© 2018 by the authors. Licensee MDPI, Basel, Switzerland. This article is an open access article distributed under the terms and conditions of the Creative Commons Attribution (CC BY) license (http://creativecommons.org/licenses/by/4.0/).

International Journal of
Molecular Sciences

MDPI

Article

Structure/Function Analysis of Cotton-Based Peptide-Cellulose Conjugates: Spatiotemporal/Kinetic Assessment of Protease Aerogels Compared to Nanocrystalline and Paper Cellulose

J. Vincent Edwards [1,*], Krystal Fontenot [1], Falk Liebner [2], Nicole Doyle nee Pircher [2], Alfred D. French [1] and Brian D. Condon [1]

[1] Southern Regional Research Center, USDA, New Orleans, LA 70124, USA;
    krystal.fontenot@ars.usda.gov (K.F.); Al.French@ars.usda.gov (A.D.F.); brian.condon@ars.usda.gov (B.D.C.)
[2] University of Natural Resources and Life Sciences Vienna, Konrad-Lorenz-Straße 24,
    A-3430 Tulln an der Donau, Austria; falk.liebner@boku.ac.at (F.L.); nicole.pircher@boku.ac.at (N.D.n.P.)
* Correspondence: vince.edwards@ars.usda.gov; Tel.: +1-504-286-4360

Received: 2 January 2018; Accepted: 25 February 2018; Published: 13 March 2018

Abstract: Nanocellulose has high specific surface area, hydration properties, and ease of derivatization to prepare protease sensors. A Human Neutrophil Elastase sensor designed with a nanocellulose aerogel transducer surface derived from cotton is compared with cotton filter paper, and nanocrystalline cellulose versions of the sensor. X-ray crystallography was employed along with Michaelis–Menten enzyme kinetics, and circular dichroism to contrast the structure/function relations of the peptide-cellulose conjugate conformation to enzyme/substrate binding and turnover rates. The nanocellulosic aerogel was found to have a cellulose II structure. The spatiotemporal relation of crystallite surface to peptide-cellulose conformation is discussed in light of observed enzyme kinetics. A higher substrate binding affinity (K_m) of elastase was observed with the nanocellulose aerogel and nanocrystalline peptide-cellulose conjugates than with the solution-based elastase substrate. An increased K_m observed for the nanocellulosic aerogel sensor yields a higher enzyme efficiency (k_{cat}/K_m), attributable to binding of the serine protease to the negatively charged cellulose surface. The effect of crystallite size and β-turn peptide conformation are related to the peptide-cellulose kinetics. Models demonstrating the orientation of cellulose to peptide O6-hydroxymethyl rotamers of the conjugates at the surface of the cellulose crystal suggest the relative accessibility of the peptide-cellulose conjugates for enzyme active site binding.

Keywords: nanocellulose; protease sensor; human neutrophil elastase; peptide-cellulose conformation; aerogel

1. Introduction

Nanocellulose-Based Protease Sensors

The sophistication and sensitivity of point of care diagnostic detectors underlies much of the progress in the rapidly growing field of biosensors as is found with point of care protease sensors [1]. The sensor's transducer serves to ground and facilitate transformation of the sensor molecule (in this study, a peptide fluorophore) into a signature signal, which in this study is a fluorescence response indicating proteolytic activity. However, the composition of the transducer surface material and its interaction at the biomolecular interface (protease binding on the transducer surface) is of equal value. Another important consideration is the detectors interface with the biological environment, which can complicate and dampen the sensitivity of detection. This is especially the case when considering the design of in situ protease sensors interfaced with dressings to detect harmful levels of proteolytic enzymes.

Proteolytic enzymes including, matrix metalloproteases (MMPs) and serine proteases such as the neutrophil serine protease (human neutrophil elastase, HNE), are responsible for proteolytic degradation of growth factors [2] and the extracellular matrix (ECM) proteins in chronic wounds. In general, acute wounds have normal levels of MMPs and HNE [3–5], which facilitates the clearance of cellular debris. However, chronic wounds have elevated levels of MMPs (0.1–0.2 U/mL) and HNE (0.02–0.1 U/mL) depending on the type, e.g., diabetic, venous, pressure, and arterial ulcers [5]. Thus, elevated concentrations of MMPs and HNE delay the healing process, and are also considered biomarkers for chronic wound treatment evaluation [6–8].

Biosensor materials should be compatible with the complexity of the bio-system they are applied to. This is especially relevant to the chronic wound environment, which varies in exudative fluid levels, bacterial burden, and biochemical diversity. Biocompatibility is imparted to resist nonselective surface adsorption by lipids, proteins, polysaccharides, cellular debris, and breakdown of the transducer surface. Due to cellulosic's hydrophilic properties, protein binding on the transducer surface for sensor detection is relatively attenuated when compared with more hydrophobic materials.

We have recently examined the relative contributions of cellulosic and nanocellulosic sensor transducer surfaces to detection sensitivity of the neutrophil protease, HNE, demonstrating that different forms of cellulose and nanocellulose materials have varying surface area, porosity, and biocompatible properties affording sensor/transducer surface functionality [9–14].

Notably it is worth considering that nanocellulose is a highly crystalline biopolymer with a hydrophilic and high specific surface area that possesses reactive hydroxyls, which can be derivatized to covalently append a wide range of biologically active molecules. Here, the focus is on comparison of the relative structure/function contributions of three different cotton transducer surfaces (nanocellulose crystals (CNC), nanocellulosic aerogels (NA), and cellulose filter paper (CFP) (Figure 1), each having unique properties contributing to protease sensor sensitivity, yet varying features of sensor and dressing applicability. The sensor molecule is a tripeptide substrate of elastase. Accordingly, we examine here a nanocellulosic aerogel material in comparison with two other cellulosic and nanocellulosic designs for protease-lowering activity, and sensor/transducer surface performance with the aim of understanding how the molecular and physical properties of the materials may provide benefit for a dressing/sensor platform useful to simultaneously test and treat targeted elevated protease levels in chronic wounds.

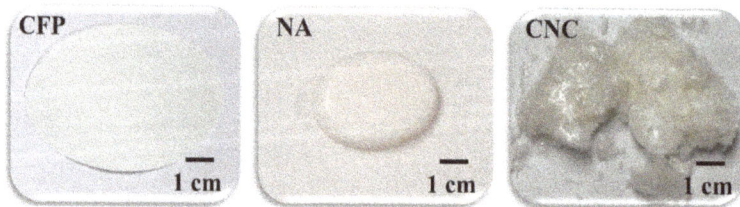

Figure 1. Images of the cellulosic (CFP) and nanocellulosic (NA and CNC) transducers.

In this study, we evaluate the relative structure/function contributions of the protease substrate, which is immobilized to a nanocellulosic aerogel. The structure function approach analyzes the enzyme kinetic turnover rate and binding affinity of the enzyme to structural findings of the crystal structure that influences the immobilized peptide in a β turn conformation on the cellulose crystallite surface in its interaction with the protease. Accordingly, the structure/function relations of two nanocellulosic and a cellulosic sensor analogs are contrasted, and also demonstrate interesting differences in protease enzymology.

2. Results

2.1. Structure/Function and Physical Property Considerations

A depiction of the physical structure of the materials of this study is shown in Figure 1. The intended application of the nanocellulose aerogel sensor is as an interface with chronic wound dressings to detect protease levels in situ. The functional properties of a protease sensor are determined by the composition of the transducer surface, the bioactivity of the enzyme substrate attached, and how the substrate is immobilized. A cellulose-based transducer surface, to which the peptide protease substrate is covalently attached, may be characterized for its crystalline structure and physical property composition [13,15]. In this study, conjugation of the tripeptide elastase substrate (Suc-Ala-Pro-Ala-AMC) to the cellulosic and nanocellulosic transducers via the amide bond formation between the carboxyl terminus of the tripeptide and the reactive hydroxyls on the transducer surfaces afforded the peptide-cellulose conjugates, pCFP, pNA and pCNC, which has previously been outlined [9,12].

2.2. Structure/Function Relations of the Protease Activity at a Molecular Level

2.2.1. Cellulose Crystal Structure

X-ray crystallography of the different transducer surfaces was undertaken and used to compare the relative contributions of cellulose structure to the peptide-cellulose conjugates and structural relationships of the cellulosic transducer surfaces to enzyme kinetics.

Figure 2 shows patterns from the X-ray diffraction (XRD) analyses. In the case of the CFP and the CNC, which are cotton based materials, the patterns were typical cellulose I_β patterns. The pattern from the aerogel transducer, seemingly a mostly amorphous material, is also shown in Figure 2A. That pattern was subjected to Rietveld analyses to do what was reasonable to confirm the initial diagnosis of amorphous material. The results of the Rietveld analysis are presented in Figure 2B.

Table 1 shows the larger values of Segal CrI (Crystallinity Index) and crystallite size for the CFP, compared to CNC. Apparently, the acid hydrolysis of the cotton reduced somewhat the size of the crystallites.

Table 1. The crystallinity index values, crystallite size values, and crystallite model for the cellulosic filter paper (CFP), cotton nanocrystalline cellulose (CNC), and nanocellulosic aerogel (NA) transducer.

Name	Segal CrI (%)	Crystallite Size (Å)	Crystallite Model	Layers Molecules	Surface Charge (mV)
CFP	87	69.3	18	144	−10.2
NC	79	54.3	14	98	−68
NA I_β	-	26	6	24	−19
NA II	-	65	16	122	−19
NA amorphous	-	19	5	24	−19

A more careful visual analysis of the NA pattern indicates a mixture of cellulose I, II and amorphous structures. [16]. The indications are the small peak at about 12° 2θ (indicating II) as well as the convex intensity in the area of 15°–17° 2θ which indicated some contributions from cellulose I. The presence of some crystalline material was supported by the poor fit of a calculated pattern from very small cellulose II crystals. Earlier, Nam et al. [17] found that such patterns with a very broad peak closely resembled exhaustively ball-milled cotton. Contributions of calculated patterns from a small cellulose I crystal improved the fit, as did one from a relatively large cellulose II crystal. Ultimately, a close fit was obtained between the calculated and experimental patterns, but it was necessary to include a substantial contribution from preferred orientation to obtain the fit depicted in

Figure 2B. Although the analysis indicated 44% of the small cellulose I crystallites, 48% of amorphous as modeled by small cellulose II crystallites, and 8% of large crystallites of II, these values should be taken as indications of the presence of three phases, not precise values. In addition, it is notable that crystallinity index values, i.e., Segal CrI (shown in Table 1) for CFP and CNC, are consistent with previously determined values for filter paper and cotton nanocrystalline cellulose. Such analyses are difficult; to get a visually satisfactory fit, 17 variables were used in the refinement with the most severe being the March–Dollase compensation for preferred orientation applied to the cellulose I component. This distorted the cellulose I pattern in Figure 2A. Because of the limited, somewhat noisy data, the obtained fit is not likely to be unique; other models could also give plausible analyses.

It is noted that these values for the composition were calculated from the exported data from the MAUD (materials analysis using diffraction) Rietveld software, not the values reported during the analysis. However, indications available are that there is considerably more surface area available for the NA samples than for the CFP and CNC materials because of the mostly smaller crystallites of NA.

The presence of the cellulose II crystallites is in good agreement with a previous study comparing the nanomorphology of various types of cellulose II aerogels obtained by dissolution-coagulation of cotton linters using different cellulose solvent systems. For calcium thiocyanate octahydrate/lithium chloride, i.e., the solvent used to prepare the three-dimensional NA scaffolds in this study, small angle X-ray spectroscopy had revealed considerably high cellulose II crystallinity (ca. 50%) and a crystal diameter of 46 ± 1 Å [18]. Notably, the crystallite size impacts the sensitivity of the sensor by increasing the specific surface area of the sensor which is correlated with sensor sensitivity [12].

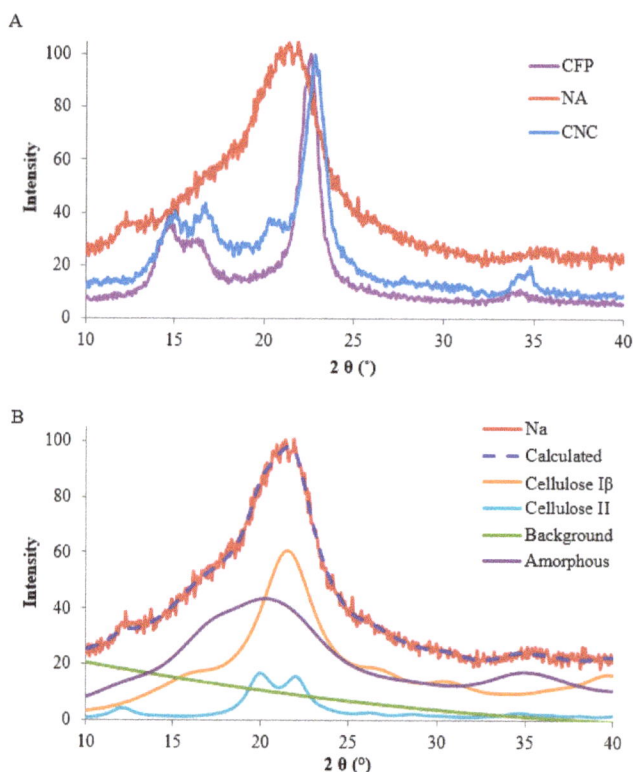

Figure 2. X-ray observed diffraction patterns (**A**) of the CFP, NA, and NC transducers; and (**B**) the calculated pattern of the NA as analyzed using the Rietveld method.

The crystallite models (shown in Figure 3) that are based on the respective cellulose crystallite sizes reveal the relative number of cellulose layers and chains which are CFP (18 layers and 144 chains), NA (12 layers and 78 chains), and CNC (14 layers and 98 chains). The number of chains (36, 28, and 24) on the 1–10 and 110 surfaces for the CFP, CNC, and NA, respectively, indicate half of the primary alcohol groups on those surface chains are in the interior structure of the cellulose or nanocellulose materials [11]. As discussed above, this structural feature effects the available primary hydroxyls for covalent attachment of the sensor molecule.

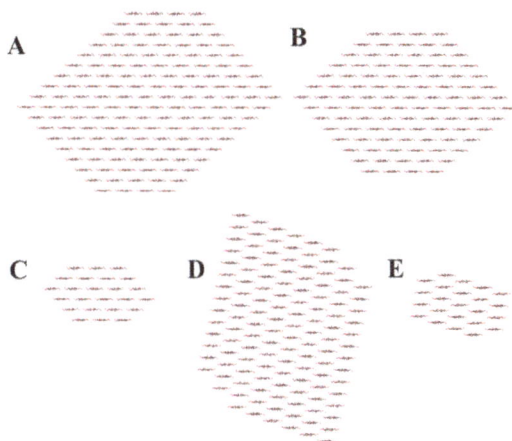

Figure 3. Models of the cotton cellulose and nanocellulose transducer materials based on the X-ray diffraction patterns. (**A**) CFP; (**B**) CNC; and (**C–E**) cellulose I, cellulose II, and amorphous contributions to the diffraction pattern for NA.

2.2.2. Peptide Conformation Consideration Based on Circular Dichroism

Here, we provide experimental evidence for secondary structure in the peptide portion of the conjugate. As shown in Figure 4, circular dichroism spectra of the peptide portion of the conjugate, i.e., Suc-Ala-Pro-Ala-AMC, are consistent with a β-turn prediction. Reflected in the CD are spectral minima at 200 (lit 208) and 214 (lit 222) nm, which are typical for alpha helices, indicate a propensity for type I β-turn formation [19], and with strong evidence for the type I β-turn formation as previously characterized for Pro-Ala [20].

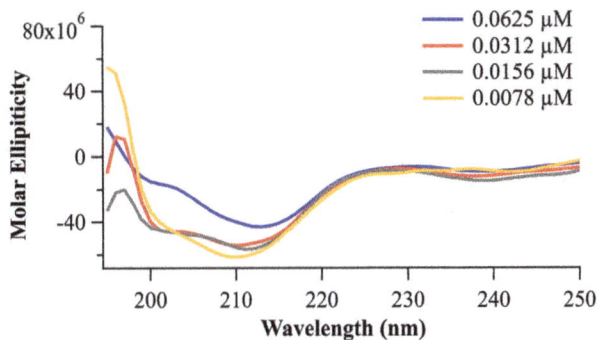

Figure 4. Circular Dichroism of the tripeptide (Suc-APA-AMC) at various concentrations ranging from 0.0625 to 0.0078 μM.

2.2.3. Bioactivity and Kinetic Evaluation of Peptide-Cellulose Conjugates with Elastase

Here, the sensor's ability to detect protease activity and the sensor functionality itself is determined through protease assays typically used to assess the enzyme/substrate recognition and relevant substrate binding and enzyme kinetics. In this study, it is applied to immobilized enzyme substrates i.e., the peptide-cellulose conjugates. The elastase activities of the tripeptide-conjugated sensor analogs, were assessed with succinamidyl-Ala-Pro-Ala-amidylcoumadin peptide-cellulose ester immobilized on the transducer surfaces through a synthesis that has previously been reported [11,13]. The bioactive response was assessed through fluorescence resulting from the protease catalyzed release of the COOH-terminal amino methyl coumarin fluorophore. The results of the bioactivity are shown in Figure 5 by way of reaction progress curves of each sensor's reaction with elastase at a protease concentration of 0.5 U/mL HNE. The relative response curves and detection sensitivity are apparent from the reaction progress and sensitivity curves.

The relative kinetic rates of the pCFP, pNA, and pCNC biosensors with HNE was assessed using traditional Michaelis–Menten kinetic analysis. Table 2 lists the kinetic values for the HNE peptide-cellulose substrates as attached in pCFP, pNA, and pCNC. The kinetic values refer to both the binding affinity of the conjugated enzyme substrate and the turnover rate or rate of product formation; a measurement of the rate of formation of hydrolyzed fluorophore (AMC) from the COOH-terminus of the bound peptide.

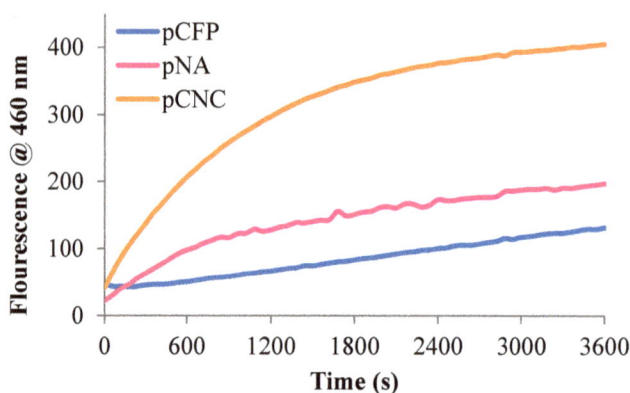

Figure 5. The response curves of 2 mg of pCFP (blue), pNA (pink), and pCNC (green) biosensor upon detection of 7-amino-4-methylcoumarin released with HNE at 0.5 U/mL substrate hydrolysis at 37 °C.

Table 2. Kinetic parameters for human neutrophil elastase (HNE) catalyzed hydrolysis of the tripeptide in solution and on the sensor.

Name	k_{cat} (s^{-1})	K_m (μM)	k_{cat}/K_m (M$^{-1} \cdot$s^{-1})	V_{max} (s^{-1})	Corr. Coeff. (Correlation Coefficient)
Suc-APA-AMC	2.56	781.4	3272.33	2.17	0.9515
pCFP	0.1201	2.150	5,5860.5	0.1021	0.8645
pNA	1.67	202	8267.33	1.42	0.8935
pCNC	0.7732	23.07	3,3515.39	0.6572	0.9956

3. Discussion

3.1. Structural Features of Peptide-Cellulose Conjugate

Figure 6 shows a model of the crystallite-based structure of the peptide cellulose conjugate representative of the ones reported here. To contrast the properties of the nanocellulosic aerogel-based sensor, we include two other cellulose-based sensors that have similarities and differences. Each of the three forms of cellulose and nanocellulose materials evaluated serves as a transducer surface of the protease sensor. Thus, assessment of the relative transducer surface properties that enable sensor protease detection sensitivity was undertaken in the context of contrasting the nanocellulose aerogel-based sensor with peptide-cellulose conjugate analogs of two other distinctly different cellulose transducer surfaces, i.e., porous cotton-based cellulose filter paper substrate (CFP) and cotton nanocellulose crystals (CNC).

As discussed above, during the generation of the NA, a portion of the cellulose is converted to cellulose II [21]. In agreement with earlier reports of several authors [22], we found that different from all other common cellulose solvent systems, cellulose bodies coagulated from respective solutions in calcium thiocyanate octahydrate suffered from distinctly less shrinkage when subjected to solvent exchange and/or scCO$_2$ drying. This can be explained by a remaining fraction of non-dissolved cellulose I acting as templating and reinforcing scaffold for coagulated cellulose II in the sense of an all-cellulose composite aerogel [23]. It is also worth noting that cellulose II hydrate has been shown to have a larger unit cell than the dehydrated form [24], which is relevant to a further observation here on the influence of nanocellulose aerogel swelling on detection sensitivity and dressing properties.

The crystallite models (Figure 3) which are based on the respective cellulose crystallite sizes reveal the relative number of cellulose layers and chains which are CFP (18 layers and 144 chains), NA (12 layers and 78 chains), and CNC (14 layers and 98 chains). The number of chains (36, 28, and 24) on the 1–10 and 110 surfaces for the CFP, CNC, and NA, respectively indicate half of the primary alcohol groups on those surface chains are in the interior structure of the cellulose or nanocellulose materials [11]. As discussed above, this effects the available primary hydroxyls for covalent attachment of the sensor molecule.

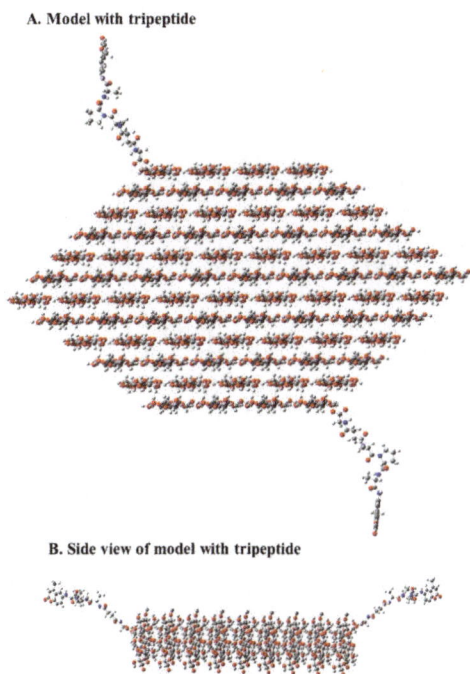

Figure 6. Models of pNA on the X-ray diffraction patterns and Rietveld calculated pattern. The pNA peptide model (**A**) Peptide-cellulose conjugate of aerogel sensor modeled in a single crystal structure with a width of 58.5 Å (78 chains and 12 layers). The side view (**B**) is an image of the same model as in A, but along the longer perimeter of the model.

Previously, we have discussed the role of the elastase substrate's peptide secondary structural conformation relative to the peptide-nanocellulosic conjugate's interaction with the protease from a computational perspective [11] and the potential interaction of the peptide with the crystallite surface. In this regard, the amino acid sequence of this study, Ala-Pro-Ala, has been shown to favor a β-turn conformation [25], which is the smallest element of secondary conformation possible for a protein or peptide. It is also worth noting that β-turns tend to form in more hydrophobic and proteinaceous environments. Thus, the formation of a β-turn in the conjugates of this study may be further enabled by the anionic transducer surface, and upon binding with the protease, which also has a high affinity for the transducer surface, and is likely to populate the transducer surface by way of its positive charge.

3.2. Sensor Bioactivity and Kinetics

The bioactivity of the sensor is dependent on the protease's affinity and interaction with the immobilized substrate and its conjugate surface, which in this study is principally crystalline cellulose. However, we have previously characterized the property of HNE binding to its immobilized substrate from a perspective of adsorption isotherms of elastase binding to the immobilized peptide substrate [26] on crosslinked gels. An assessment of elastase-substrate kinetics and Langmuir adsorption isotherms demonstrated that the protein adsorption at the solid–liquid interface of substrate-bound ethoxyacrylate gels was relatively low i.e., the ratio of maximal adsorbed enzyme to bound ligand was <1%. Thus, in this study the transducer surfaces are negatively charged, and adsorb the HNE marker protein (a positively charged protein) on their crystallite surfaces by way of ionic binding. Accordingly, the negative charge of the transducer surfaces of this study are listed in Table 1, and vary from −10 to −68 mv in order of increasing specific surface area. Moreover, the enhanced HNE binding that these types of transducer surfaces afford is reflected in the Michaelis-Menten kinetics for the immobilized elastase substrate.

An understanding of the relative peptide conformation and crystal structure of the sensor's transducer surface as presented above gives a framework in which understanding enzyme binding and the turnover rates of the cellulose peptide conjugate is possible. Moreover, a perspective of the kinetic turnover and binding affinity between the enzyme and the peptide-cellulose analogs, which possess the enzyme substrate, lends itself to an understanding of the mechanism of action of the sensor with the enzyme (HNE).

The 1.5-fold higher k_{cat} values observed for the peptide substrate in solution compared with the pNA conjugate reflects a generally slower rate of product formation when the enzyme substrates are attached to the nanocellulosic aerogel. This is understandable in light of the two-phase reaction (due to the insolubility of cellulose and nanocellulose in the assay solution) that is occurring between elastase and the peptide bound to the transducer surface, and is also observed for pCFP and pCNC. The lower V_{max} value, which is the maximum reaction rate mediated by the enzyme, observed for the pNA conjugate is consistent with the enzyme turnover rate decreasing in the conjugates.

On the other hand, the enzyme-substrate affinity or ability of the substrate to bind to the enzyme active site, as reflected in the K_m values, was approximately 4-fold greater with the pNA than in solution. Thus, the higher HNE affinity for the tripeptide substrate pNA also gives rise to a higher enzyme efficiency as seen by the higher k_{cat}/K_m assigned to pNA; it is 2.5-fold higher than the k_{cat}/K_m for the analogous enzyme substrate assessed in solution. These increases in binding can be understood in light of the crystallite size, specific surface area, and negative charge, as discussed above.

As shown previously, the macropores of the paper sensor allow passage of the serine protease to the tripeptide substrate attached within the interior structure whereas mesopores as found in the nanocellulose aerogel would not allow passage of the protease or facilitate reaction with the peptide ligand which has a polar surface area of 2 nm^2 (note the peptide substrate may be covalently bonded within the nanocellulose aerogel interior due to its relatively small size) [12]. Thus, access of HNE to the interior of the nanocellulose aerogel may be restricted especially since the pore size of pNA is 30 times less than the polar surface area of elastase (309 nm^2) based on considerations of the material's structure under non-aqueous condition. This may explain the relatively low binding affinity as measured by K_m of the nanocellulose aerogel compared with the paper and nanocrystal. Nonetheless, the turnover rate on the nanocellulose aerogel exceeded that of the paper and nanocrystal. This suggests that increasing the pore size of the aerogel to a volume sufficient for passage of the HNE may result in an increase in the overall enzyme efficiency.

3.3. Spatiotemporal Considerations of the Peptide-Cellulose Conjugate/Elastase Interaction

Spatiotemporal relationships of the peptide-cellulose conjugate to the protease are influenced by charge, protein and peptide-cellulose conjugate structure, and how these components of sensor structure and function come together in time and space. Thus, based on the kinetic properties of the sensors as discussed above it can be inferred that there are strong molecular interactions between elastase and the peptide cellulose conjugates. Specific molecular influences on sensor function are determined by the β turn peptide conformation, the frequency and orientation of peptide-cellulose conjugates on the crystallite surfaces, and the magnitude of the negative charge on the transducer surface.

The negatively charged cellulose surface would favor binding to elastase at a region of the protein where a positively charged cluster occurs and is predominated by arginine residues of HNE. Part of this strongly cationic cluster borders on the S1 active site of the serine protease [27]. Thus, it is feasible that the active site of elastase is brought into close proximity of the cellulose surface through charge transfer interactions between the protein and crystallite surface.

Since the β-turn peptide conformation and its specific resident peptide sequence is required for optimal binding in the oxyanion hole of the active site of the serine protease, stabilization of the conjugate by a negatively charged more hydrophobic surface may increase the enzyme active site binding and in part play a role for the improved K_m values observed for the nanocellulose aerogel as well as paper and nanocrystal forms when compared with the K_m determined in the solution phase.

On the other hand, it is important to recognize that the two different structures of cellulose I and cellulose II also play a role in the sensor activity by virtue of the orientation they confer to the conjugate peptide at the O6 rotamer of cellulose's hydroxymethyl, which is where the peptide is linked to the cellulose. The O6 hydroxymethyl rotamer influences the direction and orientation of the peptide with respect to the cellulose crystallite floor as depicted in Figure 7. For example the conformation of hydroxymethyls in cellulose II has been shown to be predominantly gauche/trans (gt O6) (rotamer oriented from the O5 oxygen through the C4 carbon of anyhydroglucose residues), and conversely the tg O6 rotamer predominates in cellulose I [28]. However, it is worth noting that far less work has been done on cellulose II nanocrystal shapes when they are involved in aerogels compared with cellulose I, and as discussed earlier the coagulation phase of the nanocellulose aerogel preparation may present a unique juxtaposition of minor cellulose polymorphs. Thus, it is likely that the peptide attaches to a gg O6 in both cellulose I and II since the O6 rotamer of the conjugate is not tethered by a hydrogen bond as are the hydroxymethyls in the cellulose crystalline lattice. This is consistent with our examination of a model of a cellulose II crystal where it appears that the top and bottom surfaces have the gt O6 atoms directed to the interior of the crystallite, and may not be favored for the peptide to attach to due to steric hindrance. Accordingly, the gg O6 hydroxymethyl rotamer would orient the peptide outward to the plane of the cellulose crystallite. Thus, we present a minimized model of the peptide cellulose analog in Figure 7 at a hydroxymethyl gg O6 rotamer orientation.

To further examine the relationship of the peptides orientation to the crystal surface the different possible combinations of the peptide-cellulose relation to the crystal surface are shown in Figure 8 and the relative distances of the scissile bond of the sensor peptide (the bond that HNE hydrolyzes upon binding and recognition of the enzyme active site) from the cellulose surface plane are listed in Table 3. The 1–10 and 110 surface planes form two sides of the cellulose crystal in both cellulose I and II. As shown the gg O6 rotamer of the 110 surface plane is considerably closer to the surface than found in the 1–10 surface plane. The orientation of hydroxymethyls in the 110 and 1–10 surface plane is predominantly equatorial [29]. In the case of the gt and tg O6 rotamers the proximity of the peptide to the surface planes is reversed to that observed for the gg O6 rotamer. Nonetheless the gg O6 rotamer on the 110 surface plane of the aerogel may account for the lower binding affinity due to increased steric hindrance to binding of the gg O6 110 surface plane rotamer. However, based on the distances measured for all of the combinations modeled on both surface planes it appears that there is a fairly even variation in the distribution of peptide orientations with respect to the crystal surface, and neither surface the 110 or 1–10 surface would predominately influence enzyme turnover from the standpoint of peptide orientation. On the other hand, the 110 surface plane is more hydrophobic due to exposure of fewer hydroxyls on its surface. Thus, the contribution of this feature to enzyme binding and turnover is not clear other than to point out that elastase binding tends to favor more hydrophobic substrates, and it is likely that the relative polarity and charge of the surfaces and the proximity of the peptide together have a pronounced influence on the peptide-cellulose affinity and turnover of the enzyme.

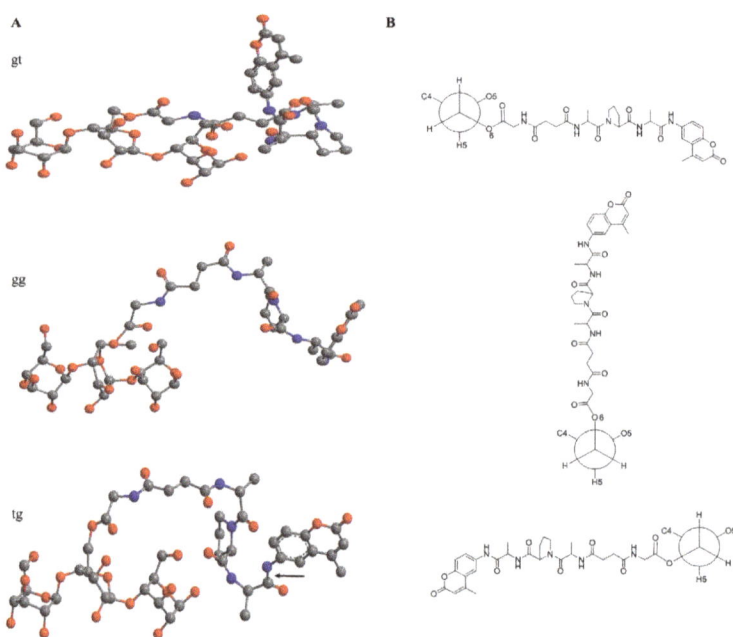

Figure 7. The conformational structures of the tripeptide substrate anchored onto a cellotriose unit at the hydroxymethyl gt O6 rotamer, gg O6 rotamer, and tg O6 rotamer orientation of the cellulose or nanocellulose sensors in: (**A**) minimized model; or (**B**) Newman projections. The black arrow shown for the tg O6 rotamer shows where the HNE protease cleaves the fluorophore. Note: gray is carbon, blue is nitrogen, red is oxygen.

Table 3. Distance of AMC (7-amino-4-methylcoumarin) tripeptide substrate from the five-chain 110 or 1–10 surface of the cellulose floor.

Name	Distance from AMC to Cellulose Floor (Å)
110 gg tripeptide	10.7
110 gt tripeptide	22.8
110 tg tripeptide	18.4
1 m 10 gg tripeptide	23.2
1 m 10 gt tripeptide	12.2
1 m10 tg tripeptide	9.9

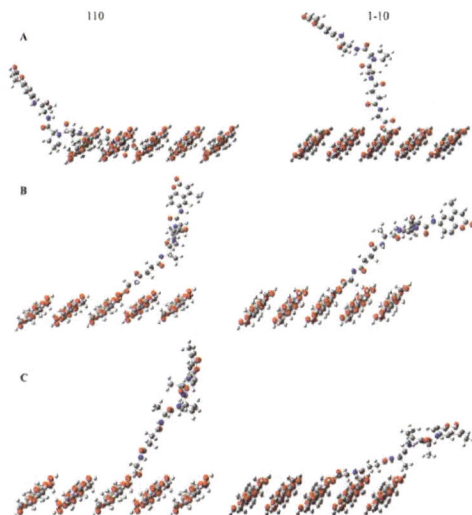

Figure 8. The tripeptide substrate anchored on the five-chain 110 or 1–10 surface of cellulose in the following rotamer positions: (**A**) gauche-gauche; (**B**) gauche-trans; and (**C**) trans-gauche.

4. Materials and Methods

4.1. General

Potassium chloride (KCl), sodium chloride (NaCl), hydrochloric acid (HCl), monosodium phosphate (NaH_2PO_4), trifluoroethanol (TFE), and sodium hydroxide (NaOH) were purchased from VWR (Radnor, PA, USA). Promogran Prisma™ (Systagenix, San Antiono, TX, USA) was purchased from the distributor HighHealthTide. Human neutrophil elastase (HNE) was purchased as a salt-free lyophilized solid from Athens Research and Technology (Athens, GA, USA) without further purification. The peptide n-succinyl-Ala-Pro-Ala-7-amino-4-methylcoumarin (Suc-Ala-Pro-Ala-AMC) was purchased from BACHEM (Torrance, CA, USA) without further purification.

4.2. Synthesis of Nanocellulosic Aerogel and Generation of Biosensors

The nanocellulosic aerogel was prepared and characterized as previously reported by Edwards et al. [12]. The esterification, and peptide immobilization of the cellulosic print cloth (CPC), cotton nanocellulosic crystals (CNC), and nanocellulosic aerogel (NA) were previously reported by Edwards et al. [9,12]. In summary, glycine-esterified cellulose and nanocellulose substrates were prepared by reacting the glycine α–amino functionality with the COOH-terminal group of Suc-Ala-Pro-Ala-AMC by way of a carbodiimide-mediated reaction. Following a simple work-up of rinsing and drying small representative samples of the active peptide component biosensors were cleaved from the matrices by adding a mixture of TFA/water/triisopropylsilane (95/2.5/2.5) to the cellulose-peptide conjugates for three hours, whereupon it was diluted with water (1:10) and submitted for ESI-LC/MS. The intact sequence of the peptide component of the cellulose-peptide conjugate (COOH-Glycine-NH2-Alanine-Proline-Ala-7-amino-4-methylcoumarin) was confirmed for its molecular weight.

4.3. Response and Sensitivity Assay

The fluorescence response and sensitivity assay of the cellulosic and nanocellulosic materials employed was previously outlined by Fontenot et al. [10]. Briefly, duplicates of ~2 mg of each biosensor were placed into a 96 well plate and 100 µL of a phosphate buffer solution (PBS) was added. A standard curve of the tripeptide substrate (1 to 0.0156 µmol/mL) was added to the 96 well plate with a total volume of 100 µL. One well containing only PBS was also included in the standard curve. To start the reaction for the response assay, 50 µL human neutrophil elastase (HNE) was added to the wells containing the biosensors and the standard curve for a total volume of 150 µL. For the sensitivity assay, 50 µL of HNE ranging from 2 to 0.0156 U/mL was added to the standard curve and to the biosensors to provide a total volume of 150 µL, from which the lowest sensitivity detection limit was determined. Fluorescent measurements commenced at 37 °C for 1 hour and measured at 1-min intervals while shaking for 3 s prior to measurement using a Biotech Synergy HT with a tungsten halogen lamp and photomultiplier detection (Winooski, VT, USA). The fluorescence measurements were acquired at 360 nm excitation and 460 nm emission. Fluorescent raw data were processed in Microsoft Excel 2013. The wound like fluid herein comprises of HNE in PBS.

4.4. Surface Charge

The investigation of the surface charge (zeta potential, ζ) values of the cellulose filter paper (CFP) was performed using an Anton Paar Surpass (Ashland, VA, USA). The electrophoretic mobility of the samples was measured using a cylindrical cell. All measurements were performed with a 1 mM KCl solution in deionized water at 21–23 °C. A 0.1M NaOH solution was used to adjust the KCl solution to pH 10.3 for FP. The following parameters were used for titration on the Anton Paar Surpass: aqueous solution of 0.100 M HCl, desired pH increment difference 0.200, volume increments 0.020 mL, pH minimum 2.5, and the pH maximum were slightly higher than the pH of the adjusted KCl solution. The flow rate ranged between 50 and 150 mL/min with a maximum pressure of 300 mbar (for the pressure and rinse program).

4.5. Kinetics Assay

The protocol of the fluorescence response assay was used for the fluorogenic kinetics assay; however, the biosensor concentration varied in duplicate concentrations between 0.5 and 4 mg (dissolved in 100 µL PBS). To start the reaction, 50 µL of 0.5 U/mL of human neutrophil elastase was added to both the free peptide substrate solution used to prepare the standard curve and for the biosensors, providing a total volume of 150 µL.

Fluorescent measurements were conducted as described for the response and sensitivity assay, however, at 20 s intervals for a period of 10 min. The data were imported into GraphPad Prism software (6, GraphPad La Jolla, CA, USA) and analyzed as follows: create XY columns for sample data with enzyme kinetics, non-linear regression for curve fit, enzyme kinetics-substrate vs. velocity, and Michaelis–Menten to determine K_m and V_{max} values.

4.6. Circular Dichroism

Circular dichroism (CD) studies were performed using a Jasco J-815 spectrometer with Spectra Manager 2 Software suite (Jasco, Easton, MD, USA). The CD measurements were carried out using 1 mm path length quartz cell. Peptide solutions were prepared at various concentrations in pH 7.4 PBS (0.5 M NaCl and 0.1 M NaH_2PO_4) containing 10% triflouroethanol (TFE). All spectra correspond to an average of three measurements, which were corrected using the baseline obtained for peptide free solutions. All data were converted to molar ellipticity and smoothed using a binomial function.

4.7. X-ray Diffraction (XRD)

All samples analyzed in the laboratory of Prof. John Wiley at the University of New Orleans Chemistry Department. The CFP, CNC, and NA samples were pressed into pellets and scanned in the θ–2θ reflection mode with a Philips X'pert powder diffractometer using Cu K radiation and a graphite monochromator. The data were obtained in a 2θ scale from 5° and 40° with a step size of 0.05 at 2.5 s per step with an overall run time of 30 min. No background corrections were made. The crystallinity indices (CrI) of the matrices were determined by subtracting the minimum intensity near 18° 2-θ (I_{AM}) from the maximum intensity (at 22.625°–725° 2-θ, I_{200}) and dividing the difference by the maximum intensity (I_{200}) (Equation (1)) [30]. The crystallite size was calculated from the XRD patterns using the Scherrer formula [31] (Equation (2)). The terms of Equation (2) are shape factor (K) of 1, copper K radiation average wavelength (λ) is 1.5418 Å, the full width at half maximum of the (200) peak (β) in radians, and the peak position divided by 2 of the (200) peak (θ) [32]. The crystallite models were drawn using Mercury software 3.5.1 (The Cambridge Crystallographic Data Centre Cambridge, UK), dividing the crystallite sizes from the Scherer equation calculations by the d_{200} interplanar spacing to get the number of chain layers. The tripeptide cellulose analogs were optimized using Chem3DPro 13.0 MM2 minimization energy calculations.

$$CrI = \frac{I200 - IAM}{I200} \times 100 \tag{1}$$

$$t = \frac{K * \lambda}{\beta * COS(\theta)} \tag{2}$$

4.8. Rietveld Refinement Method and Crystallite Models from the NA Diffraction Pattern

The Rietveld method was only used to refine the NA diffraction patterns in order to determine a crystallite size. The standard model diffraction patterns for cellulose I and cellulose II (cif) were based on work by Nishiyama et al. and more recently by French et al. [28,33]. Materials Analysis Using diffraction (Maud) 2.55 software [34] was used to apply the Rietveld method in order to determine the calculated diffraction patterns based on cellulose I, cellulose II, and amorphous patterns in order to resolve the broad overlapping cellulose I and cellulose II peaks and determine the crystallite size. Several parameter were modified in the Maud software during the refining process for the NA: the scale factor, quadratic background, and unit cell length a (layer I: the phase atom); cellulose I β: cell length a, scale factor, crystal size, and cell angle gamma (March-Dollase model: background and crystal size); and cellulose II: cell length a, scale factor, crystal size, and cell angle gamma (March-Dollase model: utilizing background and crystal size).

4.9. Molecular Modeling Studies

Computational studies minimized the energy of the tripeptide substrate anchored on the five-chain 110 or 1–10 surfaces of cellulose in the following rotamer positions: (A) gauche-gauche; (B) gauche-trans; and (C) trans-gauche [11]. The peptide substrate models were built using GaussView 5.0.9 and optimized using a semiempirical Hamiltonian method (PM3) [35] contained in Gaussian 09 revision A.02 molecular orbital software (Gaussian inc., Wallingford, CT, USA).

5. Conclusions

We evaluated cotton-based cellulosic filter paper, nanocellulosic crystals, and nanocellulosic aerogel transducers conjugated to a tripeptide substrate using: (i) X-ray diffraction to determine their diffraction pattern, cellulose orientation, crystallite size, and crystallite model; (ii) Circular dichroism for assessment of the tripeptide substrate conformation, which revealed a β-turn confirmation; (iii) the bioactivity properties of the sensors via fluorescence response and kinetic properties that indicated the sensors provide a faster response to human neutrophil elastase compared to the free tripeptide substrate; and (iv) computational assessment of the cellulose tripeptide substrate analog in different orientations. Therefore, these materials, in particular, the nanocellulosic aerogel, have promising sensitivity, selectivity, and a faster kinetics response for detecting human neutrophil elastase, and are a promising sensor interface for a multilayered chronic wound dressing motif. Furthermore, the nanocellulosic sensors are adaptable for the detection of other biomarkers of clinical interest as well, via substitution of the biomolecule.

Acknowledgments: This project was supported by the U.S. Department of Agriculture. The authors wish to thank Lawson Gary, Chief Operating Officer, T. J. Beall, Co. Inc., for the donation of True Cotton™ samples. Furthermore, the support of project CAPBONE by the Austrian Science Fund is thankfully acknowledged.

Author Contributions: J. Vincent Edwards conceived of the project, guided and mentored the experimental design and is the principle author of the manuscript. Krystal Fontenot designed the experiments, synthesized all peptide modified aerogel derivatives, performed all of the experiments, analyzed the data, conducted literature searches, designed the graphics and helped with a significant part of the writing of the manuscript. Nicole Doyle nee Pircher and Falk Liebner prepared the unmodified aerogels and contributed to editing the manuscript. Alfred D. French analyzed and interpreted X-ray crystallography data, and assisted with the molecular modeling. Brian D. Condon provided funding and management that contributed reagents/materials, analysis tools and personnel resources to the project.

Conflicts of Interest: The authors declare no conflict of interest.

Abbreviations

AMC	7-amino-4-methylcoumarin
HNE	human neutrophil elastase
CFP	(cellulosic filter paper)
NA	nanocellulosic aerogel
CNC	cotton nanocrystalline cellulose
pCFP	peptide-cellulose conjugate on filter paper
pNA	peptide-cellulose conjugate on NA
pCNC	peptide-cellulose conjugate on NC
k_{cat}	enzyme turnover rate
K_m	enzyme binding affinity constant
g	gauche conformation
t	*trans* conformation
MAUD	materials analysis using diffraction

References

1. Dargaville, T.R.; Farrugia, B.L.; Broadbent, J.A.; Pace, S.; Upton, Z.; Voelcker, N.H. Sensors and imaging for wound healing: A review. *Biosens. Bioelectron.* **2013**, *41*, 30–42. [CrossRef] [PubMed]
2. International Consensus. *The Role of Proteases in Wound Diagnostics*; An Expert Working Group Review; Wounds International: London, UK, 2011; pp. 1–13.
3. Yager, D.R.; Chen, S.M.; Ward, S.I.; Olutoye, O.O.; Diegelmann, R.F.; Kelman Cohen, I. Ability of chronic wound fluids to degrade peptide growth factors is associated with increased levels of elastase activity and diminished levels of proteinase inhibitors. *Wound Repair Regen.* **1997**, *5*, 23–32. [CrossRef] [PubMed]
4. Serena, T.E. Development of a novel technique to collect proteases from chronic wounds. *Adv. Wound Care* **2014**, *3*, 729–732. [CrossRef] [PubMed]
5. Serena, T.E.; Cullen, B.M.; Bayliff, S.W.; Gibson, M.C.; Carter, M.J.; Chen, L.; Yaakov, R.A.; Samies, J.; Sabo, M.; DeMarco, D.; et al. Defining a new diagnostic assessment parameter for wound care: Elevated protease activity, an indicator of nonhealing, for targeted protease-modulating treatment. *Wound Repair Regen.* **2016**, *24*, 589–595. [CrossRef] [PubMed]
6. Vandenbroucke, R.E.; Libert, C. Is there new hope for therapeutic matrix metalloproteinase inhibition? *Nat. Rev. Drug Discov.* **2014**, *13*, 904–927. [CrossRef] [PubMed]
7. Synder, R.J.; Driver, V.; Fife, C.E.; Lantis, J.; Peirce, B.; Serena, T.; Weir, D. Using a diagnostic tool to identify elevated protease activity levels in chronic and stalled wounds: A consensus panel discussion. *Ostomy Wound Manag.* **2011**, *57*, 36–46.
8. Moore, K.; Huddleston, E.; Stacey, M.C.; Harding, K.G. Venous leg ulcers—The search for a prognostic indicator. *Int. Wound J.* **2007**, *4*, 163–172. [CrossRef] [PubMed]
9. Edwards, J.V.; Fontenot, K.R.; Haldane, D.; Prevost, N.T.; Condon, B.D. Human neutrophil elastase peptide sensors conjugated to cellulosic and nanocellulosic materials: Part I, synthesis and characterization of fluorescent analogs. *Cellulose* **2016**, *23*, 1283–1295. [CrossRef]

10. Fontenot, K.R.; Edwards, J.V.; Haldane, D.; Graves, E.; Citron, M.S.; Prevost, N.T.; French, A.D.; Condon, B.D. Human neutrophil elastase detection with fluorescent peptide sensors conjugated to cellulosic and nanocellulosic materials: Part II, structure/function analysis. *Cellulose* **2016**, *23*, 1297–1309. [CrossRef]
11. Edwards, J.V.; Prevost, N.T.; French, A.D.; Concha, M.; Condon, B.D. Kinetic and structural analysis of fluorescent peptides on cotton cellulose nanocrystals as elastase sensors. *Carbohydr. Polym.* **2015**, *116*, 278–285. [CrossRef] [PubMed]
12. Edwards, J.V.; Fontenot, K.R.; Prevost, N.T.; Pircher, N.; Liebner, F.; Condon, B.D. Preparation, characterization and activity of a peptide-cellulosic aerogel protease sensor from cotton. *Sensors* **2016**, *16*, 1789. [CrossRef] [PubMed]
13. Edwards, J.V.; Bopp, A.; Graves, E.; Condon, B. *In Vitro Hemostatic, Hydrogen Peroxide Production and Elastase Sequestration Properties of Nonwoven Ultra Clean Greige Cotton Dressing*; Wound Healing Society: Orlando, FL, USA; Wound Repair and Regeneration: Orlando, FL, USA, 2013; p. A21.
14. Edwards, J.V.; Fontenot, K.R.; Prevost, N.T.; Haldane, D.; Nicole, P.; Liebner, F.; French, A.D.; Condon, B.D. Protease biosensors based on peptide-nanocellulose conjugates: From molecular design to dressing interface. *Int. J. Med. Nano Res.* **2016**. [CrossRef]
15. Edwards, J.V.; Prevost, N.; French, A.D.; Concha, M.; DeLucca, A.; Wu, Q. Nanocellulose-based biosensors: Design, preparation, and activity of peptide-linked cotton cellulose nanocrystals having fluorimetric and colorimetric elastase detection sensitivity. *Engineering* **2013**, *5*, 20–28. [CrossRef]
16. Reiner, R.S.; Rudie, A.W. Process scale-up of cellulose nanocrystal production to 25 kg per batch at the forest products laboratory. In *Production and Applications of Cellulose Nanomaterials*; Postek, M.T., Moon, R.J., Rudie, A.W., Bilodeau, M.A., Eds.; TAPPI Press Inc.: Peachtree Corners, GA, USA, 2013; pp. 21–24.
17. Nam, S.; French, A.D.; Condon, B.D.; Concha, M. Segal crystallinity index revisited by the simulation of X-ray diffraction patterns of cotton cellulose Iβ and cellulose II. *Carbohydr. Polym.* **2016**, *135*, 1–9. [CrossRef] [PubMed]
18. Pircher, N.; Carbajal, L.; Schimper, C.; Bacher, M.; Rennhofer, H.; Nedelec, J.-M.; Lichtenegger, H.C.; Rosenau, T.; Liebner, F. Impact of selected solvent systems on the pore and solid structure of cellulose aerogels. *Cellulose* **2016**, *23*, 1949–1966. [CrossRef] [PubMed]
19. Manning, M.C.; Illangasekare, M.; Woody, R.W. Circular dichroism studies of distorted alpha-helices, twisted β-sheets, and β turns. *Biophys. Chem.* **1988**, *31*, 77–86. [CrossRef]
20. Gierasch, L.M.; Deber, C.M.; Madison, V.; Niu, C.-H.; Blout, E.R. Conformations of (X-L-Pro-Y)2 cyclic hexapeptides. Preferred β-turn conformers and implications for β Turns in proteins. *Biochemistry* **1981**, *20*, 4730–4738. [CrossRef] [PubMed]
21. Liebner, F.; Pircher, N.; Schimper, C.; Haimer, E.; Rosenau, T. Aerogels: Cellulose-based. In *Encyclopedia of Biomedical Polymers and Polymeric Biomaterials*; CRC Press: Boca Raton, FL, USA, 2015; pp. 37–75.
22. Hoepfner, S.; Ratke, L.; Milow, B. Synthesis and characterisation of nanofibrillar cellulose aerogels. *Cellulose* **2008**, *15*, 121–129. [CrossRef]
23. Gindl, W.; Keckes, J. All-cellulose nanocomposite. *Polymer* **2005**, *46*, 10221–10225. [CrossRef]
24. Kobayashi, K.; Kimura, S.; Togawa, E.; Wada, M. Crystal transition from cellulose II hydrate to cellulose II. *Carbohydr. Polym.* **2011**, *86*, 975–981. [CrossRef]
25. Liang, G.B.; Rito, C.J.; Gellman, S.H. Thermodynamic analysis of β-turn formation in Pro-Ala, Pro-Gly, and Pro-Val model peptides in methylene chloride. *J. Am. Chem. Soc.* **1992**, *114*, 4440–4442. [CrossRef]
26. Edwards, J.V.; Caston-Pierre, S.; Bopp, A.F.; Goynes, W. Detection of human neutrophil elastase with peptide-bound cross-linked ethoxylate acrylate resin analogs. *J. Pept. Res.* **2005**, *66*, 160–168. [CrossRef] [PubMed]
27. Navia, M.A.; McKeever, B.M.; Springer, J.P.; Lin, T.Y.; Williams, H.R.; Fluder, E.M.; Dorn, C.P.; Hoogsteen, K. Structure of human neutrophil elastase in complex with a peptide chloromethyl ketone inhibitor at 1.84-a resolution. *Proc. Natl. Acad. Sci. USA* **1989**, *86*, 7–11. [CrossRef] [PubMed]
28. Nishiyama, Y.; Langan, P.; Chanzy, H. Crystal structure and hydrogen-bonding system in cellulose Iβ from synchrotron X-ray and neutron fiber diffraction. *J. Am. Chem. Soc.* **2002**, *124*, 9074–9082. [CrossRef] [PubMed]
29. Yamane, C.; Aoyagi, T.; Ago, M.; Sato, K.; Okajima, K.; Takahashi, T. Two Different Surface Properties of Regenerated Cellulose due to Structural Anisotropy. *Polym. J.* **2006**, *38*, 819–826. [CrossRef]
30. Segal, L.; Creely, J.J.; Martin, A.E.; Conrad, C.M. An empirical method for estimating the degree of crystallinity of native cellulose using the X-ray diffractometer. *Text. Res. J.* **1959**, *29*, 786–794. [CrossRef]

31. Scherrer, P. Bestimmung der Inneren Struktur und der Größe von Kolloidteilchen Mittels Röntgenstrahlen. In *Kolloidchemie ein Lehrbuch*; Springer: Berlin/Heidelberg, Germany, 1912; pp. 387–409.

32. French, A.; Santiago Cintrón, M. Cellulose polymorphy, crystallite size, and the segal crystallinity index. *Cellulose* **2013**, *20*, 583–588. [CrossRef]

33. French, A.D. Idealized powder diffraction patterns for cellulose polymorphs. *Cellulose* **2014**, *21*, 885–896. [CrossRef]

34. Lutterotti, L. Total pattern fitting for the combined size-strain-stress-texture determination in thin film diffraction. *Nucl. Instrum. Methods Phys. Res. Sect. B Beam Interact. Mater. Atoms* **2010**, *268*, 334–340. [CrossRef]

35. Stewart, J.J.P. Mopac: A semiempirical molecular orbital program. *J. Comput. Aided Mol. Des.* **1990**, *4*, 1–103. [CrossRef] [PubMed]

© 2018 by the authors. Licensee MDPI, Basel, Switzerland. This article is an open access article distributed under the terms and conditions of the Creative Commons Attribution (CC BY) license (http://creativecommons.org/licenses/by/4.0/).

International Journal of
Molecular Sciences

MDPI

Article

Effect of Surfactant Type and Sonication Energy on the Electrical Conductivity Properties of Nanocellulose-CNT Nanocomposite Films

Sanna Siljander [1,*], Pasi Keinänen [1], Anna Räty [1], Karthik Ram Ramakrishnan [1], Sampo Tuukkanen [2], Vesa Kunnari [3], Ali Harlin [3], Jyrki Vuorinen [1] and Mikko Kanerva [1]

[1]   Laboratory of Materials Science, Tampere University of Technology, FI-33720 Tampere, Finland;
       pasi.keinanen@tut.fi (P.K.); anna.raty@tut.fi (A.R.); karthik.ramakrishnan@tut.fi (K.R.R.);
       jyrki.vuorinen@tut.fi (J.V.); mikko.kanerva@tut.fi (M.K.)
[2]   BioMediTech, Tampere University of Technology, FI-33720 Tampere, Finland; sampo.tuukkanen@tut.fi
[3]   VTT Research Center, FI-02044 Espoo, Finland; vesa.kunnari@vtt.fi (V.K.); ali.harlin@vtt.fi (A.H.)
*   Correspondence: sanna.siljander@tut.fi; Tel.: +358-50-3-555-777

Received: 15 May 2018; Accepted: 15 June 2018; Published: 20 June 2018

Abstract: We present a detailed study on the influence of sonication energy and surfactant type on the electrical conductivity of nanocellulose-carbon nanotube (NFC-CNT) nanocomposite films. The study was made using a minimum amount of processing steps, chemicals and materials, to optimize the conductivity properties of free-standing flexible nanocomposite films. In general, the NFC-CNT film preparation process is sensitive concerning the dispersing phase of CNTs into a solution with NFC. In our study, we used sonication to carry out the dispersing phase of processing in the presence of surfactant. In the final phase, the films were prepared from the dispersion using centrifugal cast molding. The solid films were analyzed regarding their electrical conductivity using a four-probe measuring technique. We also characterized how conductivity properties were enhanced when surfactant was removed from nanocomposite films; to our knowledge this has not been reported previously. The results of our study indicated that the optimization of the surfactant type clearly affected the formation of freestanding films. The effect of sonication energy was significant in terms of conductivity. Using a relatively low 16 wt. % concentration of multiwall carbon nanotubes we achieved the highest conductivity value of 8.4 S/cm for nanocellulose-CNT films ever published in the current literature. This was achieved by optimizing the surfactant type and sonication energy per dry mass. Additionally, to further increase the conductivity, we defined a preparation step to remove the used surfactant from the final nanocomposite structure.

Keywords: nanocellulose; carbon nanotubes; nanocomposite; conductivity; surfactant

1. Introduction

Conductive composite materials with micrometer and nanoscale fillers, like metallic powders, carbon black, graphite and carbon fibers, are used in many applications, such as antistatic films and electromagnetic interference (EMI) shielding. Electrical conductivity of 0.01 S/cm or higher is required for the composite to be considered conductive, while materials with lower conductivity can be used as antistatic and semiconducting materials. One of the drawbacks with most fillers is that the filler content ratio needs to be as high as 50 wt. % to achieve the percolation threshold (i.e., the critical concentration of filler that corresponds to the sharp rise of conductivity). However, this high filler content ratio might lead to a decrease in the resultant composite's mechanical properties [1,2]. Nanomaterials, such as carbon nanotubes (CNTs) and graphene, play a role in the development of future composite materials. For example, CNTs and graphene have been used to toughen matrix polymers [3], to adjust barrier properties of nanocomposite films [4], and to form hierarchical reinforcements [5]. It is possible to attain the percolation threshold in the insulating polymer matrix at a low CNT concentration due to their excellent electrical, mechanical and thermal properties.

Individual CNTs are part of a group of the strongest and most conductive nanomaterials known [6]. Additionally, CNTs can carry higher current density than any other known material, with its highest measured value being 109 A/cm^2 [7,8]. However, to obtain an ideal conductive network, the carbon nanotubes have to be well separated and homogenous dispersion should be maintained in the final product. Without efficient dispersion, filler aggregates act as defect sites, which leads to lower mechanical performance [9,10]. As the most abundant polymer on earth, cellulose is a promising and well-known material that can be used as a matrix in nanocomposites.

Cellulose is environmentally conscious, low-cost, strong, dimension-stable, non-melting, non-toxic and is a non-metal matrix. The interest towards nanoscale cellulose has increased during the past few years because of its inherent properties, including its good mechanical properties, which are better than those of the respective source biomass material [11]. Cellulose-based micro-/nanofibrils (MFC/NFC) can be extracted from various types of plant fibers using mechanical forces, chemical treatments, enzymes or combinations of these. The most typical approach, however, is to apply wood pulp and mechanical methods such as homogenization, microfluidization, microgrinding and cryocrushing. Finally, after fibrillation, the width of NFC is typically between 5 and 20 nm, with a length of several micrometers. Nanocellulose (NFC) has hydroxyl groups in its structure and is therefore associated with high aspect ratio and strong hydrogen bonds formed between nanocellulose fibers [12]. These bonds enhance mechanical properties and enable the formation of free standing films. A combination of CNTs and cellulose I provides a conductive nanocomposite network. CNT-cellulose composites have been reported to be used as supercapacitor electrodes [13,14], electromagnetic interference shielding devices [15], chemical vapor sensors [16], water sensors [17,18], and pressure sensors [19].

There are different manufacturing methods for the fabrication of CNT-cellulose nanocomposites, but all the methods typically include (1) a phase of dispersing CNTs into a solution, and (2) an impregnation phase into the cellulose substrates (e.g., paper, filter paper) [15,16,20–23]. Alternatively, the dispersion can be used as a wet component with bacterial cellulose [24,25], with cellulose I and regenerated cellulose fibers [13,18,26] or in an aerogel form [17]. The processing of nanocellulose in an aqueous medium is the most common way due to its tendency to react with water, and strong affinity to itself and hydroxyl group containing materials [12]. Chen et al. [27] showed that NFCs and CNTs can form a three-dimensional conductive network structure in a gel-film morphology to achieve high electrical conductivity.

The properties of the nanocellulose-CNT composites are affected by the quality of CNT dispersion, amount of structural and oxidative defects in the graphitic structure of the CNTs, the aspect ratio of the CNTs after the disaggregate treatment, the strength of the matrix, and the interactions between the CNTs and the cellulose matrix. [28] The key challenge in numerous industrial applications is to achieve uniform and stable CNT dispersion. The homogenization phase is vital to maximize the excellent mechanical, electrical and thermal properties of the CNTs and the eco-friendly, strong and low-cost nanocellulose matrix. This is particularly important in the case of submicron- or nanometer-sized particles. In these scales, the surface chemistry plays an important role, managing the particle dispersion within the final product [29]. CNT dispersions are challenging because as the surface area of particles increases, the attractive forces between the aggregates [29] and the high aspect ratio enable the entanglement and bundling of CNTs [30]. There are two phenomena that affect CNT dispersions: nanotube morphology and the forces between the tubes. Entanglement of CNTs occurs due to tube morphology, as well as molecular forces, high aspect ratio, and high flexibility. Dispersing these entangled aggregates is difficult without damaging the nanotubes. Both CNT and aggregate size are expected to play a crucial role in the achieved level of electrical conductivity [31].

Two typical dispersion methods for CNTs include high shear mixing and pure sonication [13,15,16,19–21,24,32]. Sonication is based on ultrasonic waves that generate microscopic bubbles or inertial cavitation, which produces a shearing action. This results in liquid and suspended particles becoming intensely agitated. Another common technique is to use a centrifuge in one of the processing steps to extract the unwanted agglomerates from the supernatant, but this additional phase takes time and effort and affects the concentration of dispersed particles in the dispersion. In general, sonication is superior to shear mixing, especially for low-viscosity systems [33], where conventional mixing does not create high enough strain rates to disintegrate the CNT aggregates.

Another issue in the manufacturing of films using NFC is the shrinkage and distortion of the structure because of faster evaporation rate on surface than the mass transport of moisture within the material. When strong enough gradient occurs, film distortions emerge because of local stresses [34,35].

One widely used method for CNT dispersion is the non-covalent method. In this method, chemical moieties are adsorbed onto the surface of CNTs, the CNTs are non-covalently dispersed in a water medium, and the resultant mixture is sonicated in the presence of the moieties, namely surfactants. Surfactants are a group of organic compounds that have a hydrophilic head and a hydrophobic tail, and they are commonly used as detergents, wetting agents, emulsifiers, foaming agents and dispersants. The advantage of the non-covalent method lies in the fact that it does not deteriorate the electronic structure of the CNTs' graphitic shells, maintaining their high electrical conductivity. Good dispersion can be achieved by having a mixture of both nanocellulose and carbon nanotubes with the help of surfactants, as the surfactants lower the interfacial free energy between the particles. Table 1 lists information about surfactants and their properties used in this study.

Table 1. Surfactants used in this study.

Product Name	Triton™ x-100	Pluronic® F-127	CTAB
Type	Non-ionic	Non-ionic Polymeric	Cationic
Name	Octylphenol Ethoxylate	Poloxamer	Hexadecyltri-methylammonium bromide
Chemical Structure		CH_3 $H(OCH_2CH_2)_x(OCH_2CH)_y(OCH_2CH_2)_zOH$	CH_3 $CHBr^-$ CH_3 CH_3
Critical Micelle Concentration	0.2–0.9 mM (20–25) °C	950–1000 ppm (25 °C)	0.92 mM (20–25) °C
HLB value	13.5	22	10

In the current literature, there are several different types of surfactants used for dispersing nanocellulose and carbon nanotubes. Choosing a surfactant type for effective dispersion of nanotubes through surfactant adsorption is complicated, as the results in the published literature often give contradictory results. For example, some researchers [29] have suggested that ionic surfactants are preferable for creating aqueous dispersions. However, the non-ionic surfactant Triton X-100 was shown to be a better surfactant than the anionic surfactant SDS, which was attributed to the π-π stacking ability of the former. The quality of the NFC-CNT dispersion is dependent on the nature of the surfactant, the concentration and the type of interactions between the surfactant and dispersing particles [36]. It has been stated that, for dispersing CNTs it is preferable for the surfactant to have a relatively high HLB (hydrophilic-lipophilic balance) value [29]. This assumption was proven false in our previous study [37]. Not only are the surfactant's nature and energy carried into the dispersed system, but the concentration of the surfactant also has a crucial role in the dispersion process [38]. Too high a surfactant concentration may negatively affect conductivity properties by blocking off the charge transport through the CNT network [39]. In addition, a low surfactant concentration can cause re-aggregation, because a sufficient amount is required to cover CNT surfaces to prevent re-aggregation [39,40]. It has been shown that an efficient CNT dispersion is only possible when the surfactant concentration is above the critical micelle concentration (CMC) value [41–44]. In some cases, the surfactant concentration is reported to be higher than the (CMC), but no micelle structures are observed in the dispersion. Presumably, most of the surfactant has been adsorbed onto the surface of the CNTs [40]. In other cases, surfactants can prefer surfactant-surfactant interactions over spreading on the CNT surface [45]. It has also been reported that dispersing agents can form stable dispersions below and equal to their CMC limit [46–49]. Moreover, it has been noted that commonly, the best results can be reached with a concentration of 0.5 CMC and that any further increase in the concentration of surfactant has only a minor effect [48].

The ISO 14887:2000(E) standard can be used to determine prospective dispersing agents for both cellulose and carbon. We can categorize nanocellulose and CNTs as solids. In that case, when using water as liquid, the category of suitable dispersing agent would be a poly ethylene-oxide (PEO)/alcohol for CNT and PEO/poly propylene oxide (PPO) copolymer for nanocellulose. The standard also provides information about commercial surfactants that fall into the mentioned categories. PEO/PPO copolymer is a suitable surfactant for nanocellulose. The standard denotes that a commercial equivalent is Pluronic®. In the case of CNTs, one example of alkyl phenoxy PEO ethanol dispersing agent is Triton™.

The typical approach to the manufacturing of conductive cellulose-CNT films has been to increase CNT weight percentage without optimizing the dispersion procedure or the used surfactants. Also, the effect of the particular ratios of the cellulose, CNT and surfactant toward each other has not been fully investigated. Even though ultrasonication is widely used for the dispersion and stabilization of CNTs, there is not a standard procedure for the sonication process, and different research groups have applied different sonication treatments to their samples. Sonication can cause chemical functionalization but it can also cause defects and breakage of CNTs [1,50–52]. This will further affect the performance of CNT-based materials and their applications. It has been found in the current literature that sonication parameters such as sonicator type, sonication time and temperature control vary significantly, with reported sonication times ranging from 2 min with tip sonication to 20 h for bath sonication. Dassios et al. [53] attempted to optimize the sonication parameters for the dispersion of MWCNTs in an aqueous solution. Two critical questions concerning the homogeneity of aqueous suspensions of carbon nanotubes by ultrasonic processing were identified; namely, the dependence of dispersion quality on the duration and intensity of sonication and the identification of the appropriate conditions for retaining the highly desirable initial aspect ratio of the free-standing tubes in the dispersed state. Fuge et al. [54] studied the effect of different ultrasonication parameters (time, amplitude) on undoped and nitrogen-doped MWCNTs in aqueous dispersions and found a nearly linear decrease of the arithmetic mean average in MWCNT length with increasing ultrasonication time.

The aim of this study was to optimize the conductivity of NFC-CNT nanocomposite films using a minimum amount of processing steps (e.g., without centrifugal processing of dispersion or pressing of the film), materials and chemicals. In this paper, NFC and multiwall carbon nanotubes (MWCNT) were used to prepare composite films and study the effect of the sonication energy and surfactant type on the electrical conductivity of the nanocomposite. In addition, we investigated the removal of the surfactant from the nanocomposites and the subsequent effect on the electrical conductivity. To our knowledge this is a novel approach and has not been reported previously. The conductivity properties of the nanocomposites were studied as a function of the used sonication energy amount, as well as with and without the presence of surfactant.

2. Results

The impact of sonication energy on electrical conductivity was one of the processing parameters with the highest interest in this study. This was due to the lack of previous research in the current literature. Also, our results show that the surfactant type and sonication energy play a major role in achieving excellent conductivity. In addition to the previously mentioned parameters, removal of surfactant can enhance conductivity values toward levels never seen or reported.

Overall, the shelf-life of the sonicated dispersion samples was significantly long, since samples remained unchanged before the film preparation. Also, sedimentation was not detected, based on the fact that conductivity values were at the same level when measured from both sides of the nanocomposite films. The appearance of the sonicated dispersion samples was identical; however, the consistency and visually inspected viscosity varied with increasing sonication energy. This was observed with Triton X-100 and Pluronic F-127 samples but not in cetyl trimethylammonium bromide (CTAB) surfactant-containing dispersions.

2.1. Conductivity of NFC-CNT Nanocomposite Films

Electrical conductivity of the even and uniform centrifugally cast films was measured using the four-probe measuring technique. With this method, it is possible to minimize the contact resistances and thus provide more accurate conductivity measurements than for the commonly used two-terminal measurement. The sheet resistances of prepared and cut nanocomposite films (size 30 mm × 30 mm) were measured using a four-point probe setup made in-house and a multimeter (Keithley 2002, Tektronix, Inc., Beaverton, OR, USA) in four-wire mode. The probes were placed in line, with equal 3 mm spacing. The four-probe setup is described elsewhere in detail [55]. The conductivity measurements were carried out using a 1 mA current and voltage was measured. Measurements were taken before and after removal of surfactant.

The selection of the most functional surfactant was an important aspect in this study. This selection was determined based on the sheet resistance measurements. The effect of different surfactants and sonication energy on conductivity is shown in Figure 1. From the conductivity diagrams, the effect of surfactant type can be visually observed and estimated.

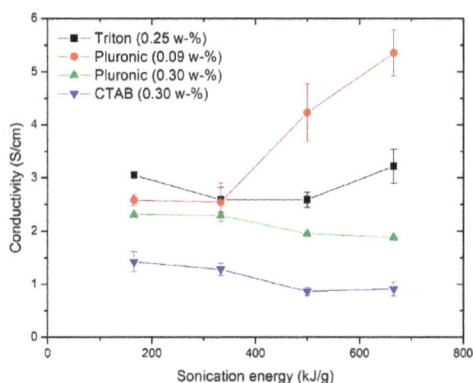

Figure 1. Conductivity of nanocomposite films processed using surfactants Triton X-100, Pluronic F-127 and cetyl trimethylammonium bromide. Concentration in weight percentages.

According to the standard, our assumption was that non-ionic surfactants would be the most promising surfactants. This was clearly the case, since the films made with surfactants Triton X-100 and Pluronic F-127 outperformed the films made with ionic surfactant CTAB.

Visual observations made with CTAB aqueous dispersion samples after sonication indicated that these samples did not gelate even with a higher amount of sonication energy per dry mass (666 kJ/g). This suggests that the dispersion process may not have been entirely successful, since samples had different consistencies and visually separate particles. The ionic surfactant (CTAB) was used to manufacture films at a 1 to 1 ratio of dry mass content of NFC and CNT. The conductivity of films processed using CTAB decreased as the sonication increased from almost 1.5 S/cm to less than 0.90 S/cm. The conductivity diagram of these films was different in its nature; the highest values were measured with the lowest amount of sonication energy.

Based on the standard Pluronic F-127, surfactant should be compatible with cellulosic materials. The first set of Pluronic F-127 nanocomposite films were done with a 1 to 1 ratio to dry mass content (0.30 wt. %). Results show that conductivity is decreasing as a function of sonication energy. Based on this finding, another set of films was manufactured using a surfactant concentration below the CMC limit (0.09 wt. %). Conductivity results for this set of samples show higher conductivity values than films manufactured using a surfactant concentration higher than the CMC value (0.30 wt. %). Using Pluronic F-127 surfactant, the highest conductivity for nanocomposite films was achieved using a sonication energy of 666 kJ/g. When comparing values of films below and above CMC value the difference is sensational 5.36 S/cm (0.09 wt. %) versus 1.88 S/cm (0.30 wt. %).

For Triton, the highest conductivity value of 3.37 S/cm was achieved with 666 kJ/g sonication energy. It should be noted that almost the same conductivity result (3.02 S/cm) was achieved using just 166 kJ/g of sonication energy.

2.2. Effect of Surfactant Removal

Conductivity measurements were also carried out after the removal of the surfactant used in the dispersing phase. Triton X-100 and Pluronic F-127 films were acetone treated and CTAB films were treated with ethanol. It can be clearly seen in Figure 2 that removal of surfactant has a strong effect. Removal of surfactant from films made with CTAB increased the conductivity significantly; the maximum conductivity was 3.02 S/cm for 166 kJ/g sonication energy. However, the films expressed a decrease in conductivity at sonication energy similar to the films with the surfactants present. For Pluronic F-127 films (Figure 2c) made below the CMC limit of the surfactant, the removal of surfactant did not have a significant effect on the conductivity. Even though there was no clear trend, the film with surfactant had somewhat higher conductivity than the one where surfactant was not present. For Pluronic F-127 (0.30 wt. %), the shape of the diagram differed from other previous sets. The films initially exhibited a decrease in conductivity as a function of increasing sonication energy, and the highest values were measured for the samples sonicated at the least energy, but also for the highest amount of sonication energy. For the samples with high conductivity, the removal of the non-ionic surfactant increased the conductivity. The highest value obtained from these measurements was 2.88 S/cm for a sonication energy of 166 kJ/g. A dramatic increase was observed in conductivity values of films manufactured with surfactant Triton X-100 (Figure 2a): the conductivity increases from approximately 3.0 S/cm to a value of 8.42 S/cm when the non-ionic surfactant was removed (sample was sonicated at 666 kJ/g). It should be noted that the films containing the surfactant did not exhibit as strong a sensitivity to increasing sonication energy (as those without surfactant).

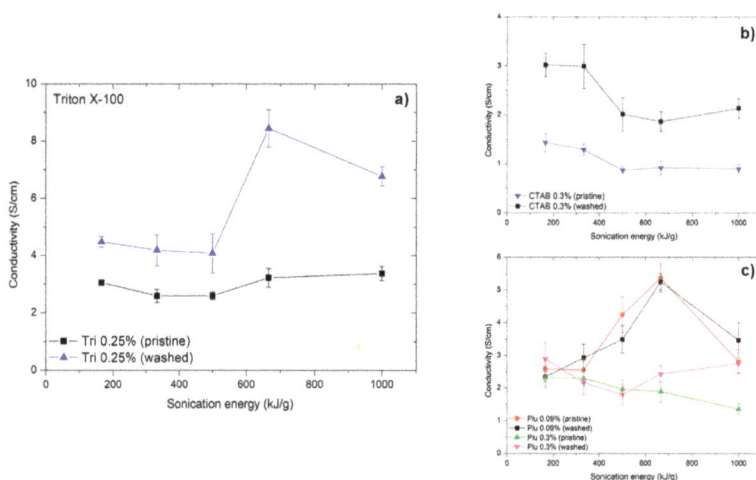

Figure 2. Conductivity of nanocomposite films before and after (**a**) Triton X-100 surfactant removal, (**b**) CTAB surfactant removal, and (**c**) Pluronic surfactant removal.

It is well known that surfactants can plasticize the structure of composites and interfere with conductivity properties by situating themselves at the interface between the conductive particles and matrix. This phenomenon was demonstrated when the properties of the nanocomposite films were compared in this study. Firstly, there was a clear increase in the conductivity of the films processed using the Triton X-100 and CTAB surfactants due to the removal of the surfactant. The diagrams (pristine vs. washed) were similar in their trend, and a clear increase in terms of conductivity was observed. When using the surfactant Pluronic F-127 for processing, a clear conclusion could not be made because the conductivity diagrams did not show a corresponding, monotonic trend due to surfactant removal. However, the results showed that, when the surfactant is present in the film structure, the effect of interference by Pluronic F-127 (concentration below CMC) on the electrical conductivity is at its minimum.

2.3. Comparison to Previous Results

When comparing our nanocomposite film's conductivity results to previous studies, we found that our results were superior to reported values. In Figure 3 are illustrated electrical conductivity results from studies that have used native NFC and manufactured homogenous nanocomposites from it. For non-ionic surfactants, the highest conductivity value found was 0.022 S/cm at a 10 wt. % CNT loading [17]. In our study, the highest value was 8.42 S/cm after removing Triton X-100 and, likewise, 5.35 S/cm with Pluronic F-127 still present in the film.

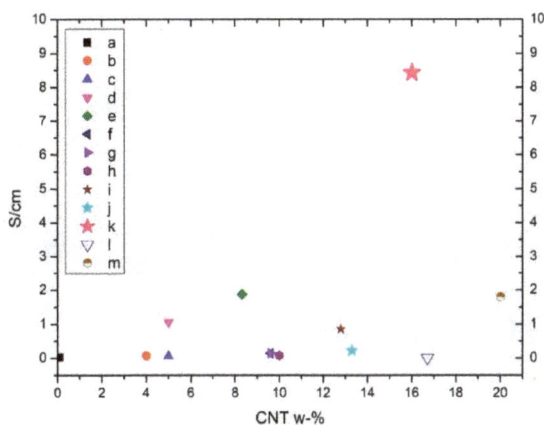

Figure 3. Comparison of obtained electrical conductivity of NFC-CNT nanocomposite films from the current literature. Pink star (letter k) refers to our data (Triton X-100), while other letters refer to a [9], b [56], c [57], d [58], e [22], f [9], g [24], h [57], i [59], j [15], l [32] and m [27].

Huang et al. [57] reported the results of a multiphase process which was used to accomplish a conductivity of 0.072 S/cm using MWCNT-doping at 10 wt. % and 0.056 S/cm with 5 wt. % doping with cotton linters and CTAB as a surfactant. CTAB surfactant was also used with bacterial nanocellulose and CNTs, where the conductivity was 0.027 S/cm (MWCNT 0.1 wt. %) [9]. Also, Yoon et al. [24] used bacterial cellulose as a matrix and obtained conductivity of 0.14 S/cm with 9.6 wt. % MWCNT loading. Electrical conductivity of TEMPO-oxidized cellulose films with 16.7 wt. % concentration of MWCNTs was 0.001 S/cm, which is lower than the conductive material limit [32]. For chitosan-cellulose-CNT membranes, Xiao et al. [56] accomplished conductivity of 0.062 S/cm with a 4 wt. % content of MWCNTs. By using comparable materials, but by applying a filtering method, Yamakawa et al. [58] obtained a 1.05 S/cm electrical conductivity with a 5 wt. % MWCNT loading and

Chen et al. 1.8 S/cm with 20 wt. % MWCNT. They were able to increase the conductivity to a value of 5.02 S/cm using a chemical alkali treatment.

In addition, studies about manufacturing conductive cellulose composites via coating cellulosic filter paper with a CNT dispersion have revealed rather good results, but the consistency of the materials is not homogeneous—not exactly an integral composite. Lee et al. [15] achieved conductivity of 1.11 S/cm using 13.3 wt. % MWCNT. Mondal et al. [59] reported conductivity values after using a dipping method, and they reached 0.85 S/cm with a 12.8 wt. % carbon nanofiber (CNF) content. Fugetsu et al. [22] manufactured conductive cellulose-based composites using a traditional paper making process with 8.32 wt. % CNT concentration and, finally, a conductivity of 1.87 S/cm was obtained.

2.4. Characterization of Nanocomposite Structure

The surface and cross-section of the films processed using surfactant Triton X-100 was studied with SEM. Images were taken with Zeiss ULTRAPlus scanning electron microscope (SEM). The effect of sonication as well as the removal of surfactant were studied with the SEM images shown in Figure 4. Two samples were specifically chosen for this inspection: 166 kJ/g and 666 kJ/g sonication energy films containing surfactant (Figure 4a,c) and after removal of surfactant Triton X-100 by washing them in acetone (Figure 4b,d).

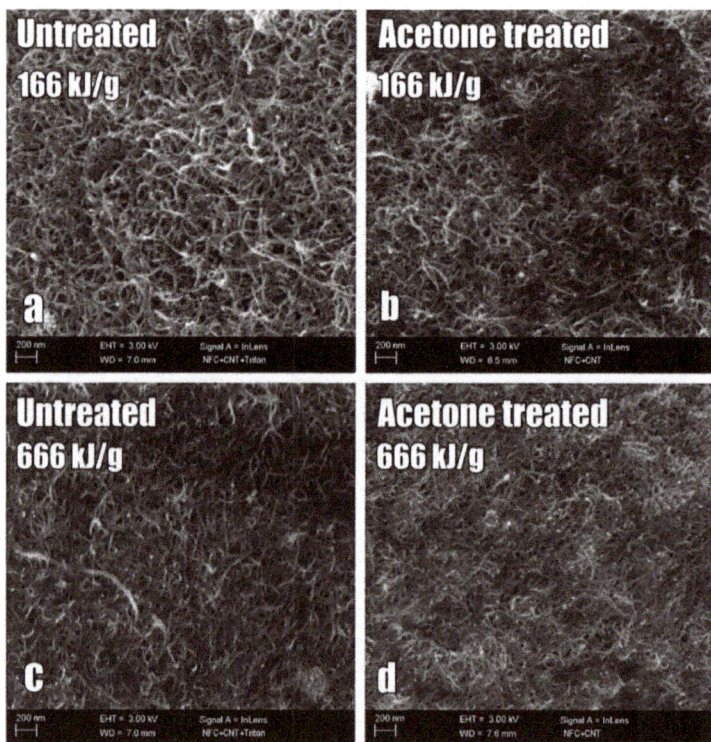

Figure 4. SEM imaging of NFC-CNT nanocomposite films surface (166 and 666 kJ/g) containing surfactant (**a,c**) and after removal of surfactant Triton X-100 by washing them in acetone (**b,d**).

In the top left (Figure 4a) image, some clusters of CNTs are present in the 166 kJ/g sonication energy sample, but not in the higher sonication energy 666 kJ/g sample. Both films have abundant amount of CNTs in the surface. Here, the 166 kJ/g film has a more porous structure than the 666 kJ/g film. In addition, the CNTs form a more consistent network in the 666 kJ/g film after washing the surfactant away (Figure 4d).

The SEM images of the sonicated samples in Figure 4 showed that there were clusters present in the 166 kJ/g sonicated film, while the higher sonicated energy sample did not have similar kinds of clusters. This indicates that, for lower sonication energies, non-dispersed particles remain in the films. This is not preferred, since the purpose is to achieve good dispersion of all the particles in the dispersion and in the films manufactured. This is an indication that the sonication process and amount of energy used affect the extent of the dispersion of particles.

The through-thickness structure of the films was also generally studied using polished cross-sections of the films embedded in epoxy. The cross-section in Figure 5 shows the CNT ends (bright contrast spots) and their even distribution in the film (500 kJ/g), (Triton system) through the thickness. A slightly layered structure can be observed and concluded as a result of dispersion flow during the casting.

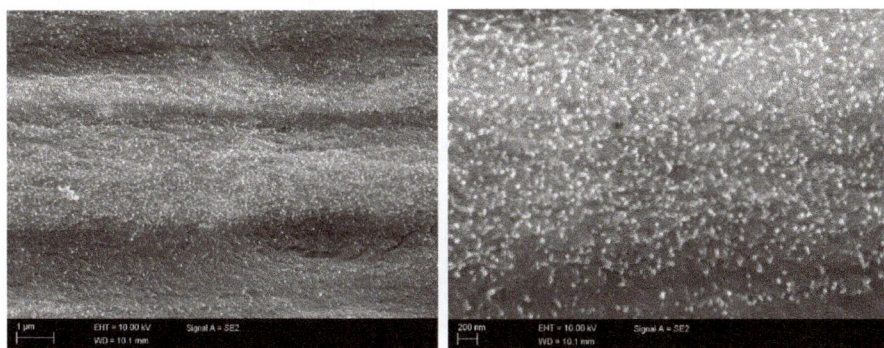

Figure 5. SEM imaging of the nanocomposite film (Triton X-100, 500 kJ/g) cross-section when embedded in epoxy: **Left** side: overall structure; **Right** side: magnification in the center of the film.

3. Discussion

Ultrasonication is a widely used process to manufacture aqueous CNT dispersions. However, how much it changes the properties of dispersed particles and the medium is often overlooked. It is known that sonication can, for example, generate hydrogen peroxide from water, degrade carbon nanotubes and ultimately destroy them. Therefore, it is important that the sonication process is optimized in terms of time and power. It also needs to be noted that the dispersion process is not linear, but follows an S-curve where temporal development of dispersion quality is related to quantity of un-dispersed solid.

Due to the re-agglomeration tendency of carbon nanotubes, it is necessary to use dispersion agents, i.e., surfactants, in manufacturing aqueous dispersions. If these dispersions are later used in conductive films, it is preferable to remove the surfactant to improve CNT network formation. Although carbon nanotubes are excellent conductors, CNT networks are not. This is due their high intertubular contact resistance. The contact points act essentially as a tunneling junction for electrons that is very sensitive to distance. The efficacy of surfactants is based on acting as a spacer between tubes, so any additional distance in a conductive network is detrimental to the conductivity itself.

4. Materials and Methods

Three-component systems containing nanocellulose, carbon nanotubes and surfactants are used in the strong, ecologically conscious nanocomposite films of this study. The CNTs add functionality to the nanocellulose matrix and the surfactant enables percolation network to build and maximize conductivity properties.

In this study, the nanocellulose (NFC) production was based on mechanical disintegration of bleached hardwood kraft pulp (BHKP). First, dried commercial BHKP produced from birch was soaked in water at approximately 1.7 wt. % concentration and dispersed using a high-shear Ystral dissolver for 10 min at 700 rpm. The chemical pulp suspension was predefined in a Masuko grinder (Supermasscolloider MKZA10-15J, Masuko Sangyo Co., Tokyo, Japan) at 1500 rpm and fluidized with six passes through a Microfluidizer (Microfluidics M-7115-30, Microfluidics Corp., Newton, MA, USA) using 1800 MPa pressure. The final material appearance of NCF was a viscous and translucent gel.

Multiwall carbon nanotubes (MWCNT, Nanocyl 7000, Nanocyl SA., Sambreville, Belgium) were purchased from Nanocyl Inc. and the product was used in the state it was in when received. This type of nanotubes is produced via catalytic chemical vapor deposition (CCVD). Concentration of CNTs was kept constant at 16 wt. % in the nanocomposite films, so the effects of sonication energy and surfactant type to the conductivity properties are more visible.

Three surfactants were chosen based on their ionic nature and standard: non-ionic Triton X-100 and Pluronic F-127, and anionic cetyl triammonium bromide (CTAB). Surfactants were purchased from Sigma-Aldrich (Merck KGaA, Darmstadt, Germany). The surfactants were diluted in deionized water to form solutions with variating dissolutions (1, 2.5, 10 wt. %).

Preparation of NFC-CNT Aqueous Dispersion

The NFC and CNT were sonicated simultaneously and after sonication no centrifuge was used so that the preparation of aqueous dispersions could be achieved using a minimum amount of processing phases. NFC-CNT aqueous dispersions with a total volume of 80 mL were prepared. One set contained NFC (0.25 wt. %), CNTs (0.05 wt. %), deionized water and one of the selected surfactants (Triton X-100, 0.25 wt. %, Pluronic F-127, 0.09 wt. % and 0.3 wt. % and CTAB 0.30 wt. %). Details about preparation produce are showed as Figure 6.

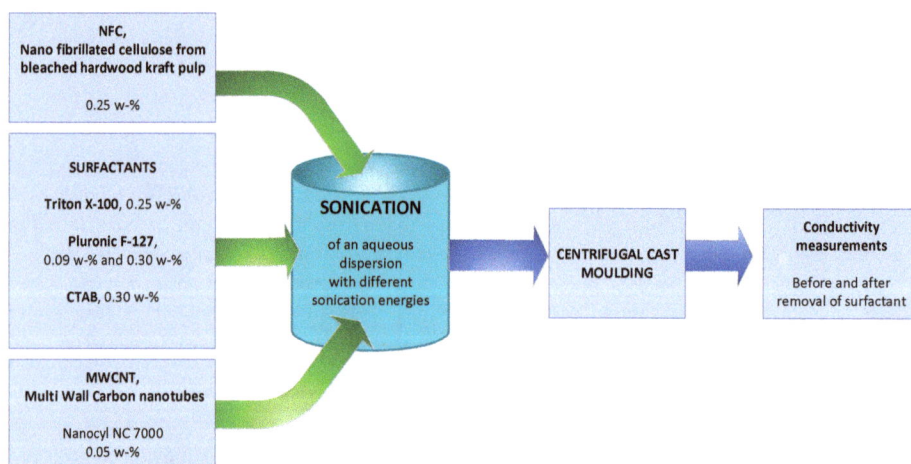

Figure 6. Preparation procedure of NFC-CNT dispersion and nanocomposite films.

The total dry mass for all the dispersions was 0.24 g. The sonication of the dispersion samples was performed with a tip horn (ø 12.7 mm) sonicator Q700 (QSonica LLC., Newton, CT, USA) in 100 mL glass beakers. The sonication amplitude of vibration (50%) was kept constant. The power output remained between 60 and 70 W for every sonication. The system included a water bath to keep samples cool during the sonication so that temperature would not rise above 30 °C. The water bath was cooled by circulating cooling glycerol through a chiller (PerkinElmer C6 Chiller, PerkinElmer Inc., Waltham, MA, USA). Samples were sonicated for four different amounts of energies per dry mass, respectively 166, 333, 500 and 666 kJ/g, which corresponded to energies of 40, 80, 120 and 160 kJ. Unsonicated samples manufactured using all three surfactants were not homogenous, and this is why film formation was unsuccessful and not analyzed.

5. Conclusions

The typical approach to the manufacturing of conductive cellulose-CNT films has been to increase CNT weight percentage without optimizing the dispersion procedure. In this study, NFC and multiwall carbon nanotubes (MWCNT) were used to prepare composite films using a minimum number of processing phases (e.g., no centrifugal dispersion or pressing of the film were used), materials and chemicals. The amount of CNTs was 0.05 wt. % in dispersion and 16 wt. % in the film after the evaporation of water in ratio to dry mass content of NFC and CNT. The effect of surfactant type (Triton X-100, Pluronic F-127 and CTAB) and sonication energy on the electrical conductivity of NFC-CNT nanocomposite films was investigated to identify optimal processing conditions for high conductivity of the nanocomposite. A conductivity of 5.36 S/cm was achieved by using Pluronic F-127 surfactant and 666 kJ/g of sonication energy. In addition, removal of the surfactant from film and its effect on the electrical conductivity was studied. A dramatic increase in conductivity values from approximately 3.0 S/cm to a value of 8.42 S/cm was observed for films manufactured with surfactant Triton X-100. Conductivity diagrams of the nanocomposite films show that sonication affects the electrical performance of the films. SEM images of sonicated samples showed that the films sonicated at 166 kJ/g have a more porous structure than the films sonicated at higher energy. The imaging also showed that the CNTs form a more consistent network with a combination of high sonication energy and surfactant removal. It can be concluded that the following parameters significantly affect the conductivity of NFC-CNT nanocomposite films:

(a) Surfactant type
(b) Surfactant concentration
(c) Sonication energy
(d) Removal of the used surfactant
(e) Film processing technique

To summarize, we manufactured nanocomposite films with exemplary conductivity in comparison to reported research and this was achieved by optimizing processing parameters and materials. Further research on the surfactant types and concentration can lead to better dispersion of the CNTs and therefore even higher conductivity.

Author Contributions: S.S. and P.K. conceived and designed the experiments; S.S. and A.R. performed the experiments and analyzed the data; V.K. and A.H. contributed materials; S.S., A.R., P.K., S.T. and K.R.R. wrote the paper. M.K. and J.V. coordinated the project aims in accordance to publication specific actions and delegation.

Funding: This research received no external funding.

Acknowledgments: This work was funded by Tekes (Finnish Funding Agency for Innovation) through a strategic opening entitled Design Driven Value Chains in the World of Cellulose (DWoC 2.0). We acknowledge the contributions of Jarmo Laakso and Essi Sarlin for SEM imaging.

Conflicts of Interest: The authors declare no conflict of interest.

References

1. Huang, J.C. Carbon black filled conducting polymers and polymer blends. *Adv. Polym. Technol.* **2002**, *21*, 299–313. [CrossRef]
2. Ma, P.C.; Siddiqui, N.A.; Marom, G.; Kim, J.K. Dispersion and functionalization of carbon nanotubes for polymer-based nanocomposites: A review. *Compos. Part A Appl. Sci. Manuf.* **2010**, *41*, 1345–1367. [CrossRef]

3. Pereira, C.; Nóvoa, P.J.R.O.; Calard, V.; Forero, S.; Hepp, F.; Pambaguian, L. Characterization of Carbon Nanotube Papers Infused with Cyanate-Ester Resin. In Proceedings of the ICCM International Conference on Composite Materials, Edinburgh, UK, 27–31 July 2009.

4. Layek, R.K.; Das, A.K.; Park, M.J.; Kim, N.H.; Lee, J.H. Enhancement of physical, mechanical, and gas barrier properties in noncovalently functionalized graphene oxide/poly(vinylidene fluoride) composites. *Carbon N. Y.* **2015**, *81*, 329–338. [CrossRef]

5. Palola, S.; Sarlin, E.; Kolahgar Azari, S.; Koutsos, V.; Vuorinen, J. Microwave induced hierarchical nanostructures on aramid fibers and their influence on adhesion properties in a rubber matrix. *Appl. Surf. Sci.* **2017**, *410*, 145–153. [CrossRef]

6. Hamedi, M.M.; Hajian, A.; Fall, A.B.; Hkansson, K.; Salajkova, M.; Lundell, F.; Wgberg, L.; Berglund, L.A. Highly conducting, strong nanocomposites based on nanocellulose-assisted aqueous dispersions of single-wall carbon nanotubes. *ACS Nano* **2014**, *8*, 2467–2476. [CrossRef] [PubMed]

7. Tuukkanen, S.; Streiff, S.; Chenevier, P.; Pinault, M.; Jeong, H.J.; Enouz-Vedrenne, S.; Cojocaru, C.S.; Pribat, D.; Bourgoin, J.P. Toward full carbon interconnects: High conductivity of individual carbon nanotube to carbon nanotube regrowth junctions. *Appl. Phys. Lett.* **2009**, *95*, 113108. [CrossRef]

8. Haghi, A.K.; Thomas, S. *Carbon Nanotubes: Theoretical Concepts and Research Strategies for Engineers*; Apple Academic Press: Waretown, NJ, USA, 2015.

9. Jung, R.; Kim, H.-S.; Kim, Y.; Kwon, S.-M.; Lee, H.S.; Jin, H.-J. Electrically Conductive Transparent Papers Using Multiwalled Carbon Nanotubes. *J. Polym. Sci. Part B Polym. Phys.* **2008**, *46*, 1235–1242. [CrossRef]

10. Haghi, A.K.; Zaikov, G.E. *Advanced Nanotube and Nanofiber Materials*; Nova Science Publishers, Inc.: New York, NY, USA, 2012; ISBN 978-1-62-081201-3.

11. Hoeng, F.; Denneulin, A.; Bras, J. Use of nanocellulose in printed electronics: A review. *Nanoscale* **2016**, *8*, 13131–13154. [CrossRef] [PubMed]

12. Gardner, D.J.; Oporto, G.S.; Mills, R.; Samir, M.A.S.A. Adhesion and surface issues in cellulose and nanocellulose. *J. Adhes. Sci. Technol.* **2008**, *22*, 545–567. [CrossRef]

13. Kuzmenko, V.; Naboka, O.; Haque, M.; Staaf, H.; Göransson, G.; Gatenholm, P.; Enoksson, P. Sustainable carbon nanofibers/nanotubes composites from cellulose as electrodes for supercapacitors. *Energy* **2015**, *90*, 1490–1496. [CrossRef]

14. Lehtimäki, S.; Tuukkanen, S.; Pörhönen, J.; Moilanen, P.; Virtanen, J.; Honkanen, M.; Lupo, D. Low-cost, solution processable carbon nanotube supercapacitors and their characterization. *Appl. Phys. A Mater. Sci. Process.* **2014**, *117*, 1329–1334. [CrossRef]

15. Lee, T.W.; Lee, S.E.; Jeong, Y.G. Carbon nanotube/cellulose papers with high performance in electric heating and electromagnetic interference shielding. *Compos. Sci. Technol.* **2016**, *131*, 77–87. [CrossRef]

16. Yun, S.; Kim, J. Multi-walled carbon nanotubes-cellulose paper for a chemical vapor sensor. *Sens. Actuators B Chem.* **2010**, *150*, 308–313. [CrossRef]

17. Qi, H.; Liu, J.; Pionteck, J.; Pötschke, P.; Mäder, E. Carbon nanotube-cellulose composite aerogels for vapour sensing. *Sens. Actuators B Chem.* **2015**, *213*, 20–26. [CrossRef]

18. Qi, H.; Mäder, E.; Liu, J. Unique water sensors based on carbon nanotube-cellulose composites. *Sens. Actuators B Chem.* **2013**, *185*, 225–230. [CrossRef]

19. Wang, M.; Anoshkin, I.V.; Nasibulin, A.G.; Korhonen, J.T.; Seitsonen, J.; Pere, J.; Kauppinen, E.I.; Ras, R.H.A.; Ikkala, O. Modifying native nanocellulose aerogels with carbon nanotubes for mechanoresponsive conductivity and pressure sensing. *Adv. Mater.* **2013**, *25*, 2428–2432. [CrossRef] [PubMed]

20. Oya, T.; Ogino, T. Production of electrically conductive paper by adding carbon nanotubes. *Carbon N. Y.* **2008**, *46*, 169–171. [CrossRef]

21. Hu, L.; Choi, J.W.; Yang, Y.; Jeong, S.; La Mantia, F.; Cui, L.-F.; Cui, Y. Highly conductive paper for energy-storage devices. *Proc. Natl. Acad. Sci. USA* **2009**, *106*, 21490–21494. [CrossRef] [PubMed]

22. Fugetsu, B.; Sano, E.; Sunada, M.; Sambongi, Y.; Shibuya, T.; Wang, X.; Hiraki, T. Electrical conductivity and electromagnetic interference shielding efficiency of carbon nanotube/cellulose composite paper. *Carbon N. Y.* **2008**, *46*, 1256–1258. [CrossRef]

23. Imai, M.; Akiyama, K.; Tanaka, T.; Sano, E. Highly strong and conductive carbon nanotube/cellulose composite paper. *Compos. Sci. Technol.* **2010**, *70*, 1564–1570. [CrossRef]

24. Yoon, S.H.; Jin, H.J.; Kook, M.C.; Pyun, Y.R. Electrically conductive bacterial cellulose by incorporation of carbon nanotubes. *Biomacromolecules* **2006**, *7*, 1280–1284. [CrossRef] [PubMed]

25. Toomadj, F.; Farjana, S.; Sanz-Velasco, A.; Naboka, O.; Lundgren, P.; Rodriguez, K.; Toriz, G.; Gatenholm, P.; Enoksson, P. Strain sensitivity of carbon nanotubes modified cellulose. *Procedia Eng.* **2011**, *25*, 1353–1356. [CrossRef]

26. Liu, Y.; Liu, D.; Ma, Y.; Sui, G. Characterization and properties of transparent cellulose nanowhiskers-based graphene nanoplatelets/multi-walled carbon nanotubes films. *Compos. Part A Appl. Sci. Manuf.* **2016**, *86*, 77–86. [CrossRef]

27. Chen, C.; Mo, M.; Chen, W.; Pan, M.; Xu, Z.; Wang, H.; Li, D. Highly conductive nanocomposites based on cellulose nanofiber networks via NaOH treatments. *Compos. Sci. Technol.* **2018**, *156*, 103–108. [CrossRef]

28. Hilding, J.; Grulke, E.; George Zhang, Z.; Lockwood, F. Dispersion of Carbon Nanotubes in Liquids. *J. Dispers. Sci. Technol.* **2003**, *24*, 1–41. [CrossRef]

29. Vaisman, L.; Wagner, H.D.; Marom, G. The role of surfactants in dispersion of carbon nanotubes. *Adv. Colloid Interface Sci.* **2006**, *128–130*, 37–46. [CrossRef] [PubMed]

30. Rastogi, R.; Kaushal, R.; Tripathi, S.K.; Sharma, A.L.; Kaur, I.; Bharadwaj, L.M. Comparative study of carbon nanotube dispersion using surfactants. *J. Colloid Interface Sci.* **2008**, *328*, 421–428. [CrossRef] [PubMed]

31. Bai, J.B.; Allaoui, A. Effect of the length and the aggregate size of MWNTs on the improvement efficiency of the mechanical and electrical properties of nanocomposites—Experimental investigation. *Compos. Part A Appl. Sci. Manuf.* **2003**, *34*, 689–694. [CrossRef]

32. Salajkova, M.; Valentini, L.; Zhou, Q.; Berglund, L.A. Tough nanopaper structures based on cellulose nanofibers and carbon nanotubes. *Compos. Sci. Technol.* **2013**, *87*, 103–110. [CrossRef]

33. Huang, Y.Y.; Terentjev, E.M. Dispersion and rheology of carbon nanotubes in polymers. *Int. J. Mater. Form.* **2008**, *1*, 63–74. [CrossRef]

34. Gimåker, M.; Östlund, M.; Östlund, S.; Wågberg, L. Influence of beating and chemical additives on residual stresses in paper. *Nord. Pulp Pap. Res. J.* **2011**, *26*, 445–451. [CrossRef]

35. Baez, C.; Considine, J.; Rowlands, R. Influence of drying restraint on physical and mechanical properties of nanofibrillated cellulose films. *Cellulose* **2014**, *21*, 347–356. [CrossRef]

36. Rosen, M.J. *Surfactants and Interfacial Phenomena*; John Wiley & Sons, Inc.: Hoboken, NJ, USA, 2004; ISBN 978-0-47-147818-8.

37. Keinänen, P.; Siljander, S.; Koivula, M.; Sethi, J.; Vuorinen, J.; Kanerva, M. Optimized dispersion quality of aqueous carbon nanotube colloids as a function of sonochemical yield and surfactant/CNT ratio. *Heliyon* **2018**, in press.

38. Blanch, A.J.; Lenehan, C.E.; Quinton, J.S. Optimizing Surfactant Concentrations for Dispersion of Single-Walled Carbon Nanotubes in Aqueous Solution. *J. Phys. Chem. B* **2010**, *114*, 9805–9811. [CrossRef] [PubMed]

39. Yu, J.; Grossiord, N.; Koning, C.E.; Loos, J. Controlling the dispersion of multi-wall carbon nanotubes in aqueous surfactant solution. *Carbon N. Y.* **2007**, *45*, 618–623. [CrossRef]

40. Islam, M.F.; Rojas, E.; Bergey, D.M.; Johnson, A.T.; Yodh, A.G. High weight fraction surfactant solubilization of single-wall carbon nanotubes in water. *Nano Lett.* **2003**, *3*, 269–273. [CrossRef]

41. Utsumi, S.; Kanamaru, M.; Honda, H.; Kanoh, H.; Tanaka, H.; Ohkubo, T.; Sakai, H.; Abe, M.; Kaneko, K. RBM band shift-evidenced dispersion mechanism of single-wall carbon nanotube bundles with NaDDBS. *J. Colloid Interface Sci.* **2007**, *308*, 276–284. [CrossRef] [PubMed]

42. Sun, Z.; Nicolosi, V.; Rickard, D.; Bergin, S.D.; Aherne, D.; Coleman, J.N. Quantitative Evaluation of Surfactant-stabilized Single-walled Carbon Nanotubes: Dispersion Quality and Its Correlation with Zeta Potential. *J. Phys. Chem. C* **2008**, *112*, 10692–10699. [CrossRef]

43. Maillaud, L.; Zakri, C.; Ly, I.; Pénicaud, A.; Poulin, P. Conductivity of transparent electrodes made from interacting nanotubes. *Appl. Phys. Lett.* **2013**, *103*, 263106. [CrossRef]

44. Bai, Y.; Lin, D.; Wu, F.; Wang, Z.; Xing, B. Adsorption of Triton X-series surfactants and its role in stabilizing multi-walled carbon nanotube suspensions. *Chemosphere* **2010**, *79*, 362–367. [CrossRef] [PubMed]

45. Calvaresi, M.; Dallavalle, M.; Zerbetto, F. Wrapping nanotubes with micelles, Hemimicelles, and cylindrical micelles. *Small* **2009**, *5*, 2191–2198. [CrossRef] [PubMed]

46. Geng, Y.; Liu, M.Y.; Li, J.; Shi, X.M.; Kim, J.K. Effects of surfactant treatment on mechanical and electrical properties of CNT/epoxy nanocomposites. *Compos. Part A Appl. Sci. Manuf.* **2008**, *39*, 1876–1883. [CrossRef]

47. Bystrzejewski, M.; Huczko, A.; Lange, H.; Gemming, T.; Büchner, B.; Rümmeli, M.H. Dispersion and diameter separation of multi-wall carbon nanotubes in aqueous solutions. *J. Colloid Interface Sci.* **2010**, *345*, 138–142. [CrossRef] [PubMed]

48. Angelikopoulos, P.; Gromov, A.; Leen, A.; Nerushev, O.; Bock, H.; Campbell, E.E.B. Below the CMC. *J. Phys. Chem. C* **2010**, *114*, 2–9. [CrossRef]

49. Bonard, J.; Stora, T.; Salvetat, J.; Maier, F.; Stockli, T.; Duschl, C.; De Heer, W.A.; Forró, L.; Châtelain, A. Purification and Size-Selection of Carbon Nanotubes. *Adv. Mater.* **1997**, *9*, 827–831. [CrossRef]

50. Lu, P.; Hsieh, Y. Lo Multiwalled carbon nanotube (MWCNT) reinforced cellulose fibers by electrospinning. *ACS Appl. Mater. Interfaces* **2010**, *2*, 2413–2420. [CrossRef] [PubMed]

51. Rossell, M.D.; Kuebel, C.; Ilari, G.; Rechberger, F.; Heiligtag, F.J.; Niederberger, M.; Koziej, D.; Erni, R. Impact of sonication pretreatment on carbon nanotubes: A transmission electron microscopy study. *Carbon N. Y.* **2013**, *61*, 404–411. [CrossRef]

52. Yang, D.; Rochette, J.-F.; Sacher, E. Functionalization of Multiwalled Carbon Nanotubes by Mild Aqueous Sonication. *J. Phys. Chem. B* **2005**, *109*, 7788–7794. [CrossRef] [PubMed]

53. Dassios, K.G.; Alafogianni, P.; Antiohos, S.K.; Leptokaridis, C.; Barkoula, N.M.; Matikas, T.E. Optimization of sonication parameters for homogeneous surfactant assisted dispersion of multiwalled carbon nanotubes in aqueous solutions. *J. Phys. Chem. C* **2015**, *119*, 7506–7516. [CrossRef]

54. Fuge, R.; Liebscher, M.; Schröfl, C.; Oswald, S.; Leonhardt, A.; Büchner, B.; Mechtcherine, V. Fragmentation characteristics of undoped and nitrogen-doped multiwalled carbon nanotubes in aqueous dispersion in dependence on the ultrasonication parameters. *Diam. Relat. Mater.* **2016**, *66*, 126–134. [CrossRef]

55. Rajala, S.; Tuukkanen, S.; Halttunen, J. Characteristics of piezoelectric polymer film sensors with solution-processable graphene-based electrode materials. *IEEE Sens. J.* **2015**, *15*, 3102–3109. [CrossRef]

56. Xiao, W.; Wu, T.; Peng, J.; Bai, Y.; Li, J.; Lai, G.; Wu, Y.; Dai, L. Preparation, structure, and properties of chitosan/cellulose/multiwalled carbon nanotube composite membranes and fibers. *J. Appl. Polym. Sci.* **2013**, *128*, 1193–1199. [CrossRef]

57. Huang, H.D.; Liu, C.Y.; Zhang, L.Q.; Zhong, G.J.; Li, Z.M. Simultaneous reinforcement and toughening of carbon nanotube/cellulose conductive nanocomposite films by interfacial hydrogen bonding. *ACS Sustain. Chem. Eng.* **2015**, *3*, 317–324. [CrossRef]

58. Yamakawa, A.; Suzuki, S.; Oku, T.; Enomoto, K.; Ikeda, M.; Rodrigue, J.; Tateiwa, K.; Terada, Y.; Yano, H.; Kitamura, S. Nanostructure and physical properties of cellulose nanofiber-carbon nanotube composite films. *Carbohydr. Polym.* **2017**, *171*, 129–135. [CrossRef] [PubMed]

59. Mondal, S.; Ganguly, S.; Das, P.; Bhawal, P.; Das, T.K.; Nayak, L.; Khastgir, D.; Das, N.C. High-performance carbon nanofiber coated cellulose filter paper for electromagnetic interference shielding. *Cellulose* **2017**, *24*, 5117–5131. [CrossRef]

© 2018 by the authors. Licensee MDPI, Basel, Switzerland. This article is an open access article distributed under the terms and conditions of the Creative Commons Attribution (CC BY) license (http://creativecommons.org/licenses/by/4.0/).

International Journal of
Molecular Sciences

MDPI

Article

Degradation Studies Realized on Natural Rubber and Plasticized Potato Starch Based Eco-Composites Obtained by Peroxide Cross-Linking

Elena Manaila [1], Maria Daniela Stelescu [2] and Gabriela Craciun [1,*]

[1] Electron Accelerators Laboratory, National Institute for Laser, Plasma and Radiation Physics,
    409 Atomistilor Street, 077125 Magurele, Romania; elena.manaila@inflpr.ro
[2] National R&D Institute for Textile and Leather—Leather and Footwear Research Institute,
    93 Ion Minulescu Street, 031215 Bucharest, Romania; dmstelescu@yahoo.com
*   Correspondence: gabriela.craciun@inflpr.ro; Tel.: +40-21-457-4346

Received: 2 August 2018; Accepted: 18 September 2018; Published: 20 September 2018

Abstract: The obtaining and characterization of some environmental-friendly composites that are based on natural rubber and plasticized starch, as filler, are presented. These were obtained by peroxide cross-linking in the presence of a polyfunctional monomer used here as cross-linking co-agent, trimethylolpropane trimethacrylate. The influence of plasticized starch amount on the composites physical and mechanical characteristics, gel fraction and cross-link density, water uptake, structure and morphology before and after accelerated (thermal) degradation, and natural (for one year in temperate climate) ageing, was studied. Differences of two orders of magnitude between the degradation/aging methods were registered in the case of some mechanical characteristics, by increasing the plasticized starch amount. The cross-link density, water uptake and mass loss were also significant affected by the plasticized starch amount increasing and exposing for one year to natural ageing in temperate climate. Based on the results of Fourier Transform Infrared Spectroscopy (FTIR) and cross-link density measurements, reaction mechanisms attributed to degradation induced by accelerated and natural ageing were done. SEM micrographs have confirmed in addition that by incorporating a quantity of hydrophilic starch amount over 20 phr and by exposing the composites to natural ageing, and then degradability can be enhanced by comparing with thermal degradation.

Keywords: natural rubber; plasticized starch; polyfunctional monomers; physical and mechanical properties; cross-link density; water uptake

1. Introduction

Natural rubber (NR) is renewable, non-toxic, has excellent physical properties and due to its low price is the most used elastomer worldwide in industry or in a variety of applications in which the final products are in contact with food or potable water [1]. The most common physical–chemical treatment of rubber is curing (cross-linking, vulcanization) by sulphur, peroxides, ultraviolet light, electron beam, and microwave irradiation, but sulphur and peroxide curing systems still remain the most desirable. The application of sulphur systems leads to the formation of sulphidic cross-links between elastomer chains [2]. In peroxide curing, high thermal stabile C–C bonds are formed. Therefore, peroxide cured elastomers exhibit high-temperature ageing resistance and low compression that is set at high temperatures. There are some disadvantages when compared peroxide to sulfur cured systems, such as low scarce safety and worse dynamic and elastic properties of vulcanizates [3]. But, the cross-linking with peroxides can be effectively improved by the use of co-agents [4–6], because they are able to boost peroxide efficiency by suppressing side reactions to a large extent, like chain scission and disproportionation [3,7], or by the formation of co-agents bridges between polymer chains as extra cross-links [8,9]. Trimethylolpropane trimethacrylate (TMPT), is a polyfunctional monomer that is used as co-agent [10,11] in order to improve the cross-linking process because is able to increase the rate state of cure and, as a consequence, to improve the physical properties of the processed material [8,12]. Anyway, the NR is used in form of mixtures, which generally contain active fillers, plasticizers, cross-linking agents, and other ingredients that give different characteristics to the final product [12,13]. New environmental-friendly elastomeric materials can be obtained by the use of natural fillers in NR and other rubber blends instead of hazardous active fillers, such as silica or carbon black, which are very well known for the harmful effects on human health [14]. So, it is preferable to replace the classics with other compatible types of filler that should maintain, or even improve, the mechanical and usable properties of NR or other rubber products. Because of the increased interests in replacement of non-renewable rubber materials with some based on components originated from natural resources, the use of natural fillers is considered as being a promising solution [15,16]. Even though their nature is known to be hydrophilic, starches are considered among the most promising available natural biopolymers and may successful candidate for the development of novel composite materials based on NR [17,18]. These are cheap, abundant available, biodegradable, recyclable, renewable, and present thermoplastic behavior [19–21]. Thermoplastic properties of starch are obtained by the disruption and plasticization of native starch with plasticizers agents (glycerol, water, and other polyols) [21–23].

The goal of the paper is to present the obtaining and characterization of a new environmental-friendly elastomeric composite that is based on NR and plasticized starch (PS). The cross-linking method was by the use of peroxide in the presence of the trimethylolpropane trimethacrylate (TMPT). The elastomeric composite behavior under accelerated (thermal) aging and natural aging in temperate climate for one year was investigated, because it is very well known that starch-based plastics have some drawbacks, including limited long term stability caused by water absorption, ageing caused degradation, poor mechanical properties and bad processability. Also, the influence of PS amount on the physical and mechanical properties, cross-linking density rate and behavior in aqueous environment, before and after aging was studied. The novelty of the present study consists in the replacing of conventional fillers (silica, carbon black) with an environmental friendly filler (plasticized starch) in order to obtain composites having improved cross-linking density and mechanical properties, and over these, a high degradation degree in natural environment.

2. Results and Discussion

2.1. Physical and Chemical Characteristics

The results of mechanical tests that have been made on unfilled NR and NR filled with PS composites (NR–PS), before and after aging are summarized in Figure 1. An improvement of mechanical properties before aging, excepting 100% Modulus (Figure 1c), due to the plasticized starch introduction, can be observed.

2.1.1. Mechanical Properties of Unfilled and Filled NR before Aging

As it can be seen from Figure 1a, hardness has slowly increased with the PS amount in blend increasing. This is due the reinforcement effect of PS, which incorporated in the NR matrix has conducted to a reduction of plasticity and flexibility of rubber chains and the composite become more rigid [14,24]. Irrespective of the PS amount added, the increase in hardness did not exceed 12.5% for unfilled NR. The results that are presented in Figure 1b show that the unfilled NR has exhibited a higher elastic response when compared to that of NR-PS samples. The composite having 10 phr of PS has a better elastic behavior than unfilled NR. The higher dynamic stiffness of the samples containing over 20 phr of PS can be the reason of the elasticity decreases [25]. The same results were obtained for 100% Modulus Figure 1c. The results presented in Figure 1d show that the tensile strength, which strongly depends on effective and uniform stress distribution, increases as the amount of PS increase, up to 20 phr. As the PS amount still increases, the tensile strength started to decrease by about

45%, because the PS tend to agglomerate, leading to insufficient wetting of filler and bad interface with the matrix. So, the PS acts as flows limiting the tensile strength. Also, another reason may be connected with the presence of some voids trapped in the composite during processing, so a specific attention has to be accorded to the preparation technique when is dealing with high filler loading in order to assure its well dispersion [26,27].

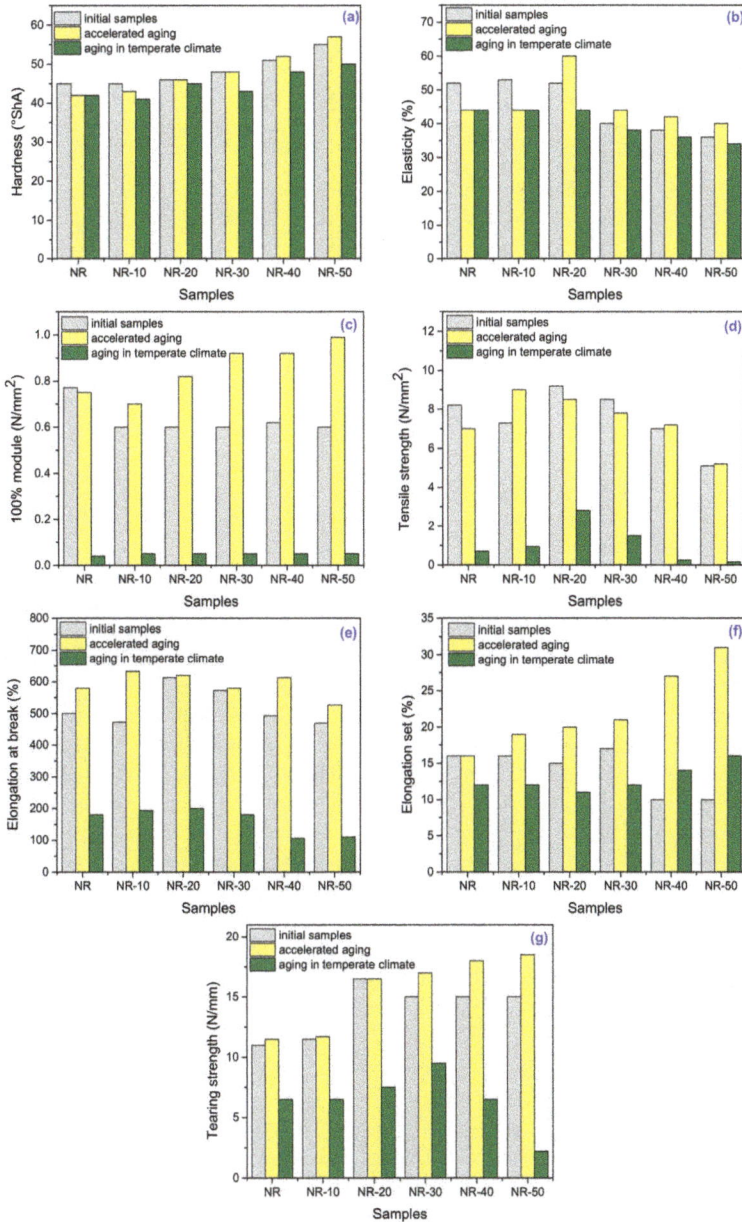

Figure 1. The Hardness (**a**), Elasticity (**b**), 100% Modulus (**c**), Tensile strength (**d**), Elongation at break (**e**), Elongation set, and (**f**) Tearing strength (**g**) variations as a function of PS amount and aging method.

As it can be seen from Figure 1e, elongation at break has the same behavior as tensile strength on the entire PS concentration range. The decrease highlighted over 20 phr of PS, is the consequence of the appeared restriction in the movement of molecular chains, which has lead on a negative effect upon the sample ductility [14,28,29]. Figure 1f shows that the elongation set has decreased with the PS amount in blend increasing, a fact that indicates also an increase in cross-link density. The decrease in residual elongation shows that the sample is vulcanized and thus returns to its original shape easily [14,24]. Figure 1g shows also the same variation trend of tearing strength as tensile strength and elongation at break. So, we can conclude that the PS has a reinforcing effect on NR when it is loaded up to 20 phr.

2.1.2. Mechanical Properties of Unfilled and Filled NR after Aging

Mechanical Properties of Unfilled and Filled NR after Accelerated (Thermal) Ageing

A specimen from each sample was subjected to thermal ageing in an air-circulating oven at 70 °C for 168 h, in order to evaluate the rubber compound properties before and after aging. As it can be seen from Figure 1a–g and Table 1, after the thermal aging, mechanical properties are modified for all the tested composites.

Table 1. Percentage modifications of the mechanical properties of unfilled and filled natural rubber (NR) after thermal aging.

Mechanical Property (%)	Sample Type (PS Loading) (phr of PS at 100 phr of NR)					
	NR	NR-10	NR-20	NR-30	NR-40	NR-50
Hardness	−6.67	−4.44	0	0	+1.96	+3.64
Elasticity	−15.38	−16.98	+15.38	+10.00	+10.53	+11.11
100% Modulus	−2.60	+16.67	+36.67	+53.33	+53.33	+65.00
Tensile strength	−14.63	+23.29	−7.61	−8.24	+2.86	+1.96
Elongation at break	+16.33	+33.83	+1.14	+1.22	+24.34	+12.13
Elongation set	0.00	+18.73	+33.33	+23.53	+170.00	+210.00
Tearing strength	+4.55	+1.74	0	+13.33	+20.00	+23.33

The resistance of rubber based composites to thermal aging is considered as being an essential requirement for better service performance. The increasing of elasticity, 100% modulus, tensile strength, elongation at break, tensile set, and tearing strength of NR-PS composites is due to the presence of peroxide free radicals that were not involved in the cross-linking reactions and that lead to the formation of few new cross-links during thermal aging. The phenomenon is well known as post-curing during aging [30,31]. On the other hand, elasticity, 100% modulus and tensile strength of the unfilled NR have diminished after thermal aging due to the post-curing during aging by which excessive cross-links were formed [30,32]. It seems that the PS presence, even in the small amount of 10 phr, has delayed the formation of excessive cross-links in NR-PS as compared with the unfilled NR.

Mechanical Properties of Unfilled and Filled NR after 1 Year of Natural Ageing in Temperate Climate

Another specimen from each sample was subjected to natural ageing for one year in temperate climate between March 2017 and March 2018, in Bucharest, Romania. In order to determine the effect of outdoor exposure, samples were suspended in vertical position on a special dryer. The influence of the natural environment (heat, cold, frost, sunlight, oxygen, moisture, precipitations) upon the degradation was evaluated by comparing the samples mechanical properties before and after aging processes.

From the results presented in Figure 1a–g and Table 2, it can be seen the important changes suffered by all mechanical properties of unfilled and filled NR composites, after one year of natural ageing in temperate climate staying. The negative modifications of mechanical properties clearly show the sample degradation during the natural process. So, 100% modulus and tensile strength present an important falling-off that indicates structural changes in rubber chains due to the cross-links dissociation.

Table 2. Percentage modifications of the mechanical properties of unfilled and filled NR after one year of natural ageing in temperate climate.

Mechanical Property (%)	Sample Type (PS Loading) (phr of PS at 100 phr of NR)					
	NR	NR-10	NR-20	NR-30	NR-40	NR-50
Hardness	−6.67	−8.89	−2.17	−10.42	−5.88	−9.09
Elasticity	−15.38	−16.98	−15.38	−5.00	−5.26	−5.56
100% Modulus	−94.81	−91.67	−91.67	−91.67	−91.67	−91.67
Tensile strength	−91.64	−86.99	−69.57	−82.35	−96.29	−96.86
Elongation at break	−64.00	−59.20	−67.37	−68.59	−78.50	−76.60
Elongation set	−25.00	−25.00	−26.67	−29.41	+40.00	+60.00
Tearing strength	−40.91	−43.48	−54.55	−36.67	−56.67	−85.33

As a consequence, the composites lose the elastic properties and the ability to act as an effective matrix material to transmit stress [30,33]. An important decrease of elongation at break was also observed. This is as a consequence of both cross-linking and scission reactions, which negative modify the elastic nature of rubber chains [30,34,35].

2.1.3. Gel Fraction and Cross-Link Density

Figure 2a,b presents the results regarding the gel fraction and cross-link density investigations before and after aging. In Table 3, the percentage changes of these parameters for unfilled and filled NR after one year of natural ageing in temperate climate are presented.

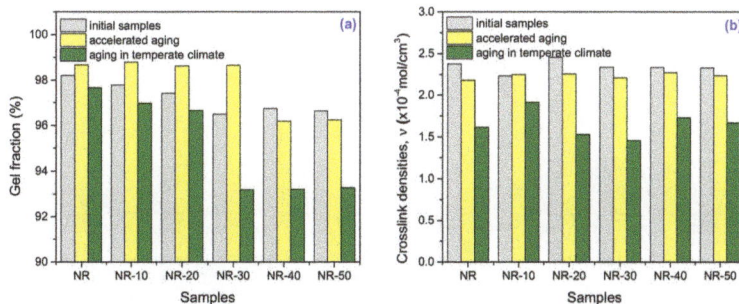

Figure 2. Gel Fraction (**a**) and Cross-link density (**b**) variations as a function of plasticized starch (PS) amount and aging method.

Table 3. Percentage modifications of gel fraction and cross-link densities of unfilled and filled NR after 1 year of natural ageing in temperate climate.

Sample Type (PS Loading)	Gel Fraction		Cross-Link Density	
	Thermal Aging	Natural Aging for 1 Year	Thermal Aging	Natural Aging for 1 Year
NR	+0.47	−0.56	−8.28	−32.13
NR-10	+1.05	−0.83	+0.71	−14.44
NR-20	+1.26	−0.78	−8.07	−37.81
NR-30	+2.24	−3.42	−5.47	−37.83
NR-40	−0.57	−3.55	−2.60	−26.03
NR-50	−0.41	−3.48	−3.99	−28.46

As it can be seen from Figure 2a,b, before being subjected to aging tests, the introduction of PS has induced a decrease of samples gel fraction and a not so significant modification of cross-link density. Continuing the loading with PS over 20 phr, the cross-link density remains on a path. The slight variations of cross-link density may be explained by the changing in phase structure character of NR in which the PS was introduced. But, for most applications, the cross-link density should not be so high and it must be sufficient to give the rubber mechanical integrity so that it can bear loads and present deformation recovery. A high cross-link density immobilized the polymer chains, fact that lead to a hard and brittle rubber [36,37].

Figure 2a,b and Table 3 show that after the accelerated (thermal) aging, an increase in the gel fraction is observed for both unfilled and filled NR composites up to the loading with 30 phr. Over this, the gel fraction has decreases and remains on a path. The cross-link density was more affected by the thermal aging. But, after one year of natural degradation in temperate climate, all the samples gel fractions and cross-link densities have strongly decreased. NR samples have showed a reduction of 32.13%, while NR-PS samples of 37.81% for the loading of 20 phr and 37.83% for 30 phr, respectively. The results are in agreement with those that were obtained in mechanical tests and prove a clear degradation of the composites.

2.2. Structural and Morphological Characteristics

2.2.1. Spectral Characterization by Fourier Transform Infrared Spectroscopy Analysis

In order to investigate structural modification of filled NR as compared with unfilled NR before and after aging, the spectral characterization by Fourier Transform Infrared Spectroscopy (FTIR) in the range of 600–4000 cm^{-1}, was done. The results are presented in Figure 3a–c.

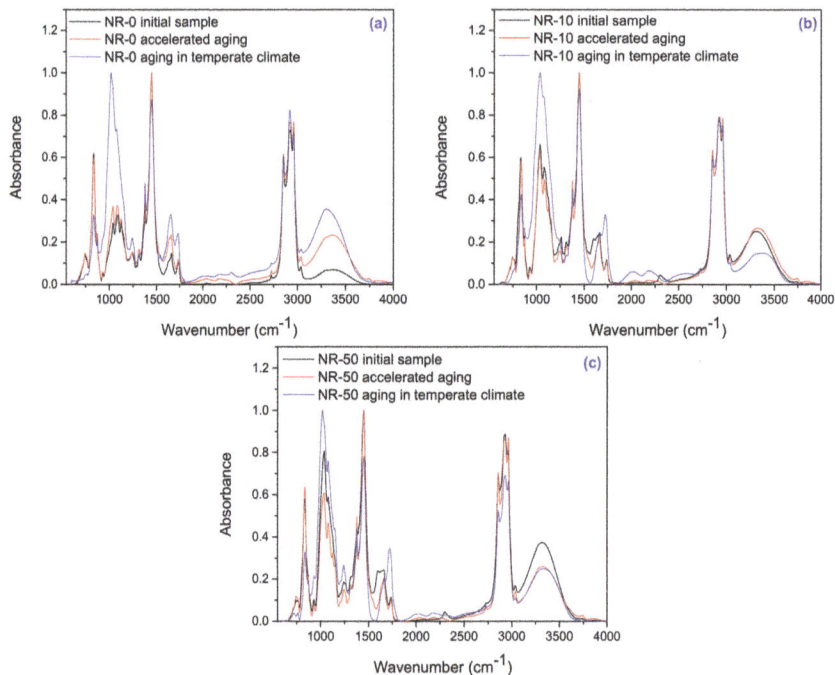

Figure 3. Infrared spectra in the range of 600–4000 cm^{-1} for samples unfilled (**a**), filled with 10 phr PS (**b**) and with 50 phr PS (**c**).

The assignments of the main bands in the NR and NR-PS before and after aging are summarized in Table 4. A comparative analysis between the FTIR spectra of NR (Figure 3a) and NR-PS (Figure 3b,c) highlights notable differences (Table 4).

Table 4. Characteristic infrared bands observed in NR and NR-PS spectra.

Wave Number (cm^{-1})	Assignment
740–760	C–O–C ring vibration from starch or deformation vibration of R$_2$C=CH–R groups from NR
833	=CH out-of-plane bending vibration from NR rubber
870	C(1)–H(α) bending vibration from starch
930–925	skeletal mode vibrations of α-(1-4) glycosidic linkage (C–O–C) from starch
1034–1038	C–O stretching vibration in C–O–H and C–O–C in the anhydrous glucose ring from starch
1080–1086	C–O–C stretching vibration that indicate the grafting of PS on NR
1125–1126	C–O stretching of C–O–C (from starch) or of alcohols >HC–OH resulted from the degradation
1240–1260	carbonyl ((>C=O) and hydroxyl (–OH) compound resulted from the degradation
1310–1315	bending vibration of C–H and C–O groups of aromatic rings (starch)
1370–1380	–CH$_3$ asymmetric deformation of NR
1440–1450	–CH$_2$– deformation vibration from NR or –CH$_2$– symmetric bending vibration from starch
1655–1665	–C=C– stretching vibration in the NR structure or may be due to absorbed water or carboxylate or conjugated ketone (>C=O) resulted from the degradation
1710–1740	the fatty acid ester groups existing in NR or carbonyl group (>C=O) from ketone (R$_2$C=O) or aldehyde (RCOH) resulted from the oxidative degradation
2852–2854	–CH$_2$– symmetric stretching vibration of NR
2919–2927	–CH$_2$– asymmetric stretching vibration of NR
2958–2960	–CH$_3$ asymmetric stretching vibration of NR
3030–3040	=CH– stretching vibration of –CH=CH$_2$ group from NR
3300–3380	N–H stretching vibration of amide groups from the existing proteins in NR or from OH-stretching vibration (–OH as a result of the degradation by oxidation)

Thus, the band that appeared at 1081 cm^{-1} (C–O–C) indicates the grafting of PS on NR [38]. Also, the band at 930–925 cm^{-1} can be attributed to the skeletal mode vibrations of α-(1-4) glycosidic linkage (C–O–C) and the one between 1100 cm^{-1} and 1030 cm^{-1} is characteristic of the anhydrous glucose ring C–O stretch [39,40]. The absorption bands at 3380 cm^{-1} that appear in NR-0 before aging treatment (Figure 3a) were identified to the proteins from NR [41]. After degradation, the spectra indicate that the intensity of the broad bend near 3380 cm^{-1} has increases. The broad band at 3380 cm^{-1} may be due formation of hydroxyl group (–OH) as a result of the degradation by oxidation [38,42]. It can be seen that after 1 year of natural ageing in temperate climate, all these bands have increased in intensity. The decreasing of *cis*-1,4 double bonds number in the polyisoprene chain at 833 cm^{-1}, the formation of hydroxyl group at 3380 cm^{-1}, the appearance of ketone and aldehyde groups between 1736–1722 cm^{-1}, and the increasing of glycosidic linkage at 930–925 cm^{-1} may be interpreted as consequences of the degradation induced by the accelerated and natural aging processes to which samples have been subjected. It can be seen that after natural ageing, all these bands have increased in intensity. All of these, correlated with the decreasing of *cis*-1,4 double bonds number in the polyisoprene chain at 870–830 cm^{-1}, the formation of hydroxyl group at 3380 cm^{-1}, the appearance of ketone and aldehyde groups between 1736–1722 cm^{-1} and the increasing of glycosidic linkage at 930–925 cm^{-1} may be interpreted as consequences of the degradation. Also, the formation of carbonyl or hydroxyl bonds (>C=O and –OH) and carboxylate or conjugated ketone (RCOOH and R$_2$C=O) is demonstrated by the occurrence of the absorption bands between 1260–1400 cm^{-1} and 1550–1690 cm^{-1}, respectively [43]. The cross-linking degree and molecular masses decreasing are due to the cleavage of the macromolecules, as demonstrated by the appearance of the absorption bands at 1370–1380 cm^{-1}, 2850–2880 cm^{-1}, and 2950–2980 cm^{-1} that correspond to –CH$_3$ groups. The slight increasing of the absorption bands at 1640–1660 cm^{-1} and 800–900 cm^{-1} due to the number of double bonds increasing, as well as the modification of absorption bands at 800–900 cm^{-1}, 1650–1680 cm^{-1}, and 3010–3040 cm^{-1} that correspond to the changes in the degree of substitution of carbon atoms of the double bond are connected with the degradation process [43].

2.2.2. Mechanisms of Natural Degradation Reactions for NR and NR-PS

In temperate climate, the main ageing processes in NR are the oxidative and thermal-oxidative degradation (photo-degradation) and UV/ozone degradation [36,44].

The initial step of oxidative and thermal-oxidative degradation consists in free-radicals formation on the NR chain by hydrogen abstraction.

The propagation of oxidative degradation takes place in several stages as it can be seen from Figure 4. The propagation first step is the reaction of a free radical with an oxygen molecule (O$_2$) to form a peroxy radical (NR–O–O•), which then abstracts a hydrogen atom from another polymer chain to form a hydroperoxide (NR–O–O–H). The hydroperoxide splits then into two new free radicals, (NR–O• and •OH) that abstract another hydrogen and from other polymer chains.

The termination of the reaction is achieved by the recombination of two radicals or by disproportionation/hydrogen abstraction, as it can be seen from Figure 5.

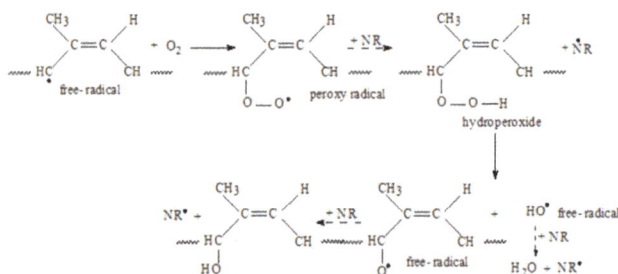

Figure 4. Propagation step of oxidative degradation.

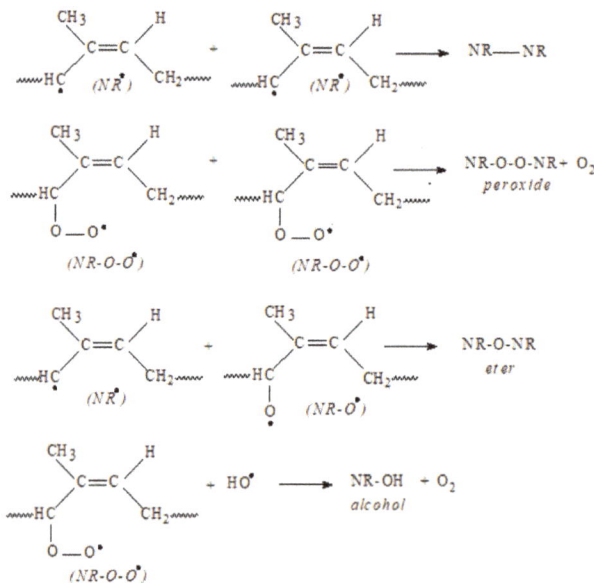

Figure 5. Termination of oxidative degradation by free radical recombination.

The results of these reactions are the polymer enbrittlement and cracking. On the other hand, termination by chain scission, as presented in Figure 6, results in the decrease of the molecular weight leading to softening of the polymer and reduction of the mechanical properties [44].

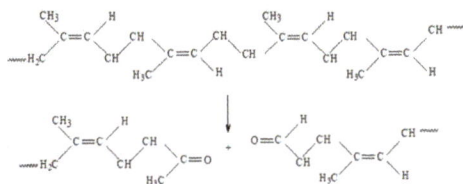

Figure 6. Termination of oxidative degradation by chain scission.

In photo-degradation, if free radicals are directly produced by UV radiation, then all of the subsequent reactions are similar to those of thermal-oxidative degradation, including chain scission, cross-linking, and secondary oxidation. Photo-oxidation can also occur by breaking the bonds that are created between NR and PS (Figure 7). Starch may form other radicals by the cleavage of a glycosidic bond and from β-fragmentation of an oxygen-centred radical resulting from cleavage of a glycosidic bond [45].

Sunlight and ozone rapidly attack the unprotected polymers and can significantly reduce them service life [44]. Ozone, being more corrosive then the molecular oxygen will attack the polymer directly at the carbon-carbon double bond. As it can be seen from Figure 8, the ozone molecule attaches itself to the double bond creating a C_2O_3 ring. The chain scission results in products containing carbonyl groups.

Ozone degradation of vulcanized NR exhibits a very specific cracking, through which it is formed a hard surface layer [46]. The reaction with ozone, leads to chain scission of the polymer chain and the formation of polymeric peroxides that can also increase the rate of oxidative aging [36,47]. As those processes are related to substantial modifications of the macromolecular backbone, substantial damage in mechanical properties are expected, even at low rate of conversion (less than 0.1%) [36,37]. This was observed for all samples aged for one year in temperate climates, for both mechanical properties and cross-linking density.

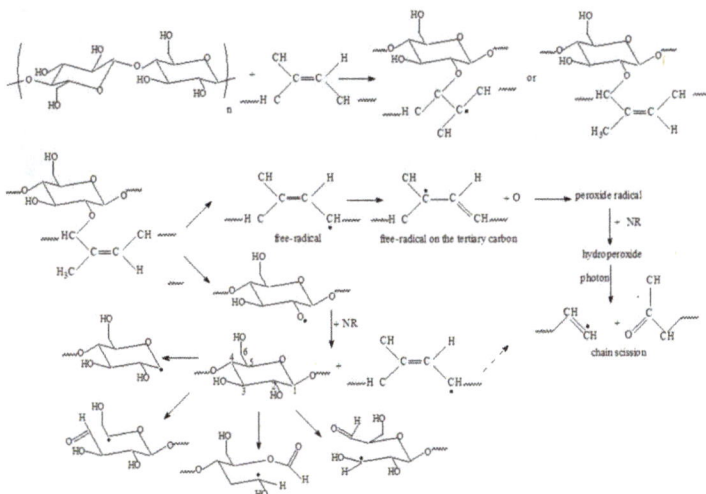

Figure 7. The photo-oxidation reaction on NR—starch bond.

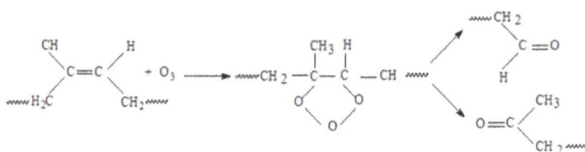

Figure 8. The ozone attack on NR chain.

2.2.3. SEM Analysis

The morphological changes of unfilled and filled NR samples before and after one year of natural ageing in temperate climate were evaluated by surfaces SEM analysis. Before and after ageing, samples were immersed in toluene, in order to remove any split fragments or un-reacted materials. The results are presented in Figures 9 and 10.

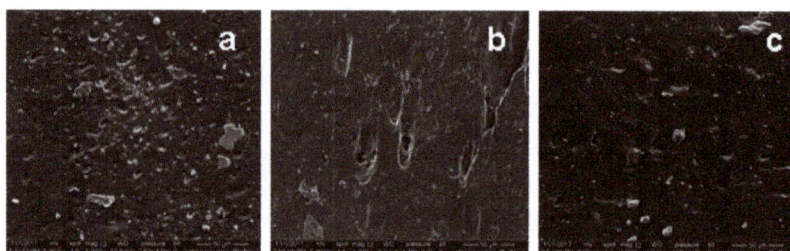

Figure 9. SEM micrographs of (**a**) NR, (**b**) NR-10, and (**c**) NR-50 samples, before one year of natural ageing in temperate climate, at magnification of 1000.

From Figure 9, it can be seen that, before aging, the NR and NR-PS samples surface looks smooth presenting only with small imperfections that can be caused by the presence of some impurities remained even after the immersion in toluene.

Figure 10. SEM micrographs of (**a**) NR, (**b**) NR-10, and (**c**) NR-50 samples, after one year of natural ageing in temperate climate, at magnification of 1000.

From Figure 10, the NR and NR-PS samples degradation after 1 year of natural ageing in temperate climate can be observed. The surfaces of all samples are rough, extensive surface cracks are formed, leading the appearance similar to a mosaic pattern. As the PS amount has increased, the roughness of surface sample also increased and more micro cracks are formed. The surface roughness can be attributed to the decreasing of elasticity, 100% modulus and tensile strength, as the filler loading was increased, but also to the dissociation of existing cross-links or to the structural changes in natural rubber chains [48]. In addition to the oxidative and ozone degradation, the outdoor thermal stress and the dust presence must be taken into account for the contribution to the cracks formation on the composite surface [48–50].

2.3. Water Uptake and Mass Loss

Water uptake tests were done before and after ageing in order to demonstrate the contribution of the strongly hydrophilic polar groups that appeared in the FTIR spectra of the composite, to the natural degradation. The results are presented in Figure 11a–d. As it can be seen, the water absorption of NR-PS composites is strongly dependent on the PS amount, increasing with the PS content due to the hydrophilic nature of starch and the greater interfacial area between the starch and the NR matrix.

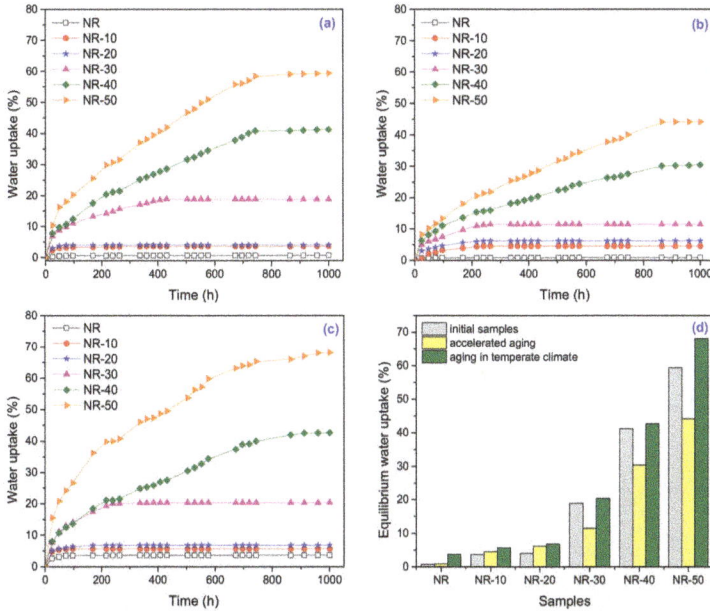

Figure 11. Water uptake (**a**) before aging, (**b**) after the accelerated ageing, (**c**) after one year of natural ageing in temperate climate, and (**d**) at equilibrium before and after ageing.

Before ageing, the smallest absorptions were obtained for PS loadings up to 20 phr. Over this, the water uptake percentages were of 41.19% and 59.31% for 40 phr and 50 phr, respectively (Figure 11a). The post-curing phenomenon, after the thermal aging, has lead to the decrease of water uptake (Figure 11b). When the composites have been exposed to high temperature for longer period (7 days at 70 °C), the oxygen molecules from the air have been diffused into the surface, but some of them were immediately consumed by oxidation reactions to produce cross-linking, scission of the rubber chains or cross-links, and combination with molecular chains of rubber [51]. So, we can conclude that the water uptake is related with the cross-link densities of the composites. After 1 year of natural ageing in temperate climate, the water uptake percentages are increased, as it can be seen from Figure 11c,d. The phenomenon theory is talking about the difficulty of solvent molecule to penetrate the carbon linkages (C–C) because of the strong bonding and high rigidity [51,52]. But, in our study, the results show that the water molecules have easy penetrated the NR, the water uptake before and after one year of natural ageing in temperate climate being of 0.71% and 3.68%, respectively. In the same, the NR-PS composites have presented a great increase in water uptake percentages as comparing with the samples before and after thermal ageing, this indicating the decreasing of cross-link density and increasing of chain scissions. The disruption of C–C bonds in NR leads mainly to chain scission, while the disruption of C–O bonds from NR-PS causes the scission of grafting bonds [53]. Since the C–C and C–O bonding energies are comparable (346 vs. 358 kJ = mol), both phenomena can equally occur in degradation. Also, the appearance of some changes in the composites structure supported by FTIR results (decreasing of *cis*-1,4 double bonds number in the polyisoprene chain, the formation of hydroxyl group, the appearance of ketone and aldehyde groups and the increasing of glycosidic linkage) that can be interpreted as being consequences of the natural degradation, can also explain the increases in water absorption. From Figure 11d, clearly the results demonstrate the superiority of the natural ageing over thermal ageing. Also, it can be seen that high PS loadings create conditions for stronger degradation. Before and after immersion in water, all of the samples were weighed in order to determine the mass loss. The results are presented in Table 5.

Table 5. Mass lose before and after thermal and natural ageing.

Samples Codes	Mass Loss (%)		
	Before Ageing	After Accelerated Ageing	After 1 Year of Natural Ageing
NR	0.256	0.263	0.282
NR-10	0.263	0.286	0.293
NR-20	0.309	0.409	0.417
NR-30	1.059	0.963	1.095
NR-40	2.083	1.031	2.619
NR-50	2.971	1.126	5.960

From Table 5 it can be seen that NR does not presents significant mass losses before and after accelerated and natural ageing, higher being however after one year in temperate climate (0.026%). The filled NR has shows small mass differences, but increased with the PS amount, before and after ageing, the highest being of 2.99% for the case of NR-50 aged for one year in temperate climate. It should be noted that the mass losses for all of the samples loaded with PS over 30%, were smaller after thermal ageing than before. These results are the consequence of the post-curing effect and are in perfect agreement with those obtained after the mechanical properties evaluation. The results presented in Table 5 for NR and NR-PS after one year of natural ageing in temperate climate, show that higher PS content increases mass loss and enhances the degradation kinetics due to the starch hydrophilic nature, which leads to the moisture retaining. The higher is the starch content in the polymer, the higher is the moisture content that renders faster degradation. This can be correlated with the polymer sample's gross morphology, which was observed to be physically changed, the surface being roughened over the degradation period as it can be seen in SEM micrographs from Figure 10 [54].

3. Materials and Methods

3.1. Materials and Samples Preparation

The raw materials that were used in the experiments were as follows: (a) Natural rubber (NR) for pharmaceutical use, Crep from Sangtvon Rubber Ltd. (Nakhon Si Thammarat, Thailand), in the form of white rubber sheets (Mooney viscosity of 67.64 ML1+4 at 100 °C, volatile materials content of 0.5%, nitrogen content of 0.45%, percentage of ash of 0.25%, impurities content of 0.026%); (b) Soluble potato starch produced by Lach-Ner Ltd. (Neratovice, Czech Republic), water insoluble substances 0.28%; loss on drying 16.9%, easily biodegradable: BOD_5 −0.6 g/g and COD −1.2 mg/g); (c) Glycerine from SC Chimreactiv SRL (Bucharest, Romania) (free acidity 0.02%, density 1.26 g/cm³, purity 99.5%); (d) IPPD antioxidant (4010 NA) *N*-isopropyl-*N*-phenyl-phenylene diamine from Dalian Richon Chem Co. Ltd. (Dalian, China), 98% purity, molecular mass: 493.6374 g/mol; (e) Peroxide Perkadox, 40 dibenzoyl peroxide, from AkzoNobel Chemicals (Deventer, The Netherlands) (density 160 g/cm³, 3.8% active oxygen content, 40% peroxide content, pH 7); and, (f) TMPT DL 75 Luvomaxx, trimethylolpropane trimethacrylate as polyfunctional monomer from Lehmann&Voss&Co (Hamburg, Germany) (22% ash, pH 9.2, density 1.36 g/cm³, 75 ± 3% active ingredient).

Composites that are based on NR and PS have been obtained according with the recipes that are presented in Table 6. Mixtures were cross-linked with peroxide in the presence of TMPT; a polyfunctional monomer was used here as curing co-agent.

Table 6. The recipes used for composites obtaining.

Ingredients (phr)	Mixtures Codes					
	NR	NR-10	NR-20	NR-30	NR-40	NR-50
Natural rubber (NR)	100	100	100	100	100	100
Starch	0	10	20	30	40	50
Glycerine	0	6	12	18	24	30
Peroxyde	8	8	8	8	8	8
TMPT	3	3	3	3	3	3
Antioxidant	1	1	1	1	1	1

PS was obtained by mixing at 70 °C, starch (50%), water (20%), and glycerine (30%) for 15 min at 50–100 rpm until the homogeneity was attended. After, the homogeneous mixture has been left to rest for 1 h, then being introduced in the oven at 80 °C for 22 h and at 110 °C for another 2 h. Finally, has been left to cool down for at least 16 h in a dry place.

The blends were prepared on an electrically roller mixer. The constituents were added in the following sequences and amounts: NR that has been mixed in the roller mixer for 2 min, PS and glycerine (mixing time between 5 and 30 min), antioxidant (mixing time: 1 min), peroxide, and TMPT (mixing time: 1 min). After all of the ingredients were added, they were homogenized for another 2 min and then removed from the roller mixer in the form of a sheet. The sheets have been cured using moulds and vulcanization press in order to obtain rubber plates with the sizes of 150 × 150 × 2 mm^3 being required for die punching test specimens. The compression temperature in the moulding machine was kept constant at 160 °C, for 20 min at a pressure of 300 kN. Cooling time was 10 min at 25 °C and 300 kN.

One specimen from each composite that have been obtained as above was subjected to accelerated ageing (thermal ageing) in an air-circulating oven at 70 °C for 168 h and another one to natural ageing for one year in temperate climate. Physical and mechanical properties, cross-linking density rate, behavior in aqueous environment, and structural and morphological investigations were done before and after ageing.

3.2. Laboratory Tests

3.2.1. Mechanical Characteristics Determining

Hardness, elasticity, 100% Modulus, tensile strength, elongation, and tearing strength were measured. Hardness was measured according to ISO 7619-1/2011 on 6 mm thick samples, while using a hardness tester. Elasticity (the rebound resilience) was evaluated according to ISO 4662/2009 also on 6 mm thick samples, while using the Schob test machine. Tensile strength and tearing strength tests were carried out with a Schopper strength tester at testing speed of 460 mm/min, using dumbbell shaped specimens according to ISO 37/2012, and angular test pieces (Type II) according to EN 12771/2003, respectively.

3.2.2. Sol-Gel Analysis

The sol-gel analysis, were performed on the cross-linked composites in order to determine the mass fraction of insoluble NR resulting from the network-forming cross-linking process. Samples having known mass were swollen in toluene for 72 h in order to remove any split fragments and un-reacted materials, and they were dried in air for six days and then in a laboratory oven at 80 °C for 12 h. Finally, samples were re-weighed and the gel fraction was calculated, as follows:

$$Gel_{fraction} = \frac{m_s}{m_i} \times 100 \tag{1}$$

where m_s and m_i are the mass of the dried sample after extraction and the initial mass of the sample, respectively [55,56].

3.2.3. Cross-Link Density Determining

The samples cross-link density was determined on the basis of equilibrium solvent-swelling measurements in toluene at 23–25 °C, by application of the modified Flory-Rehner equation for tetra functional networks. Samples having thicknesses of 2 mm were initially weighed (m_i) and immersed in toluene for 72 h. The swollen samples were dried to remove the solvent excess and weighed (m_g) being covered, in order to avoid toluene evaporation during weighing. Traces of solvent and other small molecules were eliminated by drying in air for six days and then in an oven at 80 °C for 12 h. Finally, the samples were weighed for the last time (m_s), and volume fractions of polymer in the samples at equilibrium swelling v_{2m} were determined from swelling ratio G, as follows:

$$v_{2m} = \frac{1}{1+G} \tag{2}$$

$$G = \frac{m_g - m_s}{m_s} \times \frac{\rho_e}{\rho_s} \tag{3}$$

where, ρ_e and ρ_s are the densities of samples and solvent (0.866 g/cm^3 for toluene), respectively.

Densities were determined by hydrostatic weighing method, according to SR ISO 2781/2010. The cross-linking densities v, were determined from measurements in a solvent, while using the Flory–Rehner relationship:

$$v = -\frac{Ln(1 - v_{2m}) + v_{2m} + \chi_{12}v_{2m}^2}{V_1 \left(v_{2m}^{1/3} - \frac{v_{2m}}{2} \right)} \tag{4}$$

where, V_1 is the molar volume of solvent (106.5 cm^3/mol for toluene), v_{2m} is the volume fraction of polymer in the sample at the equilibrium swelling, and χ_{12} is the Flory-Huggins polymer-solvent interaction term (the value of and χ_{12} is 0.393 for toluene) [55,56].

3.2.4. Structural and Morphological Measurements

Structural changes of NR and NR-PS before and after ageing, were highlighted by FTIR measurements that have been done while using the TENSOR 27 (Bruker, Bremen, Germany) FTIR spectrophotometer by ATR measurement method. The spectra were obtained from 30 scans mediation, realized in absorption in the range of 4000–600 cm^{-1}, with a resolution of 4 cm^{-1}.

Morphological measurements of NR and NR-PS, also before and after ageing, were done on the samples surfaces while using the scanning electron microscope FEI/Phillips (FEI Company, Hillsboro, OR, USA). For this, the samples were placed on an aluminum mount, sputtered with gold palladium, and then scanned at an accelerating voltage of 30 kV.

3.2.5. Mechanisms of Degradation Reactions

Based on the results that were obtained by FTIR measurements, reaction mechanisms attributed to degradation induced by natural ageing were done. Generally, degradation can be induced by heat (thermal degradation), oxygen (oxidative and thermal-oxidative degradation), light (photo-degradation), and weathering (generally UV/ozone degradation) [44,45].

3.2.6. Water Uptake Tests

The water uptake tests were done in accordance with SR EN ISO 20344/2004 in order to study the water absorption on NR and NR-PS before and after ageing. For this, the samples were dried in a laboratory oven at 80 °C for 3 h and then were cooled at room temperature in desiccators before weighing. Water absorption tests were conducted by immersing the samples in distilled water in beaker and then maintaining at room temperature (25 ± 2 °C). After immersion, the samples were taken out from the water at periodic intervals and the wet surfaces were quickly dried while using a clean dry cloth or tissue paper before weighing. Absorption was calculated from the weight difference. The percentage of samples weight gaining was measured at different intervals of time. The water uptake was calculated, as follows:

$$\text{Water uptake } (\%) = \frac{m_t - m_i}{m_i} \times 100 \tag{5}$$

where, m_t is the weight of the sample immersed in water at time t and m_i is the initial weight of the oven-dried specimen.

Note: For mechanical, sol-gel analysis, cross-link determining, rubber-filler interaction, and water uptake tests before and after ageing, five samples were taken in work and the results are the averages of these five measurements.

4. Conclusions

A new elastomeric composite that is based on natural rubber and plasticized potato starch was obtained by the peroxide cross-linking method in the presence of the cross-linking co-agent trimethylolpropane trimethacrylate. The composite behavior before and after accelerated (thermal) aging and natural degradation in temperate climate for 1 year was investigated in terms of physical and mechanical characteristics. So, an improvement of mechanical properties, excepting 100% Modulus, due to the plasticized starch introduction, was observed before ageing. Also, by the increasing of the plasticized starch amount in the composite, excepting hardness, all other mechanical characteristics have started to decrease around the loading of 20 phr. Thus, we can say that the plasticized starch loading up to 20 phr has a reinforcing effect on natural rubber. By comparing the mechanical properties of the samples after ageing, we have observed the appearance of the post-curing during aging effect, reflected in minor modifications after the accelerated thermal ageing and notable negative modifications due to the natural process. These results are in accordance with those that were obtained by studying the gel fraction and cross-link density. Structural investigations through the FTIR technique before and after thermal and natural ageing were done. The intensity increasing of some bands corresponding to the formation of polar groups, such as carbonyl and hydroxyl between 1400–1260 cm^{-1}, carboxylate or conjugated ketone between 1690–1550 cm^{-1}, and aldehyde groups between 1736–1722 cm^{-1} were attributed to the degradation, sustaining the process efficiency, and are in a perfect accordance with the termination phase of reaction mechanisms that were achieved. Morphological investigations through the SEM technique have showed that, after aging, the plasticized starch amount increasing has been reflected in both surface roughness and more micro cracks appearance also. The water uptake and mass loss tests that have been done before and after ageing also have demonstrated that the natural degradation is favored by the addition of plasticized starch in the composite due to the appearance of some strongly hydrophilic groups from natural rubber, plasticized starch, and composite also. In the process of replacing conventional fillers, as silica and carbon black, with some natural fillers in order to obtain composites that are highly degradable in natural environment, starch can be considered as being a solution.

Author Contributions: The authors contribution was as fallows: conceptualization and methodology, M.E. and S.M.D.; Investigations, M.E., C.G. and S.M.D.; Writing-Original Draft Preparation, M.E.; Writing-Review & Editing, C.G.

Funding: This research was funded by Romanian Ministry of National Education and Scientific Research through NUCLEU Program, grant numbers 26 N/14.03.2016, PN 16.34.01.01/2016-2017 and PN 18.13.01.01./2018.

Acknowledgments: The authors want to thank to Dumitru Marius Grivei for the given support for SEM investigations.

Conflicts of Interest: The authors declare no conflict of interest.

References

1. Craciun, G.; Manaila, E.; Stelescu, M.D. New Elastomeric Materials Based on Natural Rubber Obtained by Electron Beam Irradiation for Food and Pharmaceutical Use. *Materials* **2016**, *9*, 999. [CrossRef] [PubMed]
2. Kruželák, J.; Sýkora, R.; Hudec, I. Sulphur and peroxide vulcanisation of rubber compounds-overview. *Chem. Pap.* **2016**, *70*, 1533–1555. [CrossRef]
3. Kruželák, J.; Sýkora, R.; Hudec, I. Peroxide vulcanization of natural rubber. Part II: Effect of peroxides and co-agents. *J. Polym. Eng.* **2015**, *35*, 21–29. [CrossRef]
4. Henning, S.K.; Boye, W.M. Fundamentals of Curing Elastomers with Peroxides and Coagents II: Understanding the Relationship Between Coagent and Elastomer. *Rubber World* **2009**, *240*, 31–39.
5. Rajan, R.; Varghese, S.; George, K.E. Role of coagents in peroxide vulcanizatin of natural rubber. *Rubber Chem. Technol.* **2013**, *86*, 488–502. [CrossRef]
6. Vieira, E.R.; Mantovani, J.D.; de Camargo Forte, M.M. Comparison between peroxide/coagent cross-linking systems and sulfur for producing tire treads from elastomeric compounds. *J. Elastom. Plast.* **2013**, *47*, 347–359. [CrossRef]
7. Bucsi, A.; Szocs, F. Kinetics of radical generation in PVC with dibenzoyl peroxide utilizing high-pressure technique. *Macromol. Chem. Phys.* **2000**, *201*, 435–438. [CrossRef]
8. Alvarez Grima, M.M. Novel Co-Agents for Improved Properties in Peroxide Cure of Saturated Elastomers. Ph.D. Thesis, University of Twente, Enschede, The Netherlands, 16 February 2007.
9. Drobny, J.G. *Ionizing Radiation and Polymers: Principles, Technology and Applications*, 1st ed.; Elsevier Health Sciences; William Andrew: Norwich, NY, USA, 2012; pp. 88–91. ISBN 978-1-4557-7881-.
10. Dikland, H.G.; Ruardy, T.; Van der Does, L.; Bantjes, A. New coagents in peroxide vulcanization of EPM. *Rubber Chem. Technol.* **1993**, *66*, 693–711. [CrossRef]
11. Thitithammawong, A.; Uthaipan, N.; Rungvichaniwat, A. The effect of the ratios of sulfur to peroxide in mixed vulcanization systems on the properties of dynamic vulcanized natural rubber and polypropylene blends. *Songklanakarin J. Sci. Technol.* **2012**, *34*, 653–662. Available online: http://rdo.psu.ac.th/sjstweb/journal/34-6/0597-0721-34-6-653-662.pdf (accessed on 1 August 2018).
12. Manaila, E.; Craciun, G.; Stelescu, M.D.; Ighigeanu, D.; Ficai, M. Radiation vulcanization of natural rubber with polyfunctional monomers. *Polym. Bull.* **2014**, *71*, 57–82. [CrossRef]
13. Stelescu, M.D.; Manaila, E.; Craciun, G.; Dumitrascu, M. New Green Polymeric Composites Based on Hemp and Natural Rubber Processed by Electron Beam Irradiation. *Sci. World J.* **2014**, *684047*. [CrossRef] [PubMed]
14. Manaila, E.; Stelescu, M.D.; Craciun, G.; Ighigeanu, D. Wood Sawdust/Natural Rubber Ecocomposites Cross-Linked by Electron Beam Irradiation. *Materials* **2016**, *9*, 503. [CrossRef] [PubMed]
15. Datta, J. Effect of Starch Fillers on the Dynamic Mechanical Properties of Rubber Biocomposite Materials. *Polym. Compos.* **2015**, *23*, 109–112. Available online: http://www.polymerjournals.com/pdfdownload/1189816.pdf (accessed on 26 June 2018). [CrossRef]
16. Datta, J.; Rohn, M. Structure, thermal stability and mechanical properties of polyurethanes, based on glycolysate from polyurethane foam waste prepared with use of 1,6-hexanediol as a glycol. *Polimery* **2008**, *53*, 871–875.
17. Liu, C.; Shao, Y.; Jia, D. Chemically modified starch reinforced natural rubber composites. *Polimer* **2008**, *49*, 2176–2181. [CrossRef]
18. Mente, P.; Motaung, T.E.; Hlangothi, S.P. Natural Rubber and Reclaimed Rubber Composites–A Systematic Review. *Polym. Sci.* **2016**, *2*, 1–19. [CrossRef]
19. Lomeli Ramírez, M.G.; Satyanarayana, K.G.; Iwakiri, S.; Bolzon de Muniz, G.; Tanobe, V.; Flores-Sahagun, T.S. Study of the properties of biocomposites. Part, I. Cassava starch-green coir fibers from Brazil. *Carbohydr. Polym.* **2011**, *86*, 1712–1722. [CrossRef]
20. Mali, S.; Grossmann, M.V.E.; García, M.A.; Martino, M.N.; Zaritzky, N.E. Antiplasticizing effect of glycerol and sorbitol on the properties of cassava starch Films. *Braz. J. Food Technol.* **2008**, *11*, 194–200. Available online: http://bj.ital.sp.gov.br/artigos/html/busca/pdf/v11n3a6107a.pdf (accessed on 26 June 2018).

21. Gaspar, M.; Benko, Z.; Dogossy, G.; Reczey, K.; Czigany, T. Reducing water absorption in compostable starch-based plastics. *Polym. Degrad. Stab.* **2005**, *90*, 563–569. [CrossRef]

22. Mathew, A.P.; Dufresne, A. Plasticized waxy maize starch: Effects of polyols and relative humidity on material properties. *Biomacromolecules* **2002**, *3*, 1101–1108. [CrossRef] [PubMed]

23. Van der Burg, M.C.; Van der Woude, M.E.; Janssen, L.P.B.M. The influence of plasticizer on extruded thermoplastics starch. *J. Vinyl Addit. Technol.* **1996**, *2*, 170–174. [CrossRef]

24. Ahmed, K.; Nizami, S.S.; Raza, N.Z.; Mahmood, K. Effect of micro-sized marble sludge on physical properties of natural rubber composites. *Chem. Ind. Chem. Eng. Q.* **2013**, *19*, 281–293. [CrossRef]

25. Thongsang, S.; Sombatsompop, N. Dynamic Rebound Behavior of Silica/Natural Rubber Composites: Fly Ash Particles and Precipitated Silica. *J. Macromol. Sci. Part B* **2007**, *46*, 825–840. [CrossRef]

26. Chen, R.S.; Ahmad, S.; Ab Ghani, M.H.; Salleh, M.N. Optimization of High Filler Loading on Tensile Properties of Recycled HDPE/PET Blends Filled with Rice Husk. *AIP Conf. Proc.* **2014**, *1614*, 46–51. [CrossRef]

27. Nourbakhsh, A.; Baghlani, F.F.; Ashori, A. Nano-SiO_2 filled rice husk/polypropylene composites: Physico-mechanical properties. *Ind. Crop. Prod.* **2011**, *33*, 183–187. [CrossRef]

28. Ahmed, K. Hybrid composites prepared from Industrial waste: Mechanical and swelling behavior. *J. Adv. Res.* **2015**, *6*, 225–232. [CrossRef] [PubMed]

29. Kukle, S.; Gravitis, J.; Putnina, A.; Stikute, A. The effect of steam explosion treatment on technical hemp fibres. In Proceedings of the 8th International Scientific and Practical Conference, Rezekne, Latvia, 20–22 June 2011; pp. 230–237.

30. Hanafi, I.; Muniandy, K.; Othman, N. Fatigue life, morphological studies, and thermal aging of rattan powder-filled natural rubber composites as a function of filler loading and a silane coupling agent. *BioResources* **2012**, *7*, 841–858. Available online: http://ojs.cnr.ncsu.edu/index.php/BioRes/article/view/BioRes_07_1_0841_Ismail_MO_Fatigue_Morpholog_Thermal_Aging_Rattan_Composite/1352 (accessed on 25 June 2018).

31. Abdul Kader, M.; Bhowmick, A.K. Acrylic rubber–fluorocarbon rubber miscible blends: Effect of curatives and fillers on cure, mechanical, aging, and swelling properties. *J. Appl. Polym. Sci.* **2003**, *89*, 1442–1452. [CrossRef]

32. Rattansom, N.; Prasertsri, S. Relationship among Mechanical properties, heat ageing resistance, cut growth behaviour and morphology in natural rubber: Partial replacement of clay with various type of carbon black at similar hardness level. *Polym. Test.* **2009**, *28*, 270–276. [CrossRef]

33. Ismail, H.; Ishiaku, U.S.; Azhar, A.A.; Mohd Ishak, Z.A. A comparative study of the effect of thermo-oxidative aging on the physical properties of rice husk ash and commercial fillers in epoxidized natural rubber compounds. *J. Elastom. Plast.* **1997**, *29*, 270–289. [CrossRef]

34. Azura, A.R.; Ghazali, S.; Mariatti, M. Effects of the filler loading and aging time on the mechanical and electrical conductivity properties of carbon black filled natural rubber. *J. Appl. Polym. Sci.* **2008**, *110*, 747–752. [CrossRef]

35. Khanlari, S.; Kokabi, M. Thermal stability, aging properties, and flame resistance of NR-based nanocomposite. *J. Appl. Polym. Sci.* **2010**, *119*, 855–862. [CrossRef]

36. Martins, A.F.; Visconte, L.L.Y.; Schuster, R.H.; Boller, F.; Nunes, H.C.R.; Nunes, R.C.R. Ageing Effect on Dynamic and Mechanical Properties of NR/Cel II Nanocomposites. *Kautsch. Gummi Kunstst.* **2004**, *57*, 446–451.

37. Somers, A.E.; Bastow, T.J.; Burgar, M.I.; Forsyth, M.; Hill, A.J. Quantifying rubber degradation using NMR. *Polym. Degradr. Stab.* **2000**, *70*, 31–37, . [CrossRef]

38. Riyajan, S.-A.; Sasithornsonti, Y.; Phinyocheep, P. Green natural rubber-g-modified starch for controlling urea release. *Carbohydr. Polym.* **2012**, *89*, 251–258. [CrossRef] [PubMed]

39. Fang, J.M.; Fowler, P.A.; Tomkinson, J.; Hill, C.A.S. The preparation and characterization of a series of chemically modified potato starches. *Carbohydr. Polym.* **2002**, *47*, 245–252. [CrossRef]

40. Mu, T.-H.; Zhang, M.; Raad, L.; Sun, H.-N.; Wang, C. Effect of α-Amylase Degradation on Physicochemical Properties of Pre-High Hydrostatic Pressure-Treated Potato Starch. *PLoS ONE* **2015**, *10*, e0143620. [CrossRef] [PubMed]

41. Eng, A.H.; Tanaka, Y.; Gan, S.N. FTIR studies on amino groups in purified Hevea rubber. *J. Nat. Rubber Res.* **1992**, *7*, 152–155.

42. Kim, I.-S.; Lee, B.-W.; Sohn, K.-S.; Yoon, J.; Lee, J.-H. Characterization of the UV Oxidation of Raw Natural Rubber Thin Film Using Image and FT-IR Analysis. *Elastom. Compos.* **2016**, *51*, 1–9. [CrossRef]
43. Coates, J. Interpretation of Infrared Spectra, A Practical Approach. In *Encyclopedia of Analytical Chemistry*, 1st ed.; Meyers, R.A., Ed.; John Wiley & Sons Ltd.: Chichester, UK, 2000; pp. 10815–10837. Available online: https://pdfs.semanticscholar.org/9203/59b562f615d9f68f8e57b6b6b505aa213174.pdf (accessed on 20 June 2018).
44. Polymer Properties Database. Available online: http://polymerdatabase.com/polymer%20chemistry/Thermal%20Degradation.html (accessed on 12 June 2018).
45. Alberti, A.; Bertini, S.; Gastaldi, G.; Iannaccone, N.; Macciantelli, D.; Torri, G.; Vismara, E. Electron beam irradiated textile cellulose fibres: ESR studies and derivatisation with glycidyl methacrylate (GMA). *Eur. Polym. J.* **2005**, *41*, 1787–1797. [CrossRef]
46. Connors, S.A. Chemical and Physical Characterization of the Degradation of Vulcanized Natural Rubber in the Museum Environment. Master's Thesis, Queen's University, Kingston, ON, Canada, 1998.
47. Palmas, P.; Le Campion, L.; Bourgeisat, C.; Martel, L. Curing and thermal ageing of elastomers as studied by H-1 Broadband and C-13 high-resolution solid-state NMR. *Polymer* **2001**, *42*, 7675–7683. [CrossRef]
48. Muniandy, K.; Hanafi, I.; Othman, N. Studies on natural weathering of rattan powder filled natural rubber composites. *BioResources* **2012**, *7*, 3999–4011. Available online: http://ojs.cnr.ncsu.edu/index.php/BioRes/article/view/BioRes_07_3_3999_Muniandy_Natural_Weathering_Rattan_Powder_Rubber/1653 (accessed on 20 May 2018).
49. Datta, R.N. Rubber-curing systems. In *Current Topics in Elastomers Research*; Bhowmick, A.K., Ed.; CRC Press: Boca Raton, FL, USA, 2008; ISBN 9781420007183.
50. Noriman, N.Z.; Ismail, H. The effects of electron beam irradiation on the thermal properties, fatigue life and natural weathering of styrene butadiene rubber/recycledacrylonitrile–butadiene rubber blends. *Mater. Des.* **2011**, *32*, 3336–3346. [CrossRef]
51. Rohana Yahya, Y.S.; Azura, A.R.; Ahmad, Z. Effect of Curing Systems on Thermal Degradation Behaviour of Natural Rubber (SMR CV 60). *J. Phys. Sci.* **2011**, *22*, 1–14. Available online: http://web.usm.my/jps/22-2-11/22.2.1.pdf (accessed on 26 June 2018).
52. Azura, A.R.; Muhr, A.H.; Thomas, A.G. Diffusion and reactions of oxygen during ageing for conventionally cured natural rubber vulcanisate. *Polym. Plast. Technol. Eng.* **2006**, *45*, 893–896. [CrossRef]
53. Pimolsiriphol, V.; Saeoui, P.; Sirisinha, C. Relationship among Thermal Ageing Degradation, Dynamic Properties, Cure Systems, and Antioxidants in Natural Rubber Vulcanisates. *Polym. Plast. Technol. Eng.* **2007**, *46*, 113–121. [CrossRef]
54. Hoque, M.E.; Ye, T.J.; Yong, L.C.; Dahlan, K.Z.M. Sago Starch-Mixed Low-Density Polyethylene Biodegradable Polymer: Synthesis and Characterization. *J. Mater.* **2013**, 365380. [CrossRef]
55. Arroyo, M.; Lopez-Manchado, M.A.; Herrero, B. Organo-montmorillonite as substitute of carbon black in natural rubber compounds. *Polymer* **2003**, *44*, 2447–2453. [CrossRef]
56. Chenal, J.M.; Chazeau, L.; Guy, L.; Bomal, Y.; Gauthier, C. Molecular weight between physical entanglements in natural rubber: A critical parameter during strain-induced crystallization. *Polymer* **2007**, *48*, 1042–1046. [CrossRef]

© 2018 by the authors. Licensee MDPI, Basel, Switzerland. This article is an open access article distributed under the terms and conditions of the Creative Commons Attribution (CC BY) license (http://creativecommons.org/licenses/by/4.0/).

International Journal of
Molecular Sciences

MDPI

Article

Impact of pH Modification on Protein Polymerization and Structure–Function Relationships in Potato Protein and Wheat Gluten Composites

Faraz Muneer [1], Eva Johansson [1], Mikael S. Hedenqvist [2], Tomás S. Plivelic [3] and Ramune Kuktaite [1,*]

[1] Department of Plant Breeding, Swedish University of Agricultural Sciences, Box 101, SE-23053 Alnarp, Sweden; faraz.muneer@slu.se (F.M.); eva.johansson@slu.se (E.J.)
[2] KTH Royal Institute of Technology, School of Engineering Sciences in Chemistry, Biotechnology and Health, Fibre and Polymer Technology, SE-10044 Stockholm, Sweden; mikaelhe@kth.se
[3] MAX-IV Laboratory, Lund University, Box 118, SE-22100 Lund, Sweden; tomas.plivelic@maxiv.lu.se
* Correspondence: ramune.kuktaite@slu.se; Tel.: +46-40-41-5337

Received: 12 November 2018; Accepted: 20 December 2018; Published: 24 December 2018

Abstract: Wheat gluten (WG) and potato protein (PP) were modified to a basic pH by NaOH to impact macromolecular and structural properties. Films were processed by compression molding (at 130 and 150 °C) of WG, PP, their chemically modified versions (MWG, MPP) and of their blends in different ratios to study the impact of chemical modification on structure, processing and tensile properties. The modification changed the molecular and secondary structure of both protein powders, through unfolding and re-polymerization, resulting in less cross-linked proteins. The β-sheet formation due to NaOH modification increased for WG and decreased for PP. Processing resulted in cross-linking of the proteins, shown by a decrease in extractability; to a higher degree for WG than for PP, despite higher β-sheet content in PP. Compression molding of MPP resulted in an increase in protein cross-linking and improved maximum stress and extensibility as compared to PP at 130 °C. The highest degree of cross-linking with improved maximum stress and extensibility was found for WG/MPP blends compared to WG/PP and MWG/MPP at 130 °C. To conclude, chemical modification of PP changed the protein structures produced under harsh industrial conditions and made the protein more reactive and attractive for use in bio-based materials processing, no such positive gains were seen for WG.

Keywords: wheat gluten; potato protein; chemical pre-treatment; structural profile; tensile properties

---

1. Introduction

Wheat gluten (WG) and potato protein (PP) are industrial side-streams originating from wheat and potato processing into ethanol and potato starch, respectively. Both protein-rich by-products are currently mostly utilized in baking (e.g., WG) and animal feed industries (PP). In addition, both of these proteins have attractive properties as raw materials for non-food applications such as in bio-based plastics [1,2]. During the production processes of WG and PP, high drying temperature is utilized, and for PP, an acid mediated coagulation step is also included. Both procedures are known to denature the chemical and structure-determining bonds in the protein and may also result in ionization of amino, carboxyl and phenolic groups [3] and reduced protein functionality [2,4,5]. In addition, for PP being treated at a low pH ≈ 4 during industrial processing, causing definite changes in chemical and physical properties to occur. The chemical processes resulting in intra- and intermolecular changes of the protein limit their processing window, thereby decreasing their suitability for processing into bio-based materials [2,6]. Similarly for WG, the relatively high temperature during drying negatively impacts the proteins and results in aggregated gluten with limited functionality [5]. An alternative and added-value opportunity is to use chemical tools to retain the WG and PP processing ability, both in their present or modified form, for benefit of industry, the consumer and society, thereby contributing to the circular bio-economy. Manipulating the chemical nature of industrial WG and PP would create new possibilities and contribute to the development of alternative uses and an added-value of these proteins for different applications. Development of new bio-based materials from WG and PP proteins is one such promising application.

Un-plasticized WG and PP are difficult to process as their glass transition temperature (T_g) is located close to their thermal degradation temperature [2,7–9]. Thus, easy processing requires the addition of chemical agents or polyol-based plasticizers that depress T_g and widen the processing window [2,4,10,11]. In general, proteins perform better in terms of their structure–functional properties when they are processed at basic pH [12,13]. The basic pH is known to denature and unfold the proteins and thereby expose sulfhydryl and hydrophobic protein sections, which open-up for new protein interactions when processed [14].

Chemical additives, such as NaOH, NH$_4$OH or urea, create basic conditions for the protein, resulting in changes of secondary and supramolecular protein structures which correlate with improved functional properties of processed materials [7,9,10,12,13,15–17]. However, previous studies have mainly focused on additives or chemical modifiers being added directly to the protein where the blend was immediately processed, thereby leaving a short reaction time for additives to interact with the protein. Although, pre-treatment of the proteins with chemical modifiers in order to modify the protein and later use it for processing has not been evaluated before. This has left a gap for further investigations on the impact of such chemical pre-treatment on protein structure and also functional properties of the final product. In this study, we pre-treated the WG and PP with NaOH in a solution to achieve a basic pH and to promote protein unfolding and induce reactivity. The pH modification was aimed to promote new protein–protein interactions, that could positively impact the mechanical performance of the protein films and composites. The objective was also to understand the effect of the chemical modification on protein molecular and secondary structures, processing behavior and mechanical performance of composites with varying WG and PP ratios. Increased understanding of protein–protein interactions and chemistry, as well as structure-function relationships of proteins in a blend can offer novel opportunities to create versatile and attractive performance of films and composites.

2. Results and Discussion

2.1. Protein Extractability in pH-Modified Protein Powders, Films and Composites

In this study, SE-HPLC was used to assess protein polymerization behavior in the pH modified protein powders and high temperature processed films (Figure 1). The modified potato protein (MPP) and modified wheat gluten (MWG) powders showed greater amounts of extractable polymeric proteins (HMw and LMw) compared to the non-modified samples, PP and WG (Figure 1a–d). This was particularly evident for the HMw proteins extracted during 2Ex and 3Ex extractions (Figure 1b,d). The results indicated, that the alkaline pH induced unfolding in the studied protein and made the proteins more extractable.

For protein films and composites processed at two temperatures, 130 and 150 °C, variation in protein extractability was observed. The processing of the protein powders into films at high temperature induced a higher degree of protein cross-linking and promoted the formation of larger protein aggregates and polymers, thereby decreasing their extractability during SE-HPLC analysis as has previously been reported in processed protein based materials [2,4,12,16,18–20]. In addition, possible interactions of the proteins studied with residual starches (wheat starch from wheat gluten fraction and potato starch from potato protein fraction) could have potentially occurred through non-covalent reactions such as, hydrogen and van der Waals-bonding between the protein and reactive hydroxyl groups from starch. Although those interactions were minor and did not substantially impact protein cross-linking.

For the pressed materials in this study an increase in protein extractability was seen for protein materials pressed at 150 °C, as compared to those pressed at 130 °C (Figure 1e). This increase in extractability in films pressed at the higher temperature (150 °C), was mainly due to an increase in LMw proteins being extracted as a consequence of protein degradation (except, for MWG, in which increase in HMw proteins at higher temperature was also observed) and lack of protein–protein interactions. In addition, proteins from films produced from the chemically modified proteins were in general more easily extractable, with the exception of MPP pressed at 130 °C (compare MPP with PP pressed at 130 °C; Figure 1e).

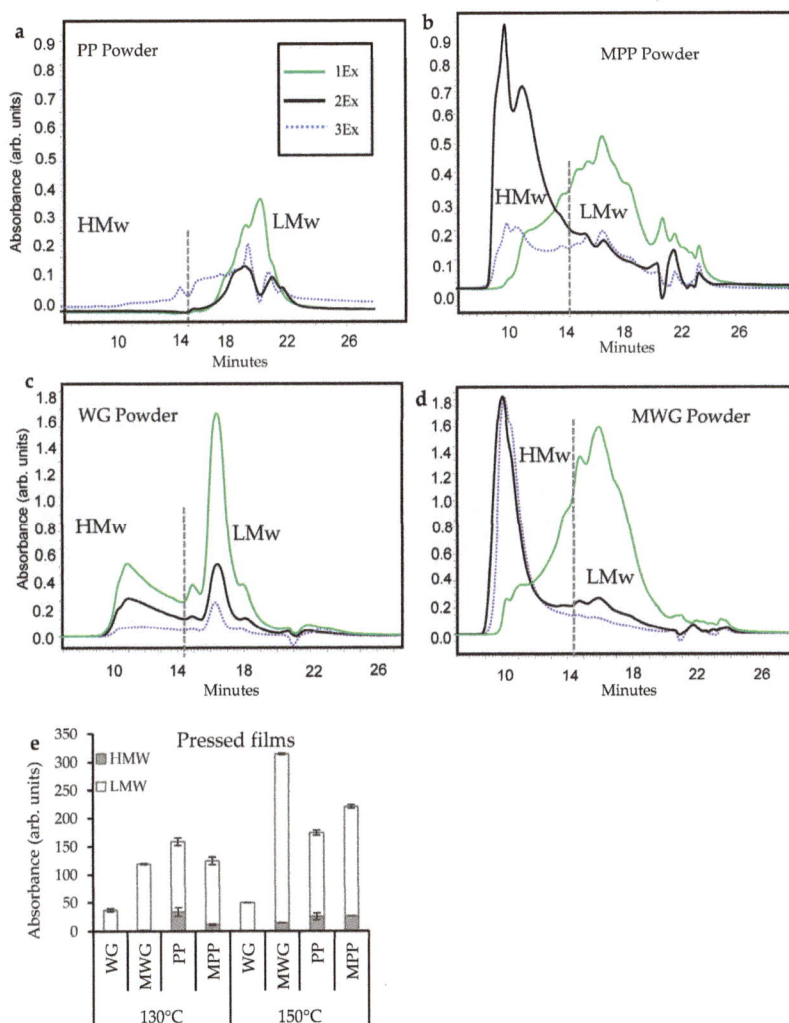

Figure 1. Representative chromatograms of protein molecular size distribution of potato protein powder (PP) (**a**), modified potato protein powder (MPP) (**b**), wheat gluten (WG) (**c**) and modified wheat gluten (MWG) powder (**d**), and the total protein extractability of films from the 3 extractions of non-modified and pH-modified proteins pressed at 130 °C and 150 °C (**e**).

The protein extractability results clearly suggests that the basic pH-modification of protein powders contributed to a protein network unfolding, decrease in protein–protein interactions and de-polymerization of protein aggregates being present in the non-modified protein powder. The unfolding, decrease of protein–protein interactions and de-polymerization of the proteins after modification resulted in an increased extractability of the proteins. The fact that the extractability increased primarily as HMw proteins were extracted during 2Ex and 3Ex extraction steps (where sonication energy is used), suggests that primarily protein aggregates/polymers were affected by the pH adjustment (chemical modification). However, the majority of the protein extraction taking place at 2Ex and 3Ex indicated the presence of some of the cross-links, although with lower degree of intermolecular disulphide bonds, while hydrogen bonds and non-covalent protein interactions might still be in place [21,22].

Compression molding of the powder into films resulted in excessive cross-linking and re-formation of intermolecular disulphide bonds for all samples, as this has been previously reported for wheat gluten and potato protein materials [2,11,16,19,23]. In particular, most of the HMw proteins in all the hot pressed samples ended up in forming large polymers, which were not extractable in the films after pressing. It is also known from previous studies that compression molding of films under alkaline conditions leads to increased protein–protein interactions with the formation of disulphide cross-links in WG, hydrogen bonding in WG and PP and also the formation of isopeptide bonds such as dehydroalanin, lanthionine and lysinoalanine [7,16,24–26]. In addition, alkaline pH and the use of cross-linking enzymes, such as transglutaminase, for protein pre-treatment can further introduce new protein cross-linking opportunities, as has been shown in grass pea flour films [27]. Also, newly sourced transglutaminase species that have been shown to positively impact certain functional behavior of proteins in wheat dough [28,29] should further be explored in protein-based materials.

In this study, the proteins were found to be less polymerized after hot pressing at 150 °C than at 130 °C, most likely due to protein degradation, with HMw proteins breaking into LMw fragments [8,23]. In terms of nano-structure for unmodified WG and PP samples pressed at 130 and 150 °C, few differences due to high temperature were observed in the main scattering broad peak, d_1 and for the scattering distance d_2, studied by SAXS, for both types of proteins (supporting information, Figure S1). For the PP film, the broad peak (d_1) became "sharper", while d_2 intensity decreased with increasing temperature. While for the WG film, at 150 °C the scattering distance, d_2, disappeared, indicated a less complex structural morphology compared to the previously observed structural morphologies in WG films pressed at 130 °C [5]. A possible explanation is a decrease in polymerization and an increase in protein breakdown as shown by SE-HPLC (Figure 1e). Regarding the PP film morphology, intensity of the d_2 peak was higher at 130 °C than at 150 °C, indicating more structural complexity, although without any specific peak ratio [2]. This change in peak position and intensity may be affected by protein polymerization behavior, which was observed at different pressing temperatures. Regarding the MWG and MPP films morphology, the MPP showed a clear peak, d_1, indicating similar structural morphology observed previously for the PP films pressed at 130 °C [2] (supporting information, Figure S2). While MWG showed d_1 and d_2, scattering distances, that can be referred to some undefined, though poorly hierarchically arranged structural entities (as compared to previously observed hexagonal arrangement of WG proteins). Thus, modification may not have considerably changed the nano-structure of MPP, while MWG nanostructure was not improved.

Previous study of WG has indicated that a less polymerized structure in the powder used for processing contributes to a higher degree of re-polymerization, chemical flexibility for proteins to interact and cross-link during processing, and the availability of chemically reactive sites for disulphide bond formation [5]. Although, the described effect could not be seen for the WG proteins in this study, indicating the alkali modification either leads to irreversible changes in the protein conformation (negative impact) or to chemical reactions not involving disulphide bond formation.

2.2. Protein Extractability in Pressed Composites

Among the pressed composites the highest protein extractability was clearly seen for the MWG/MPP samples at 150 °C, almost independent of the ratio of the two modified proteins within the composite (Figure 2). These results corresponded well with the high protein extractability found for the modified proteins pressed individually at 150 °C (compare Figure 1e), although protein extractability was even higher in the composites than in the pressed films of individual proteins. Thus, mixing of the two modified proteins did not contribute to increased protein polymerization as compared to protein polymerization for the modified proteins separately. Similarly, a high proportion of MWG (50 and 70%) in the composites pressed at 130 °C resulted in increased protein extractability as compared to the films from each of the individually modified proteins (Figure 2a compared with Figure 1). The extractability was also increased with the increase in temperature from 130 to 150 °C and with higher MWG content in the blend (Figure 2). An increased protein extractability with increase in MWG content (at 130 °C) could be due to the fact that MWG has a more complex protein molecular structure and chemistry, compared to PP, and was more denatured and depolymerized in the pH-modification process compared to PP (see MWG and MPP powder profiles; Figure 1b,d). Therefore, an MWG increase in the MWG/MPP blends (i.e., the 70/30 sample) resulted in a decrease in polymerization and lack of disulphide cross-link formation compared to 30/70 and 50/50 samples. However, using a 150 °C pressing temperature reduced the protein polymerization (increased protein extractability) for all MWG/MPP compositions compared to the individual modified proteins (compare Figures 1 and 2b). This increase in protein extractability at higher temperature may be due to the lack of disulphide and irreversible cross-links between the protein chains and also due to breakdown of HMw protein into LMw fragments.

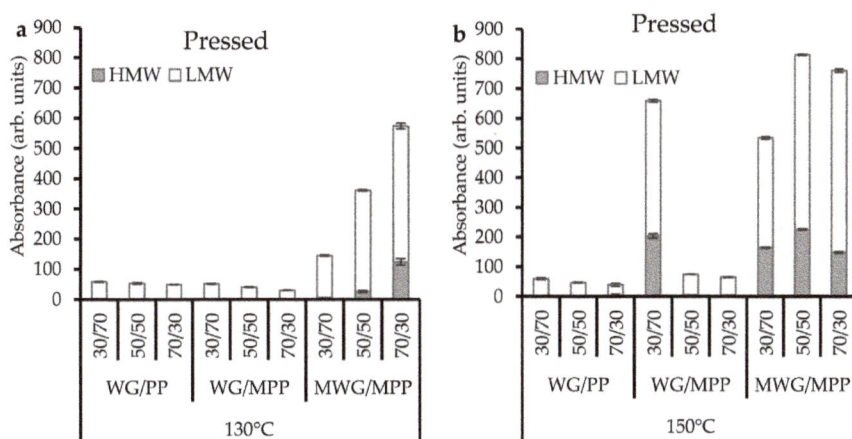

Figure 2. Total protein extractability of WG/PP, WG/MPP and MWG/MPP samples pressed at 130 °C (a) and 150 °C (b), as studied by SE-HPLC.

For the composites based on non-modified proteins (WG/PP), such an increase in protein extractability (as seen for modified samples) was not seen (compare Figures 1 and 2); instead a slight decrease in protein extractability was observed, especially at high WG ratios (70/30). The non-modified (as received) WG and PP have a low extractability due to the formation of HMw protein aggregates during their industrial processing [2,5]. Therefore, their further processing at high temperatures favored the formation of non-reducible protein cross-links and also incorporation of the LMw fragments in larger protein networks, thus reducing their extractability [8,22,30]. This suggests that composites of non-modified proteins produced in this study did show reduced protein extractability as reported for individual protein based materials produced previously [2,19].

As for the WG (non-modified) composites with different ratios of MPP, the protein extractability was in a similar range to that of WG/PP films at both pressing temperatures (compare Figures 1 and 2) except for WG/MPP 30/70 at 150 °C with surprisingly higher amounts of extractable proteins. Hence, higher pressing temperature (as 150 °C) increased protein polymerization and decreased extractability when ≤ 50 *wt*.% of MPP was used. No such effect of temperature and amount used of MPP was seen in the samples pressed at 130 °C, suggesting that high temperature may be one of the main driving factors in determining the protein polymerization.

Interestingly, the lowest protein extractability (highest degree of protein polymerization) was found for the 70/30 WG/MPP composite film at 130 °C, and for the 70/30 WG/PP film at 150 °C (Figure 2). Thus, to increase protein polymerization in WG based films, modified potato protein (MPP) is beneficial at 130 °C, while non-modified PP is beneficial to have at higher temperature. Increased protein polymerization in 70/30 WG/MPP could be explained by a higher degree of protein cross-linking of WG at 130 °C as has been reported in previous studies [5,19,31]. However, an increased protein polymerization in WG/PP 70/30 sample at 150 °C suggests formation of new protein–protein interactions in a composite, because the extractability of these blends decreased slightly more when compared to both non-modified proteins pressed at 150 °C.

2.3. Effect of Chemical Modification on Secondary Structure of Proteins

As shown in previous studies [5,10,12,31–33], the non-modified WG powder showed relatively low amounts of β-sheets, as indicated by a flat shoulder in the amide I region (1620–1625 cm^{-1}), while peaks at 1660, 1650 and 1641 cm^{-1} verified the presence of α-helices, α-helices and random coils, and unordered structures, respectively (Figure 3a). Chemical modification of the WG protein (MWG sample) resulted in a slight decrease in the intensity of peaks in the region 1640–1660 cm^{-1}, indicating that some of either α-helices, α-helices and random coils, or unordered structures became involved in formation of strong hydrogen bonded β-sheets (Figure 3a). Thus, FT-IR data indicate formation of β-sheets at the expense of either α-helices, α-helices and random coils or unordered structures, may indicate development of novel interactions (presumably via hydrogen bonds, protein-protein and peptide–protein interactions [34]), although not verified by SE-HPLC results (Figure 1).

The FT-IR spectra of PP showed a very distinct and high intensity peak at 1622 cm^{-1} verifying the presence of strongly hydrogen-bonded β-sheets. The formation of these protein structures are known to originate from the harsh treatment of the PP during industrial extraction with the use of high temperature and acidic processing conditions promoting a high degree of cross-linking [2]. The high degree of cross-linking in PP was also supported by the SE-HPLC data, showing low extractability of PP powder (Figure 1a). Furthermore, results of the SE-HPLC were supported by FT-IR data for the MPP powder sample, in general showing a lower amount of strongly hydrogen-bonded β-sheets (less aggregated structure) as reported in previous studies [4,34,35].

The pressed WG/PP blended films resulted in FT-IR spectra with a clear and high intensity peak at 1620 cm^{-1} and also several well defined peaks in the 1640–1660 and 1622 cm^{-1} region (Figure 3b,c), thereby showing structural features resembling both single powders used in the blend (Figure 3a) and indicating presence of strongly bonded β-sheet interactions, α-helices, α-helices and random coils, and unordered structures [2,5]. Including MPP in the blends (WG/MPP) for compression molding resulted in FT-IR spectra with reduced intensity and a shift to lower frequency of the major β-sheet peak (especially at 150 °C when the shoulder also became broader), and a decrease in intensity of the peaks in the region of 1640–1660 cm^{-1} (Figure 3b,c). In general protein aggregation has been correlated to an increased amount of strongly hydrogen-bonded β-sheets contributing to ordered conformation and improved tensile strength [2,4,7,10,12,31–34,36–38]. However, the lowest extractability of the proteins (indicating protein polymerization) as resolved by SE-HPLC results (Figure 2) was found for WG/MPP 70/30 at 130 °C and WG/PP 70/30 at 150 °C, not corresponding with blends showing the most intense peak for β-sheets by FT-IR (Figure 3b,c). These results suggest that the protein polymerization behavior and relationships with structural properties seems to be more complicated for compression molded composites than for individual proteins, especially while modifying of the chemical structures of the proteins used in the blends.

FT-IR spectra of MWG/MPP composites showed more differentiation between the ratios of proteins, where the strongly hydrogen-bonded β-sheets related structural peak was more pronounced at higher PP ratio and lower temperature (supporting information, Figure S3), corresponding to lower protein extractability by SE-HPLC (Figure 2). This might indicate that differences in protein polymerization were too limited among WG/PP and WG/MPP blended compression molded films to be differentiated by FT-IR in most cases, and basically only differences between separate blended protein components were seen. Thus, only for differences in protein polymerization behavior of MWG/MPP blends were large enough to be differentiated by FT-IR as also indicated by SE-HPLC, and corresponded to the characteristic relationship of SE-HPLC, FT-IR and functional properties data [12,31].

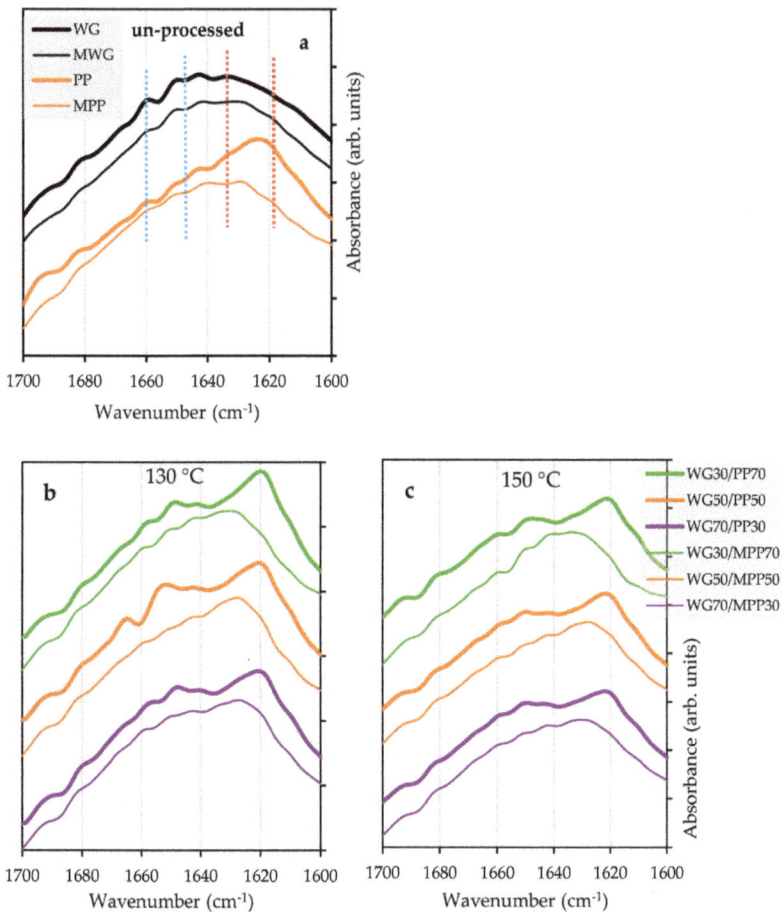

Figure 3. FT-IR spectra of individual protein powders and pressed films at 130 and 150 °C. WG, MWG, PP and MPP powders (**a**), pressed samples of WG/MPP blends along with controls (**b,c**).

2.4. Effect of Protein Modification on Mechanical Performance of the Pressed Composites

Compression molding of WG, MWG, PP, MPP and their blends, yielded samples of different colors as can be seen in Figure 4. High WG and MWG content generally resulted in a sample with light brown color (Figure 4, MWG100 and WG/MPP 70/30), while high PP and MPP content generally resulted in samples with dark brown (WG/PP 30/70) or almost black in color (MPP100), which can influence the composites' attractiveness and end-use. It is however an easy task to incorporate a pigment or dye to obtain a desired color (excluding white color). WG and PP samples showed opposite tensile performance, WG having lower strength and E-modulus but higher extensibility (elongation at break) than PP. Thus, comparing these two protein types, the tensile strength correlated positively with protein extractability. This observation was different to many previous studies, where a higher degree of cross-linking yielded higher strengths [2,5,19,22,39,40]. Here, the WG proteins most likely have a higher ability to cross-link (due to their more complex molecular structure characteristics and specific polymerization properties [8,22]) during pressing than the PP as SE-HPLC results indicate, although FT-IR shows a higher degree of β-sheets in the pressed PP which might explain their higher maximum stress and E-modulus [2].

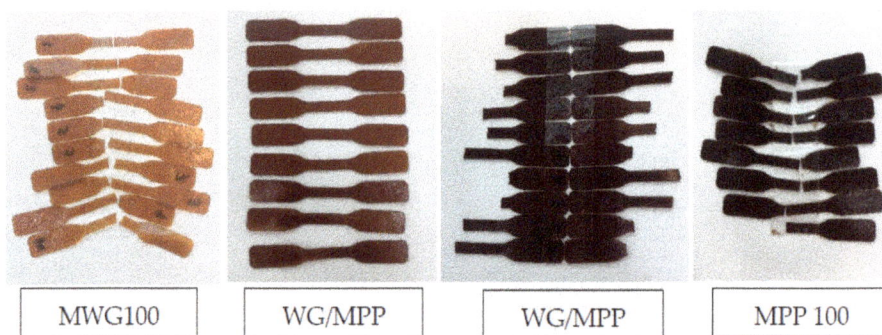

Figure 4. Representation of the range of colors of films and composites MWG100, WG/MPP 70/30, WG/MPP 30/70 and MPP100 pressed at 130 °C.

Modification of WG (MWG) resulted in lower maximum stress and E-modulus and reduced extensibility compared to WG for both evaluated pressing temperatures (Figure 5). Modification of the PP (MPP), resulted in a decrease in E-modulus for samples pressed at both temperatures and maximum stress decreased at 150 °C while the extensibility increased particularly at 130 °C (Figure 5). SE-HPLC data indicated an increase in cross-linking at 130 °C, although a decrease in β-sheet content was shown by FT-IR, which might explain the improvement in the extensibility and decrease in E-modulus for the MPP as compared to PP. In addition, a previous study reported an increase in extensibility and a decrease in modulus for PP modified with NaOH, when pressed with 30% glycerol [4]. Thus, modification of PP prior to processing (e.g., hot pressing) leads to improved tensile properties but also contributes to opportunities for using lower processing temperatures than was in previous studies ($\geq$ 150 °C) [2], thereby saving energy input during the production process.

The pressed composites of WG and PP resulted in tensile properties generally as expected from the ratios of the individually pressed proteins (Figure 5). In addition, pressed WG/MPP samples showed tensile properties generally between those of the individually pressed protein samples, although minor differences were seen among the samples. Furthermore, as also indicated by comparing FT-IR and SE-HPLC results, differences in the protein cross-linking obtained by SE-HPLC could not explain the differences in tensile properties between the WG/PP and WG/MPP blends.

The most rubber-like properties were observed for the pressed MWG material (lower stiffness than WG), which also resulted in that the MWG/MPP blends were with the most rubber-like properties (low modulus but still with appreciable extensibility) (Figure 5), thereby correlating with SE-HPLC and FT-IR data. It is noteworthy that the higher pressing temperature (150 °C) leads to lower stiffness, strength and extensibility for the modified materials and their blends (MWG, MPP and MWG/MPP) (Figure 5).

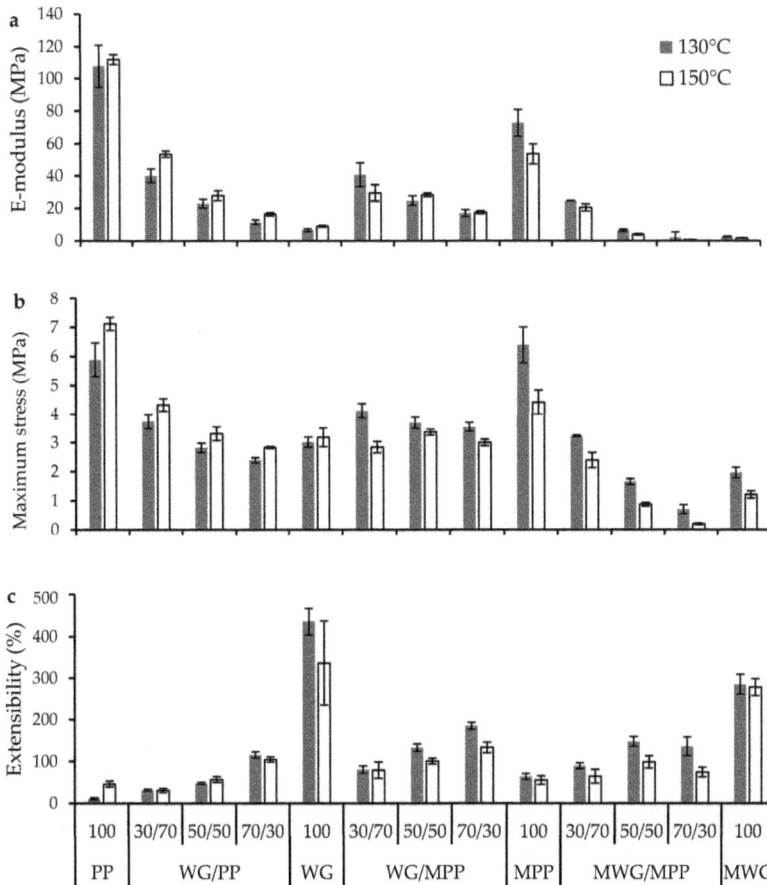

Figure 5. Mechanical properties of individual protein films and WG/PP, WG/MPP and MWG/MPP blends pressed at 130 and 150 °C; (**a**) E-modulus, (**b**) maximum stress and (**c**) extensibility.

3. Materials and Methods

3.1. Wheat Gluten and Potato Proteins

Wheat gluten with protein content of 77.7 *wt*.%, starch 5.8 *wt*.%, moisture 6.9 *wt*.% and fat 1.2 *wt*.%, was purchased from Lantmännan Reppe AB, Lidköping, Sweden. Commercial PP was supplied by Lyckeby Starch AB, Kristianstad, Sweden. The potato protein content was 82.2 *wt*.% (Dumas method, Flash 2000 NC Analyser, Thermo Scientific, USA, NX 6.25) and moisture content of 8.1% (dry basis, dried at 105 °C for 3 h). Glycerol (purity 99.5 *wt*.%, 0.5 *wt*.% water) was supplied by Karlshamns Tefac AB, Karlshamn, Sweden.

3.2. Protein Modification

Potato protein powder, 50 g was slowly dispersed in 600 mL distilled water under stirring. The suspension was adjusted to pH 10 by addition of a 5 M NaOH solution, followed by stirring for 30 min at 75 °C ($\pm$ 3 °C). The PP was well dispersed in these conditions forming a dark brown suspension. The PP suspension was then lyophilized and ground to powder using an IKA A10 grinder (IKA, Germany) and was designated as modified PP (MPP) in this study. WG was modified in the same way as the PP above and was designated as modified WG (MWG). Protein blends were prepared by dispersing various ratios of WG and PP in water followed by the pH modification step as described for the separate protein sources. Thereafter, MWG/MPP blend were lyophilized and ground to obtain powder for further sample processing.

3.3. Sample Preparation and Compression Molding

Powders from individually modified PP and WG proteins were mixed with glycerol, (70 *wt.*% protein and 30 *wt.*% glycerol) prior to compression molding and later pressed into films. Each type of protein was pressed into two films, one at 130 °C and another one at 150 °C. For blends, different ratios of WG/PP and WG/MPP (30/70, 50/50 and 70/30) were manually mixed with glycerol, (70 *wt.*% protein and 30 *wt.*% glycerol) prior to compression molding. The sample compositions are presented in Table 1. The prepared individual protein films and blends were then placed in a 0.5 mm thick aluminum frame with a 100 × 100 mm opening to control the size and thickness of the film. The film and blends were then molded for 5 min at 200 bar in a hydraulic press (Polystat 400s, Servitech, Germany) between pre-heated aluminum plates with poly(ethylene terephthalate) release films. From each of the blends two films were compression molded, one at 130 °C and another at 150 °C. Pressed films were removed from the hot press and left to cool between two room temperature aluminum plates.

Table 1. Compositions of the blends of WG, PP, MWG and MPP.

Type	Abbreviation	WG (*wt.*%)	PP (*wt.*%)	MWG (*wt.*%)	MPP (*wt.*%)
	WG/PP 30/70	30	70	—	—
Controls	WG/PP 50/50	50	50	—	—
	WG/PP 70/30	70	30	—	—
Only PP	WG/MPP 30/70	30	—	—	70
Modified	WG/MPP 50/50	50	—	—	50
	WG/MPP 70/30	70	—	—	30
MWG/MPP	MWG/MPP 30/70	—	—	30	70
Modified in	MWG/MPP 50/50	—	—	50	50
Composite	MWG/MPP 70/30	—	—	70	30

3.4. SE-HPLC to Assess Protein Polymerization in Processed Composites

The soluble amount of protein and protein size distribution of the protein powders, compression molded samples and composites were examined in this study. For this analysis, films were cut into approximately 0.2 mm pieces using a scalpel. Thereafter, 16.5 mg (± 0.05 mg) of each sample was added to 1.4 mL buffer solution (0.5% SDS, 0.05 M NaH_2PO_4, pH 6.9) in a Eppendorf tube (1.5 mL). To obtain the first extraction (1Ex) tubes were vortexed for 10 s (Whirli Vib 2, Labassco, Sweden) and then shaken for 5 min (IKA-VIBRAX VXR, IKA, Germany) at 2000 rpm. The tubes were then centrifuged (Legend Micro 17, Sorvall, Germany) for 30 min at 12,500 rpm and the supernatant was collected in HPLC vials. For the 2nd extraction (2Ex), extraction buffer (1.4 mL) was added to the pellet from 1Ex and then sonicated for 30 s at an amplitude of 5 microns using a Sanyo Soniprep 150 Ultrasonic Disintegrator (Tamro, UK) and thereafter centrifuged for 30 min and the supernatant collected in HPLC vials for analysis. The third extraction (3Ex) was similar to 2Ex, the pellet of 2Ex was used with sonication intervals of 30 + 60 + 60 s, to avoid overheating, samples were left to cool at room temperature between each sonication interval. All three extractions (1Ex, 2Ex and 3Ex) were analyzed with a Waters 2690 Separation Module connected to a Waters 996 Photodiode Array Detector (Waters, Millford, MA, USA).

An SE-HPLC column (Biosep-SEC-S 4000, Phenomenex, Torrance, CA, USA) was used for protein size distribution determination. For each sample, 20 μL was injected onto the column at an isocratic flow of 0.2 mL/min (50% acetonitrile, 0.1% TFA; 50% H_2O, 0.1% TFA). Chromatograms were obtained at 210 nm and integrated using Empower Pro software (Waters, USA). The chromatograms were divided into two sections depending on the elution time of the proteins and total area was calculated. The high molecular weight (HMw) proteins were eluted between 7–14.5 min and low molecular weight (LMw) between 14.5–28 min. HMw are referred to as polymeric proteins, and LMw as smaller molecular size proteins.

3.5. Tensile Testing

For mechanical testing the films were conditioned for at least 48 h at 23 °C and 50% relative humidity and then cut into dumbbell shaped samples (ISO 37 type 3, Elastocon, Sweden) and thereafter tested in the same conditions. Prior to testing the thickness of all the samples was measured at 5 different points (Mitutoyo IDC 112B) in the test area and the average was used for calculations. A clamp separation distance of 40 mm, crosshead speed of 10 mm/min and a 500 N load cell was used to test all samples on an Instron 5566 universal testing machine with a Bluehill software (Instron AB, Danderyd, Sweden).

3.6. Fourier Transform Infrared Spectroscopy

All the samples were dried for at least 72 h in a desiccator over silica gel prior to the analysis. FT-IR spectroscopy was carried out on all films and powders using a Spectrum 2000 FT-IR spectrometer (Perkin-Elmer inc., Norwalk, CN, USA) equipped with single reflection ATR (Golden Gate, Speac Ltd., USA). Data was Fourier self-de-convoluted using the Spectrum software with γ=2 and a smoothing factor of 70%.

4. Conclusions

Chemical modification of plant proteins, through heating at a basic pH, results in a protein with a decreased amount of intermolecular disulphide bonds and thereby a less cross-linked structure, as shown here for both WG and PP. Cross-linking of the proteins in the starting powder, when producing bio-based materials, largely impacts the properties of compression molded materials both in terms of opportunities for the proteins to further cross-link and form new bonds and structures during processing. From our previous study, mild protein extraction of WG has been shown to contribute with less cross-linked protein powder which when hot compression molded into films resulted in improved tensile properties (E-modulus and maximum stress) [5]. However, a less cross-linked structure in the protein starting material does not necessarily contribute to improved properties of processed bio-based materials. Here, a decrease in cross-linking of proteins by chemical modification, improved cross-linking (during temperature pressing) and tensile properties (E-modulus and maximum stress) of the material of PP but not for WG. Thus, using basic pH for modification of WG must contribute to irreversible changes in the protein conformation/structure and/or to chemical reactions not involving disulphide bond formation, which reduces the opportunities for the proteins to re-cross-link when processed. For PP, instead the chemical modification through basic pH contributed to the breaking of disulphide bonds which were created during industrial production of the PP starting material (the powder).

A major advantage with the chemical modification of PP was that it enabled the production of films with improved properties at a pressing temperature as low as 130 °C, thereby contributing to a lower environmental foot-print due to a reduction of energy use.

Thus, chemical modification of proteins for bio-based material production on an industrial scale may potentially help to widen the functional properties and improve the socio-economic value of the product (especially in the case of PP). Figure 6 provides a schematic summary of the important findings of this study.

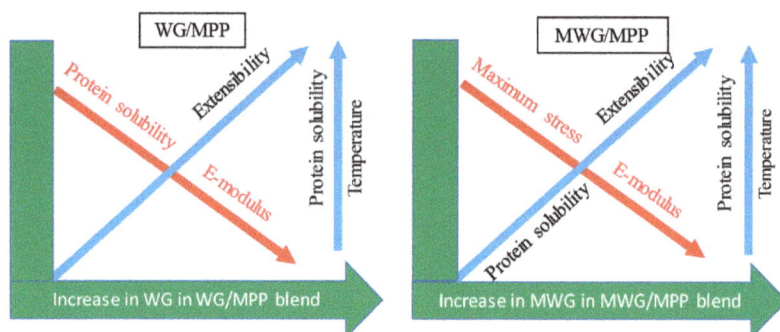

Figure 6. Schematic diagram of summarized effects of temperature and composition of the blend on protein polymerization and mechanical properties.

Supplementary Materials: Supplementary materials can be found at http://www.mdpi.com/1422-0067/20/1/58/s1.

Author Contributions: R.K. conceived and designed the experiment together with F.M., F.M. performed the experiments and wrote the manuscript with contribution from the co-authors. M.S.H. performed FT-IR measurements and contributed with critical reading of the manuscript. R.K. contributed with performance of some SAXS analysis, coordinated SAXS experiments and obtained a beamtime for SAXS analysis, while T.S.P. contributed with interpretation of SAXS results. E.J. and R.K. contributed with obtaining of partial funding for this study and with critical commenting and writing of this manuscript.

Funding: This work was funded by the Swedish Governmental research program Trees and Crops for the Future (TC4F; Vinnova) and Partnerskap Alnarp.

Acknowledgments: We acknowledge Maria Luisa-Prieto Linde for her kind assistance in the laboratory, Lyckeby Starch AB for their kind provision of the potato protein and William Newson for his kind editing of the manuscript and English.

Conflicts of Interest: The authors declare no conflict of interest.

Abbreviations

Ex	Extraction
FT-IR	Fourier Transform Infrared Spectroscopy
HMw	High molecular weight proteins
LMw	Low molecular weight proteins
MPP	Modified potato protein
MWG	Modified wheat gluten
PP	Potato protein
SE-HPLC	Size Exclusion High Performance Liquid Chromatography
T_g	Glass transition temperature
WG	Wheat gluten

References

1. Kuktaite, R.; Newson, W.R.; Rasheed, F.; Plivelic, T.S.; Hedenqvist, M.S.; Gällstedt, M.; Johansson, E. Monitoring Nanostructure Dynamics and Polymerization in Glycerol Plasticized Wheat Gliadin and Glutenin Films: Relation to Mechanical Properties. *ACS Sustain. Chem. Eng.* **2016**, *4*, 2998–3007. [CrossRef]

2. Newson, W.R.; Rasheed, F.; Kuktaite, R.; Hedenqvist, M.S.; Gällstedt, M.; Plivelic, T.S.; Johansson, E. Commercial potato protein concentrate as a novel source for thermoformed bio-based plastic films with unusual polymerisation and tensile properties. *RSC Adv.* **2015**, *5*, 32217–32226. [CrossRef]

3. Neurath, H.; Greenstein, J.P.; Putnam, F.W.; Erickson, J.A. The chemistry of protein denaturation. *Chem. Rev.* **1944**, *34*, 157–265. [CrossRef]

4. Du, Y.; Chen, F.; Zhang, Y.; Rempel, C.; Thompson, M.R.; Liu, Q. Potato protein isolate-based biopolymers. *J. Appl. Polym. Sci.* **2015**, *132*. [CrossRef]

5. Rasheed, F.; Hedenqvist, M.S.; Kuktaite, R.; Plivelic, T.S.; Gällstedt, M.; Johansson, E. Mild gluten separation–A non-destructive approach to fine tune structure and mechanical behavior of wheat gluten films. *Ind. Crops Prod.* **2015**, *73*, 90–98. [CrossRef]

6. Van Koningsveld, G.A.; Gruppen, H.; de Jongh, H.H.J.; Wijngaards, G.; van Boekel, M.A.J.S.; Walstra, P.; Voragen, A.G.J. Effects of pH and Heat Treatments on the Structure and Solubility of Potato Proteins in Different Preparations. *J. Agric. Food Chem.* **2001**, *49*, 4889–4897. [CrossRef] [PubMed]

7. Ullsten, N.H.; Cho, S.-W.; Spencer, G.; Gallstedt, M.; Johansson, E.; Hedenqvist, M.S. Properties of Extruded Vital Wheat Gluten Sheets with Sodium Hydroxide and Salicylic Acid. *Biomacromolecules* **2009**, *10*, 479–488. [CrossRef] [PubMed]

8. Pommet, M.; Morel, M.-H.; Redl, A.; Guilbert, S. Aggregation and degradation of plasticized wheat gluten during thermo-mechanical treatments, as monitored by rheological and biochemical changes. *Polymer* **2004**, *45*, 6853–6860. [CrossRef]

9. Olabarrieta, I.; Cho, S.W.; Gällstedt, M.; Sarasua, J.R.; Johansson, E.; Hedenqvist, M.S. Aging properties of films of plasticized vital wheat gluten cast from acidic and basic solutions. *Biomacromolecules* **2006**, *7*, 1657–1664. [CrossRef]

10. Ture, H.; Gallstedt, M.; Kuktaite, R.; Johansson, E.; Hedenqvist, M.S. Protein network structure and properties of wheat gluten extrudates using a novel solvent-free approach with urea as a combined denaturant and plasticiser. *Soft Matter* **2011**, *7*, 9416–9423. [CrossRef]

11. Rasheed, F.; Newson, W.R.; Plivelic, T.S.; Kuktaite, R.; Hedenqvist, M.S.; Gällstedt, M.; Johansson, E. Macromolecular changes and nano-structural arrangements in gliadin and glutenin films upon chemical modification: Relation to functionality. *Int. J. Biol. Macromol.* **2015**, *79*, 151–159. [CrossRef] [PubMed]

12. Kuktaite, R.; Plivelic, T.S.; Cerenius, Y.; Hedenqvist, M.S.; Gällstedt, M.; Marttila, S.; Ignell, R.; Popineau, Y.; Tranquet, O.; Shewry, P.R.; et al. Structure and Morphology of Wheat Gluten Films: From Polymeric Protein Aggregates toward Superstructure Arrangements. *Biomacromolecules* **2011**, *12*, 1438–1448. [CrossRef] [PubMed]

13. Ullsten, N.H.; Gällstedt, M.; Spencer, G.M.; Johansson, E.; Marttila, S.; Ignell, R.; Hedenqvist, M.S. Extruded High Quality Materials From Wheat Gluten. *Polym. Renew. Resour.* **2010**, *1*, 173–186. [CrossRef]

14. Gennadios, A.; Brandenburg, A.H.; Weller, C.L.; Testin, R.F. Effect of pH on properties of wheat gluten and soy protein isolate films. *J. Agric. Food Chem.* **1993**, *41*, 1835–1839. [CrossRef]

15. Kuktaite, R.; Türe, H.; Hedenqvist, M.S.; Gällstedt, M.; Plivelic, T.S. Gluten Biopolymer and Nanoclay-Derived Structures in Wheat Gluten–Urea–Clay Composites: Relation to Barrier and Mechanical Properties. *ACS Sustain. Chem. Eng.* **2014**, *2*, 1439–1445. [CrossRef]

16. Newson, W.R.; Kuktaite, R.; Hedenqvist, M.S.; Gällstedt, M.; Johansson, E. Effect of Additives on the Tensile Performance and Protein Solubility of Industrial Oilseed Residual Based Plastics. *J. Agric. Food Chem.* **2014**, *62*, 6707–6715. [CrossRef]

17. Kowalczyk, D.; Baraniak, B. Effects of plasticizers, pH and heating of film-forming solution on the properties of pea protein isolate films. *J. Food Eng.* **2011**, *105*, 295–305. [CrossRef]

18. Domenek, S.; Morel, M.-H.; Bonicel, J.; Guilbert, S. Polymerization kinetics of wheat gluten upon thermosetting. A mechanistic model. *J. Agric. Food Chem.* **2002**, *50*, 5947–5954. [CrossRef]

19. Gällstedt, M.; Mattozzi, A.; Johansson, E.; Hedenqvist, M.S. Transport and Tensile Properties of Compression-Molded Wheat Gluten Films. *Biomacromolecules* **2004**, *5*, 2020–2028. [CrossRef]

20. Muneer, F.; Johansson, E.; Hedenqvist, M.S.; Plivelic, T.S.; Markedal, K.E.; Petersen, I.L.; Sørensen, J.C.; Kuktaite, R. The impact of newly produced protein and dietary fiber rich fractions of yellow pea (Pisum sativum L.) on the structure and mechanical properties of pasta-like sheets. *Food Res. Int.* **2018**, *106*, 607–618. [CrossRef]

21. Gontard, N.; Guilbert, S.; Cuq, J.-L. Water and Glycerol as Plasticizers Affect Mechanical and Water Vapor Barrier Properties of an Edible Wheat Gluten Film. *J. Food Sci.* **1993**, *58*, 206–211. [CrossRef]

22. Johansson, E.; Malik, A.H.; Hussain, A.; Rasheed, F.; Newson, W.R.; Plivelic, T.; Hedenqvist, M.S.; Gällstedt, M.; Kuktaite, R. Wheat Gluten Polymer Structures: The Impact of Genotype, Environment, and Processing on Their Functionality in Various Applications. *Cereal Chem. J.* **2013**, *90*, 367–376. [CrossRef]

23. Newson, W.R.; Kuktaite, R.; Hedenqvist, M.; Gällstedt, M.; Johansson, E. Oilseed Meal Based Plastics from Plasticized, Hot Pressed Crambe abyssinica and Brassica carinata Residuals. *J. Am. Oil Chem. Soc.* **2013**, *90*, 1229–1237. [CrossRef]

24. Gerrard, J.A. Protein–protein crosslinking in food: Methods, consequences, applications. *Trends Food Sci. Technol.* **2002**, *13*, 391–399. [CrossRef]

25. Lagrain, B.; Goderis, B.; Brijs, K.; Delcour, J.A. Molecular basis of processing wheat gluten toward biobased materials. *Biomacromolecules* **2010**, *11*, 533–541. [CrossRef] [PubMed]

26. Rombouts, I.; Lagrain, B.; Delcour, J.A.; Türe, H.; Hedenqvist, M.S.; Johansson, E.; Kuktaite, R. Crosslinks in wheat gluten films with hexagonal close-packed protein structures. *Ind. Crops Prod.* **2013**, *51*, 229–235. [CrossRef]

27. Giosafatto, C.; Al-Asmar, A.; D'Angelo, A.; Roviello, V.; Esposito, M.; Mariniello, L. Preparation and Characterization of Bioplastics from Grass Pea Flour Cast in the Presence of Microbial Transglutaminase. *Coatings* **2018**, *8*, 435. [CrossRef]

28. Ceresino, E.B.; de Melo, R.R.; Kuktaite, R.; Hedenqvist, M.S.; Zucchi, T.D.; Johansson, E.; Sato, H.H. Transglutaminase from newly isolated Streptomyces sp. CBMAI 1617: Production optimization, characterization and evaluation in wheat protein and dough systems. *Food Chem.* **2018**, *241*, 403–410. [CrossRef]

29. Ceresino, E.B.; Kuktaite, R.; Sato, H.H.; Hedenqvist, M.S.; Johansson, E. Impact of gluten separation process and transglutaminase source on gluten based dough properties. *Food Hydrocolloid.* **2019**, *87*, 661–669. [CrossRef]

30. Morel, M.H.; Redl, A.; Guilbert, S. Mechanism of heat and shear mediated aggregation of wheat gluten protein upon mixing. *Biomacromolecules* **2002**, *3*, 488–497. [CrossRef]

31. Rasheed, F.; Newson, W.R.; Plivelic, T.S.; Kuktaite, R.; Hedenqvist, M.S.; Gallstedt, M.; Johansson, E. Structural architecture and solubility of native and modified gliadin and glutenin proteins: Non-crystalline molecular and atomic organization. *RSC Adv.* **2014**, *4*, 2051–2060. [CrossRef]

32. Blomfeldt, T.O.J.; Kuktaite, R.; Johansson, E.; Hedenqvist, M.S. Mechanical Properties and Network Structure of Wheat Gluten Foams. *Biomacromolecules* **2011**, *12*, 1707–1715. [CrossRef] [PubMed]

33. Muneer, F.; Andersson, M.; Koch, K.; Menzel, C.; Hedenqvist, M.S.; Gallstedt, M.; Plivelic, T.S.; Kuktaite, R. Nanostructural Morphology of Plasticized Wheat Gluten and Modified Potato Starch Composites: Relationship to Mechanical and Barrier Properties. *Biomacromolecules* **2015**, *16*, 695–705. [CrossRef] [PubMed]

34. Nowick, J.S. Exploring β-sheet structure and interactions with chemical model systems. *Acc. Chem. Res.* **2008**, *41*, 1319–1330. [CrossRef] [PubMed]

35. Jackson, M.; Mantsch, H.H. The Use and Misuse of FTIR Spectroscopy in the Determination of Protein Structure. *Critical Rev. Biochem. Mol. Biol.* **1995**, *30*, 95–120. [CrossRef] [PubMed]

36. Cho, S.W.; Gällstedt, M.; Johansson, E.; Hedenqvist, M.S. Injection-molded nanocomposites and materials based on wheat gluten. *Int. J. Biol. Macromol.* **2011**, *48*, 146–152. [CrossRef] [PubMed]

37. Blomfeldt, T.O.; Kuktaite, R.; Plivelic, T.S.; Rasheed, F.; Johansson, E.; Hedenqvist, M.S. Novel freeze-dried foams from glutenin-and gliadin-rich fractions. *RSC Adv.* **2012**, *2*, 6617–6627. [CrossRef]

38. Muneer, F.; Andersson, M.; Koch, K.; Hedenqvist, M.S.; Gällstedt, M.; Plivelic, T.S.; Menzel, C.; Rhazi, L.; Kuktaite, R. Innovative Gliadin/Glutenin and Modified Potato Starch Green Composites: Chemistry, Structure, and Functionality Induced by Processing. *ACS Sustain. Chem. Eng.* **2016**, *4*, 6332–6343. [CrossRef]

39. Sun, S.; Song, Y.; Zheng, Q. Thermo-molded wheat gluten plastics plasticized with glycerol: Effect of molding temperature. *Food Hydrocolloid.* **2008**, *22*, 1006–1013. [CrossRef]

40. Muneer, F.; Johansson, E.; Hedenqvist, M.S.; Gällstedt, M.; Newson, W.R. Preparation, Properties, Protein Cross-Linking and Biodegradability of Plasticizer-Solvent Free Hemp Fibre Reinforced Wheat Gluten, Glutenin, and Gliadin Composites. *BioResources* **2014**, *9*, 5246–5261. [CrossRef]

© 2018 by the authors. Licensee MDPI, Basel, Switzerland. This article is an open access article distributed under the terms and conditions of the Creative Commons Attribution (CC BY) license (http://creativecommons.org/licenses/by/4.0/).

International Journal of
Molecular Sciences

MDPI

Article

Exploring the Structural Transformation Mechanism of Chinese and Thailand Silk Fibroin Fibers and Formic-Acid Fabricated Silk Films

Qichun Liu [1,2], Fang Wang [1,*], Zhenggui Gu [2], Qingyu Ma [3] and Xiao Hu [4,5,6,*]

1   Center of Analysis and Testing, Nanjing Normal University, Nanjing 210023, China; 40021@njnu.edu.cn
2   School of Chemistry and Materials Science, Nanjing Normal University Jiangsu, Nanjing 210023, China; guzhenggui@njnu.edu.cn
3   School of Physics and Technology, Nanjing Normal University, Nanjing 210023, China; maqingyu@njnu.edu.cn
4   Department of Physics and Astronomy, Rowan University, Glassboro, NJ 08028, USA
5   Department of Biomedical Engineering, Rowan University, Glassboro, NJ 08028, USA
6   Department of Molecular and Cellular Biosciences, Rowan University, Glassboro, NJ 08028, USA
*   Correspondence: wangfang@njnu.edu.cn (F.W.); hu@rowan.edu (X.H.); Tel.: +1-86-025-8589-8176 (F.W.); +1-856-256-4860 (X.H.)

Received: 19 September 2018; Accepted: 22 October 2018; Published: 24 October 2018

Abstract: Silk fibroin (SF) is a protein polymer derived from insects, which has unique mechanical properties and tunable biodegradation rate due to its variable structures. Here, the variability of structural, thermal, and mechanical properties of two domesticated silk films (*Chinese and Thailand B. Mori*) regenerated from formic acid solution, as well as their original fibers, were compared and investigated using dynamic mechanical analysis (DMA) and Fourier transform infrared spectrometry (FTIR). Four relaxation events appeared clearly during the temperature region of 25 °C to 280 °C in DMA curves, and their disorder degree (f_{dis}) and glass transition temperature (T_g) were predicted using Group Interaction Modeling (GIM). Compared with *Thai* (Thailand) regenerated silks, *Chin* (Chinese) silks possess a lower T_g, higher f_{dis}, and better elasticity and mechanical strength. As the calcium chloride content in the initial processing solvent increases (1%–6%), the T_g of the final SF samples gradually decrease, while their f_{dis} increase. Besides, SF with more non-crystalline structures shows high plasticity. Two α- relaxations in the glass transition region of tan δ curve were identified due to the structural transition of silk protein. These findings provide a new perspective for the design of advanced protein biomaterials with different secondary structures, and facilitate a comprehensive understanding of the structure-property relationship of various biopolymers in the future.

Keywords: silk fibroin; glass transition; DMA; FTIR; stress-strain

1. Introduction

Silk is a biopolymer with perfect biocompatibility and tunable biodegradability due to its unique protein compositions and structures [1–5]. In the past few decades, silk has been developed into variable biomaterials including tubes, sponges, hydrogels, fibers and thin films, and combined with various functional nanomaterials to provide unique properties that can be applied to biomedical, electrical, or material engineering [6–11].

Generally, different material fabrication methods can affect the multi-step structural transitions and physical properties of silk fibroin materials. For example, Philips et al. [12] compared the dissolution of silk fibroin using different ionic liquids, and demonstrated that 1-butyl-3-methylimidazolium chloride, 1-butyl-2,3-dimethylimidazolium, and 1-ethyl-3-

methylimidazolium were able to disrupt the hydrogen bonding in silk fibroin fibers. By controlling drying rate, Lu et al. [13] were able to prepare water-insoluble silk films from 9.3 mol/L LiBr aqueous solution. Tian et al. [14] added poly epoxy materials, such as formaldehyde, glutaraldehyde, and epoxy compounds into silk fibroin. They suggested that the flexibility of silk materials can be improved through the epoxy compounds, which also acted as crosslinking agents for silk fibroin proteins.

Dynamic Mechanical Analysis (DMA) is one of thermal analysis techniques, which is an advanced technique for measuring the viscoelastic change of polymeric materials during their structural relaxation. [15–20] Juan et al. [21] investigated the effect of temperature and thermal history on the mechanical properties of native silkworm and spider dragline silks by dynamic mechanical thermal analysis (DMTA). Their results showed that the DMA storage modulus and loss tangent of silk materials depend on their different chemical and physical processing methods. Wang et al. [22] also explored the variability of individual as-reeled *A. pernyi* silk fibers using DMTA. They suggested that different polar solvents could affect the tensile properties and structure of silk fibers during the quasi-static tensile tests in ethanol, air, methanol, or water. Porter et al. [23,24] assumed that spider silk's stiffness and strength attributed to the high cohesive energy density of hydrogen bonding, and the toughness attributed to the high energy absorption during post-yield deformation. Furthermore, they found that silk strength was associated with the peculiar molecular and nanoscale structure of its morphology. Kawano et al. [25] measured *Nafion* silk films with different types of solvent and cations using DMA in the controlled force mode. Their results demonstrated that silk elasticity decreased with the increase of water, methanol, ethanol, or ethanol/water mixture content in the *Nafion* film, and also decreased with increasing temperature and cation substitutions (Li^+, Na^+, K^+, Cs^+ and Rb^+).

Besides, Step-scan Differential Scanning Calorimetry (SSDSC) is a relatively new technique which is another thermal analysis technique under temperature modulation, where the temperature program comprises a periodic succession of short, heating rates, and isothermal steps; thus, the measured heat flow contains contributions which arise from the heat capacity and those due to physical transformations or chemical reactions. The total heat flow can be separated into the reversing and non-reversing components, because the reversing component is only observed on the heating part of the cycle and the non-reversing only on the isothermal. Since both the heat capacity equilibration and DSC equilibration are rapid, the C_p calculation is said to be independent of kinetic processes [26]. Therefore, through SSDSC, the "reversing heat capacity", which represents the reversible heat effect of samples within the temperature range of the modulation, such as the specific heat of samples during the glass transition region, can be measured and calculated. Hu et al. [3] used temperature modulated DSC (TMDSC) to eliminate the non-reversing thermal phenomena of the sample and measure the reversing thermal properties of the silk-tropoelastin samples. Sheng et al. [27] characterized the heat capacity, phase contents and transitions of PLA scaffold using SSDSC approach.

Bombyx Mori silkworms are domestically raised silkworms that can produce white silk fibers (e.g., from China (*Chin* silk)), or yellow silk fibers (e.g., from Thailand (*Thai* silk)), due to their different geographical and growing environments [28,29]. Derived from *Bombyx mori* cocoons, silk fibroin (SF) is a fibrous protein consisting of repeating glycine-alanine or glycine-serine peptides responsible for beta-sheet crystal structures mixed with amorphous regions [3–11]. Different silkworm species have different amino acid compositions and therefore have different crystallinity [5,23,24]. The environmental climate can also affect the mulberry leaves. Various silkworm leaves or foods may lead to differences in their cocoons, such as the color and strength. In our previous work, *Indian Antheraea mylitta*, *Philosamia ricini*, *Antheraea assamensis*, Thailand and Chinese *Bombyx mori* mulberry (*Thai*, *Chin*) silk films have been successfully regenerated from the aqueous solution [30]. Moreover, it was found that Chin and Thai silk fibroin films can be regenerated through a calcium chloride-formic acid ($CaCl_2$/FA) solution system [31]. It was demonstrated that Ca^{2+} ions could interact with the silk structure, and change their glass transition temperature, specific heat, and thermal stability [32]. In this work, the DMA technique was, for the first time, used to explore and compare the structure and mechanical property of these two kinds of silk fibers (Thai, Chin), and also combined with SSDSC and FTIR technologies to investigate these properties and transformation mechanism of their protein films regenerated from the FA solution with a changing CaCl$_2$ content (1%~6%). In addition, a theory developed by Porter et al. [24] Group Interaction Modeling (GIM), was used to investigate and verified the relationship between the stability and structure of regenerated silk materials during the glass transition temperature region (T_g). This work also explained the impact of CaCl$_2$ content to the dynamic mechanical properties of two domesticated silks comprehensively. These comparative studies are important for the design of advanced silk-based materials with tunable structures and properties.

2. Results and Discussion

2.1. Dynamic Mechanical Analysis of the Degumming Silk Fiber

Figure 1 shows the storage modulus E' and loss factor tan δ curves of CRS and TRS natural fibers with the change of temperature at five frequencies (1 Hz, 2 Hz, 5 Hz, 10 Hz, and 20 Hz), respectively. Four peaks were observed in E' curves of Chin silk fibroin fiber sample under all frequencies (Figure 1a), which were assigned to the protein relaxation of γ, β, α_c and α at 25 °C to 280 °C, respectively. As the frequency increases from 1 Hz to 20 Hz, the transition peak moves slightly to a higher temperature, since the time of the molecular segment relaxation and movement is inversely proportional to the frequency intensity. According to the equivalent principle of time and temperature, the increased frequency is equivalent to the shortened relaxation time of the material. Therefore, when the material is tested at a higher frequency, the transition peak of the segment movement could move to a higher temperature [33,34]. Different frequencies of 1 Hz to 20 Hz have the same effect of dynamic thermomechanical property on the silk fibroin. Therefore, in the remaining studies, we will only discuss experimental phenomena with a frequency of 1 Hz. In the 1 Hz force-controlled E' curve (solid line), the γ-relaxation endear at about 41.64 °C due to the molecular motion of the silk protein side chain and the initial evaporation of free water molecules from silk. The β-relaxation at 96.12 °C could not be precisely assigned, but may be related to the molecular motion of silk fibroin after complete evaporation of water, or to the pendant group of the silk polymers (e.g., Ardhyananta et al. [35] has pointed out that the pendant group of the polysiloxanes could affect the thermal and mechanical properties). This can be confirmed by previous findings [36,37] that regenerative silk usually contains 5–10% (w/w) bound water, which significantly affect the thermal properties of silk. The water content of our samples in this work was around 6 wt.% (Table 1) measured by thermogravimetric analysis (TG), which has been discussed previously [29]. The α-relaxation at about 235.34 °C is associated with the glass transition of the silk protein noncrystalline structure due to the segmental motion of the silk protein backbone. Notably, Um et al. [38] pointed out that an α_c-relaxation above 260 °C might occur after the α-relaxation in silk proteins fabricated from the aqueous solution. In our present work, this relaxation appeared around 269.92 °C for CRS and 272.39 °C for TRS. Besides, during the heating scan, the tan δ curves also show clearly three peaks (39.91 °C, 92.72 °C, and 214.25 °C), corresponding to the γ, β, and α-relaxation (Figure 1b). However, the peak of α_c-relaxation did not appear obviously in the tan δ curve. The same phenomena can be also found from the TRS fiber sample. For the TRS fiber, three transition events can be observed at 43.82 °C, 109.07 °C, and 238.48 °C in the E' curve at 1 Hz (Figure 1c, solid line), which correspond to γ, β, α relaxations, respectively. The final transition peak at 272.39 °C (Figure 1c) is belonged to α_c-relaxation. Furthermore, in the tan δ curve (Figure 1d, solid line), γ, β and α-relaxation peaks were observed at 42.11 °C, 100.55 °C, and 218.92 °C in the range of 25~280 °C, respectively.

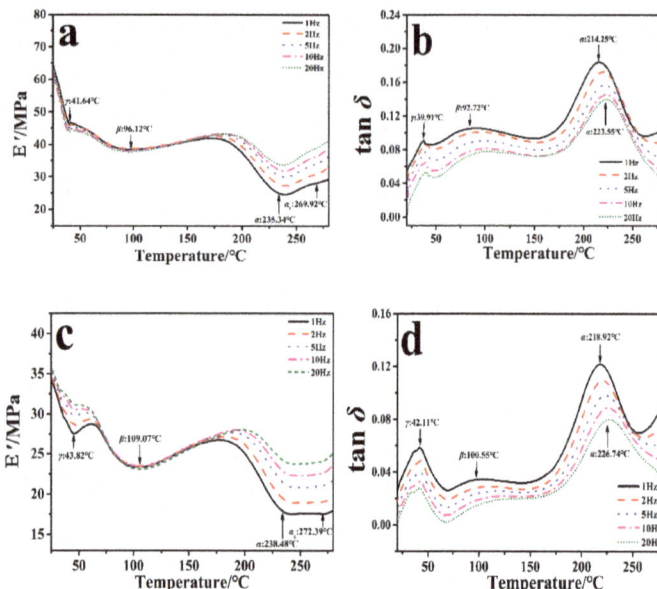

Figure 1. Dynamic mechanical analysis of the storage modulus (*E′*) curves of Chin silk fibroin fibers (**a**) and Thai silk fibroin fibers (**c**); and tan *δ* curves of Chin silk fibroin fibers (**b**) and Thai silk fibroin fibers (**d**) at five different frequencies: 1 Hz (solid line), 2 Hz (dash line), 5 Hz (dot line), 10 Hz (dash dot line), and 20 Hz (short dot line).

Table 1. The water content of different SF film samples by Thermogravimetric Analysis (TG) *.

Sample	Water Content/%	Sample	Water Content/%
CRS	9.70	TRS	7.26
CSF-1.0	3.23	TSF-1.0	1.26
CSF-1.5	2.97	TSF-1.5	1.04
CSF-2.0	2.05	TSF-2.0	1.10
CSF-3.0	1.79	TSF-3.0	1.20
CSF-4.0	1.35	TSF-4.0	0.52
CSF-6.0	1.16	TSF-6.0	0.51

* All of the numbers have an error bar of ± 3%.

Table 2 compared *E′* and tan *δ* of these two kinds of regenerated silk fibers (CRS and TRS) in the relationship of *γ*, *β*, *α*, and *α*$_c$ under the condition of 1Hz frequency. During the *γ*-relaxation, CRS sample has a storage modulus of 18.33 MPa and tan *δ* of 10.45 at 41.64 °C, while TRS sample has a lower storage modulus of 7.48 MPa and a loss factor of 7.21 at 43.82 °C. In the storage modulus *E′* curve, the peak temperatures of the CRS sample (41.64 °C, 96.12 °C, 235.34 °C, and 269.92 °C) were 2.18~12.95 °C lower than the TRS sample (43.82 °C, 109.07 °C, 238.48 °C, and 272.39 °C) (Table 2). In tan *δ* curve, the peak temperatures of the CRS sample (39.91 °C, 92.72 °C, and 214.25 °C) were also 2.20~4.67 °C lower than those of TRS sample (42.11 °C, 100.55 °C, and 218.2 °C) (Table 2). Moreover, the storage modulus *E′* of TRS sample under the 1 Hz (7.48 MPa, 11.71 MPa, 23.54 MPa, and 25.73 MPa) were also lower than those of the CRS sample (18.33 MPa, 26.28 MPa, 40.20 MPa and 38.34 MPa) during the *γ*, *β*, *α* and *α*$_c$ transitions, respectively (Table 2). These results indicated that *Chin* white silk fiber (CRS) can dehydrate more easily and have more disorder in its structure than the yellow *Thai* TRS fiber, and *Chin* silk molecular chains can be moved more easily when heated. Besides, it will possess a higher degree of viscous deformation, stronger damping, and faster energy dissipation than the TRS sample. Meanwhile, this might also imply that the CRS fiber has more elasticity and stiffness.

Table 2. The experimental parameters of degummed Chinese (CRS) and Thailand (TRS) *B. Mori* silk fibers obtained from DMA analysis *

Sample	CRS/TRS			
Attribution	γ	β	α	α_c
E′/MPa	18.33/7.48	26.28/11.71	40.20/23.54	38.34/25.73
Tan δ	10.45/7.21	6.24/4.72	23.49/19.36	N/A
$T_{E'}$/°C	41.64/43.28	96.12/109.07	235.34/238.48	269.92/272.39
$T_{tan\delta}$/°C	39.91/42.11	92.72/100.55	214.25/218.92	N/A
$\Delta T_{E'}$/°C	2.18	12.95	3.14	2.47
$\Delta T_{tan\delta}$/°C	2.20	7.83	4.67	N/A

* E' and tan δ represent the storage modulus and integral loss factor of CRS and TRS samples at 1 Hz frequency respectively in DMA tensile mode. $T_{E'}$ and $T_{tan\delta}$ represent the peak temperatures for the storage modulus and the loss factor curves, which are corresponding to the protein relaxation of γ, β, α and α_c. $\Delta T_{E'}$ and $\Delta T_{tan\delta}$ represent the peak temperature differences between TRS and CRS samples at γ, β, or α-relaxation regions. The E' and Tan δ values have an error bar of ± 0.3, the $T_{E'}$ and $T_{tan\delta}$ values have an error bar of ± 0.5 °C.

Born et al. [39] pointed out that the thermally induced vitreous transition of silk fibroin was proposed to the derive non-cooperative or cooperative movements of the skeleton segments in the non-crystalline or disordered regions of silk structure, when the intermolecular forces pass through a maximum or the intermolecular rigidity tends to zero. The transition condition is known as Born's elastic instability criterion, which focuses on the stiffness or mobility of the bonds perpendicular to the axis of interaction instead of the bonds along the axis interaction. Quantitatively, Porter's Group Interaction Modeling (GIM) theory provided a relationship between the properties and structure of polymeric materials, and the expression between structural parameters and T_g can be presented in Equation (1):

$$T_g^c = 0.224y + 0.0513\frac{E_{coh}}{N} \tag{1}$$

where the T_g^c is the theoretical glass transition temperature at a reference rate of 1 Hz, which can be written in terms of several parameters: (1) the temperature of skeletal mode vibrations, y; (2) the cohesive energy, E_{coh}; and (3) the skeletal degrees of freedom, N [24].

The degrees of freedom N, in the GIM frame is defined as the number of normal vibration skeletons at the axis of the polymer backbone. For detailed calculation of E_{coh} for each peptide base, reference can be made to Porter's work [24]. Wang et al. [22] and Guan et al. [40] studied the cohesive energy E_{coh} and degree of freedom N of each group of China silk from *Bomby Mori cocoon* according to the data in Table 3. *B. mori* silk' E_{coh} was calculated as: E_{coh} = 24.3 (contribution of the peptide base) + 0 × 47.5% (contribution of glycine -H) + 4.5 × 31.7% (Alanine contribution -CH$_3$) + 10.8 × 15.8% (Serine -CH$_2$-OH contribution) + 35.8 × 5% (tyrosine contribution -CH$_2$-Ph-OH) = 29.2 (kJ·mol^{-1}).

Table 3. Calculated GIM parameters based on peptide group contributions and amino acid (AA) sequences for *B Mori* silks [24].

Peptide	Structure	E_{coh}/kJ·mol^{-1} (without H-Bond)	Degrees of Freedom N	Molar Fraction as Counted in AA Sequence *B Mori*
Peptidebase(-R group)	-C-CO-NH-	24.3	6	0
Glycine(G)	-H	0	0	47.5%
Alanine(A)	-CH$_3$	4.5	2	31.7%
Serine(S)	-CH$_2$-OH	10.8	3	15.8%
Glutamine(Q)	-CH$_2$-CH$_2$-CO-NH$_2$	28.8	5	0
Tyrosine(Y)	-CH$_2$-Ph-OH	35.8	4	5%
Leucine(L)	-CH$_2$-C(CH3)$_2$	18	4	0
Arginine(R)	-CH$_2$-CH$_2$-CH$_2$-NH-C(NH$_2$)$_2$	45	7	0
Average	*B. Mori*	29.2	6	/

While the CRS and TRS samples in our study were grown in different regions (China and Thailand) and have different thermal properties [28–30], where they all came from *Bombyx Mori* silkworm species. Therefore, we considered that these two kinds of silk have the same cohesive energy E_{coh} and degree of freedom N.

Meanwhile, in our research, the experimental value of T_g was defined as the temperature from fitted gaussian peak position on the tan δ curve during the glass transition, as shown in Figure 1b,d and Table 4. Using the GIM method and the structural parameters in Table 3, we have $E_{coh} = 29.2$ kJ·mol^{-1}, $N = 6$, and $y = 241$ °C. The value of y was common to all structures due to the same average group molecular weight [23]. Therefore, the theoretical value of $T_g{}^c$ of CRS or TRS fiber sample was calculated to 82.4 °C, without considering the contribution from hydrogen-bonds. This theoretical value $T_g{}^c$ was much lower than that of DMA observation (about 218 °C). If one hydrogen bond per peptide group was taken, the calculated result T_g from Equation (1) would become 157.6 °C, which is close to the lower limit of the silk's experimental T_g. Guan et al. [40] argued that the molecular structures responsible for the glass transition of silk do not have a singular form, but a probability spectrum with several favored forms, e.g., one or two hydrogen bonds per peptide. Thus, the experimental T_g temperatures of 214.25 °C for Chin silk fibroin fibers in Figure 1b and of 218.92 °C for Thai silk fibroin fibers in Figure 1d are the results of the averaged hydrogen-bonding density contributed by hydrogen-bonding forms with different probabilities, respectively. Vollrath et al. [23] believed that if one or two hydrogen bonds were adopted, the molecular structure in the silk would have a 70% chance of 2 hydrogen bonds (H-bonds). Hydrogen bond energy took 10 kJ·mol^{-1} as an average of N-H···O and N-H···N forms, respectively. The higher T_g implied more hydrogen bonds existed among amide groups of silks, through which highly oriented molecular structure and the number of hydrogen bonds have impact on the cohesive energy directly. In *B. Mori* silk sample, if two hydrogen bonds per peptide were counted, an additional energy of 20 kJ·mol^{-1} would be added to the peptide base value of 29.2 kJ·mol^{-1}, which gives the final average E_{coh} of 49.2 kJ·mol^{-1} for each characteristic segment. As a result, the theoretical value $T_g{}^c$ of silk is 243.1 °C through Equation (1) calculation, which is close to the upper limit of experimental temperature of T_g at 214.25 °C in tan δ curve and at 235.34 °C in E' curve for CRS fiber sample, while at 218.92 °C in tan δ curve, and at 238.48 °C in E' curve for TRS fiber sample.

Table 4. GIM parameters used for T_g and the degree of structural disorder f_{dis} calculations b*

Sample	Group	H-Bonds	E_{coh}/kJ·mol^{-1}	N	$T_g{}^c$/°C	T_g/°C	Tan δ^c	Tan δ	f_{dis}
Chin	$G_{0.475}A_{0.317}S_{0.158}Y_{0.005}$	1	39.2	6	157.6	214.25	56	23.49	0.63
		2	49.2	6	243.1		70		0.51
Thai	$G_{0.475}A_{0.317}S_{0.158}Y_{0.005}$	1	39.2	6	157.6	218.92	56	19.36	0.52
		2	49.2	6	243.1		70		0.42

* The number of H-bonds per peptide group is 1 or 2. Cohesive energy E_{coh} is the sum of energy from hydrogen bonds and the peptide base. N is the degrees of freedom. $T_g{}^c$ is the theoretical glass transition temperature calculated from Equation (1), in which y is set as 241 °C for all cases. T_g represents the experimental glass transition temperature from tan δ curve at α relaxation. The theoretical Tan δ^c was calculated from Equation (2), which represents the energy dissipation for 100% degrees of structural disorder. Tan δ is the integral loss factor at α relaxation in DMA curve, and f_{dis} is the predicted degree of structural disorder by using Equation (3).

Tan δ curves from DMA can be further used to determine the structural change of regenerated silk. As previously introduced, the order-disorder distribution can avoid the complicated assignments of secondary structures, which could be used to effectively predict the macroscopic properties of silk materials. First, a structural parameter, f_{dis}, is defined as the degree of structural disorder, which is the molar fraction of the non-crystalline structures that are responsible for glass transition. The degree of structural disorder approximates an averaged structural parameter of heterogeneous nano-structures in a mean-field homogeneous micro and macroscopic morphology. For these two kinds of silk, their f_{dis} values obtained from amino acid sequence analysis are listed in Table 4.

Equation (2) from GIM model described the quantitative relationship of cumulative tan δ (over the transition temperature range) with the structural parameters of the interactive group, E_{coh} and N. A quick calculation of tan δ^c using Equation (2) for *B. Mori* silk is shown in Table 4, which appeared much greater than the experimental values of tan δ. Therefore, for semi-crystalline silks, the degree of structural disorder f_{dis} was introduced into the equation, and adapted as a new form, as presented in Equation (3). The function of factor f_{dis} is easy to understand as only the motions of the disordered structure could be activated during the glass transition and contribute to the experimentally measured tan δ.

The coefficient (2/3) in Equation (3) was used to correct the experimental effect of the uniaxial tensile mode in DMA measurement, because the molecular structures subjected to the static stress of the tensile direction could not be relaxed as the motions along this direction are restrained. As a result, the probability of molecular motions of the overall disordered structure through glass transition was reduced by one dimension, or a factor of 2/3.

$$\tan \delta^c = 0.0085 \times \frac{E_{coh}}{N} \tag{2}$$

$$\tan \delta = \frac{2}{3} \times f_{dis} \times 0.0085 \frac{E_{coh}}{N} \tag{3}$$

Equation (3) opened two avenues: First, it allowed the prediction of the cumulative loss tangent with a known degree of structural disorder. Second, it allowed the calculation of the degree of structural disorder from the theoretical tan δ^c. A quick calculation using GIM framework (Equation (2)) for *B. Mori* silk showed that tan δ^c was in the numerical range of 56–70, which represented the energy dissipation of 100% degrees of structural disorder. The apparent discrepancy between the experimental cumulative tan δ and the theoretical tan δ^c during the glass transition for native *B. Mori* silks clearly suggested that crystalline or ordered structure existed in our silk samples. This phenomenon was also mentioned in the work of Porter et al. [41–43] The degree of disorder f_{dis} for two silk fibers were also calculated individually by using both the number of hydrogen bonds and the cohesive energy from the Equation (3) (Table 4). Simultaneously, apparent discrepancy between the experimental tan δ and the theoretical tan δ^c in glass transition region of silks appeared by using the GIM framework. Guan et al. and Porter et al. also reported the same phenomenon [21,24]. By comparing the tan δ of these two kinds of regenerated *B. Mori* silk, the degree of disorder f_{dis} was deduced: 0.63 for one H-bond and 0.51 for two H-bonds in the CRS sample, and 0.52 for one H-bond and 0.42 for two H-bonds in the TRS sample, as listed in Table 4. The results showed that the disordered structure of silk fibroin had a significant effect on its glass transition. Additionally, the glass transition temperature (T_g) decreased with the degree of disorder (f_{dis}) increasing, since tan δ is directly associated with the f_{dis} as well as the ordered molecular structures in silk.

2.2. Structural Transformation of Chin and Thai Silk Protein Films

Glass transition temperature is a characterization temperature at which the chain segment of polymer molecules starts to move, which is related to the flexibility of polymer chains. In the glass transition region, when the semi-crystalline polymer material changes from the solid state to the flowing liquid state, the specific heat of the semi-crystalline polymer material undergoes a discontinuous mutation during the heating process [28,29]. Our previous studies on silk fibroin films by scanning electron microscopy (SEM) [30] and X-ray diffraction (XRD) [33] showed that a high $CaCl_2$ concentration can significantly reduce the silk fibril structure and micro-/nanoscale morphology in the silk film. Further, two small diffraction peaks appeared at 20.7° and 24.0° in the XRD curves of low $CaCl_2$ concentration sample (e.g., TSF-1.5), which is recognized as the silk I structure. This phenomenon implied that the β-sheet crystal content decreases with the increase of calcium chloride concentration, and higher calcium chloride concentrations may disrupt the hydrogen bonds between the silk fibroin molecular chains, which reduces the silk II content and increases the silk I content. Here, we will explore the structure transformation of silk proteins fabricated from CaCl₂–formic acid solution by using DMA, SSDSC, and FTIR results.

In general, the loss factors (tan δ) of silk fibroin membranes CSF-1.0, CSF-1.5, CSF-2.0, CSF-3.0, CSF-4.0, and CSF-6.0 all have three discontinuous events that corresponded to the protein relaxation of γ, β, and α, which have been discussed in the previous section. The γ relaxation peak around 50~60 °C with little shoulders in two side of curve implied the co-events of γ-relaxation (protein-water T_g) and the evaporation of mobile H_2O in this region [21]. The water evaporation completed around 80~140 °C, and the major molecular motion of pure silk fibroin (the α relaxation peak of tan δ curve) appeared in the range of 150–230 °C (Figure 2). The water content of various silk protein films samples contained around 0.5–3.23% bound water molecules measured by TG in our previous work [30]. Hu et al. [36]. focused on the interaction of the solid silk film with the intermolecular bound water molecules. The results showed that the silk film start to release the water molecules into air at 35 °C. As the temperature increased, more and more water evaporated and the weight of silk film decreased until about 160 °C. Above 160 °C, there is no more intermolecular water in the silk film. Based on these results, the water should have no contribution to α-relaxation. In addition, this peak also decreased gradually with the increasing of calcium chloride. To better understand the change during the T_g region, StepScan differential scanning calorimetry (SSDSC) measurement were also used to determine the heat capacity increment ($\Delta C_p{}^s$) of silk samples during the T_g. It demonstrated that the glass transition temperature ($T_g{}^s$) of SF increased with the calcium chloride concentration decreasing, e.g., from 157.30 °C to 176.88 °C with the change of calcium chloride concentration from 6.0% to 1.0%, respectively. The heat capacity increment ($\Delta C_p{}^s$) is directly proportional to the average chain mobility of proteins, which reflects the number of freely rotating bonds capable of changing the chain conformation [30]. The $\Delta C_p{}^s$ results summarized in Table 5 indicated that a non-crystalline structure exists in all samples and the SF-6.0 protein chains have the highest fraction of the non-crystalline structure, while the average chain mobility and non-crystalline fraction in the SF-1.0 sample are the lowest. Therefore, with the CaCl₂ concentration increasing, the $T_g{}^s$ decreased while $\Delta C_p{}^s$ increased, which indicated that content of non-crystalline structures in SF is increasing with the increase of CaCl₂ content.

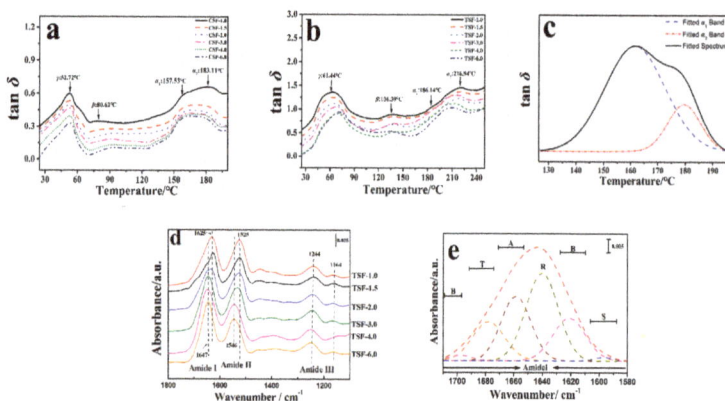

Figure 2. Tan δ curves of CSF (**a**) and TSF (**b**) from 1.0% (Solid); 1.5 (Dash); 2.0 (Dot); 3.0 (Dash Dot); 4.0 (Short Dot); 6.0 (Dash Dot Dot) CaCl₂/FA; (**c**) Curve fitting example of the T_g region (sample TSF-3.0). The fitted peaks are shown as Dash (α₁ Peak) and Dash Dot Dot (α₂ Peak); (**d**) FTIR absorbance spectra of TSF samples from different CaCl₂ conctration solutions in the range of 1100–1800 cm^{-1}; (**e**) Curve fitting example of the amide I region (sample TSF-3.0) in FTIR spectra. The fitted peaks were shown as short dash-dotted lines, and assigned as side chains (S), β-sheets (B), random coils (R), α-helixes (A), and turns (T).

Table 5. SSDSC and DMA analysis of CSF and TSF samples in their glass transition region [30] *.

Sample	[CaCl$_2$]/ wt %	$T_g{}^s$/°C	ΔC_p/ J·g^{-1}·°C^{-1}	$T_{g\text{-}\alpha1}$/°C	Content in α_1 Region/%	Tan $\delta_{\text{-}\alpha1}$	$f_{\text{dis-}\alpha1}$	$T_{g\text{-}\alpha2}$/°C	Content in α_2 Region/%	Tan $\delta_{\text{-}\alpha2}$	$f_{\text{dis-}\alpha2}$
	1.0	176.88	0.1667	157.53	32.87	6.65	0.18	177.11	67.13	7.52	0.20
	1.5	172.65	0.1721	155.49	52.56	7.15	0.19	171.47	47.44	7.92	0.21
CSF	2.0	169.04	0.1890	155.16	53.95	7.89	0.21	168.32	46.05	8.91	0.24
	3.0	167.42	0.1998	154.87	71.53	9.89	0.27	165.51	28.47	9.25	0.25
	4.0	163.21	0.2157	154.39	73.54	11.57	0.3	163.93	26.46	10.37	0.27
	6.0	157.30	0.2204	153.92	79.99	12.44	0.34	162.36	20.01	11.03	0.30
	1.0	181.54	0.1641	186.14	54.70	4.34	0.12	216.94	45.30	4.02	0.10
	1.5	177.08	0.1643	184.54	65.51	5.73	0.15	212.01	34.49	4.97	0.13
TSF	2.0	175.94	0.1644	184.31	71.14	6.52	0.18	209.27	28.86	5.56	0.14
	3.0	173.42	0.1972	183.69	75.24	7.22	0.19	206.45	24.76	6.25	0.16
	4.0	164.01	0.2136	183.08	75.72	8.93	0.23	204.85	24.28	7.67	0.20
	6.0	159.20	0.2137	182.47	76.65	10.17	0.26	200.82	23.35	8.93	0.23

* [CaCl$_2$] is the concentration of calcium chloride in the solution. $T_g{}^s$ is the glass transition temperature of SF measured by SSDSC. $T_{g\text{-}\alpha1}$ and $T_{g\text{-}\alpha2}$ represent the peak temperatures of α_1-relaxation and α_2-relaxtion from the tan δ curves, respectively. Their content values were obtained by fitting the tan δ curve in α_1 and α_2-relaxtion regions using Gaussian peaks. Tan $\delta_{\text{-}\alpha1}$ and Tan $\delta_{\text{-}\alpha2}$ are the integral loss factor at α relaxations. $f_{\text{dis-}\alpha1}$ and $f_{\text{dis-}\alpha2}$ are the predicted degree of structural disorder at α_1-relaxation and α_2-relaxation by the GIM model, respectively. The $T_g{}^s$, $T_{g\text{-}\alpha1}$, and $T_{g\text{-}\alpha2}$ have an error bar of ±0.5 °C. The Content in α_1 region and Content in α_2 region have an error bar of ±3%. The Tan $\delta_{\text{-}\alpha1}$, Tan $\delta_{\text{-}\alpha2}$, $f_{\text{dis-}\alpha1}$ and $f_{\text{dis-}\alpha2}$ have an error bar of ±0.05.

In our previous study [30], we found that the concentration of CaCl$_2$/FA could significantly affect the secondary structures of silk. Since DMA technique is more sensitive than the SSDSC technique for the glass transition measurement, two relaxation peaks (α_1 and α_2) can be found in the glass transition region of tan δ curve from 125 °C to 195 °C (Figure 2). For the CSF-1.0 sample, the two transition peaks appeared at 157.53 °C ($T_{g\text{-}\alpha1}$) and 177.11 °C ($T_{g\text{-}\alpha2}$), which are associated with α_1 and α_2- relaxation (Figure 2a, solid line), respectively. For the TSF-1.0 sample, the α_1 and α_2-relaxation transitions appeared at 186.14 °C ($T_{g\text{-}\alpha1}$) and 216.94 °C ($T_{g\text{-}\alpha2}$), respectively. To quantify the percentage of these two peaks, the tan δ curves were curve fitted using Gaussian peaks in the glass transition region of 125–195 °C.

Figure 2c showed an example of curve fitted tan δ curves from the TSF-3.0 sample (dashed lines). Table 5 summarized the fitted percentage of each peak in T_g region for all silk samples. It shows that the TSF-1.0 film has 54.70% α_1 relaxion at 186.14 °C ($T_{g\text{-}\alpha1}$), and 45.30% α_2-relaxation at 216.94 °C ($T_{g\text{-}\alpha2}$). While the TSF-6.0 sample has 76.65% α_1 relaxion at 182.47 °C and 23.35% α_2 relaxion at 200.82 °C. This suggests that as the concentration of calcium chloride increased, both peaks shifted to lower temperatures (Table 5), but the content of α_1 relaxion increased, while the content of α_2 relaxion decreased. To better understand the secondary structures of our materials, a FTIR deconvolution method was performed to quantify the percentage of the secondary structures in all silk samples [28]. Figure 2d showed the main characteristics of protein structures of all six silk films in the FTIR spectra. With the concentration of CaCl$_2$ increasing, the center of the absorption band in Amide I region shifted gradually from 1625 cm^{-1} (beta-sheet structure, TSF-1.0) to 1647 cm^{-1} (random coil structure, TSF-6.0), while the peak in Amide II region also shifted gradually from 1525 cm^{-1} (TSF-1.0) to 1546 cm^{-1} (TSF-6.0). Moreover, for the Amide-III and FTIR fingerprinting regions, Thai silk sample showed same characteristic peaks and two obvious peaks at 1244 cm^{-1} and 1164 cm^{-1} [30]. Figure 2e shows an example of the Amide I region of TSF-3.0 sample with the fitted FTIR peaks shown as dashed lines. The peak positions and their related secondary structures were assigned as side chains (S), β-sheets (B), random coil (R), α-helix (A), and turns (T) [30]. We found that the content of helix structures in TSF samples could increase significantly by increasing the concentration of CaCl$_2$. For example, with the increase of CaCl$_2$ concentration, the percentage of β-sheet structures gradually decreased from 26.60% in TSF-1.0 film to 7.07% in the TSF-6.0 sample, while the percentage of random coils decreased from 56.07% to 46.43%, while α-helixes content increased from 8.59% to 28.39%, and the percentage of turns also increased from 7.11% to 17.74%. The similar trend of α-relaxation temperatures and the content of secondary structures in the silk samples were also observed in CSF samples. Table 6 summarized the percentage of secondary structures for each CSF and TSF sample. These results implied that α_1 and α_2- relaxation events in tan δ curve are associated with the change of protein secondary structures. Um et al. [38] found that there is α_c-relaxation after the major α-relaxation in water-regenerated silk materials, which is related to the cooling crystallization of silk proteins. In our silk samples regenerated from $CaCl_2$/FA system, with the decrease of β-sheets in the silk films, the α_1 content increased, while α_2 percentage decreased, and both of the α_1 and α_2 peaks shifted to a lower temperature. Therefore, we can claim that random coils or other non-crystalline structures may be contributed to the transition of α_1-relaxation which was equaled to the α-relaxation, while the change of α-helix structure may be related to α_2-relaxation which was the α_c-relaxation found previously [38].

Table 6. Percentage of secondary structures in CSF and TSF samples obtained by FTIR structural analysis [28].

	CSF					TSF				
[CaCl₂] (wt %)	β-Sheet/%	Turns/%	Side Chains%	α-Helix/%	Random Coils/%	β-Sheet/%	Turns/%	Side Chains/%	α-Helix/%	Random Coils/%
1.0	23.32	6.03	1.33	10.07	59.25	26.60	7.11	1.63	8.59	56.07
1.5	20.25	7.59	1.07	11.39	59.70	24.76	8.02	1.15	9.96	56.11
2.0	13.56	12.05	0.91	13.78	59.70	14.70	14.31	1.19	12.54	57.26
3.0	8.09	14.13	0.85	18.79	58.14	10.92	16.61	0.94	18.22	53.31
4.0	8.17	15.41	0.63	25.66	50.13	8.25	17.05	0.79	21.83	52.08
6.0	6.83	15.89	0.57	30.26	46.45	7.07	17.74	0.37	28.39	46.43

Similarly, DMA experiment also showed that $T_{g\text{-}\alpha1}$ and $T_{g\text{-}\alpha2}$ of the CSF samples were lower than those of the TSF samples (Figure 2 and Table 5) at the same calcium chloride content. Both FTIR and DMA results indicate that lower β-sheet structures and higher non-crystalline structures can be obtained in silk films regenerated from the solution with higher calcium chloride concentration.

Base on the above work, the relationship between degree of disorder f_{dis} and the T_g were discussed and the influence of calcium chloride on structure of these silk samples was further explored in this section. It was supposed that the amount of hydrogen bonding effects of calcium chloride on different silk fibroin films were identical (H-bonds = 1; $N = 6$; $E_{coh} = 39.2$ kJ·mol^{-1}, $y = 241$ °C), the degree of disorder $f_{dis\text{-}\alpha1}$ and $f_{dis\text{-}\alpha2}$ of two kinds of regenerated silk fibroin were calculated from GIM Equation (3) and is shown in Table 5, as well as the tan $\delta_{\text{-}\alpha1}$, tan $\delta_{\text{-}\alpha2}$, $T_{g\text{-}\alpha1}$, and $T_{g\text{-}\alpha2}$ from DMA test. The increasing trend of $f_{dis\text{-}\alpha1}$ and $f_{dis\text{-}\alpha2}$ indicated that more concentration of calcium chloride could induce the β-sheet transform into the random coil in the secondary structure of silk fibroin and more non-crystalline structure form. Simultaneously, in the same concentration of calcium chloride, CSF samples showed more disorder degree than that of TSF samples, e.g., 0.20 for CSF-1.0 sample, 0.10 for TSF-1.0 sample at α_1-relaxation. The results suggested that more β-sheet and crystalline structures in TSF sample than those in CSF sample.

2.3. Stress-Strain Study of CSF and TSF

The stress-strain curve gives information about the Young modulus (slope at the origin), yield point, break point, and recovery behavior of polymeric films. Meanwhile, the stress-strain curve analysis can also provide information on polymer structure, degree of crosslinking, degree of crystallization, processing conditions, and viscoelastic properties of polymer [44–46]. The stress-strain curve of the CSF sample (Figure 3a) shows that as the calcium chloride concentration increases from 1.0 to 6.0 mass%, the initial slope decreases from 7.86 to 1.83 and its stress decreases from 14.91 to 11.93 MPa at the yield point, respectively. In addition, under the stress of 10 MPa, their strains were 1.21%, 2.15%, 2.47%, 3.39%, 4.17%, and 5.88%, respectively (Table 7). Compared with the trend of change in CSF samples under the initial slope and yield stress, the trend of change in TSF samples is the same (Figure 3b). These results reveal that increasing the content of calcium chloride can increase the elongation of the sample, while its elasticity and stiffness decrease due to the formation of dominated random coils and other non-crystalline structures in silk samples [33]. Besides, with the same concentration of calcium chloride and the same stress loaded on the sample, the initial slope of CSF sample was higher than that of the TSF sample. For example, the initial slope of the

CSF-6.0 sample is 1.83 under the 10 MPa, which was higher than the initial slope of 1.52 from the TSF-6.0 sample.

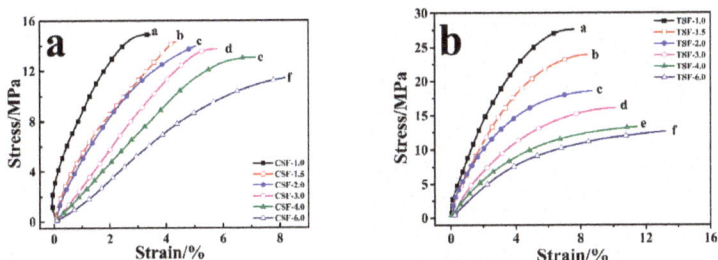

Figure 3. The stress-strain curves of CSF (**a**) and TSF (**b**) samples regenerated from 1.0% (solid square), 1.5% (hollow square), 2.0% (solid circle), 3.0% (hollow circle), 4.0% (solid triangle), and 6.0% (hollow triangle) CaCl$_2$/FA solutions.

Table 7. Mechanical properties of CSF and TSF samples measured by DMA *.

	CSF			TSF		
[CaCl$_2$]/wt %	Initial Slope	Yield Stress/MPa	Strain/%	Initial Slope	Yield Stress/MPa	Strain/%
1.0	7.86 ± 0.10	14.91 ± 1.57	1.21 ± 0.08	7.37 ± 0.24	14.28 ± 1.29	1.86 ± 0.07
1.5	7.11 ± 0.32	14.30 ± 1.38	2.15 ± 0.15	6.94 ± 0.19	13.33 ± 1.37	2.27 ± 0.10
2.0	5.48 ± 0.55	13.98 ± 1.55	2.47 ± 0.13	3.57 ± 0.36	12.45 ± 1.56	2.50 ± 0.05
3.0	2.96 ± 0.37	13.66 ± 1.29	3.39 ± 0.16	2.70 ± 0.27	11.20 ± 1.28	3.93 ± 0.12
4.0	2.16 ± 0.86	12.97 ± 1.37	4.17 ± 0.09	1.80 ± 0.58	10.17 ± 1.35	5.45 ± 0.27
6.0	1.83 ± 0.27	11.43 ± 0.98	5.88 ± 0.17	1.52 ± 0.49	8.93 ± 1.88	6.96 ± 0.16

* [CaCl$_2$] is the concentration of calcium chloride in the formic acid solution system. The initial slope is the ratio of stress to strain, which represents the elasticity of SF sample. The yield stress is the stress at the yield point in stress-strain curve. The strain value is the material elongation ratio under the stress force of 10 MPa. Every sample have to do five experiment times. Their errors or deviations were shown in each column after symbol '±'.

Similarly, the strain of the TSF-6.0 sample was 6.96%, which was greater than 5.88% of the CSF-6.0 sample. And the yield stress of CSF-6.0 sample was 11.43 MPa, which was higher than 8.93 MPa of the TSF-6.0 sample. These results indicate that the CSF sample has better elasticity than the TSF sample, while the TSF sample has better plasticity than the CSF sample. These results also imply that the average chain mobility of silk films in TSF samples is higher than that in CSF samples. Therefore, at the same CaCl$_2$ concentration, the TSF sample would contain more non-crystalline structures than the CSF sample, which has been proved by FTIR structural analysis in Table 6.

3. Materials and Methods

3.1. Materials and Preparation

The Chinese (Dandong Qiyue Trade co., LTD.) and Thailand *Bombyx mori* silk cocoons (Queen Sericulture Center, Nakornratchasima, Thailand) were first degummed into Chin regenerated silk fibroin fibers (CRS) and Thai regenerated silk fibroin fibers (TRS) according to a previously reported procedure [47]. Different amounts of calcium chloride (AR, purity 96%, West Gansu Chemical Plant, Shantou, Guangdong) were mixed with formic acid (AR, purity 88%, West Gansu Chemical Plant, Shantou, Guangdong) to form 1.0, 1.5, 2.0, 3.0, 4.0, and 6.0 mass% CaCl$_2$/FA solutions. Then, the degummed silk fibroin (SF) fibers were quickly dissolved into these CaCl$_2$/FA solutions at room temperature. The final solutions were immediately cast onto a Teflon mold (50 mm × 20 mm × 10 mm) and dried at room temperature to form SF films. After being washed (30 mins in running water) and vacuum dried to fully remove CaCl$_2$/FA solvents, the final Chin SF (CSF) and Thai SF (TSF) films were obtained. CSF and TSF samples were named with the numbers (−1.0,

−1.5, −2.0, −3.0, −4.0, and −6.0) after the sample names to indicate the initial concentration of $CaCl_2$ (1.0, 1.5, 2.0, 3.0, 4.0, and 6.0 mass%) in the solution.

3.2. Thermal and Mechanical Analyses

Sample with a dimension of $5.0 \times 2.0 \times 1.0mm^3$ was subjected to the mechanical analysis by using a Dynamic Mechanical Analyzer (Perkin-Elmer Diamond DMA, Waltham, MA, USA). Experiment was proceeded under a temperature range from 25 °C to 280 °C with a heating rate of 2 °C min^{-1} and at 1, 2, 5, 10, and 20 Hz frequencies simultaneously. The glass transition temperature (T_g) of sample was taken from the temperature at which the maximum peak of the tangent δ was exhibited. Furthermore, at the atmosphere temperature (25 °C), the stress-strain property of silk fibroin film was tested by DMA in a tensile controlled model. The tensile force increased from 5 mN to 2000 mN at a lifting speed of 50 mN min^{-1} until the sample was broken. For all experiments, at least three samples were measured under the same test conditions to check the consistency of the experimental results. Dynamic mechanical analysis (DMA) refers to the technique of measuring the dynamic modulus and mechanical loss of a specimen and its relation to temperature or frequency under programmed temperature and alternating stress. In general, a periodically varying (usually refers to the sine) force was applied to the sample to produce periodic changes in the stress. The sample will have a corresponding deformation behavior of the stress; thus, the mechanical modulus of the sample can be determined by the stress and deformation. For example, if a shear stress is applied, a shear G can be obtained, and if a type of stress is applied in tension or bending, Young's Modulus E can be obtained. DMA measurement modes include tension, compression, bending, single-cantilever, shear, and reverse. In most cases, the specimen undergoes a periodical variation of the mechanical vibration stress, which causes the corresponding vibrational strain. However, the specimen does not always respond instantaneously to the changing stress and the lags behind for a certain period of time. This mainly depends on the viscoelasticity of the specimen and the phase shift between applied stress and deformation. Therefore, sample modulus consists of both real and imaginary parts. The real part describes the sample response as the periodic stress, which is a measurement of the elasticity of the sample, called the storage modulus. The imaginary part describes the response phase shifted by 90°, which is a measurement of the mechanical energy converted to heat, called the loss modulus. The ratio of loss modulus (E'') and storage modulus (E') is called the loss factor, the phase shift tangent δ (tan δ), which is the express of sample damping performance. The storage modulus E' is proportional to the mechanical energy stored in the specimen during stress, while the loss modulus E'' represents the energy dissipated in the specimen during stress. The larger the loss modulus is, the more viscous the specimen is, and the stronger the damping of specimen is. The tan δ is independent of the geometry, therefore it can be accurately measured even if the geometry of the sample is not regular. In dynamic mechanical analysis, the force amplitude F_A and the displacement amplitude L_A are used to calculate the modulus of samples. For example, the tensile modulus or elastic modulus of the experiment E' can be obtained through the formula:

$$E' = \frac{\sigma}{\varepsilon} = \frac{F_A}{L_A} \frac{L_0}{A} = \frac{F_A}{L_A} g \tag{4}$$

$$\sigma = F_A / A \tag{5}$$

$$\varepsilon = L_0 / L_A \tag{6}$$

$$g = L_0 / A \tag{7}$$

where σ and ε are the stress and strain of the sample, respectively; F_A and L_A are the force amplitude and the displacement amplitude, respectively; L_0 is the original length of the sample; A is the unit area of the sample; g is the geometry efficiency; and F_A / L_A is the rigidity of the material [48].

Generally, materials with low storage modulus E' implies it will deform easily when applying load on it; and the tan δ represents the viscoelasticity of materials during the loading cycle. A high value of tan δ suggests a high degree of energy dissipation and a high degree of viscous deformation for this material.

Besides, the dried CRS and TRS samples were encapsulated in Al pans and heated in a differential scanning calorimeter under Step-Scan modulated (SSDSC, Diamond DSC, Perkin-Elmer, USA) at a heating rate of 3 °C/min^{-1} with a 3 °C step and isothermal time of 2 min with 25 mL min^{-1} purged dry nitrogen gas, and equipped with a refrigerated cooling system. Each sample was about 5 mg.

3.3. Fourier Transform Infrared Spectrometry

Fourier transform infrared spectra (FTIR) of silk protein film sample was obtained using a FTIR spectrometer (Nicolet-NEXUS 670, Nicolet, Madison, WI, USA), equipped with a deuterated triglycine sulfate detector and a multiple-reflection, horizontal MIRacle ATR attachment (OMNIT, using a Ge crystal, Madison, WI, USA). Spectra were recorded in the wavenumber range of 1800 to 1100 cm^{-1} with a resolution of 4 cm^{-1}, and 64 scans were applied for each measurement. Fourier self-deconvolution (FSD) of the IR spectra covering the Amide I region (1595–1705 cm^{-1}) was performed by the Nicolet Omnic software. Deconvolution was performed using Gauss line shape with a half-bandwidth of 25 cm^{-1} and a noise reduction factor of 0.3. FSD spectra were then curve-fitted by Gaussian peaks to measure the relative areas in the Amide I region.

3.4. Thermogravimetric Analysis

Thermogravimetric (TG) analysis (PerkinElmer Pyris 1, Waltham, MA, USA) was used to measure the change in the mass of the silk samples during temperature increase. The TG curves were obtained under a nitrogen atmosphere with a gas flow of 50 mL·min^{-1}. Samples of about 2–3 mg were heated from 30 to 600 °C with a heating rate of 10 °C·min^{-1}. The mass change percentages during heating were recorded.

4. Conclusions

Dynamic Mechanical Analysis (DMA) analysis is more sensitive than Differential Scanning Calorimetry (DSC), as short-range chain motion changes are easier to detect than heat capacity changes during the phase transitions of biopolymer materials. The various structures and mechanical properties of two kinds of domesticated silk fibroin (SF) films (Chinese and Thailand *B. Mori*) regenerated from formic acid-CaCl$_2$ solutions were investigated using DMA. Our study showed that by using the GIM model, the disorder degree (f_{dis}) of silk samples can be inferred from the cohesive energy (E_{coh}), the skeletal degree of freedom (N), and the loss factor (tan δ) at the glass transition region. Four disordered phase relaxation events were explored by DMA technique, with a focus on the glass transition and the degree of structural disorder. Our results illustrated that there are nearly two hydrogen bonds formed in each peptide group in the *Bombyx Mori* silk fibers. With the increase of calcium chloride concentration in the SF sample, the T_g of silk material decreases, which implied that the sample could contain more non-crystalline structures, such as random coils and helix. α_1 and α_2-relaxation events in DMA curves are both associated with the silk secondary structures. The random coils as well as other non-crystalline structures may be attributed to the change of α_1-relaxation, while α_2-relaxation could be directly associated to the α-helix to β-sheet transition. Moreover, at the same calcium chloride concentration, the CSF sample is more disordered than the TSF sample, which suggests that there are more β-sheet in the TSF sample than in the CSF sample. It also showed that the elasticity of the TSF sample is lower than that of the CSF sample, while their ductility is the opposite. Besides, SF samples prepared with lower concentrations of calcium chloride have higher elasticity, while SF samples prepared with high concentrations have better ductility. The effects of calcium chloride concentrations on the structure and mechanical properties of regenerated silk fibroin films were further investigated by comparing GIM theoretical model calculations with experimental results. These results provide us a new way to understand the structural changes and mechanical properties of different domesticated silk regenerated from acid-based solution system, which would be critical for engineering applications of silk materials. These results provide us with a new way to understand the structural changes and mechanical properties of different domesticated silk regenerated from acid-based solution system, which would be critical for engineering applications of silk materials in the future.

Author Contributions: Q.L., F.W. and X.H. conceived and designed the experiments; Q.L., F.W., Z.G., and Q.M. performed the experiments, Q.L., F.W. and X.H. analyzed the data and wrote the paper.

Funding: This work was supported by the National Natural Science Foundation of China (11474166) and the College of Natural Science Foundation of Jiangsu Province, China (15KJB150018). X.H. is supported by Rowan University Start-up Grants, and US-NSF Biomaterials Program (DMR-1809541) and Materials Eng. and Processing program (CMMI-1561966).

Conflicts of Interest: The authors declare no conflict of interest.

References

1. Bhardwaj, N.; Singh, Y.P.; Devi, D.; Kandimalla, R.; Kotoky, J.; Manda, B.B. Potential of silk fibroin/chondrocyte constructs of muga silkworm Antheraea assamensis for cartilage tissue engineering. *J. Mater. Chem. B* **2016**, *4*, 3670–3684. [CrossRef]
2. Yu, S.; Yang, W.; Chen, S.; Chen, M.; Liu, Y.; Shao, Z.; Chen, X. Floxuridine-loaded silk fibroin nanospheres. *RSC Adv.* **2014**, *4*, 18171–18177. [CrossRef]
3. Hu, X.; Wang, X.; Rnjak, J.; Weiss, A.S.; Kaplan, D.L. Biomaterials derived from silk-tropoelastin protein systems. *Biomaterials* **2010**, *31*, 8121–8131. [CrossRef] [PubMed]
4. Hu, X.; Shmelev, K.; Sun, L. Regulation of Silk Material Structure by Temperature-Controlled Water Vapor Annealing. *Biomacromolecules* **2011**, *12*, 1686–1696. [CrossRef] [PubMed]
5. Jao, D.; Xue, Y.; Medina, J.; Hu, X. Protein-Based Drug-Delivery Materials. *Materials* **2017**, *10*, 517. [CrossRef] [PubMed]
6. Kambe, Y.; Murakoshi, A.; Urakawa, H.; Kimura, Y.; Yamaoka, T. Vascular induction and cell infiltration into peptide-modified bioactive silk fibroin hydrogels. *J. Mater. Chem. B* **2017**, *5*, 7557–7571. [CrossRef]
7. Aytemiz, D.; Sakiyama, W.; Suzuki, Y.; Nakaizumi, N.; Tanaka, R.; Ogawa, Y.; Takagi, Y.; Nakazawa, Y.; Asakura, T. Small-diameter silk vascular grafts (3 mm diameter) with a double-raschel knitted silk tube coated with silk fibroin sponge. *Adv. Healthc. Mater.* **2013**, *2*, 361–368. [CrossRef] [PubMed]
8. Shahbazi, B.; Taghipour, M.; Rahmani, H.; Sadrjavadi, K.; Fattahi, A. Preparation and characterization of silk fibroin/oligochitosan nanoparticles for siRNA delivery. *Colloids Surf. B* **2015**, *136*, 867–877. [CrossRef] [PubMed]
9. Tesfaye, M.; Patwa, R.; Kommadath, R.; Kotecha, P.; Katiyar, V. Silk nanocrystals stabilized melt extruded poly (lactic acid) nanocomposite films: Effect of recycling on thermal degradation kinetics and optimization studies. *Thermochim. Acta* **2016**, *643*, 41–52. [CrossRef]
10. Yin, Z.; Wu, F.; Xing, T.; Yadavalli, V.K.; Kundu, S.C.; Lu, S. A silk fibroin hydrogel with reversible sol-gel transition. *RSC Adv.* **2017**, *7*, 24085–24096. [CrossRef]
11. Hu, X.; Park, S.H.; Gil, E.S.; Xia, X.X.; Weiss, A.S.; Kaplan, D.L. The Influence of Elasticity and Surface Roughness on Myogenic and Osteogenic-Differentiation of Cells on Silk-Elastin Biomaterials. *Biomaterials* **2011**, *32*, 8979–8989. [CrossRef] [PubMed]
12. Phillips, D.M.; Drummy, L.F.; Conrady, D.G.; Fox, D.M.; Naik, R.R.; Stone, M.O. Dissolution and Regeneration of Bombyx mori Silk Fibroin Using Ionic Liquids. *J. Am. Chem. Soc.* **2004**, *126*, 14350–14361. [CrossRef] [PubMed]
13. Lu, Q.; Hu, X.; Wang, X.; Kluge, J.A.; Lu, S. Water-Insoluble Silk Films with Silk I Structure. *Acta Biomater.* **2010**, *6*, 1380–1387. [CrossRef] [PubMed]
14. Tian, L.; Chen, Y.; Min, S. Research on Cytotoxicity of Silk Fibroin Gel Materials Prepared with Polyepoxy Compound. *J. Biomed. Eng.* **2007**, *24*, 1309–1313.
15. Crawford, D.M.; Escarsega, J.A. Dynamic mechanical analysis of novel polyurethane coating for military applications. *Thermochim. Acta* **2000**, *357*, 161–168. [CrossRef]

16. Yin, B.; Hakkarainen, M. Core–shell nanoparticle–plasticizers for design of high-performance polymeric materials with improved stiffness and toughness. *J. Mater. Chem.* **2011**, *21*, 8670–8677. [CrossRef]

17. Khandaker, M.S.K.; Dudek, D.M.; Beers, E.P.; Dillard, D.A.; Bevan, D.R. Molecular modeling of the elastomeric properties of repeating units and building blocks of resilin, a disordered elastic protein. *J. Mech. Behav. Biomed. Mater.* **2016**, *61*, 110–121. [CrossRef] [PubMed]

18. Mahdi, E.M.; Tan, J.C. Dynamic molecular interactions between polyurethane and ZIF-8 in a polymer-MOF nanocomposite: Microstructural, thermo-mechanical and viscoelastic effects. *Polymer* **2016**, *97*, 31–43. [CrossRef]

19. Saba, N.; Jawaid, M.; Alothman, O.Y.; Paridah, M.T. A review on dynamic mechanical properties of natural fibre reinforced polymer composites. *Constr. Build. Mater.* **2016**, *106*, 149–159. [CrossRef]

20. Nalyanya, K.M.; Migunde, O.P.; Ngumbu, R.G.; Onyuka, A.; Rop, R.K. Thermal and dynamic mechanical analysis of bovine hide. *J. Therm. Anal. Calorim.* **2016**, *121*, 1–8. [CrossRef]

21. Guan, J.; Porter, D.; Vollrath, F. Thermally induced changes in dynamic mechanical properties of native silks. *Biomacromolecules* **2013**, *14*, 930–937. [CrossRef] [PubMed]

22. Wang, Y.; Guan, J.; Hawkins, N.; Porter, D.; Shao, Z. Understanding the variability of properties in Antheraea pernyi silk fibres. *Soft Matter* **2014**, *10*, 6321–6331. [CrossRef] [PubMed]

23. Vollrath, F.; Porter, D. Spider silk as a model biomaterial. *Appl. Phys. A* **2006**, *82*, 205–212. [CrossRef]

24. Porter, D.; Vollrath, F.; Shao, Z. Predicting the mechanical properties of spider silk as a nanostructured polymer. *Eur. Phys. J. E* **2005**, *16*, 199–206. [CrossRef] [PubMed]

25. Kawano, Y.; Wang, Y.; Palmer, R.; Aubuchon, A.; Steve, R. Stress-Strain Curves of Nafion Membranes in Acid and Salt Forms. *Polímeros* **2002**, *12*, 96–101. [CrossRef]

26. Pyda, M.; Wunderlich, B. Reversing and nonreversing heat capacity of poly (lactic acid) in the glass transition region by TMDSC. *Macromolecules* **2005**, *38*, 10472–10479. [CrossRef]

27. Sheng, S.J.; Hu, X.; Wang, F.; Ma, Q.Y.; Gu, M.F. Mechanical and thermal property characterization of poly-L-lactide (PLLA) scaffold developed using pressure-controllable green foaming technology. *Mat. Sci. Eng. C.-Mater.* **2015**, *49*, 612–622. [CrossRef] [PubMed]

28. Wang, F.; Wolf, N.; Rocks, E.M.; Vuong, T.; Hu, X. Comparative studies of regenerated water-based Mori, Thai, Eri, Muga and Tussah silk fibroin films. *J. Therm. Anal. Calorim.* **2015**, *122*, 1069–1076. [CrossRef]

29. Mazzi, S.; Zulker, E.; Buchicchio, J.; Anderson, B.; Hu, X. Comparative thermal analysis of Eri, Mori, Muga, and Tussar silk cocoons and fibroin fibers. *J. Therm. Anal. Calorim.* **2014**, *116*, 1337–1343. [CrossRef]

30. Wang, F.; Yu, H.Y.; Gu, Z.G.; Si, L.; Liu, Q.C.; Hu, X. Impact of calcium chloride concentration on structure and thermal property of Thai silk fibroin films. *J. Therm. Anal. Calorim.* **2017**, *130*, 1–9. [CrossRef]

31. Wang, F.; Chandler, P.; Oszust, R.; Sowell, E.; Graham, Z.; Ardito, W.; Hu, X. Thermal and structural analysis of silk–polyvinyl acetate blends. *J. Therm. Anal. Calorim.* **2017**, *127*, 923–929. [CrossRef]

32. Hu, X.; Lu, Q.; Sun, L.; Cebe, P.; Wang, X.Q.; Zhang, X.H.; Kaplan, D.L. Biomaterials from Ultrasonication-Induced Silk Fibroin−Hyaluronic Acid Hydrogels. *Biomacromolecules* **2010**, *11*, 3178–3188. [CrossRef] [PubMed]

33. Zhu, Z.; Jiang, C.; Cheng, Q.; Zhang, J.; Guo, S. Accelerated aging test of hydrogenated nitrile butadiene rubber using the time-temperature-strain superposition principle. *RSC Adv.* **2015**, *5*, 90178–90183. [CrossRef]

34. Butaud, P.; Placet, V.; Klesa, J.; Ouisse, M.; Foltête, E.; Gabrion, X. Investigations on the frequency and temperature effects on mechanical properties of a shape memory polymer (Veriflex). *Mech. Mater.* **2015**, *8*, 50–60. [CrossRef]

35. Ardhyananta, H.; Kawauchi, T.; Ismail, H.; Takeichi, T. Effect of pendant group of polysiloxanes on the thermal and mechanical properties of polybenzoxazine hybrids. *Polymer* **2009**, *25*, 5959–5969. [CrossRef]

36. Hu, X.; Kaplan, D.; Cebe, P. Effect of Water on Thermal Properties of Silk Fibroin. *Thermochim. Acta* **2007**, *461*, 137–144. [CrossRef]

37. Motta, A.; Fambri, L.; Migliaresi, C. Regenerated silk fibroin films: Thermal and dynamic mechanical analysis. *Chem. Phys.* **2002**, *203*, 1658–1665. [CrossRef]

38. Um, I.C.; Kim, T.H.; Kweon, H.Y.; Chang, S.K.; Park, Y.H. A comparative study on the dielectric and dynamic mechanical relaxation behavior of the regenerated silk fibroin films. *Macromol. Res.* **2009**, *17*, 785–790. [CrossRef]

39. Born, M. Thermodynamics of crystals and melting. *J. Chem. Phys.* **1939**, *7*, 591–603. [CrossRef]

40. Guan, J.; Wang, Y.; Mortimer, B.; Holland, C.; Shao, Z.; Porter, D.; Vollrath, F. Glass transitions in native silk fibres studied by dynamic mechanical thermal analysis. *Soft Matter* **2016**, *12*, 5926–5936. [CrossRef] [PubMed]

41. Porter, D.; Vollrath, F. Water mobility denaturation and the glass transition in proteins. *Biochim. Biophys. Acta* **2012**, *1824*, 785–791. [CrossRef] [PubMed]

42. Chen, F.; Porter, D.; Vollrath, F. Silk cocoon (Bombyx mori): Multi-layer structure and mechanical properties. *Acta Biomater.* **2012**, *8*, 2620–2627. [CrossRef] [PubMed]

43. Fu, C.; Porter, D.; Chen, X.; Vollrath, F.; Shao, Z. Understanding the Mechanical Properties of Antheraea Pernyi Silk-From Primary Structure to Condensed Structure of the Protein. *Adv. Funct. Mater.* **2015**, *21*, 729–737. [CrossRef]

44. Qin, G.; Hu, X.; Cebe, P.; Kaplan, D.L. Mechanism of resilin elasticity. *Nat. Commun.* **2012**, *3*, 1003–1013. [CrossRef] [PubMed]

45. Roberts, D.R.T.; Holder, S.J. Mechanochromic systems for the detection of stress, strain and deformation in polymeric materials. *J. Mater. Chem.* **2011**, *21*, 8256–8268. [CrossRef]

46. Hu, X.; Kaplan, D.; Cebe, P. Determining beta-sheet crystallinity in fibrous proteins by thermal analysis and infrared spectroscopy. *Macromolecules* **2006**, *39*, 6161–6170. [CrossRef]

47. Yu, H.Y.; Wang, F.; Liu, Q.C.; Ma, Q.Y.; Gu, Z.G. Structure and Kinetics of Thermal Decomposition Mechanism of Novel Silk Fibroin Films. *Acta Phys.-Chim. Sin.* **2017**, *33*, 344–355.

48. Sgreccia, E.; Chailan, G.F.; Khadhraoui, M.; Vona, M.L.D.; Knauth, P. Mechanical properties of proton-conducting sulfonated aromatic polymer membranes: Stress–strain tests and dynamical analysis. *J. Power Sources* **2010**, *195*, 7770–7775. [CrossRef]

© 2018 by the authors. Licensee MDPI, Basel, Switzerland. This article is an open access article distributed under the terms and conditions of the Creative Commons Attribution (CC BY) license (http://creativecommons.org/licenses/by/4.0/).

International Journal of
Molecular Sciences

MDPI

Article

Characterization of *Ecklonia cava* Alginate Films Containing Cinnamon Essential Oils

Su-Kyoung Baek, Sujin Kim and Kyung Bin Song *

Department of Food Science and Technology, Chungnam National University, Daejeon 34134, Korea;
sukyoungb@cnu.ac.kr (S.-K.B.); pppink32@gmail.com (S.K.)
* Correspondence: kbsong@cnu.ac.kr; Tel.: +82-42-821-6723; Fax: +82-42-825-2664

Received: 27 October 2018; Accepted: 8 November 2018; Published: 10 November 2018

Abstract: In this study, *Ecklonia cava* alginate (ECA) was used as a base material for biodegradable films. Calcium chloride ($CaCl_2$) was used as a cross-linking agent, and various concentrations (0%, 0.4%, 0.7%, and 1.0%) of cinnamon leaf oil (CLO) or cinnamon bark oil (CBO) were incorporated to prepare active films. The ECA film containing 3% $CaCl_2$ had a tensile strength (TS) of 17.82 MPa and an elongation at break (E) of 10.36%, which were higher than those of the film without $CaCl_2$. As the content of essential oils (EOs) increased, TS decreased and E increased. Addition of CLO or CBO also provided antioxidant and antimicrobial activities to the ECA films. The antioxidant activity of the ECA film with CBO was higher than that of the film containing CLO. In particular, the scavenging activities of the 2,2-diphenyl-1-picrylhydrazyl (DPPH) and 2,2'-azino-bis (3-ethylbenzothiazoline-6-sulphonic acid) (ABTS) radicals in the ECA film containing 1% CBO were 50.45% and 99.37%, respectively. In contrast, the antimicrobial activities against *Escherichia coli* O157:H7, *Salmonella* Typhimurium, *Staphylococcus aureus*, and *Listeria monocytogenes* were superior in the ECA films with CLO. These results suggest that ECA films containing CLO or CBO can be applied as new active packaging materials.

Keywords: active packaging materials; alginate films; antimicrobial agents; antioxidant activity; biodegradable films; essential oils

---

1. Introduction

In recent years, environmental pollution caused by plastic waste has worsened, and interest in eco-friendly packaging materials has increased. In particular, biodegradable films have been studied as new packaging materials in the food industry [1–3]. As biodegradable film materials, carbohydrates, proteins, and lipids are used individually or in combination.

Seaweeds can grow in abundance, and they are an excellent alternative to replace limited land resources [2]. Among them, *Ecklonia cava* is a brown alga present mainly in the East Asian region [4], and it contains approximately 26% alginate by dry weight [5]. Alginate, which is mainly extracted from brown algae, is non-toxic, relatively inexpensive, and biocompatible, and it can be considered as a base material for biodegradable films [3]. However, *Ecklonia cava* alginate (ECA) has not been studied as a biodegradable film.

Alginate is composed of guluronic acid and mannuronic acid, and their ratio affects alginate film properties. When the proportion of guluronic acid is high, the alginate film is hard and fragile. On the contrary, when the proportion of mannuronic acid is high, elastic and soft gels can be formed [6]. Ionic bonds can be formed by the interactions between divalent cations such as calcium and the carboxyl groups of guluronic acid, and the alginate polymer forms an egg-box structure surrounding the divalent cations [7,8], resulting in films with rigid structures by cross-linking. Therefore, the incorporation of calcium chloride ($CaCl_2$) in alginate films can be effective in improving the physical properties of the films.

Active packaging can protect foods during storage and extend their shelf life by the addition of active materials to packaging materials [9]. Essential oils (EOs), classified as "Generally Recognized as Safe" by the Food and Drug Administration, are added as natural active materials to produce active packaging [1]. Among EOs, cinnamon oil is extracted from various parts of cinnamon (*Cinnamomum verum*) such as leaf, bark, flower, and root, but cinnamon leaf oil (CLO) and cinnamon bark oil (CBO) are commonly used as EOs. The major compounds in CLO and CBO are eugenol and trans-cinnamaldehyde, respectively, which contribute to antioxidant and antimicrobial activities [10,11]. CLO and CBO can be added to ECA films as natural active materials to produce active packaging films.

Therefore, this study aimed to prepare alginate films extracted from *Ecklonia cava* with improved physical properties and to characterize the films containing various concentrations of CLO or CBO.

2. Results and Discussion

2.1. Attenuated Total Reflectance-Fourier Transform Infrared (ATR-FTIR) Analysis of ECA

To investigate the molecular characteristics of ECA, ATR-FTIR analysis was performed and the results are presented in Figure 1. A sharp peak at 3600–3500 cm^{-1} indicates the presence of free OH groups [12], but it was not observed in the FTIR spectrum of ECA. The broad band at 3500–3100 cm^{-1} was observed owing to the vibration of the hydroxyl group [12,13]. In the case of alginate, the stretching of protonated carboxyl groups appeared at approximately 1730 cm^{-1}. When monovalent ions (sodium) displaced the proton, peaks on the polymeric backbone of alginate appeared at approximately 1600 cm^{-1} and 1400 cm^{-1} by asymmetric and symmetric stretching vibration of the carboxylate group of sodium alginate [8]. In this study, the bands were observed at 1607.01 cm^{-1} and 1426.74 cm^{-1}. These results are in accordance with the data found in previous reports on alginate [7,8]. The band at 1027.94 cm^{-1} represented the stretching vibration of C-O, and the band at 880.16 cm^{-1} was related to the C1-H deformation vibration of β-mannuronic acid residues [7]. In general, the content of guluronic acid (G) and mannuronic acid (M) in alginate varies, depending upon the alginate species. Since G and M have characteristic bands at 1025 cm^{-1} and 1100 cm^{-1}, respectively, an approximate M/G ratio can be predicted by FTIR analysis [14], and the M/G ratio of ECA was estimated to be lower than 1 in the present study.

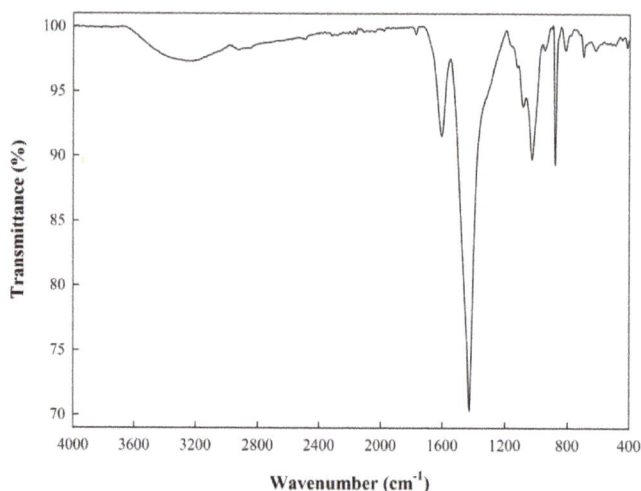

Figure 1. FTIR spectra of *Ecklonia cava* alginate (ECA).

2.2. Physical Properties of ECA Films with CaCl$_2$

The concentration of the cross-linking agent and the type and concentration of the plasticizer have been optimized based on preliminary experiments. In this study, CaCl$_2$ was incorporated as a cross-linking agent, and various concentrations of CaCl$_2$ (1%, 2%, 3%, 5%, and 7%) were added to the ECA films. The physical properties of the ECA films prepared with varying contents of CaCl$_2$ are shown in Table 1. The thickness and tensile strength (TS) of the ECA films with added CaCl$_2$ increased compared to those of films without CaCl$_2$. In particular, the TS of the ECA film containing 3% CaCl$_2$ was significantly improved from 10.49 MPa to 17.82 MPa compared to that of the ECA film without CaCl$_2$. The incorporation of CaCl$_2$ increased the TS because of the formation of more rigid polymers via interactions between the calcium ions and carboxyl groups of alginate [15]. Zactiti and Kieckbusch [16] reported that increasing calcium concentration resulted in a thicker "egg-box" structure and the formation of a stronger network. Although the addition of CaCl$_2$ improved the TS of the ECA film, it did not affect water vapor permeability (WVP). Similar results have been reported in the literature [8,17]. In the present study, the concentration of CaCl$_2$ was optimized to 3% based on the mechanical properties of the film.

Table 1. Mechanical properties of the *Ecklonia cava* alginate (ECA) films containing different amounts of CaCl$_2$.

CaCl$_2$ (%)	Thickness (μm)	Tensile Strength (MPa)	Elongation at Break (%)	Water Vapor Permeability ($\times 10^{-9}$ g/m s Pa)
0	34.60 ± 0.86 [d]	10.49 ± 1.38 [d]	9.30 ± 0.89 [bc]	1.89 ± 0.13 [a]
1	35.00 ± 0.81 [cd]	14.78 ± 2.74 [b]	9.90 ± 0.56 [ab]	1.86 ± 0.16 [a]
2	35.96 ± 1.42 [bc]	12.99 ± 1.65 [bc]	9.96 ± 0.52 [ab]	1.85 ± 0.15 [a]
3	36.68 ± 0.61 [b]	17.82 ± 1.05 [a]	10.36 ± 0.27 [a]	1.78 ± 0.02 [a]
5	37.12 ± 0.64 [b]	15.26 ± 1.80 [b]	9.69 ± 0.60 [ab]	1.83 ± 0.10 [a]
7	38.80 ± 1.20 [a]	10.79 ± 1.55 [cd]	8.85 ± 0.49 [c]	1.74 ± 0.07 [a]

Means ± S.D., $n = 5$, [a-d]: any means in the same column followed by different letters are significantly ($p < 0.05$) different by Duncan's multiple range test.

2.3. Physical Properties of ECA Films Containing EOs

EOs were added to ECA films at different concentrations (0.4%, 0.7%, and 1.0%), and Table 2 shows the physical properties of ECA films incorporated with CLO or CBO. The addition of EOs increased the film thickness and elongation at break (E) values, whereas the TS decreased. In particular, the addition of CLO or CBO at 0.7% into the ECA films increased E from 10.36% to 18.25% or 18.65%, respectively, resulting in flexible ECA films. EOs can interfere with the interactions between alginate and calcium ions, thereby improving the flexibility of the films by reducing the intermolecular forces along the polymer chain. Benavides et al. [15] found that, as the content of oregano EOs increased in alginate films, TS decreased and E increased because oregano EOs could act as a plasticizer and increase flexibility and chain mobility. These findings are consistent with those in previous studies on other alginate films containing EOs [18,19].

In general, the addition of hydrophobic EOs causes a decrease in WVP in the films [1]. It has been reported that the incorporation of EOs into various films improved the WVP [20–22]. On the contrary, the WVP of ECA films containing CLO or CBO increased as the contents of EOs increased in the present study. This difference can be explained by the formation of pores via changes in the internal structure of the film matrix. Previously, it was reported that the WVP of alginate films incorporated with garlic oil increased as a result of expansion by the intermolecular interactions in the film matrix [19]. Similarly, the WVP of alginate films with other EOs was not improved [18,23]. Atarés et al. [24] also reported that the WVP of the films was not easily improved by the addition of hydrophobic materials, although the incorporation of lipids in the microstructure of the films affected the WVP.

Table 2. Physical properties of the ECA films containing essential oils (EOs).

EOs (%)		Thickness (μm)	Tensile Strength (MPa)	Elongation at Break (%)	Water Vapor Permeability ($\times 10^{-9}$ g /m s Pa)
Control	0	36.68 ± 0.61 [d]	17.82 ± 1.05 [a]	10.36 ± 0.27 [c]	1.78 ± 0.02 [d]
CLO	0.4	40.12 ± 0.73 [c]	17.40 ± 0.87 [ab]	17.94 ± 2.59 [ab]	2.01 ± 0.05 [c]
	0.7	44.12 ± 0.67 [b]	15.10 ± 1.88 [cd]	18.25 ± 3.32 [ab]	2.35 ± 0.06 [b]
	1.0	49.80 ± 0.60 [a]	14.94 ± 1.22 [cd]	17.28 ± 1.40 [ab]	2.56 ± 0.05 [a]
CBO	0.4	40.40 ± 1.14 [c]	16.02 ± 0.80 [bc]	16.92 ± 1.73 [ab]	1.99 ± 0.03 [c]
	0.7	44.08 ± 0.61 [b]	15.45 ± 1.32 [c]	18.65 ± 2.06 [a]	2.30 ± 0.07 [b]
	1.0	49.68 ± 0.46 [a]	13.58 ± 0.92 [d]	15.61 ± 1.77 [b]	2.54 ± 0.09 [a]

Means ± S.D., $n = 5$, [a-d]: any means in the same column followed by different letters are significantly ($p < 0.05$) different by Duncan's multiple range test. CLO: cinnamon leaf oil; CBO: cinnamon bark oil.

2.4. Optical Properties of ECA Films Containing EOs

The optical characteristics of the films are important because they directly affect the appearance and quality of food products. They influence not only customer preference but also the lipid oxidation rate, which can affect the quality of foods [20]. As shown in Table 3, the incorporation of EOs decreased the values of L * and b *, whereas the value of a * increased. The total color difference (ΔE *) in ECA films with EOs increased compared with that in films without EOs. Furthermore, the ECA films gradually became opaque with increasing EO content. The opacity values of ECA films with CLO or CBO at 1.0% were 25.12 or 22.54 A/mm, respectively, whereas ECA films without EOs had an opacity of 9.77 A/mm. Similar to our results, the addition of CBO into chitosan films decreased lightness, whereas redness and opacity increased [25].

Table 3. Optical properties of the ECA films containing EOs.

EOs (%)		L *	a *	b *	ΔE *	Opacity (A/mm)
Control	0	72.03 ± 0.27 [a]	5.16 ± 0.10 [e]	49.19 ± 0.34 [ab]	-	9.77 ± 0.31 [f]
CLO	0.4	68.52 ± 0.40 [b]	7.71 ± 0.26 [d]	49.76 ± 0.32 [a]	4.39 ± 0.48 [d]	13.43 ± 0.51 [d]
	0.7	66.40 ± 0.43 [d]	8.53 ± 0.19 [b]	45.21 ± 1.08 [d]	7.75 ± 0.35 [b]	22.64 ± 0.53 [b]
	1.0	64.73 ± 1.11 [e]	8.96 ± 0.66 [a]	44.55 ± 0.77 [d]	9.48 ± 1.20 [a]	25.12 ± 0.37 [a]
CBO	0.4	68.48 ± 0.33 [b]	7.52 ± 0.19 [d]	49.00 ± 0.97 [ab]	4.37 ± 0.36 [d]	12.12 ± 0.19 [e]
	0.7	67.12 ± 1.10 [c]	8.17 ± 0.24 [c]	48.52 ± 0.81 [b]	5.86 ± 1.10 [c]	20.85 ± 0.34 [c]
	1.0	64.01 ± 0.67 [f]	8.59 ± 0.42 [b]	46.87 ± 1.29 [c]	9.11 ± 0.79 [a]	22.54 ± 0.54 [b]

Means ± S.D., $n = 10$, [a-f]: any means in the same column followed by different letters are significantly ($p < 0.05$) different by Duncan's multiple range test.

2.5. Scanning Electron Microscopy (SEM) Analysis

Cross-sectional images showing the internal structure of ECA films incorporated with CLO or CBO are presented in Figure 2. As the content of EOs increased, it was apparent that the microstructure was not compact and pores were formed. The structural change and pore formation in the film increased the WVP. Similarly, when CBO was added to gelatin films, pores were formed by aggregation caused by irregular dispersion of hydrophobic molecules as the concentration of EOs increased [26]. Zhang et al. [23] also reported that alginate films had a relatively homogeneous and smooth surface, but the addition of cinnamon EOs or soybean EOs resulted in irregular surfaces and pores. It was reported that these irregular surfaces and pores were generated by the arrangement of alginate molecules and the agglomeration of oil droplets when the films were formed.

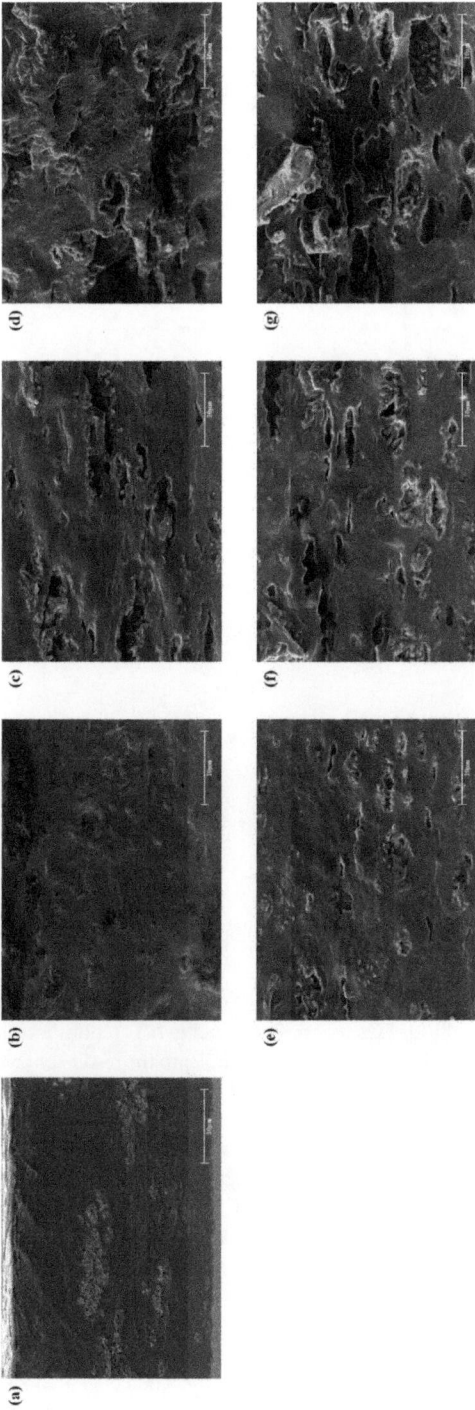

Figure 2. SEM images of the ECA films containing essential oils (EOs). (**a**) Control, (**b**) cinnamon leaf oil (CLO) 0.4%, (**c**) CLO 0.7%, (**d**) CLO 1.0%, (**e**) cinnamon bark oil (CBO) 0.4%, (**f**) CBO 0.7%, and (**g**) CBO 1.0%.

2.6. Thermal Analysis

Figure 3 shows the effect of the incorporation of CLO or CBO on the thermal stability of the ECA films using thermogravimetric analysis (TGA). The initial loss of mass appeared to be due to loss of moisture in the ECA film matrix [27]. The main reduction of mass occurred in the range of 200 °C to 300 °C, and the decrease in this region was caused by the volatilization of glycerol used as a plasticizer and the alginate polymer [28]. A loss of mass in ECA films with CLO or CBO was observed at 450 °C, probably due to the presence of highly stable substances in the EOs. At 600 °C, the residue amounts in the control, CLO 0.7%, and CBO 0.7% films were 46.93%, 40.31%, and 40.68%, respectively. The amount of residue was highest in the control, suggesting high thermal stability. The decrease in thermal stability of ECA films incorporated with EOs could be due to the structural change in the ECA film network, causing a decrease in TS by the incorporation of EOs.

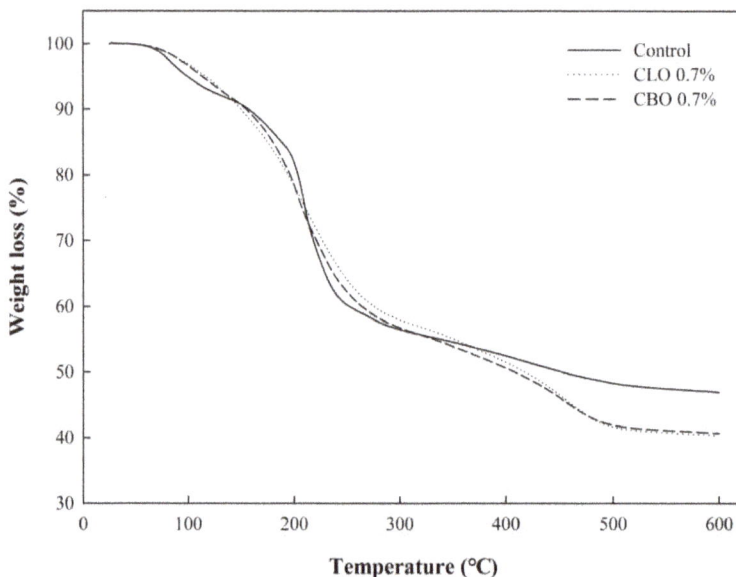

Figure 3. Thermogravimetric analysis (TGA) curves of the ECA films containing cinnamon leaf oil (CLO) and cinnamon bark oil (CBO).

The results of differential scanning calorimetry (DSC) analysis of the ECA films, the first heating scans, are presented in Figure 4. The ECA film without EOs (control) showed two T_e (endothermic peak temperature) at 86.63 (T_{e1}) and 233.65 °C (T_{e2}). This observation was similar to the results of DSC analysis of alginate having Te at 86 °C and 197 °C [29]. However, it should be noted that the T_{e2} was higher for the ECA film, probably due to the cross-linking effect of $CaCl_2$. T_{e1} reflected water loss, and the addition of CLO and CBO to the ECA film increased T_{e1} to 109.81 °C and 108.46 °C, respectively. The ECA films also showed an exothermic peak, and a second endothermic peak T_{e2} appeared at 233.65 °C in the control and at 236.66 °C and 229.14 °C for ECA films with CLO and CBO, respectively. These peaks could be attributed to recrystallization of ECA after heat induction [30]. The values of T_g (glass transition temperature) were measured in the second heating scan. In general, the T_g of alginate is around 120 °C [31]. In this study, the T_g of the control was 126.34 °C and decreased to 103.54 °C and 103.63 °C with the addition of CLO and CBO, respectively. The lower T_g was reflected by the decrease in TS due to the addition of EOs in the ECA films, thereby lowering the amount of thermal energy required to transform the ECA films from a glassy to a rubbery state. Tongnuanchan et al. [32] also observed similar results in gelatin films, where EOs interfered with the interactions of polymers and increased the mobility of the polymer by weakening the film structure and consequently lowering T_g. These results clearly support the TGA results, where the thermal stability of the ECA films with CLO or CBO did not improve compared to the control.

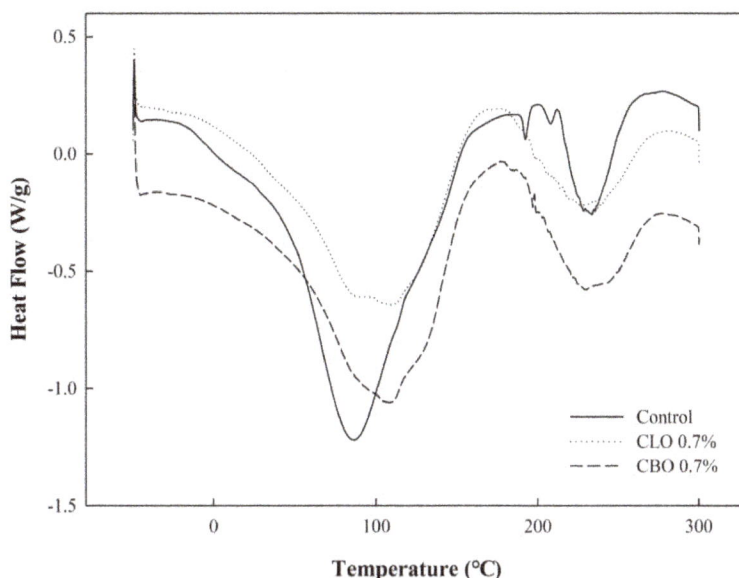

Figure 4. DSC curves of the ECA films containing CLO and CBO.

2.7. Antioxidant Activity

The antioxidant property of ECA films containing EOs was evaluated (Figures 5 and 6). As the content of CLO or CBO increased, the radical scavenging properties of 2,2-diphenyl-1-picrylhydrazyl (DPPH) and 2,2′-azino-bis (3-ethylbenzothiazoline-6-sulphonic acid) (ABTS) increased. The antioxidant activity of the ECA films with CBO was superior to that of the films with CLO. ECA films containing CLO at various concentrations did not exhibit significant differences in antioxidant activity, except for the film incorporated with 1.0% CLO. In contrast, ECA films with CBO exhibited significant differences in antioxidant activity as the EO content increased. The DPPH radical scavenging activities of ECA films containing CLO or CBO at 1.0% were 9.58% and 50.45%, respectively, and the ABTS radical scavenging activities were 55.05% and 99.37%, respectively. It is known that eugenol has higher radical scavenging activity than trans-cinnamaldehyde [33]. However, the effect of CBO on the radical scavenging activities of DPPH and ABTS was greater than that of CLO in the present study. Singh et al. [34] reported that the DPPH radical scavenging ability of CBO was better than that of CLO, even though eugenol has stronger DPPH radical scavenging ability than trans-cinnamaldehyde. These results suggest that antioxidant activity is affected not only by the major component of the EOs, eugenol or trans-cinnamaldehyde, but also by other constituents in the EOs. Overall, our results suggest that ECA films containing CBO are more suitable as antioxidant films than those containing CLO.

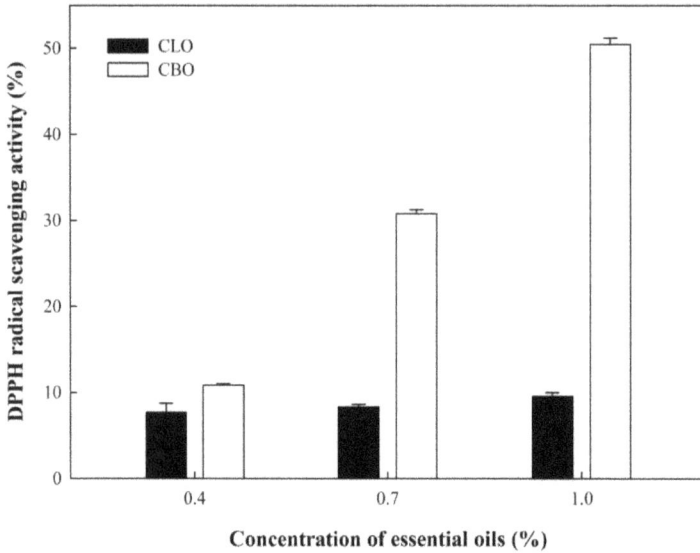

Figure 5. DPPH radical scavenging activity of the ECA films containing EOs.

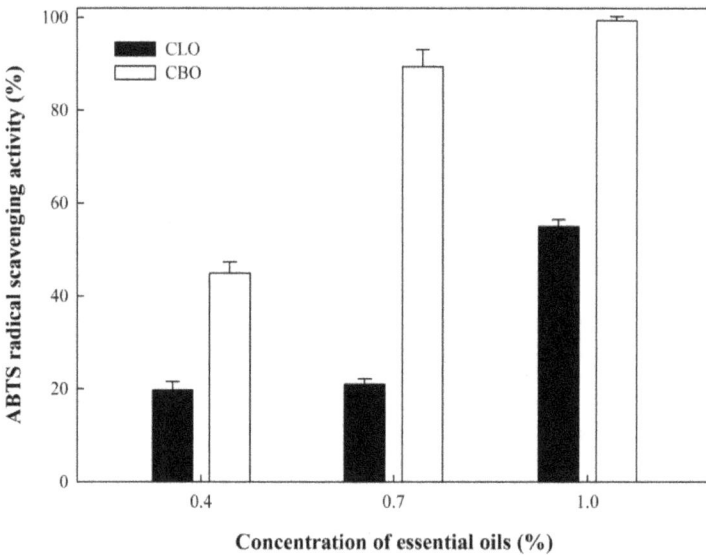

Figure 6. ABTS radical scavenging activity of the ECA films containing EOs.

2.8. Antimicrobial Activity

Table 4 shows the inhibitory effects of ECA films containing EOs against various bacteria (*Escherichia coli* O157:H7, *Salmonella* Typhimurium, *Staphylococcus aureus*, and *Listeria monocytogenes*). As expected, ECA films without CLO or CBO did not show antimicrobial activity against bacterial strains. In addition, the ECA film, containing 0.4% EO, showed no antimicrobial activity because of the low EO concentration. In contrast, ECA films containing 0.7% and 1.0% EOs exhibited antimicrobial activity, and the inhibition zones against four strains significantly increased as the amount of EOs increased. The antibacterial ability of EOs is mainly due to the presence of functional compounds, such as eugenol and trans-cinnamaldehyde, which disrupt the bacterial cell structure by penetrating cell membranes and inhibiting the production of essential enzymes in bacteria [35]. As shown in Table 3, ECA films containing CLO showed higher antimicrobial activity against four strains, when compared to the films incorporated with CBO. This is probably due to the difference in antimicrobial activity between eugenol, the main component of CLO, and trans-cinnamaldehyde, a major component of CBO [36]. Kang and Song [37] demonstrated the higher antibacterial effect of CLO (higher than that of CBO) by measuring the minimum inhibitory concentration of the EOs.

Table 4. Antimicrobial activity of the ECA films containing EOs.

EOs (%)		Inhibition Zone (mm)			
		E. coli O157:H7	*S.* Typhimurium	*S. aureus*	*L. monocytogenes*
Control	0	ND *	ND	ND	ND
CLO	0.4	ND	ND	ND	ND
	0.7	12.04 ± 0.10 [c]	11.49 ± 0.14 [c]	10.77 ± 0.07 [b]	13.17 ± 0.49 [b]
	1.0	14.50 ± 0.51 [a]	15.37 ± 0.16 [a]	14.28 ± 0.25 [a]	15.02 ± 0.17 [a]
CBO	0.4	ND	ND	ND	ND
	0.7	10.88 ± 0.08 [d]	10.77 ± 0.09 [d]	10.35 ± 0.52 [b]	12.03 ± 0.66 [c]
	1.0	13.48 ± 0.17 [b]	12.76 ± 0.22 [b]	13.93 ± 0.36 [a]	14.48 ± 0.13 [a]

Means ± S.D., $n = 5$, [a-d]: any means in the same column followed by different letters are significantly ($p < 0.05$) different by Duncan's multiple range test, * ND: not detected.

3. Materials and Methods

3.1. Materials

Ecklonia cava (Jeju-do, Korea) was processed in the laboratory, as described below. Glycerol and CaCl$_2$ were purchased from Samchun Pure Chemical Co. (Pyeongtaek-si, Gyeonggi-do, Korea). Span 80 was purchased from Yakuri Pure Chemical Co. (Osaka, Japan). CLO and CBO were obtained from Gooworl Co. (Daegu, Korea).

3.2. Alginate Extraction

Alginate extraction from *Ecklonia cava* was performed as reported previously [38,39]. Ground *Ecklonia cava* in 0.025% H$_2$SO$_4$ solution (1:50, w/v) was stirred for 3 h. After filtration, the residue was washed with distilled water and filtered again. The residue in 3% Na$_2$CO$_3$ solution (1:50, w/v) was heated at 60 °C for 2 h. Distilled water (1:25, w/v) was added and the mixture was centrifuged at 5000× g for 10 min. Methanol (95%, 1:50, v/v) was added to the supernatant and kept at 4 °C for 1 h. After centrifugation (10,000× g, 20 min), the supernatant was removed and the precipitate was lyophilized to obtain ECA.

3.3. ATR-FTIR Spectroscopy

FTIR spectrometry (Vertex 80v, Bruker, Billerica, MA, USA) was used to obtain ATR-FTIR spectra of ECA. Analysis was performed with 16 scans for each spectrum in the region between 4000 cm^{-1} and 400 cm^{-1}.

3.4. Preparation of ECA Films Containing EOs

To prepare the film-forming solution, ECA (1.5%), glycerol (0.3 g/g ECA), and CaCl$_2$ (0.03 g/g ECA) were dispersed in distilled water, followed by stirring at 70 °C for 30 min until complete dissolution was achieved. After homogenization (12,000 rpm, 3 min), various concentrations (0.4%, 0.7%, and 1.0% of total volume) of EOs and Span 80 (25% of EOs) were incorporated in the solution. Next, the mixture was homogenized at 12,000 rpm for 7 min, sonicated for 8 min, and degassed for

5 min. The filtered film solution (16 mL) was uniformly spread onto Petri dishes (90 mm diameter) and dried at 25 °C for 18 h.

3.5. Physical Properties

A micrometer (No. 7327, Mitutoyo Co., Kawasaki, Japan) was used to measure the thickness of each film. TS and E of the ECA films containing CLO or CBO were evaluated with a Testometric machine (Model 250-2.5 CT, Testometric Co., Lancashire, UK). The films were cut into a rectangular shape (2.54 cm × 10 cm) for the measurement of their physical properties such as TS and E. The initial grip distance in the instrument was 50 mm, and its cross-head speed was set at 50 mm/min. The TS of the films, which is the maximum stress to withstand before breakage and is measured as the highest point of stress–strain curve, was obtained by dividing the maximum force (N), applied when the film was broken, by the cross-sectional area of the film. In addition, the E (%) was calculated by dividing the maximum length (mm) at the moment of film rupture by the initial grip distance (mm), followed by multiplying with 100. The method used for WVP measurement was previously described by Lee and Song [40]. Before the measurement, the water content of the ECA films was controlled by conditioning in a thermos-hygrostat (25 °C, 50% relative humidity) for 24 h.

3.6. Optical Properties

To characterize the color of the ECA films containing CLO or CBO, a colorimeter (CR-400M, Minolta, Tokyo, Japan) was used. According to the CIELAB color scale, the L *, a *, and b * values of the prepared ECA films were evaluated on a white standard plate with L *, a *, and b * values of 97.37, −0.14, and 2.00, respectively. The difference in color between the ECA film without EO and those containing EO was indicated as ΔE *. The method described by Yang et al. [41] was used to obtain the opacity of the ECA films with a spectrophotometer (UV-2450, Shimadzu Co., Kyoto, Japan).

3.7. SEM Analysis

To investigate the microstructure of the ECA films, SEM analysis was conducted. Before the analysis, the ECA films containing CLO or CBO were fractured in liquid nitrogen and coated with a platinum layer. SEM images of cross sections were acquired using an ion beam scanning electron microscope (LYRA3 WMU, TESCAN, Brno, Czech Republic) under an accelerating voltage of 5 kV. All images were displayed at a magnification of 5000×.

3.8. Thermal Property Analysis

To analyze the thermal properties of the ECA films, TGA and DSC were conducted. TGA was carried out with a thermogravimetric analyzer (Mettler Toledo, Columbus, OH, USA). The films (3.8 ± 0.1 mg) were sealed in a sample pan and heated in the temperature range of 25–600 °C at a rate of 10 °C/min under nitrogen flow at 50 mL/min. DSC measurements were carried out using a DSC1 (Mettler Toledo, Schwerzenbach, Switzerland). Each sample was scanned from −50 °C to 300 °C at a rate of 10 °C/min. After the first scan was completed, the second scan was conducted in the same way.

3.9. Antioxidant Activity

To examine the antioxidant property of the ECA films, DPPH and ABTS were used. The method described by Shojaee-Aliabadi et al. [42] was used to measure the DPPH radical scavenging activities of the ECA films. The film solution (1%) for this measurement was prepared by diluting the ECA film (0.1 g) in distilled water (10 mL) and shaking in an incubator for 30 min. Next, the solution was mixed with the DPPH solution at a ratio of 1:39 and stored in the dark for 1 h, followed by measuring the absorbance at 517 nm. The ABTS radical scavenging activities of the ECA films were also determined [43]. Before the test, ABTS solution was prepared by blending 7 mM ABTS and 2.45 mM potassium persulfate (2:1). After keeping the solution in the dark for 16 h, ethanol was added to dilute the solution until it had an absorbance value of 0.7 at 734 nm. The ABTS solution was then mixed with the film solution (1%). After being stored in the dark for 10 min, the supernatant of the mixture was separated by centrifugation (3000 rpm, 5 min) and the absorbance was measured at 734 nm.

3.10. Antimicrobial Activity

The disc diffusion method was applied to confirm the antibacterial property of the ECA films. *E. coli* O157:H7 (NCTC 12079, ATCC 43889) and *S.* Typhimurium (KCTC 2421, ATCC 14028) were inoculated into tryptic soy broth. *S. aureus* (KCTC 1621, ATCC 10537) and *L. monocytogenes* (KCTC 13064, ATCC 15313) were cultured in brain heart infusion broth. Culture cocktails were prepared by combining each bacterial strain. The bacterial solutions were diluted with peptone water (0.1%) and placed on Mueller Hinton Agar (MHA) medium. The ECA film-forming solutions (80 µL) were impregnated into 8 mm paper discs and dried for 30 min. After drying, the discs were put onto the inoculated MHA medium. Thereafter, it was incubated at 37 °C for 24 h, and the size of the inhibition zone was measured with a digimatic caliper (Model 500-181-20, Mitutoyo Corp., Kawasaki, Japan).

3.11. Statistical Analysis

The SAS program (version 9.4, SAS Institute Inc., Cary, NC, USA) was used for statistical analysis of the experimental results. All values are presented as mean ± standard deviations, and $p < 0.05$ indicates statistically significant differences. All experiments were repeated at least 5 times.

4. Conclusions

New biodegradable films were prepared using ECA, which has not yet been studied, as a base material in this study. The physical properties of the ECA films were improved with the use of $CaCl_2$ as a cross-linking agent, and 3% of $CaCl_2$ was found to be optimal. CLO and CBO were incorporated to develop active packaging films and, as a result, have influenced the various properties of the ECA film. As the content of EOs increased, the TS decreased, while E increased. The incorporation of CLO or CBO into the films resulted in antioxidant activities, and the ECA films with CBO showed stronger DPPH and ABTS radical scavenging activities than did the ECA films containing CLO. Antimicrobial activities against various bacteria were confirmed, and higher antimicrobial activity was found in the ECA film containing CLO. Therefore, our studies demonstrate that the ECA films containing CLO or CBO can be applied as new active packaging materials in the food industry.

Author Contributions: S.-K.B. performed the experiments, analyzed the data, and wrote the original draft. S.-K.B. and S.K. carried out the discussion of the work results. K.B.S. supervised the study and finally revised the manuscript.

Funding: This study was supported by the National Research Foundation of Korea.

Conflicts of Interest: The authors declare no conflict of interest.

References

1. Atarés, L.; Chiralt, A. Essential oils as additives in biodegradable films and coatings for active food packaging. *Trends Food Sci. Technol.* **2016**, *48*, 51–62. [CrossRef]
2. Khalil, H.A.; Tye, Y.Y.; Saurabh, C.K.; Leh, C.P.; Lai, T.K.; Chong, E.W.N.; Syakir, M.I. Biodegradable polymer films from seaweed polysaccharides: A review on cellulose as a reinforcement material. *Express Polym. Lett.* **2017**, *11*, 244–265. [CrossRef]
3. Tavassoli-Kafrani, E.; Shekarchizadeh, H.; Masoudpour-Behabadi, M. Development of edible films and coatings from alginates and carrageenans. *Carbohydr. Polym.* **2016**, *137*, 360–374. [CrossRef] [PubMed]
4. Wijesinghe, W.A.J.P.; Jeon, Y.J. Exploiting biological activities of brown seaweed *Ecklonia cava* for potential industrial applications: A review. *Int. J. Food Sci. Nutr.* **2012**, *63*, 225–235. [CrossRef] [PubMed]

5.   Park, Y.H. Seasonal variation in the chemical composition of brown algae with special reference to alginic acid. *Korean J. Fish. Aquat. Sci.* **1969**, *2*, 71–82.

6.   Jost, V.; Reinelt, M. Effect of Ca^{2+} induced crosslinking on the mechanical and barrier properties of cast alginate films. *J. Appl. Polym. Sci.* **2018**, *135*, 45754. [CrossRef]

7.   Fawzy, M.A.; Gomaa, M.; Hifney, A.F.; Abdel-Gawad, K.M. Optimization of alginate alkaline extraction technology from *Sargassum latifolium* and its potential antioxidant and emulsifying properties. *Carbohydr. Polym.* **2017**, *157*, 1903–1912. [CrossRef] [PubMed]

8.   Sellimi, S.; Younes, I.; Ayed, H.B.; Maalej, H.; Montero, V.; Rinaudo, M.; Nasri, M. Structural, physicochemical and antioxidant properties of sodium alginate isolated from a Tunisian brown seaweed. *Int. J. Biol. Macromol.* **2015**, *72*, 1358–1367. [CrossRef] [PubMed]

9.   Kuorwel, K.K.; Cran, M.J.; Orbell, J.D.; Buddhadasa, S.; Bigger, S.W. Review of mechanical properties, migration, and potential applications in active food packaging systems containing nanoclays and nanosilver. *Compr. Rev. Food Sci. F.* **2015**, *14*, 411–430. [CrossRef]

10.   Ranasinghe, P.; Pigera, S.; Premakumara, G.S.; Galappaththy, P.; Constantine, G.R.; Katulanda, P. Medicinal properties of 'true'cinnamon (*Cinnamomum zeylanicum*): A systematic review. *BMC Complement Altern. Med.* **2013**, *13*, 275. [CrossRef] [PubMed]

11.   Sangal, A. Role of cinnamon as beneficial antidiabetic food adjunct: A review. *Adv. Appl. Sci. Res.* **2011**, *2*, 440–450.

12.   Mahmood, A.; Bano, S.; Kim, S.G.; Lee, K.H. Water–methanol separation characteristics of annealed SA/PVA complex membranes. *J. Memb. Sci.* **2012**, *415*, 360–367. [CrossRef]

13.   Falkeborg, M.; Paitaid, P.; Shu, A.N.; Pérez, B.; Guo, Z. Dodecenyl succinylated alginate as a novel material for encapsulation and hyperactivation of lipases. *Carbohydr. Polym.* **2015**, *133*, 194–202. [CrossRef] [PubMed]

14.   Pereira, L.; Sousa, A.; Coelho, H.; Amado, A.M.; Ribeiro-Claro, P.J. Use of FTIR, FT-Raman and 13 C-NMR spectroscopy for identification of some seaweed phycocolloids. *Biomol. Eng.* **2003**, *20*, 223–228. [CrossRef]

15.   Benavides, S.; Villalobos-Carvajal, R.; Reyes, J.E. Physical, mechanical and antibacterial properties of alginate film: Effect of the crosslinking degree and oregano essential oil concentration. *J. Food Eng.* **2012**, *110*, 232–239. [CrossRef]

16.   Zactiti, E.M.; Kieckbusch, T.G. Potassium sorbate permeability in biodegradable alginate films: Effect of the antimicrobial agent concentration and crosslinking degree. *J. Food Eng.* **2006**, *77*, 462–467. [CrossRef]

17.   Zactiti, E.M.; Kieckbusch, T.G. Release of potassium sorbate from active films of sodium alginate crosslinked with calcium chloride. *Packag. Technol. Sci.* **2009**, *22*, 349–358. [CrossRef]

18.   Han, Y.; Yu, M.; Wang, L. Physical and antimicrobial properties of sodium alginate/carboxymethyl cellulose films incorporated with cinnamon essential oil. *Food Packag. Shelf Life* **2018**, *15*, 35–42. [CrossRef]

19.   Pranoto, Y.; Salokhe, V.M.; Rakshit, S.K. Physical and antibacterial properties of alginate-based edible film incorporated with garlic oil. *Food Res. Int.* **2005**, *38*, 267–272. [CrossRef]

20.   Abdollahi, M.; Rezaei, M.; Farzi, G. Improvement of active chitosan film properties with rosemary essential oil for food packaging. *Int. J. Food Sci. Technol.* **2012**, *47*, 847–853. [CrossRef]

21.   Shojaee-Aliabadi, S.; Hosseini, H.; Mohammadifar, M.A.; Mohammadi, A.; Ghasemlou, M.; Hosseini, S.M.; Khaksar, R. Characterization of κ-carrageenan films incorporated plant essential oils with improved antimicrobial activity. *Carbohydr. Polym.* **2014**, *101*, 582–591. [CrossRef] [PubMed]

22.   Tongnuanchan, P.; Benjakul, S.; Prodpran, T. Properties and antioxidant activity of fish skin gelatin film incorporated with citrus essential oils. *Food Chem.* **2012**, *134*, 1571–1579. [CrossRef] [PubMed]

23.   Zhang, Y.; Ma, Q.; Critzer, F.; Davidson, P.M.; Zhong, Q. Physical and antibacterial properties of alginate films containing cinnamon bark oil and soybean oil. *LWT Food Sci. Technol.* **2015**, *64*, 423–430. [CrossRef]

24.   Atarés, L.; De-Jesús, C.; Talens, P.; Chiralt, A. Characterization of SPI-based edible films incorporated with cinnamon or ginger essential oils. *J. Food Eng.* **2010**, *99*, 384–391. [CrossRef]

25.   Peng, Y.; Li, Y. Combined effects of two kinds of essential oils on physical, mechanical and structural properties of chitosan films. *Food Hydrocoll.* **2014**, *36*, 287–293. [CrossRef]

26.   Wu, J.; Sun, X.; Guo, X.; Ge, S.; Zhang, Q. Physicochemical properties, antimicrobial activity and oil release of fish gelatin films incorporated with cinnamon essential oil. *Aquac. Fish.* **2017**, *2*, 185–192. [CrossRef]

27.   Yadav, M.; Rhee, K.Y.; Park, S.J. Synthesis and characterization of graphene oxide/carboxymethylcellulose/alginate composite blend films. *Carbohydr. Polym.* **2014**, *110*, 18–25. [CrossRef] [PubMed]

28. Shankar, S.; Wang, L.F.; Rhim, J.W. Preparations and characterization of alginate/silver composite films: Effect of types of silver particles. *Carbohydr. Polym.* **2016**, *146*, 208–216. [CrossRef] [PubMed]

29. Soares, J.P.; Santos, J.E.; Chierice, G.O.; Cavalheiro, E.T.G. Thermal behavior of alginic acid and its sodium salt. *Eclética Química* **2004**, *29*, 57–64. [CrossRef]

30. Pongjanyakul, T.; Priprem, A.; Puttipipatkhachorn, S. Investigation of novel alginate–magnesium aluminum silicate microcomposite films for modified-release tablets. *J. Control Release.* **2005**, *107*, 343–356. [CrossRef] [PubMed]

31. Naidu, B.V.K.; Sairam, M.; Raju, K.V.; Aminabhavi, T.M. Thermal, viscoelastic, solution and membrane properties of sodium alginate/hydroxyethylcellulose blends. *Carbohydr. Polym.* **2005**, *61*, 52–60.

32. Tongnuanchan, P.; Benjakul, S.; Prodpran, T. Structural, morphological and thermal behavior characterizations of fish gelatin film incorporated with basil and citronella essential oils as affected by surfactants. *Food Hydrocoll.* **2014**, *41*, 33–43. [CrossRef]

33. Sharma, U.K.; Sharma, A.K.; Pandey, A.K. Medicinal attributes of major phenylpropanoids present in cinnamon. *BMC Complement Altern. Med.* **2016**, *16*, 156. [CrossRef] [PubMed]

34. Singh, G.; Maurya, S.; Catalan, C.A. A comparison of chemical, antioxidant and antimicrobial studies of cinnamon leaf and bark volatile oils, oleoresins and their constituents. *Food Chem. Toxicol.* **2007**, *45*, 1650–1661. [CrossRef] [PubMed]

35. Sanla-Ead, N.; Jangchud, A.; Chonhenchob, V.; Suppakul, P. Antimicrobial Activity of cinnamaldehyde and eugenol and their activity after incorporation into cellulose-based packaging films. *Packag. Technol. Sci.* **2012**, *25*, 7–17. [CrossRef]

36. Nazzaro, F.; Fratianni, F.; De Martino, L.; Coppola, R.; De Feo, V. Effect of essential oils on pathogenic bacteria. *Pharmaceuticals* **2013**, *6*, 1451–1474. [CrossRef] [PubMed]

37. Kang, J.H.; Song, K.B. Inhibitory effect of plant essential oil nanoemulsions against *Listeria monocytogenes*, *Escherichia coli* O157: H7, and *Salmonella* Typhimurium on red mustard leaves. *Innov. Food Sci. Emerg. Technol.* **2018**, *45*, 447–454. [CrossRef]

38. Cho, M.; Yoon, S.J.; Kim, Y.B. The nutritional composition and antioxidant activity from *Undariopsis petseniana*. *Ocean Polar Res.* **2013**, *35*, 273–280. [CrossRef]

39. You, B.J.; Jeong, I.H.; Lee, K.H. Effect extraction conditions on bile acids binding capacity in vitro of alginate extracted from sea tangle (*Laminaria* spp.). *Korean J. Fish. Aquat. Sci.* **1997**, *30*, 31–38.

40. Lee, K.Y.; Song, K.B. Preparation and characterization of an olive flounder (*Paralichthys olivaceus*) skin gelatin and polylactic acid bilayer film. *J. Food Sci.* **2017**, *82*, 706–710. [CrossRef] [PubMed]

41. Yang, S.Y.; Lee, K.Y.; Beak, S.E.; Kim, H.; Song, K.B. Antimicrobial activity of gelatin films based on duck feet containing cinnamon leaf oil and their applications in packaging of cherry tomatoes. *Food Sci. Biotechnol.* **2017**, *26*, 1429–1435. [CrossRef] [PubMed]

42. Shojaee-Aliabadi, S.; Hosseini, H.; Mohammadifar, M.A.; Mohammadi, A.; Ghasemlou, M.; Ojagh, S.M.; Khaksar, R. Characterization of antioxidant-antimicrobial κ-carrageenan films containing *Satureja hortensis* essential oil. *Int. J. Biol. Macromol.* **2013**, *52*, 116–124. [CrossRef] [PubMed]

43. Bitencourt, C.M.; Fávaro-Trindade, C.S.; Sobral, P.J.A.; Carvalho, R.A. Gelatin-based films additivated with curcuma ethanol extract: Antioxidant activity and physical properties of films. *Food Hydrocoll.* **2014**, *40*, 145–152. [CrossRef]

© 2018 by the authors. Licensee MDPI, Basel, Switzerland. This article is an open access article distributed under the terms and conditions of the Creative Commons Attribution (CC BY) license (http://creativecommons.org/licenses/by/4.0/).

International Journal of
Molecular Sciences

MDPI

Article

Morphological Study of Chitosan/Poly (Vinyl Alcohol) Nanofibers Prepared by Electrospinning, Collected on Reticulated Vitreous Carbon

**Diana Isela Sanchez-Alvarado [1], Javier Guzmán-Pantoja [2], Ulises Páramo-García [1,*],
Alfredo Maciel-Cerda [3], Reinaldo David Martínez-Orozco [1] and Ricardo Vera-Graziano [3]**

[1]   Tecnológico Nacional de México/Instituto Tecnológico de Cd. Madero, Centro de Investigación en
       Petroquímica, Prol. Bahía de Aldhair y Av. De las Bahías, Parque de la Pequeña y Mediana Industria,
       Altamira, Tamaulipas 89600, Mexico; diana.sanchez@itsna.edu.mx (D.I.S.-A.);
       rd.martinez.orozco@gmail.com (R.D.M.-O.)
[2]   Instituto Mexicano del Petróleo, Gerencia de Transformación de Hidrocarburos, Eje Central Lázaro Cárdenas
       No. 152, Col. San Bartolo Atepehuacán, 07730 Ciudad de México, Mexico; jguzmanp@imp.mx
[3]   Instituto de Investigaciones en Materiales, Universidad Nacional Autónoma de México, Circuito Exterior
       S/N, Circuito de la Investigación Científica, Ciudad Universitaria, 04510 Ciudad de México, Mexico;
       macielal@unam.mx (A.M.-C.); graziano@unam.mx (R.V.-G.)
*   Correspondence: uparamo@itcm.edu.mx; Tel.: +52-833-357-4820

Received: 3 May 2018; Accepted: 1 June 2018; Published: 9 June 2018

Abstract: In this work, chitosan (CS)/poly (vinyl alcohol) (PVA) nanofibers were prepared by
using the electrospinning method. Different CS concentrations (0.5, 1, 2, and 3 wt %), maintaining
the PVA concentration at 8 wt %, were tested. Likewise, the studied electrospinning experimental
parameters were: syringe/collector distance, solution flow and voltage. Subsequently, the electrospun
fibers were collected on a reticulated vitreous carbon (RVC) support for 0.25, 0.5, 1, 1.5, and 2 h.
The morphology and diameter of the CS/PVA nanofibers were characterized by scanning electron
microscopy (SEM), finding diameters in the order of 132 and 212 nm; the best results (uniform fibers)
were obtained from the solution with 2 wt % of chitosan and a voltage, distance, and flow rate of
16 kV, 20 cm, and 0.13 mL/h, respectively. Afterwards, a treatment with an ethanolic NaOH solution
was performed, observing a change in the fiber morphology and a diameter decrease (117 ± 9 nm).

Keywords: chitosan; PVA; nanofibers; electrospinning

---

1. Introduction

The electrospinning technique is an efficient method for the manufacture of micro- and nanofibers;
it is also an efficient and simple method to produce homogenous fibers for various applications.
This technique relies on multiple operating parameters such as applied voltage, collector type,
temperature, relative humidity, flow rate, etc., as well as on intrinsic solution parameters like molecular
weight, polymeric concentration, viscosity, surface tension, electrical conductivity, among others.
The electrospinning process uses electrostatic forces to produce polymeric fibers. When high voltage
is applied, polymer drops are subjected to instability, forming fibers when the surface tension of the
polymer solution drop at the tip of the needle is overcome. A series of parameters must be optimized
in order to produce fibers instead of drops [1–3].

Chitosan (CS) is a biopolymer derived from chitin, often obtained from the exoskeleton of
crustaceans. It has very useful properties such as biocompatibility, biodegradability, and antimicrobial
activity; however, the production of fibers of this polymer is very difficult due to its insolubility in
organic solvents, its ionic character in solution and the formation of three-dimensional networks by
strong hydrogen bonds [4–6].

In this sense, Li and Hsieh (2006) found that by combining an anionic polymer such as poly (vinyl alcohol) (PVA) and chitosan, which is a cationic polymer, the formation of fibers was improved, which is in contrast with the combination of polymers having the same charge [7]. PVA is a biodegradable polymer that may be used to immobilize chitosan [8]. Moreover, PVA can reduce the crystallinity of chitosan [9]. The functional groups of chitosan make hydrogen bonds with PVA which lead to form defect-free nanofibers [10].

Although CS and PVA can improve the production of fibers, the right proportion to have a stable solution has to be established. In this sense, several works studying different CS/PVA ratios have been reported [11–13]. To do so, the calculation of the thermodynamic interaction parameter (X) by using the Flory–Huggins theory is necessary. This parameter represents the energy associated with the blending of polymers; depending on its value, the interaction can be classified as favorable or unfavorable.In our case, we did not consider its calculation because the CS/PVA ratio selected to carry out the electrospinning experiments produced a homogeneous solution, which we could witness in the lab by observing a miscible behavior.

Paipitak et al. [14] used the electrospinning method to produce CS/PVA nanofibers, with a glass plate as collector, and study the effect of CS concentration on the formation of nanofibers. The morphology and diameter of the CS/PVA nanocomposite fibers were analyzed by SEM, finding a fiber diameter of 100 nm. The best CS solution concentration for fiber formation was 2 wt % due to the maximum fiber yield.

According to the points described above and in order to improve the efficiency of the electrospinning method and morphology of the obtained CS/PVA nanofibers, in the present work, reticulated vitreous carbon (RVC) was used as a collector due to its high electrical conductivity and minimal reactivity. RVC is currently used in electrochemical systems such as battery and fuel cells, metal ion removal, electroanalytical analysis, sensors and synthesis of organic compounds. As it can be seen, these applications are related to the promotion of turbulence aimed at increasing the mass transfer phenomena in reactors. To the best of our knowledge, in the present research work, RVC has been used for the first time as a fiber collector in the electrospinning process. The ultimate goal, after fundamental studies, is to produce CS/PVA nanofibers, establishing the best parameters of the electrospinning technique (voltage, drop-collector distance and volumetric flow), on a highly porous support to generate turbulence and adequate contact between the materials and a solution. The obtained results encouraged us to consider the experimental procedure featured in this work for its application in the reduction of sulfur compounds in FCC (Fluid Catalytic Cracking) gasoline, where the chitosan fibers could work as adsorbents of metals such as nickel and molybdenum in an oxidative desulfurization process. The results of this study will be part of a forthcoming paper.

2. Results and Discussions

2.1. Electrospinning Method: Experimental Parameters

Preliminary electrospinning processes were carried out using different operation parameters. Not all the analyzed electrospinning parameters showed good results since in some cases, the fibers were not uniform with large nodules incorporated into them or simply no fibers were observed.

The operation parameters used to produce fibers are shown in Table 1. In order to establish the best CS/PVA proportions, four of them were experimentally tested (0.5 wt % CS/8 wt % PVA, 1 wt % CS/8 wt % PVA, 2 wt % CS/8 wt % PVA, and 3 wt % CS/8 wt % PVA). As the most uniform nanofibers were obtained with 2 wt % CS/8 wt % PVA, the other three concentrations were discarded (data not shown). Then, the optimal electrospinning conditions were obtained with a 2 wt % CS/8 wt % PVA solution, and a flow rate, distance, and voltage of 0.13 mL/h, 20 cm, and 16 kV, respectively.

Table 1. Experimentally evaluated parameters using a 2 wt % CS/8 wt % PVA solution.

2 wt. CS/8 wt % PVA Solution	Image	Flow Rate	Distance	Voltage	Observations *
	Id.	mL/h	cm	kV	
Flow rate	a	0.12	15	14	Non-uniform fibers
	b	0.13	15	14	Non-uniform fibers with nodules
Voltage	c	0.13	15	15	Non-uniform fibers with nodules
	d	0.13	15	14	Non-uniform fibers
Tip–target distance	e	0.12	20	14	Non-uniform fibers
	f	0.12	25	14	Non-uniform fibers
	g	0.13	20	16	Non-uniform fibers
	h	0.12	25	16	Non-uniform fibers
	i	0.13	20	14	Non-uniform fibers
	j	0.13	25	14	Non-uniform fibers
	k	0.13	20	16	Uniform fibers
	l	0.13	25	16	Non-uniform fibers with nodules

* Uniformity criteria were the fiber abundance and lack of nodules observed in the micrographs shown in Figure 1.
CS: chitosan; PVA: poly (vinyl alcohol).

Figure 1 shows the CS/PVA electrospun sample using a 2 wt % CS/8 wt % PVA solution, with the parameters featured in Table 1, varying flow rate, distance, and voltage during the performance of the electrospinning technique, where it is observed that the fibers are non-uniform: some fibers are thinner than others and some nodules can be seen, unlike what is observed under the optimal conditions mentioned above. These experiments were performed on a glass surface.

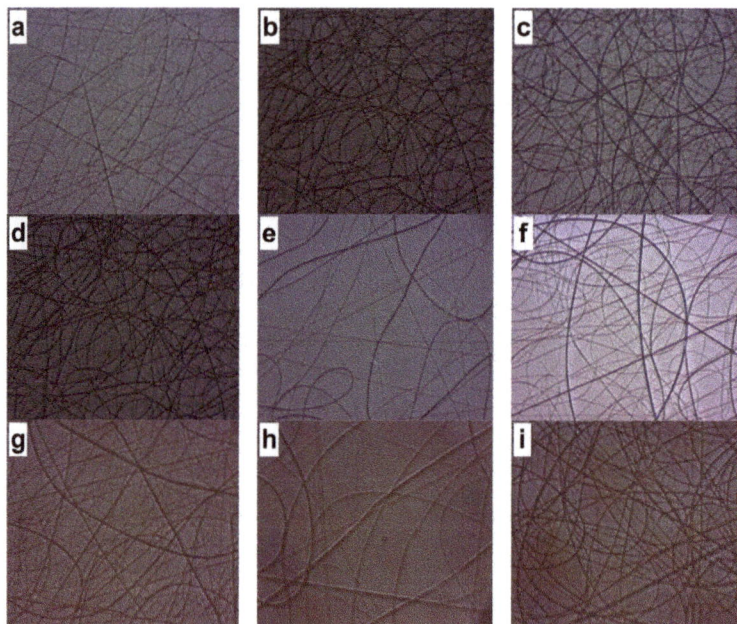

Figure 1. *Cont.*

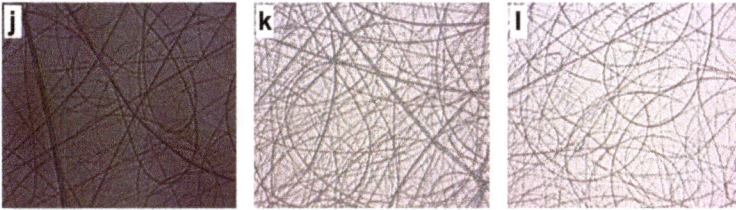

Figure 1. Micrographs of fibers produced by using a 2 wt % CS/8 wt % PVA solution (40× resolution): (**a**) 0.12 mL/h, 15 cm and 14 kV; (**b**) 0.13 mL/h, 15 cm and 14 kV; (**c**) 0.13 mL/h, 15 cm and 15 kV; (**d**) 0.13 mL/h, 15 cm and 14 kV; (**e**) 0.12 mL/h, 20 cm and 14 kV; (**f**) 0.12 mL/h, 25 cm and 14 kV; (**g**) 0.13 mL/h, 20 cm and 16 kV; (**h**) 0.12 mL/h, 25 cm and 16 kV; (**i**) 0.13 mL/h, 20 cm and 14 kV; (**j**) 0.13 mL/h, 25 cm and 14 kV; (**k**) 0.13 mL/h, 20 cm and 16 kV; (**l**) 0.13 mL/h, 25 cm and 16 kV.

Figure 2 shows a comparison of optical images for threedifferent solution compositions with optimal conditions. Non-uniform fibers are observed for 0.5 (a) and 0.1 wt % (b) of CS. For 2 wt % CS (c), uniform fibers were obtained.

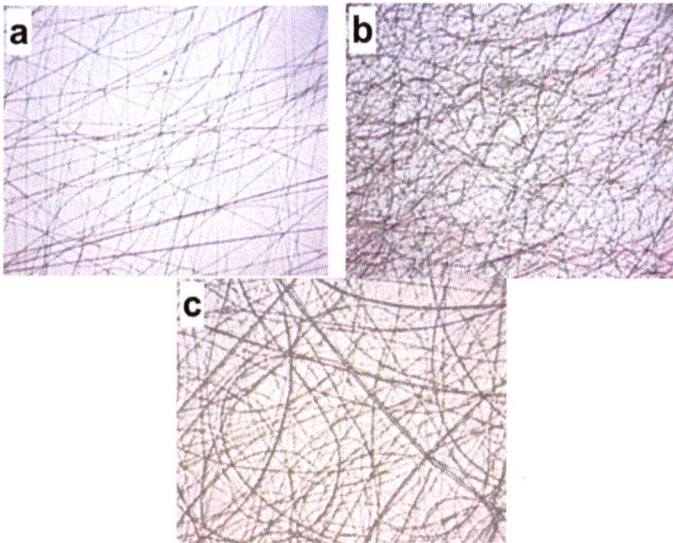

Figure 2. Micrographs of fibers produced with different CS concentrations (wt %): (**a**) 0.5; (**b**) 1; and (**c**) 2 with a voltage, distance, and flow rate of 0.13 mL/h, 20 cm, and 16 kV, respectively (40× resolution).

2.2. Electrospun CS/PVA Samples Collected on Reticulated Vitreous Carbon

Figure 3 shows the images of the electrospun fibers on the reticulated vitreous carbon (RVC) support at different times. The deposition of the fibers was observed during the spinning process, and the surface of the RVC support was modified by showing a white coloration caused by the electrospun fibers.

Int. J. Mol. Sci. **2018**, *19*, 1718

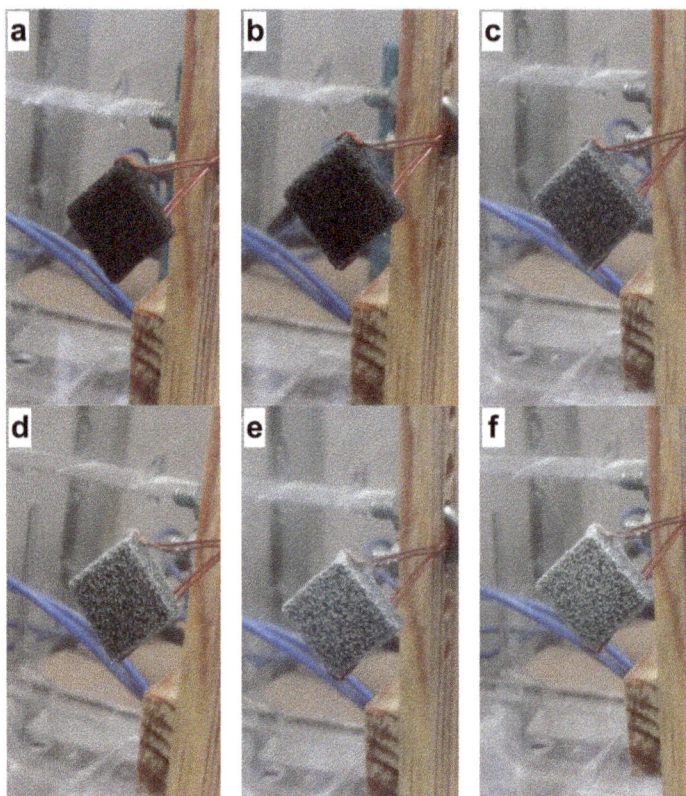

Figure 3. Electrospun fibers on the RVC support: (**a**) 0; (**b**) 15; (**c**) 30; (**d**) 60; (**e**) 90; and (**f**) 120 min.

Zargarian and Haddadi [15] indicated that the electrospinning technique can produce fibers with diameters ranging from 5 nm to 1 μm under a high-voltage electrostatic field that runs between a metal capillary syringe and an electrically grounded collector. Fibers made of polymer blends can be produced by this technique. In their study, the authors fabricated nanofibrous scaffolds of poly (ε-caprolactone)/hydroxyapatite/chitosan–PVA in order to investigate the influence of several parameters that affect the process and morphology such as concentration, solvent composition, voltage and tip-to-collector distance (TCD). In this work, the surface morphology of the nanofibrous material was studied using SEM micrographs. The distribution of fiber diameter frequencies was plotted to facilitate the optimization process. The voltage played an important role, while the TCD was largely responsible for the determination of the average diameter of the fibers.

2.3. Morphology of the RVC Support and CS/PVA Fibers on RVC

As for RVC, it is a macroporous polymeric carbon form as it can be seen in Figure 4. This material is obtained, for example, by carbonization of an expanded polymer or a vacuum removed material [16,17]. This material is a solid foam constituted by an open cell network composed of vitreous carbon, a material with high electrical and thermal conductivity. The adjective reticulated means "constructed, arranged or marked as a network or part of a network". RVC has a remarkably high vacuum volume and surface area. Its rigid structure has low density and withstands low temperatures. Figure 4 shows a SEM micrograph of unmodified RVC.

Figure 5 shows the micrographs of the electrospun fibers on RVC at different times with 2 wt % CS/8 wt % PVA, with a voltage, distance, and flow rate of 0.13 mL/h, 20 cm, and 16 kV, respectively. It is observed that the deposited fibers increased as the electrospinning time increased too; very few defects and a relatively uniform morphology can be observed.

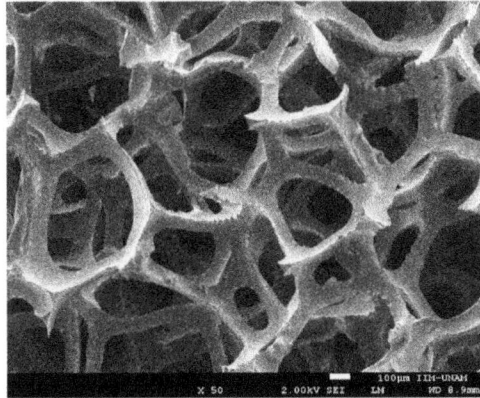

Figure 4. SEM micrograph of reticulated vitreous carbon used as support material.

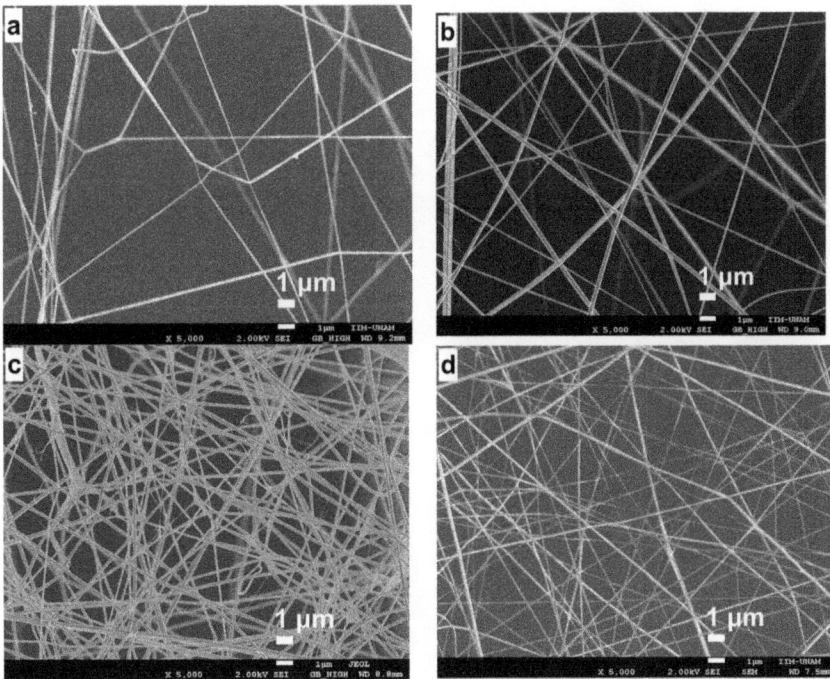

Figure 5. *Cont.*

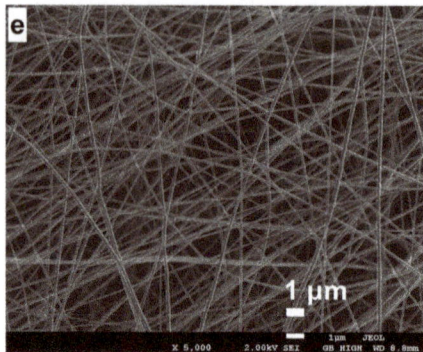

Figure 5. SEM characterization of electrospun nanofibers on a RVC support at different times: (**a**) 15; (**b**) 30; (**c**) 60; (**d**) 90; and (**e**) 120 min.

For the diameter frequency histograms, about 100 electrospun fibers were analyzed with particle counting. Figure 6 presents the histograms of the nanofibers analyzed above, where diameters from 90 to 390 nm were observed.

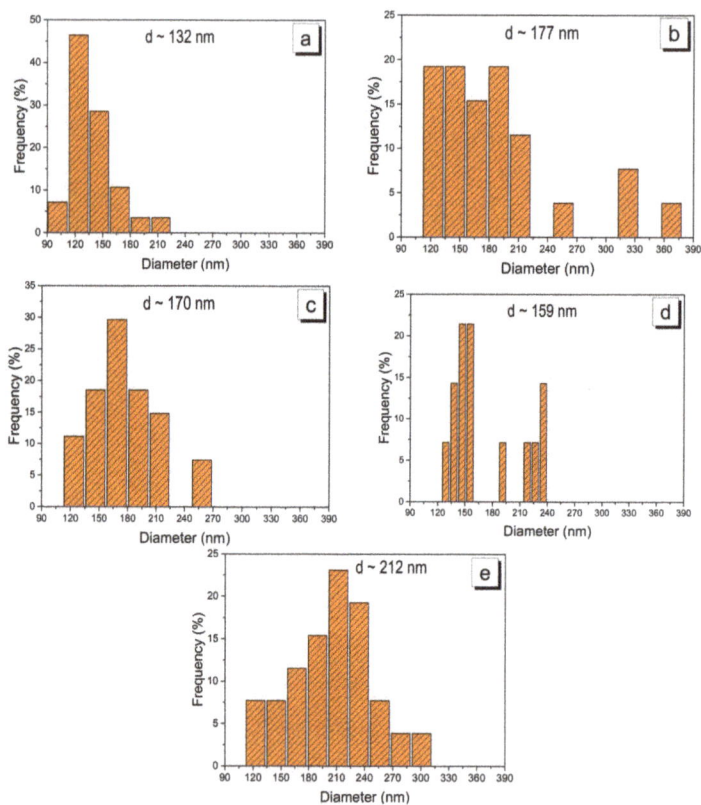

Figure 6. Histograms of nanofibers on the RVC support: (**a**) 15; (**b**) 30; (**c**) 60; (**d**) 90; and (**e**) 120 min.

Table 2 shows the average fiber diameters at electrospinning times of 15 min, 0.5, 1, 1.5 and 2 h. Diameters from 132 to 212 nm were observed with a standard deviation from ±6 to ±20 nm.

Table 2. Diameters of the electrospun nanofibers.

Parameter	15 min	30 min	1 h	1.5 h	2 h
Diameter (nm)	132	177	170	159	212
Standard deviation (nm)	±20	±9	±7	±11	±6

2.4. Stabilization of CS/PVA Nanofibers

Figure 7 shows micrographs of CS/PVA nanofibers deposited on RVC at an electrospinning time of 90 min before ((a) and (c)) and after ((b) and (d)) the ethanolic NaOH solution treatment used to stabilize the nanofibers. This treatment resulted in a fiber diameter decrease, which is in good agreement with results reported in the literature [18].

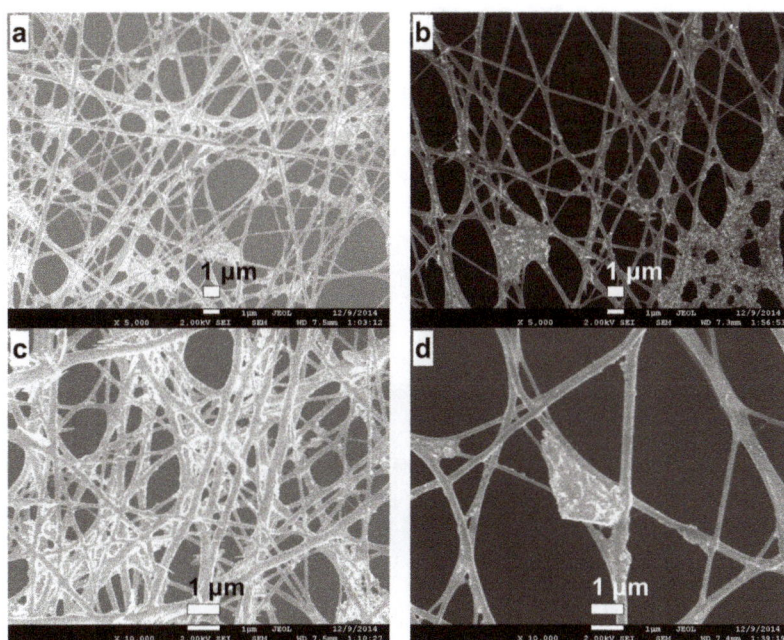

Figure 7. SEM micrographs of CS/PVA electrospun nanofibers deposited on RVC for 90 min before (**a,c**) and after (**b,d**) ethanolic NaOH solution treatment.

Table 3 presents the diameter of the electrospun fibers for 90 min before and after the alkaline treatment, which produced a diameter decrease of 44 nm.

Table 3. Diameters of the electrospun fibers before and after the alkaline treatment.

Parameter	Before	After
Diameter (nm)	159 ± 11	115 ± 9

2.5. Solution Viscosity

The viscosity of the CS/PVA solution decreased with time (weeks) from 10,500 to 410 cP, which was due to the hydrolysis of chitosan, as stated elsewhere [19,20]. Based on this fact, the electrospun nanofibers were produced by using a recently prepared solution. The results are shown in Figure 8.

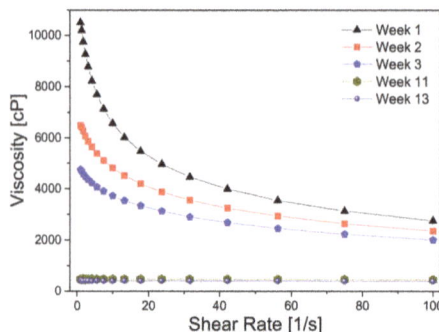

Figure 8. Viscosity variation of the 2 wt % CS/8 wt % PVA solution.

Figure 9 shows the micrographs obtained from the electrospun process on RVC, where cross ((a) and (b)) and topview ((c) and (d)) sections were taken for the SEM analysis. It is observed that the fibers are only found on the RVC surface, confirming the fact that they did not penetrate into the pores of the RVC support.

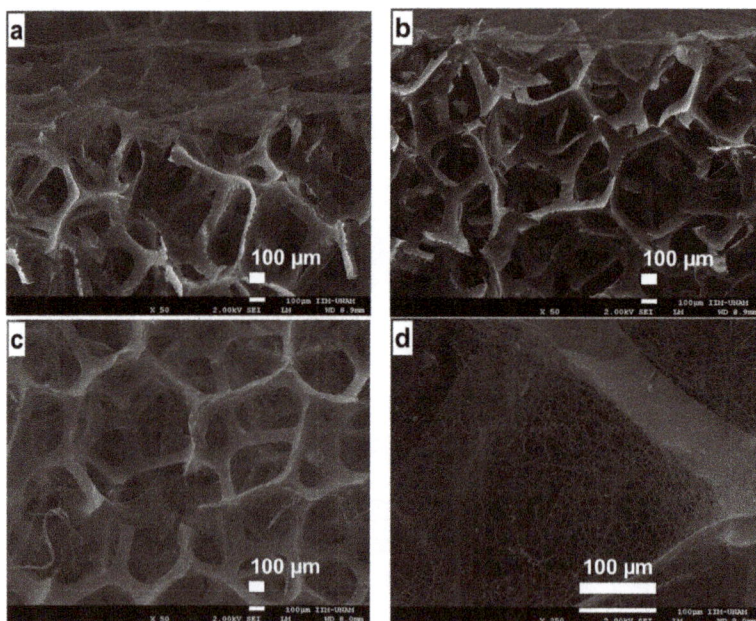

Figure 9. SEM micrographs of electrospun nanofibers deposited on RVC (90 min): (**a**,**b**) cross sections; (**c**,**d**) top-view sections.

3. Materials and Methods

3.1. Materials

CS from crab shells with deacetylation degree of 85% (lot number 91k1265) was used as received from Sigma Aldrich. Medium molecular weight PVA was purchased from Sigma Aldrich and used as received. Acetic acid (99.7%) was acquired from J.T Baker. RVC 60 ppi (normal pores per lineal inch) was obtained from Electrosynthesis, Co. Inc., Lancaster, NY, USA.

3.2. Preparation of the CS/PVA Solution

The electrospinning solutions were prepared by dissolving a maximum of 3 wt % of CS in acetic acid at 2 wt %. This solution was magnetically stirred until complete dissolution at room temperature. PVA powder was dissolved in deionized water at 8 wt % near its boiling point to facilitate complete dissolution. Subsequently, both solutions were homogenized by magnetic stirring.

In order to study the component ratio effect on the electrospinning process, the CS concentration was varied using 0.5, 1, 2, and 3 wt % while the PVA concentration remained at 8 wt % for each prepared solution.

3.3. Electrospinning Process

Electrospinning solutions were pumped using a 5 mL commercial plastic syringe and a Programmable Syringe Pump NE-1600. The studied flow rates were adjusted to 0.10, 0.12, 0.13, 0.15, and 0.20 mL/h at three different working voltages: 14, 15, and 16 kV, controlled by a high DC (Direct Current) voltage supply (Glassman High Voltage EH60). A glass plate or RVC was used as a collector, and the tip-to-collector distanceswere set at 15, 20, and 25 cm. All electrospinning procedures were carried out at room temperature (i.e., at around 23 °C). As the solution viscosity is a very important parameter, all the electrospinning experiments were carried out with recently prepared CS/PVA solutions (no more than 1 week old).

3.4. Alkaline Treatment with NaOH

The as-obtained electrospun nanofibers were immersed overnight in a saturated NaOH ethanolic solution in order to analyze the alkaline treatment effect. The treated samples were then rinsed with water and dried in air to be observed by SEM.

3.5. Viscometric Studies

The solution viscosity was measured in a rheometer (MCR Physica 301, Anton Paar) with cone and plate geometry, varying the shear rate between 1 and 100 rad/s. The measurements were carried out after 1, 2, 3, 11, and 13 weeks of their preparation. The storage temperature was 25 °C.

4. Conclusions

In the present study, CS/PVA nanofibers were manufactured by the electrospinning technique. The use of RVC allowed the formation of homogeneous fibers, compared to the fibers collected on a glass plate, with diameters ranging from 132 to 212 nm, making RVC a suitable substrate to obtain thin fibers. However, the deposit of fibers occurred only on the substrate surface. After testing CS concentrations of 0.5, 1, 2, and 3 wt %, it was found that the best solution to produce fibers was that containing 2 wt % CS and 8 wt % PVA. After evaluating electrospinning parameters such as flow rate, tip-to-collector distance, and voltage on either glass or RVC substrates, the optimum parameters for producing CS/PVA electrospun nanofibers were a voltage, distance, and flow rate of 16 kV, 20 cm, and 0.13 mL/h, respectively. , which allowed the manufacture of uniform fibers that were observed by SEM. The use of RVC improved the formation of homogeneous fibers compared to the fibers collected on the glass plate, with diameters ranging from 132 to 212 nm. In order to obtain chemically stable CS/PVA nanofibers, it was necessary to treat the nanofibers with an ethanolic NaOH solution.

Author Contributions: D.I.S.-A. performed the experimental tests. J.G.-P. designed and supervised the experimental part aimed at obtaining the polymeric materials. He also was in charge of the manuscript revision process. U.P.-G. carried out the discussion of the work overall results. A.M.-C. supervised the electrospinning experiments. R.D.M.-O. ran and interpreted the SEM experiments and results, respectively. R.V.-G. provided the infrastructure to perform the electrospinning experiments.

Acknowledgments: Diana Isela Sanchez-Alvarado acknowledges the scholarship granted by CONACYT-Mexico (269921/218106).

Conflicts of Interest: The authors declare no conflicts of interest.

References

1. Fong, H.; Reneker, D.H. *Structure Formation in Polymeric Fibers*; Hanser Gardner Publishers: Munich, Germany, 2001; pp. 225–246.
2. Huang, Z.; Zhang, Y.; Kotaki, M.; Ramakrishna, S. A review on polymer nanofibers by electrospinning and their applications in nanocomposites. *Compos. Sci. Technol.* **2003**, *63*, 2223–2253. [CrossRef]
3. Gibson, P.; Gibson, H.; Rivin, D. Transport properties of porous membranes based on electrospun nanofibers. *Colloids Surf. A* **2001**, *187–188*, 469–481. [CrossRef]
4. Desai, K.; Kit, K.; Li, J.; Zivanovic, S. Morphological and surface properties of electrospun chitosan nanofibers. *Biomacromolecules* **2008**, *9*, 1000–1006. [CrossRef] [PubMed]
5. Geng, X.; Kwon, O.; Jang, J. Electrospinning of chitosan dissolved in concentrated acetic acid solution. *Biomaterials* **2005**, *26*, 5427–5432. [CrossRef] [PubMed]
6. Homayoni, H.; Hosseini Ravandi, S.A.; Valizadeh, M. Electrospinning of chitosan nanofibers: Processing optimization. *Carbohydr. Polym.* **2009**, *77*, 656–661. [CrossRef]
7. Li, L.; Hsieh, Y.-L. Chitosan bicomponent nanofibers and nanoporous fibers. *Carbohydr. Res.* **2006**, *341*, 374–381. [CrossRef] [PubMed]
8. Kumar, K.; Tripathi, B.P.; Shahi, V.K. Crosslinked chitosan/polyvinyl alcohol blend beads for removal and recovery of Cd(II) from wastewater. *J. Hazard. Mater.* **2009**, *172*, 1041–1048. [CrossRef] [PubMed]
9. Kim, J.K.; Kim, J.Y.; Lee, Y.M.; Kim, K.Y. Properties and swelling characteristics of cross-linked poly (vinyl alcohol)/chitosan blend membrane. *J. Appl. Polym. Sci.* **1992**, *45*, 1711–1717. [CrossRef]
10. Habiba, U.; Siddique, T.A.; Talebian, S.; Lee, J.J.L.; Salleh, A.; Ang, B.C.; Afifi, A.M. Effect of deacetylation on property of electrospun chitosan/PVA nanofibrous membrane and removal of methyl orange, Fe(III) and Cr(VI) ions. *Carbohydr. Polym.* **2017**, *177*, 32–39. [CrossRef] [PubMed]
11. Jawalkar, S.; Raju, K.; Halligudi, S.; Sairam, M.; Aminabhavi, T. Molecular modeling simulation to predict compatibility of poly (vinyl alcohol) and chitosan blends: A comparison with experiments. *J. Phys. Chem. B* **2007**, *111*, 2431–2439. [CrossRef] [PubMed]
12. Kumar, N.; Prabhakar, M.; Prasad, V.; Rao, M.; Reddy, A.; Rao, C.; Subha, M. Compatibility studies of chitosan/PVA blend in 2% aqueous acetic acid solution at 30 °C. *Carbohydr. Polym.* **2010**, *82*, 251–255. [CrossRef]
13. Lewandowska, K. Viscometric Studies in dilute solution mixtures of chitosan and microcrystalline chitosan with poly (vinyl alcohol). *J. Solut. Chem.* **2013**, *42*, 1654–1662. [CrossRef] [PubMed]
14. Paipitak, K.; Pornpra, T.; Mongkontalang, P.; Techitdheera, W.; Pecharapa, W. Characterization of PVA-chitosan Nanofibers prepared by electrospinning. *Procedia Eng.* **2011**, *8*, 101–105. [CrossRef]
15. Zargarian, S.S.H.; Haddadi-Asl, V. A nanofibrous composite scaffold of PCL/hydroxyapatite-chitosan/PVA prepared by electrospinning. *Iran. Polym. J.* **2010**, *19*, 457–468.
16. Kizling, K.; Dzwonek, M.; Olszewski, B.; Bącal, P.; Tymecki, L.; Więckowska, A.; Stolarczyk, K.; Bilewicz, R. Reticulated vitreous carbon as a scaffold for enzymatic fuel cell designing. *Biosens. Bioelectron.* **2017**, *95*, 1–7. [CrossRef] [PubMed]
17. Walsh, F.C.; Arenas, L.F.; Ponce de León, C.; Reade, G.W.; Whyte, I.; Mellor, B.G. The continued development of reticulated vitreous carbon as a versatile electrode material: Structure, properties and applications. *Electrochim. Acta* **2016**, *215*, 566–591. [CrossRef]
18. Huang, X.-J.; Ge, D.; Xu, Z.-K. Preparation and characterization of stable chitosan nanofibrous membrane for lipase immobilization. *Eur. Polym. J.* **2007**, *43*, 3710–3718. [CrossRef]

19. Skaugrud, O. Chitosan-New biopolymer for cosmetics and drugs. *Drug Cosmet. Ind.* **1991**, *148*, 24–29.
20. Ravi Kumar, M.N.V. A review of chitin and chitosan applications. *React. Funct. Polym.* **2000**, *46*, 1–27. [CrossRef]

© 2018 by the authors. Licensee MDPI, Basel, Switzerland. This article is an open access article distributed under the terms and conditions of the Creative Commons Attribution (CC BY) license (http://creativecommons.org/licenses/by/4.0/).

International Journal of
Molecular Sciences

MDPI

Article

Polymeric Micelles Based on Modified Glycol Chitosan for Paclitaxel Delivery: Preparation, Characterization and Evaluation

Na Liang [1], Shaoping Sun [2,*], Xianfeng Gong [2], Qiang Li [2], Pengfei Yan [2,3,*] and Fude Cui [4]

[1] Key Laboratory of Photochemical Biomaterials and Energy Storage Materials, Heilongjiang Province, College of Chemistry & Chemical Engineering, Harbin Normal University, Harbin 150025, China; liangna528@163.com

[2] Key Laboratory of Chemical Engineering Process & Technology for High-Efficiency Conversion, College of Heilongjiang Province; School of Chemistry and Material Science, Heilongjiang University, Harbin 150080, China; gongxianfeng@sina.com (X.G.); liqianghlj100@163.com (Q.L.)

[3] Key Laboratory of Functional Inorganic Material Chemistry, Heilongjiang University, Harbin 150080, China

[4] School of Pharmacy, Shenyang Pharmaceutical University, Shenyang 110016, China; syphucuifude@163.com

* Correspondence: sunshaoping111@163.com (S.S.); yanpf@vip.sina.com (P.Y.); Tel./Fax: +86-451-8660-8616 (S.S.)

Received: 22 April 2018; Accepted: 15 May 2018; Published: 23 May 2018

Abstract: Amphiphilic polymer of α-tocopherol succinate modified glycol chitosan (TS-GC) was successfully constructed by conjugating α-tocopherol succinate to the skeleton of glycol chitosan and characterized by Fourier-transform infrared (FT-IR) and proton nuclear magnetic resonance (^{1}H-NMR). In aqueous milieu, the conjugates self-assembled to micelles with the critical aggregation concentration of 7.2×10^{-3} mg/mL. Transmission electron microscope (TEM) observation and dynamic light scattering (DLS) measurements were carried out to determine the physicochemical properties of the micelles. The results revealed that paclitaxel (PTX)-loaded TS-GC micelles were spherical in shape. Moreover, the PTX-loaded micelles showed increased particle sizes (35 nm vs. 142 nm) and a little reduced zeta potential (+19 mV vs. +16 mV) compared with blank micelles. The X-ray diffraction (XRD) spectra demonstrated that PTX existed inside the micelles in amorphous or molecular state. In vitro and in vivo tests showed that the PTX-loaded TS-GC micelles had advantages over the Cremophor EL-based formulation in terms of low toxicity level and increased dose, which suggested the potential of the polymer as carriers for PTX to improve their delivery properties.

Keywords: glycol chitosan; α-tocopherol succinate; amphiphilic polymer; micelles; paclitaxel

1. Introduction

Paclitaxel (PTX), as a powerful anti-tumor drug, has been extensively used in the clinical treatment of several solid tumors, such as refractory ovarian cancer, metastasis breast cancer, non-small cell lung cancer, Acquired Immune Deficiency Syndrome-related Kaposi's sarcoma and other cancers [1,2]. Due to its poor water solubility of approximately <2 μg/mL, PTX is currently solubilized in a 50:50 mixture of Cremophor EL (PEG-35 caster oil) and dehydrated ethanol as Taxol® (Bristol-Myers Squibb, New York City, NY, USA). However, several studies reported that Cremophor EL induced serious side effects such as hypersensitivity, nephrotoxicity, neurotoxicity, and the extraction of plasticizer from the infusion tubes [3]. In light of these drawbacks, a number of alternative preparations were investigated, including liposomes, nanocrystals, micelles, cyclodextrin complexes and PTX conjugates [4–8].

Among these formulations, polymeric micelles have been proven as promising drug delivery systems for PTX administration [9,10], because of their attractive characteristics, such as biocompatibility, high drug-loading content, small size (<200 nm) and propensity to evade scavenging by the mononuclear phagocyte system (MPS) [11,12]. Moreover, the nanoscale dimensions of polymeric micelles permit their selectively accumulation in tumor tissues due to the enhanced permeability and retention (EPR) effect, which is termed "passive targeting" [13,14]. The formation of polymeric micelles is generally considered as the self-assembly of polymeric amphiphiles. In aqueous medium, polymeric amphiphiles form the micelles consisting of the inner hydrophobic core and the outer hydrophilic shell. The hydrophobic core provides a storeroom for loading hydrophobic drugs, and the hydrophilic shell allows retaining the stability of micelles in an aqueous environment [15].

Due to its favorable properties, such as biodegradability, biocompatibility, nontoxicity and bioadhesivity, chitosan has been widely studied as a pharmaceutical carrier for drug delivery [16,17]. However, chitosan is insoluble at pH values above its pKa (6.4) in water, and this obviously limits its biomedical applications.

In recent years, glycol chitosan (GC), which possesses good solubility over a broad range of pH, has been studied to construct drug delivery systems for PTX, such as nanocrystals, nanoparticles, hydrogels and microspheres [18–21]. The hydrophobically modified GC, which could be used as micellar carriers, has been extensively studied [22,23].

α-Tocopherol is a good solvent for many hydrophobic drugs because of its excellent lipophilic nature [24]. Once it is grafted on to the backbone of GC, it may serve as the hydrophobic segment and therefore provide sufficient capacity for poorly soluble drugs.

Inspired by the above investigations, in this study, an amphiphilic polymer α-tocopherol succinate modified glycol chitosan (TS-GC) was designed for PTX delivery. The TS-GC was prepared through amide formation. The preparation, characterization and self-assembling ability of TS-GC were studied. Furthermore, the physicochemical properties, hemolysis, in vitro cytotoxicity and in vivo antitumor activity of PTX-loaded micelles were evaluated deeply.

2. Results and Discussion

2.1. Synthesis and Characterization of TS-GC

In this study, the polymer TS-GC was synthesized via the coupling reaction between carboxyl group of TS and amine group of GC in the presence of water-soluble 1-Ethyl-3-(3-dimethylaminopropyl) carbodiimide hydrochloride (EDC). Firstly, an active ester intermediate was formed between the carboxyl group of TS and EDC. Then the amino bond can easily formed by reaction of primary amino of GC with the intermediate. NHS was used to stabilize the intermediate to achieve higher yield [25]. The scheme of the reaction between GC and TS was shown in Figure 1.

^{1}H-NMR spectra confirmed the grafting of TS chain onto GC as illustrated in Figure 2. In the spectrum of TS-GC, the proton peaks of TS, including methyl (0.77–0.88 ppm) and methylene (0.99–1.54 ppm) that belonged to the protons of the long-chain alkyl group of TS were observed. Moreover, the new-emerged signals at 2.45–2.63 ppm were attributed to the methene hydrogen ($-COCH_2CH_2-$) of the succinyl group of TS [26]. These spectra proved the formation of TS-GC.

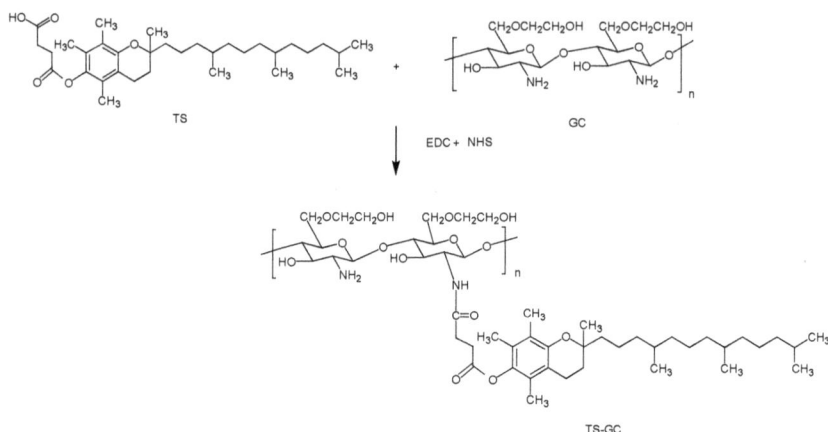

Figure 1. Synthesis of α-tocopherol succinate modified glycol chitosan (TS-GC).

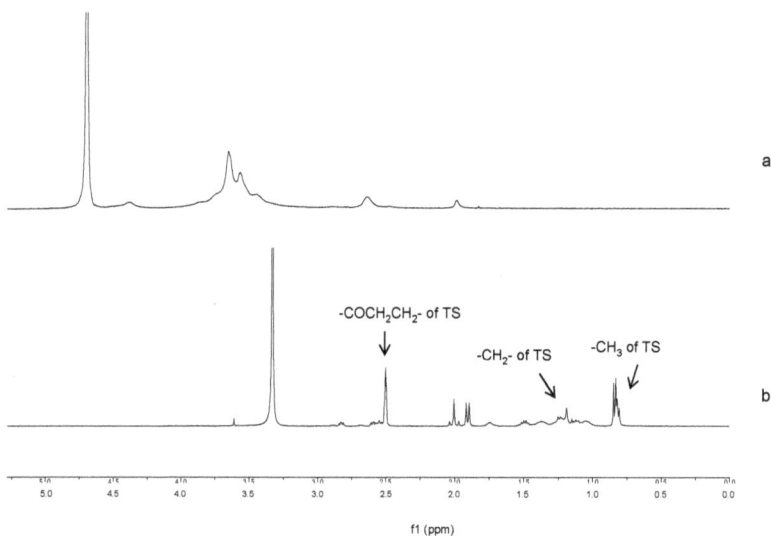

Figure 2. Proton nuclear magnetic resonance (^{1}H-NMR) spectra of (a) glycol chitosan (GC) and (b) TS-GC.

FT-IR was used to characterize the functional groups of GC before and after modification. The FT-IR spectra of GC (a), TS-GC (b), physical mixture of GC and TS (c) and TS (d) are shown in Figure 3. In curve a, GC showed characteristic signals at 3133 cm^{-1} (O–H stretch overlapped with N–H stretch), 1669 cm^{-1} (amide I band, C=O stretch of acetyl group), 1558 cm^{-1} (amide II band, N–H bending) and 1399 cm^{-1} (C–H bending), respectively. The spectrum of TS (curve d) showed characteristic peaks of C=O bond at 1754 cm^{-1} (the carbonyl of ester bond) and at 1715 cm^{-1} (carboxylic C=O). And the peak at 934 cm^{-1} was attributed to the out-of-plane bending vibrations of carboxylic C–OH. Compared with GC, the increased intensity of peaks at 1669 cm^{-1} and 1558 cm^{-1} in TS-GC (curve b) indicated the formation of amide bond. In addition, disappearance of carboxylic C=O signal at 1715 cm^{-1} and carboxylic C–OH signal at 934 cm^{-1} further confirmed full reaction of TS with GC. All the above indicated the successful introduction of TS.

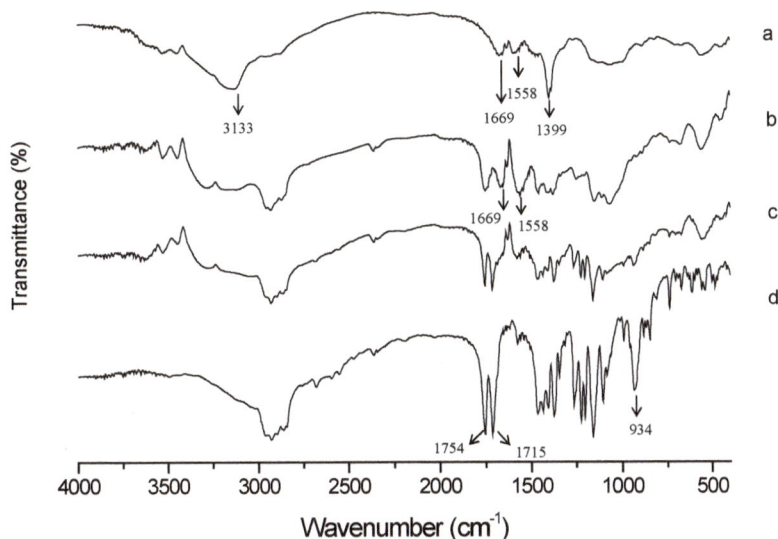

Figure 3. Fourier-transform infrared (FT-IR) spectra of (**a**) GC, (**b**) TS-GC, (**c**) physical mixture of GC and TS, and (**d**) TS.

The degree of amino substitution was calculated by measuring the amount of terminal amino groups of TS-GC with 2,4,6-trinitrobenzene sulphonic acid (TNBS reagent). The maximum absorbance of the yellow colour product was at 344 nm, and the absorbance was in proportion to the number of primary amino groups. From the calibration curve obtained with unmodified GC solution with different concentrations, the substitution degree of TS-GC in this experiment was calculated as 11.3%.

Critical aggregation concentration (CAC) plays an important role in maintaining the stability of micelles upon dilution. Only when the concentration of the polymer is higher than its CAC can micelles be formed. Polymeric micelles are generally more stable because of their markedly lower CAC. For the determination of the CAC of TS-GC, the pyrene fluorescence was employed to monitor the properties of TS-GC in solution. This method utilized pyrene's sensitivity to the local polarity of the environment. Below the CAC, the polymers only exist as single chains, and pyrene is solubilized in water. When the concentration increases to reach the critical value called CAC, polymer chains start to associate to form micelles, and pyrene partitions preferentially toward the hydrophobic cores of micelles. This leads to the increase of fluorescence intensity, and the intensity of the third peak increased significantly compared to that of the first peak. So the CAC values were calculated from curves of I_3/I_1 versus polymer concentration. The CAC was defined as the intercept of the tangents to the curve before and after the point of inflection [27]. For TS-GC, the CAC was calculated to be 7.2×10^{-3} mg/mL, which was significantly lower than that of the low molecular weight surfactants in water. This implied the TS-GC micelles could be stable and not easily dissociate upon dilution.

2.2. Preparation of TS-GC Micelles

In general, the aggregates of amphiphilic polymers can be prepared through diafiltration or sonication methods. As for diafiltration, the polymer was dissolved in co-solvent and then dialyzed against water. For TS-GC, it was not easily dissolved in the mixture of organic solvent with water, so the probe-sonication was employed. Micelle formation is a delicate balance between the attractive force that leads to the association of molecules and the repulsive force that prevents unlimited growth of the micelles [28]. It was reported that once the micelle structure was formed completely, the drugs were hardly to be incorporated into the micelles [29]. So, in this study, in order to get high encapsulation efficiency, the PTX loading occurred simultaneously with self-assembly of TS-GC under the probe-sonication treatment as stated in Section 3.5. The PTX encapsulation efficiency reached to 71.8%, and the drug loading capacity was calculated as 8.0%.

2.3. Characterization of PTX-Loaded TS-GC Micelles

2.3.1. XRD Analysis

To confirm the existence form of PTX in PTX-loaded TS-GC micelles, XRD analysis was conducted for PTX, blank micelles, their physical mixture and PTX-loaded micelles. As illustrated in Figure 4, typical intense diffraction peaks of PTX were still observed with weak intensity in the pattern obtained from the physical mixture of PTX and blank micelles, which reflected the presence of PTX crystal in the mixture. While the spectrum of lyophilized PTX-loaded micelles was similar to that of the blank micelles, and there were no diffraction peaks for PTX. It can be concluded that PTX was encapsulated in the polymeric micelles in molecular or amorphous state and there was no free drug on the surface of micelles.

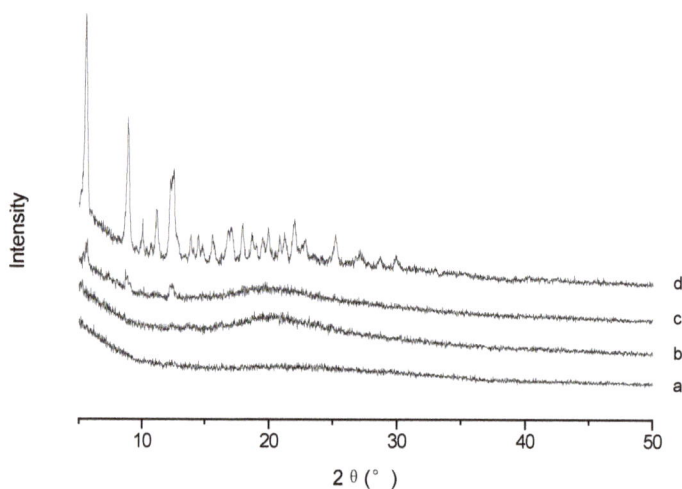

Figure 4. X-ray diffraction (XRD) spectra of (**a**) blank micelles; (**b**) paclitaxel (PTX)-loaded micelles; (**c**) physical mixture of PTX and blank micelles; and (**d**) PTX.

2.3.2. Particle Size and Zeta Potential

Size of the micelles is an important factor affecting the in vivo fate of the drug. In this study, the particle size and their distribution of the micelles were measured by dynamic light scattering method. From the result, the mean particle size of PTX-loaded micelles was 142 nm with polydispersity index (PDI) of 0.186, and it was larger than that of blank micelles (35 nm, PDI of 0.105), which indicated the encapsulation of PTX into the micelles. For both bare and PTX-loaded micelles, the size distribution was narrow. It was reasonably safe to assume an increasing accumulation of the drug in tumor tissue. Because it was reported that the small size of micelles (<200 nm) can reduce non-selective clearance by the reticuloendothelial system (RES) and show EPR effect for passive accumulation in certain tumor sites [30]. Moreover, the size-sieving may occur during the distribution process in the body, and the narrow size distribution may promote the selective accumulation at the target site. Transmission electron microscope (TEM) micrograph of PTX-loaded micelles is shown in Figure 5. It was obvious that the micelles were spherical in shape.

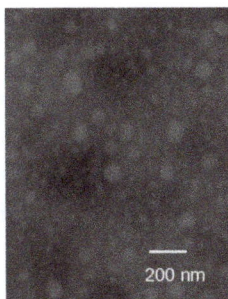

Figure 5. Transmission electron microscope (TEM) image of PTX-loaded micelles.

For a micellar solution, zeta potential can greatly influence the particle stability through the electrostatic repulsion. Relatively high surface charge could provide a repelling force between the particles, and thus increase the stability of the system. In this study, the blank micelles and PTX-loaded micelles in aqueous medium were positively charged with zeta potential of +19 mV and +16 mV, respectively, due to the presence of ionized amino groups of TS-GC distributing on the surface of micelles. It is reasonable to confirm the aqueous stability of the micelles.

2.4. In Vitro Hemolytic Test

As analogs of low molecular weight surfactants, amphiphilic polymers may solubilize lipids or be inserted into phospholipid membranes, and consequently lead to hemolysis of red blood cells following intravenous administration [31]. It is necessary to determine whether TS-GC induces hemolysis and is safe for intravenous injection. The hemolysis of PTX-loaded TS-GC micelles was compared with that of Cremophor EL-based formulation. It was observed that in the range of 10–200 μg/mL, the hemolysis of PTX-loaded TS-GC micelles was almost negligible, with only 4.2% at the concentration of 200 μg/mL, while the hemolysis induced by the Cremophor EL-based formulation increased from 0.04% to 10.9%. The results suggested that PTX-loaded micelles were not toxic to the erythrocytes.

2.5. In Vitro Cytotoxicity Study

The cytotoxicity study was conducted by 3-(4,5-dimethylthiazol-2-yl)-2,5-diphenyl tetrazolium bromide (MTT) method to assess the effectiveness of PTX-loaded TS-GC micelles. As shown in Figure 6, the cytotoxicity of PTX-loaded TS-GC micelles was similar to that of Cremophor EL-based formulation with equivalent doses in the concentration ranges used in this study. Furthermore, as the concentration and incubation time increased, PTX-loaded micelles and Cremophor EL-based formulation displayed increasing cytotoxicity. It was straightforward to understand that both concentration and incubation time played a major role in the in vitro cytotoxicity of PTX. For longer incubation periods, a larger number of cells enter the G2 and M cell cycle phases, during which PTX is more active. The findings suggested that PTX-loaded TS-GC micelles could be used as a potential PTX carrier.

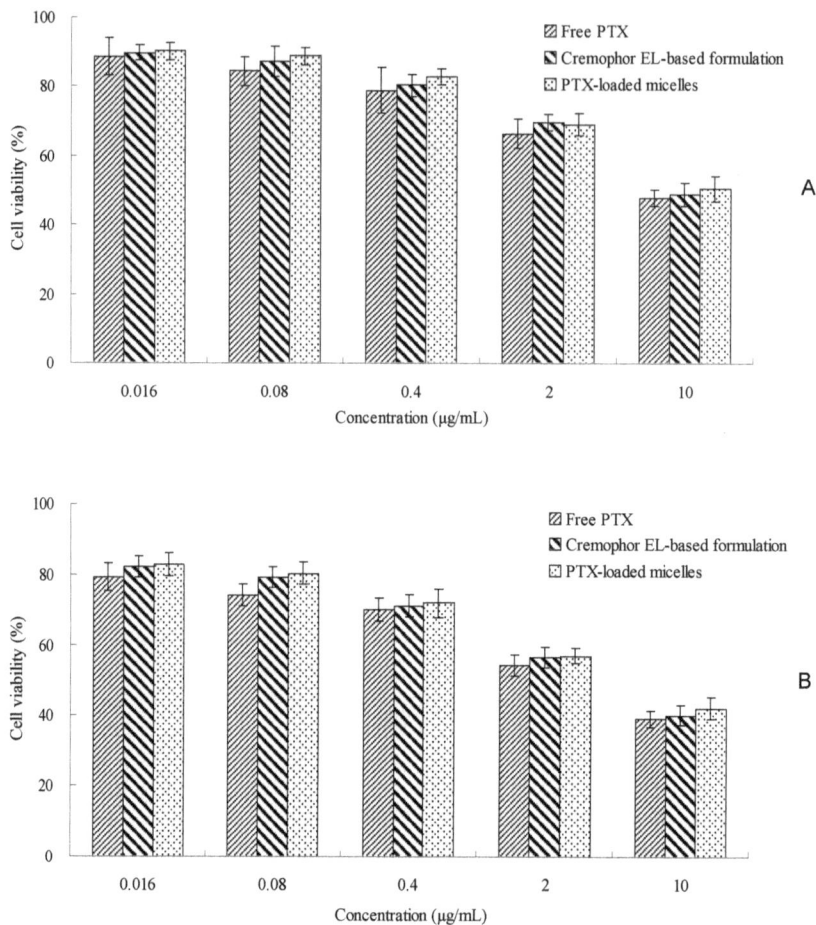

Figure 6. In vitro cytotoxicity of free PTX in dimethyl sulfoxide (DMSO), Cremophor EL-based formulation and PTX-loaded micelles against MCF-7 cells after 24 h (**A**) and 48 h (**B**) incubation (mean ± SD, n = 3).

2.6. In Vivo Antitumor Activity Study

The antitumor efficacy of PTX-loaded micelles in vivo was consistent well with the in vitro cell experiment. As shown in Figure 7, comparing with the saline control, a striking antitumor response was observed in all the treatment groups ($p < 0.05$). Furthermore, PTX-loaded micelles and Cremophor EL-based formulation had the similar antitumor efficacy at the same dose of 10 mg/kg, with tumor inhibition rate (TIR) of 72.3% and 69.6%, respectively, and the difference between the two groups was not statistically significant. The slight superior of the micelles might be explained by the increased concentration of PTX in tumor tissue due to the EPR effect, and the exact mechanism will be further studied.

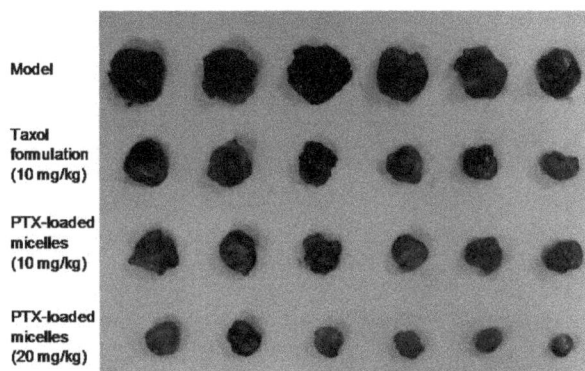

Figure 7. Photograph of tumors from each treatment group excised after intravenous injection treatment on day 13.

On the other hand, the toxic effect of PTX-loaded micelles was much less than that of the Cremophor EL-based formulation. At the high dose of 20 mg/mL, the intravenous administration of Cremophor EL-based formulation resulted in development of immediate ataxia and enhanced respiration, and 3 in 6 animals died although the injection speed was slowed down and the first aid treatment was given immediately after the symptoms appeared. In contrast, none of the mice treated with PTX-loaded micelles died throughout the study. This may be explained by the serious side effect of Cremophor EL. While for the micelles, all the components constructed TS-GC were nontoxic and biocompatible. It was obvious that the low toxicity of PTX-loaded micelles is of great value.

Taken together, these results implied that PTX-loaded TS-GC micelles not only had similar to better antitumor efficacy as Cremophor EL-based formulation, but also induced less systemic toxicity, especially at the higher dose. The PTX-loaded micelles might allow the administration of a higher dose of PTX.

3. Materials and Methods

3.1. Materials

Glycol chitosan (Mw 250 kDa, deacetylation degree of 82.7%), 2,4,6-trinitrobenzene sulfonic acid (TNBS), pyrene (purity > 99%) and 3-(4,5-dimethylthiazol-2-yl)-2,5-diphenyl tetrazolium bromide (MTT) were purchased from Sigma, St. Louis, MO, USA. 1-Ethyl-3-(3-dimethylaminopropyl) carbodiimide hydrochloride (EDC) and *N*-hydroxysuccinimide (NHS) were obtained from Shanghai Medpep Co., Ltd., Shanghai, China. α-Tocopherol succinate (TS) was kindly donated by Xinchang Pharmaceutical Co., Ltd., Shaoxing, China. Paclitaxel (PTX, purity of 99.9%) was supplied by Tianfeng Bioengineering Technology Co., Ltd., Shenyang, China. Cremophor EL was a kind gift from BASF Corp., Ludwigshafen, Germany. Dulbecco's modified Eagle's medium (DMEM), fetal bovine serum (FBS), and penicillin–streptomycin mixture were purchased from Gibco BRL, Carlsbad, CA, USA. All other chemicals and solvents were of analytical or chromatographic grade and used without further purification. Distilled water or Milli-Q water was used in all experiments.

3.2. Animals and Cell Line

The male New Zealand rabbit (weighing 2 kg) and specific pathogen-free female Kunming mice (5–6 weeks old, weighing 20–25 g) were acquired from Laboratory Animal Center of Harbin Medical University, Harbin, China. MCF-7 cells (human breast cancer cells) were obtained from American Type Culture Collection, Manassas, VA, USA. U14 cells (mouse uterine cervix carcinoma cells) were provided by the Cell Resource Center of Chinese Academy of Medical Sciences, Beijing, China. All animal procedures were performed in compliance with the animal care protocols approved by the Animal Ethics Committee of Harbin Medical University (18 May 2017, No. 201705180118).

3.3. Synthesis of α-Tocopherol Succinate Modified Glycol Chitosan (TS-GC)

α-Tocopherol succinate modified glycol chitosan (TS-GC) was synthesized by the coupling reaction between carboxyl group of TS with amine group of GC in the presence of EDC [32]. Briefly, GC (100 mg) was dissolved in 10 mL of distilled water and TS (64 mg) was dissolved in 10 mL of methanol. The TS solution was slowly added to the GC solution and followed by the addition of excessive EDC and NHS under gentle stirring at room temperature. After overnight reaction, the mixture was dialyzed against distilled water using a cellulose membrane (Viskase, Willowbrook, IL, USA, MWCO: 7000), and TS-GC was obtained by freeze-drying of the dialyzed solution.

3.4. Characterization of TS-GC

To confirm the formation of TS-GC, proton nuclear magnetic resonance (^{1}H-NMR) analysis was operated at 300 MHz using Bruker Avance spectrometer (AV-300, Bruker, Karlsruhe, Germany). GC and the conjugate were dissolved in D_2O and DMSO-d6 at the concentration of 1% (w/v), respectively.

In order to further investigate the structural changes during the synthesis process, Fourier-transform infrared (FT-IR) spectra of GC, TS, their physical mixture and the conjugate were obtained on FT-IR spectrometer (Tensor II, Bruker, Fällanden, Switzerland) in the range between 4000 and 400 cm^{-1} after compressed into KBr pellets.

The degree of substitution means the number of tocopherol groups per 100 anhydroglucose units (amino groups) of TS-GC. The amount of remaining terminal amino residues on TS-GC was measured using 2,4,6-trinitrobenzene sulfonic acid (TNBS reagent) according to TNBS method [33,34]. The numerical value was calculated from the calibration curve that obtained by GC solution.

The critical aggregation concentration (CAC) of TS-GC in aqueous milieu was measured using a fluorometer (F-2500 FL Spectrophotometer, Hitachi Ltd., Tokyo, Japan). The pyrene fluorescence was monitored to evaluate the micropolarity and hydrophobicity of the region in which it was solubilized, so as to prove the formation of the micelles [35].

3.5. Preparation of PTX-Loaded TS-GC Micelles

PTX-loaded TS-GC micelles were prepared by a probe-type ultrasonic method. Briefly, 10 mg of TS-GC was swollen in 10 mL of distilled water under gentle stirring overnight at room temperature to ensure complete dispersion. PTX solution was prepared by dissolving PTX in methanol at the concentration of 1 mg/mL. The PTX solution was added into the TS-GC solution, and the final mixture was sonicated for 10 min at 400 W in an ice bath by a ultrasonicator (JY92-II, Ningbo Scientz Biotechnology Co., Ltd., Ningbo, China). The sonication was carried out with the pulse function (turned on for 3 s and off for 2 s). The unloaded PTX was removed by centrifugation at 4000 rpm for 10 min. The resulting supernatant was lyophilized to obtain PTX-loaded TS-GC micelles. The blank micelles were prepared by the same procedure except no PTX was added.

3.6. Characterization of PTX-Loaded TS-GC Micelles

3.6.1. X-ray Diffraction (XRD) Analysis

X-ray diffraction diagrams were detected using an X-ray diffractometer (Geigerflex, Rigaku Co., Akishima, Japan) with Cu Kα radiation. Samples were scanned from 5 to 50° (2θ) at a scanning speed of 2°/min and step size of 0.02°. The X-ray system was operated at a potential of 30 kV and current of 30 mA.

3.6.2. Transmission Electron Microscopy (TEM) Observation

Morphology of the micelles was observed using a transmission electron microscope (TEM) (Jeol JEM1200EX, Tokyo, Japan) that operated at an accelerating voltage of 60 kV. For TEM, an aqueous droplet of micelles was immobilized on copper grids and negatively stained with phosphotungstate solution (2%, w/v), then dried at room temperature before observation.

3.6.3. Measurement of Particle Size and Zeta Potential

The number-weighted diameter, their distribution, and zeta potential of the micelles were measured using photon correlation spectroscopy with a Zetasizer Nano-ZS90 at 25 °C. The lyophilized samples were suspended in distilled water before measurement.

3.6.4. Determination of Drug Loading and Drug Encapsulation Efficiency

The drug loading and entrapment efficiency of the PTX-loaded micelles were determined as follows: 100 µL of PTX-loaded TS-GC micelles solution (concentration of TS-GC = 1.0 mg/mL) was centrifuged at 10,000 rpm for 10 min with ultrafilter (Vivaspin 500, MWCO 10 k, Sartorius Co., Göttingen, Germany). The unentrapped PTX amount in the ultrafiltrate (W_1) was determined using HPLC method. The HPLC system consisted of a mobile phase delivery pump (LC-10ATVP HPLC pump, Shimadzu, Kyoto, Japan) and a UV detector (SPD-10A UV/Vis detector, Shimadzu, Japan). For separation, a Diamonsil^{TM} C_{18} reverse-phase column (200 × 4.6 mm, 5 µm, Dikma technologies Inc., Beijing, China) was used. The mobile phase consisted of a mixture of acetonitrile and water (60:40, v/v). The detector wavelength, column temperature, and flow rate of the mobile phase were set at 227 nm, 30 °C and 1.0 mL/min, respectively. The injection volume of the test samples was 20 µL.

In order to measure the total PTX content (W_0) in the micelles, another 100 µL of identical PTX-loaded micelles solution was ultrasonicated in 10 mL methanol to extract PTX from the micelles. The encapsulation efficiency (EE%) and drug loading (DL%) of PTX were calculated using the following Equations (1) and (2):

$$EE\% = (W_0 - W_1)/W_0 \times 100\% \tag{1}$$

$$DL\% = (W_0 - W_1)/(W_0 - W_1 + 100) \times 100\% \tag{2}$$

where W_1 is the PTX content in the ultrafiltrate, and W_0 is the total PTX amount in the solution. The unit was microgram.

3.7. In Vitro Hemolysis Test

Hemolysis assessment of the TS-GC micelles was conducted as the method reported by Gong et al. [36]. The rabbit blood, freshly drawn from the ear vein, was centrifuged at 3000 rpm for 10 min to isolate the erythrocytes. The erythrocytes were washed with normal saline for three times to make the supernatant achromatic, and then suspended in normal saline to get a 2% (v/v) suspension. The lyophilized powder of PTX-loaded TS-GC micelles was dispersed in 0.9% NaCl, and different amounts of micelle solution were added into the tubes with 2.5 mL of 2% erythrocyte dispersion in each. Then adequate amounts of normal saline were added in every tube to obtain a final volume of 5 mL. After incubating at 37 °C for 4 h, the mixture was centrifuged at 3000 rpm for 10 min to remove intact red blood cells (RBC). The supernatant was analyzed for released hemoglobin at 540 nm using a spectrophotometer (UV-Vis Spectrophotometer Model 752, Shanghai Spectrum Instruments Co., Ltd., Shanghai, China). To obtain 0 and 100% hemolysis, 2.5 mL of saline and 2.5 mL of distilled water was added to 2.5 mL of RBC suspension, respectively. The degree of hemolysis was calculated by the following Equation (3).

$$Hemolysis\,(\%) = (A_{sample} - A_{0\%})/(A_{100\%} - A_{0\%}) \times 100\% \tag{3}$$

where A_{sample}, $A_{0\%}$, and $A_{100\%}$ are the absorbance of the samples, a solution of 0% hemolysis, and a solution of 100% hemolysis, respectively.

3.8. In Vitro Cytotoxicity

The in vitro cytotoxic activity of samples was evaluated by MTT method using MCF-7 cell line [37]. Briefly, 50 μL of MCF-7 cells growing in the logarithmic phase were cultured in Dulbecco's modified Eagle's medium (DMEM) containing 10% (*v*/*v*) fetal bovine serum, 100 IU/mL of penicillin G sodium and 100 μg/mL of streptomycin sulfate. The cells were seeded in a 96-well microtitre plate at the density of 1×10^4 cells per well and maintained in an incubator supplied with 5% CO_2 at 37 °C. After reaching 75% confluence, the cells were incubated with free drug in dimethyl sulfoxide (DMSO), Cremophor EL-based PTX formulation and PTX-loaded TS-GC micelles at the equivalent drug concentrations ranging from 0.016 to 10 μg/mL for 24 and 48 h. At designated time intervals, the medium was removed and the wells were washed with PBS for two times. Then, 10 μL of MTT (5 mg/mL in PBS) was added to the wells. After an additional 4 h of incubation, the MTT medium was aspirated off and 100 μL of DMSO was added to each well to dissolve the formazan crystals. The absorbance was measured at 570 nm with a microplate reader (Bio-Tek Instruments Inc., Vernusky, VT, USA). Untreated cells were taken as the control with 100% viability, and cells without addition of MTT were used as blank to calibrate the spectrophotometer to zero absorbance. The cytotoxicity was calculated as follows Equation (4):

$$\text{Cytotoxicity} = (A_{\text{culture medium}} - A_{\text{sample}})/A_{\text{culture medium}} \times 100\% \qquad (4)$$

where $A_{\text{culture medium}}$ and A_{sample} are the absorbance of cells incubated with culture medium and absorbance of cells exposed to the sample, respectively.

3.9. In Vivo Antitumor Activity

Specific pathogen-free Kunming mice, 5 to 6 weeks old, weighing 20–25 g were used for this study. The animals were housed six per cage in standard size cages under standard laboratory conditions (21 ± 2 °C, 12-h light: 12-h dark cycle, relative humidity of 50–60%) and allowed to access sterilized food and water freely. The mice were left to acclimatize for a week prior to the experiment.

U14 cells of the third passage in vivo were used for tumor development. Animals were inoculated with 2.0×10^6 cells (0.2 mL/mouse) subcutaneously in the armpit of right anterior limb. Three days later, the tumors were palpable, and the tumor model was established. The animals were randomized and divided into different groups (*n* = 6): (1) normal saline group (negative control); (2) Cremophor EL-based PTX formulation groups (positive control, 10 mg/kg and 20 mg/kg); and (3) PTX-loaded TS-GC micelles groups (10 mg/kg and 20 mg/kg), and the treatments were initiated. All samples were injected intravenously via the tail vein every 3 days, four times in total. The day that mice received treatment was set as day 1. At the end of the experiment, on day 13, the animals were sacrificed by cervical dislocation, and the tumors were excised, weighed and imaged. The tumor inhibition rate (TIR) of each formulation was defined as follows Equation (5):

$$\text{TIR} = (\text{tumor weight of negative control group - tumor weight of treatment group})/\text{tumorweightofnegativecontrolgroup} \times 100\% \qquad (5)$$

3.10. Statistical Analysis

Each experiment was performed in triplicate. Values were expressed as mean±standard deviation (SD). Statistical data analysis was performed using the Student's *t*-test with $p < 0.05$ as the level of significance.

4. Conclusions

In this study, a novel amphiphilic derivative of glycol chitosan was successfully synthesized by grafting α-tocopherol succinate onto the skeleton of glycol chitosan. In aqueous milieu, the conjugates provide stable self-aggregates above the CAC. Furthermore, the water-insoluble anticancer agent, PTX, was successfully encapsulated into the core of the micelles. The mean diameters of PTX-loaded micelles were about 142 nm, which was larger than blank ones. The spherical morphology of the micelles was visually confirmed by TEM. Moreover, the PTX-loaded TS-GC micelles possessed antitumor activities in vitro and in vivo, and had advantages over the commercially available Cremophor EL-based formulation in terms of low toxicity levels and increased tolerated dose. Therefore, this novel TS-GC polymer might be used as a potential carrier for PTX.

Author Contributions: N.L. and S.S. conceived and designed the experiments. N.L. performed the experiments, analyzed the data and wrote the paper; X.G. and Q.L. performed studies of in vitro cytotoxicity. F.C. and P.Y. gave us much useful advice and some pieces of guidance.

Funding: This work was funded by the National Natural Science Foundation of China (No. 51403057), Harbin Science and Technology Innovation Talents Special Fund Project (No. 2016RQQXJ097, No. 2016RQQXJ131), and the Doctoral Scientific Research Startup Foundation of Harbin Normal University (No. XKB201304).

Conflicts of Interest: The authors declare no conflict of interest.

References

1. Abou-ElNaga, A.; Mutawa, G.; El-Sherbiny, I.; Abd-ElGhaffar, H.; Allam, A.; Ajarem, J.; Mousa, S. Novel nano-therapeutic approach actively targets human ovarian cancer stem cells after xenograft into nude mice. *Int. J. Mol. Sci.* **2017**, *18*, 813. [CrossRef] [PubMed]
2. Bernabeu, E.; Cagel, M.; Lagomarsino, E.; Moretton, M.; Chiappetta, D.A. Paclitaxel: What has been done and the challenges remain ahead. *Int. J. Pharm.* **2017**, *526*, 474–495. [CrossRef] [PubMed]
3. Gelderblom, H.; Verweij, J.; Nooter, K.; Sparreboom, A. Cremophor EL: The drawbacks and advantages of vehicle selection for drug formulation. *Eur. J. Cancer* **2001**, *37*, 1590–1598. [CrossRef]
4. Liu, Y.; Zhang, B.; Yan, B. Enabling anticancer therapeutics by nanoparticle carriers: The delivery of paclitaxel. *Int. J. Mol. Sci.* **2011**, *12*, 4395–4413. [CrossRef] [PubMed]
5. Ravar, F.; Saadat, E.; Gholami, M.; Dehghankelishadi, P.; Mahdavi, M.; Azami, S.; Dorkoosh, F.A. Hyaluronic acid-coated liposomes for targeted delivery of paclitaxel, in-vitro characterization and in-vivo evaluation. *J. Control. Release* **2016**, *229*, 10–22. [CrossRef] [PubMed]
6. Hou, J.; Sun, E.; Zhang, Z.H.; Wang, J.; Yang, L.; Cui, L.; Ke, Z.C.; Tan, X.B.; Jia, X.B.; Lv, H.X. Improved oral absorption and anti-lung cancer activity of paclitaxel-loaded mixed micelles. *Drug Deliv.* **2017**, *24*, 261–269. [CrossRef] [PubMed]
7. Alani, A.W.G.; Bae, Y.; Rao, D.A.; Kwon, G.S. Polymeric micelles for the pH-dependent controlled, continuous low dose release of paclitaxel. *Biomaterials* **2010**, *31*, 1765–1772. [CrossRef] [PubMed]
8. Erdoğar, N.; Esendağlı, G.; Nielsen, T.T.; Esendağlı-Yılmaz, G.; Yöyen-Ermiş, D.; Erdoğdu, B.; Sargon, M.F.; Eroğlu, H.; Bilensoy, E. Therapeutic efficacy of folate receptor-targeted amphiphilic cyclodextrin nanoparticles as a novel vehicle for paclitaxel delivery in breast cancer. *J. Drug Target* **2018**, *26*, 66–74. [CrossRef] [PubMed]
9. Nakamura, I.; Ichimura, E.; Goda, R.; Hayashi, H.; Mashiba, H.; Nagai, D.; Yokoyama, H.; Onda, T.; Masuda, A. An in vivo mechanism for the reduced peripheral neurotoxicity of NK105: A paclitaxel-incorporating polymeric micellar nanoparticle formulation. *Int. J. Nanomed.* **2017**, *12*, 1293–1304. [CrossRef] [PubMed]
10. Zhang, T.; Luo, J.; Fu, Y.; Li, H.; Ding, R.; Gong, T.; Zhang, Z. Novel oral administrated paclitaxel micelles with enhanced bioavailability and antitumor efficacy for resistant breast cancer. *Colloids Surf. B Biointerfaces* **2017**, *150*, 89–97. [CrossRef] [PubMed]
11. Torchilin, V.P. Micellar nanocarriers: Pharmaceutical perspectives. *Pharm. Res.* **2006**, *24*, 1–16. [CrossRef] [PubMed]

12. Deshmukh, A.S.; Chauhan, P.N.; Noolvi, M.N.; Chaturvedi, K.; Ganguly, K.; Shukla, S.S.; Nadagouda, M.N.; Aminabhavi, T.M. Polymeric micelles: Basic research to clinical practice. *Int. J. Pharm.* **2017**, *532*, 249–268. [CrossRef] [PubMed]

13. Rapoport, N. Physical stimuli-responsive polymeric micelles for anti-cancer drug delivery. *Prog. Polym. Sci.* **2007**, *32*, 962–990. [CrossRef]

14. Biswas, S.; Kumari, P.; Lakhani, P.M.; Ghosh, B. Recent advances in polymeric micelles for anti-cancer drug delivery. *Eur. J. Pharm. Sci.* **2016**, *83*, 184–202. [CrossRef] [PubMed]

15. Kwon, G.S.; Okano, T. Polymeric micelles as new drug carriers. *Adv. Drug Deliv. Rev.* **1996**, *21*, 107–116. [CrossRef]

16. Yang, Y.; Wang, S.; Wang, Y.; Wang, X.; Wang, Q.; Chen, M. Advances in self-assembled chitosan nanomaterials for drug delivery. *Biotechnol. Adv.* **2014**, *32*, 1301–1316. [CrossRef] [PubMed]

17. Prabaharan, M. Chitosan-based nanoparticles for tumor-targeted drug delivery. *Int. J. Biol. Macromol.* **2015**, *72*, 1313–1322. [CrossRef] [PubMed]

18. Fu, Y.-N.; Li, Y.; Li, G.; Yang, L.; Yuan, Q.; Tao, L.; Wang, X. Adaptive chitosan hollow microspheres as efficient drug carrier. *Biomacromolecules* **2017**, *18*, 2195–2204. [CrossRef] [PubMed]

19. Yu, J.; Liu, Y.; Zhang, L.; Zhao, J.; Ren, J.; Zhang, L.; Jin, Y. Self-aggregated nanoparticles of linoleic acid-modified glycol chitosan conjugate as delivery vehicles for paclitaxel: Preparation, characterization and evaluation. *J. Biomater. Sci. Polym. Ed.* **2015**, *26*, 1475–1489. [CrossRef] [PubMed]

20. Sharma, S.; Verma, A.; Teja, B.V.; Shukla, P.; Mishra, P.R. Development of stabilized paclitaxel nanocrystals: In-vitro and in-vivo efficacy studies. *Eur. J. Pharm. Sci.* **2015**, *69*, 51–60. [CrossRef] [PubMed]

21. Zhao, L.; Zhu, L.; Liu, F.; Liu, C.; Shan, D.; Wang, Q.; Zhang, C.; Li, J.; Liu, J.; Qu, X.; et al. pH triggered injectable amphiphilic hydrogel containing doxorubicin and paclitaxel. *Int. J. Pharm.* **2011**, *410*, 83–91. [CrossRef] [PubMed]

22. Park, J.S.; Han, T.H.; Lee, K.Y.; Han, S.S.; Hwang, J.J.; Moon, D.H.; Kim, S.Y.; Cho, Y.W. N-acetyl histidine-conjugated glycol chitosan self-assembled nanoparticles for intracytoplasmic delivery of drugs: Endocytosis, exocytosis and drug release. *J. Control. Release* **2006**, *115*, 37–45. [CrossRef] [PubMed]

23. Huo, M.; Fu, Y.; Liu, Y.; Chen, Q.; Mu, Y.; Zhou, J.; Li, L.; Xu, W.; Yin, T. N-mercapto acetyl-N'-octyl-O,N''-glycol chitosan as an efficiency oral delivery system of paclitaxel. *Carbohyd. Polym.* **2018**, *181*, 477–488. [CrossRef] [PubMed]

24. Nielsen, P.B.; Müllertz, A.; Norling, T.; Kristensen, H.G. The effect of α-tocopherol on the in vitro solubilisation of lipophilic drugs. *Int. J. Pharm.* **2001**, *222*, 217–224. [CrossRef]

25. Park, C.; Vo, C.L.-N.; Kang, T.; Oh, E.; Lee, B.-J. New method and characterization of self-assembled gelatin–oleic nanoparticles using a desolvation method via carbodiimide/N-hydroxysuccinimide (EDC/NHS) reaction. *Eur. J. Pharm. Biopharm.* **2015**, *89*, 365–373. [CrossRef] [PubMed]

26. Xu, X.; Li, L.; Zhou, J.; Lu, S.; Yang, J.; Yin, X.; Ren, J. Preparation and characterization of N-succinyl-N'-octyl chitosan micelles as doxorubicin carriers for effective anti-tumor activity. *Colloids Surf. B Biointerfaces* **2007**, *55*, 222–228.

27. Băran, A.; Stîngă, G.; Anghel, D.F.; Iovescu, A.; Tudose, M. Comparing the spectral properties of pyrene as free molecule, label and derivative in some colloidal systems. *Sens. Actuators B Chem.* **2014**, *197*, 193–199. [CrossRef]

28. Astafieva, I.; Zhong, X.F.; Eisenberg, A. Critical micellization phenomena in block polyelectrolyte solutions. *Macromolecules* **1993**, *26*, 7339–7352. [CrossRef]

29. Zhang, C.; Ping, Q.; Zhang, H. Self-assembly and characterization of paclitaxel-loaded N-octyl-O-sulfate chitosan micellar system. *Colloids Surf. B Biointerfaces* **2004**, *39*, 69–75. [CrossRef] [PubMed]

30. Danhier, F.; Danhier, P.; De Saedeleer, C.J.; Fruytier, A.-C.; Schleich, N.; Rieux, A.d.; Sonveaux, P.; Gallez, B.; Préat, V. Paclitaxel-loaded micelles enhance transvascular permeability and retention of nanomedicines in tumors. *Int. J. Pharm.* **2015**, *479*, 399–407. [CrossRef] [PubMed]

31. Huo, M.; Zhang, Y.; Zhou, J.; Zou, A.; Yu, D.; Wu, Y.; Li, J.; Li, H. Synthesis and characterization of low-toxic amphiphilic chitosan derivatives and their application as micelle carrier for antitumor drug. *Int. J. Pharm.* **2010**, *394*, 162–173. [CrossRef] [PubMed]

32. Liang, N.; Sun, S.; Li, X.; Piao, H.; Piao, H.; Cui, F.; Fang, L. α-Tocopherol succinate-modified chitosan as a micellar delivery system for paclitaxel: Preparation, characterization and in vitro/in vivo evaluations. *Int. J. Pharm.* **2012**, *423*, 480–488. [CrossRef] [PubMed]

33. Bernkop-Schnürch, A.; Krajicek, M.E. Mucoadhesive polymers as platforms for peroral peptide delivery and absorption: Synthesis and evaluation of different chitosan-EDTA conjugates. *J. Control. Release* **1998**, *50*, 215–223. [CrossRef]
34. Dosio, F.; Brusa, P.; Crosasso, P.; Arpicco, S.; Cattel, L. Preparation, characterization and properties in vitro and in vivo of a paclitaxel–albumin conjugate. *J. Control. Release* **1997**, *47*, 293–304. [CrossRef]
35. Yan, M.; Li, B.; Zhao, X. Determination of critical aggregation concentration and aggregation number of acid-soluble collagen from walleye pollock (*Theragra chalcogramma*) skin using the fluorescence probe pyrene. *Food Chem.* **2010**, *122*, 1333–1337. [CrossRef]
36. Gong, J.; Huo, M.; Zhou, J.; Zhang, Y.; Peng, X.; Yu, D.; Zhang, H.; Li, J. Synthesis, characterization, drug-loading capacity and safety of novel octyl modified serum albumin micelles. *Int. J. Pharm.* **2009**, *376*, 161–168. [CrossRef] [PubMed]
37. Laskar, P.; Samanta, S.; Ghosh, S.K.; Dey, J. In vitro evaluation of pH-sensitive cholesterol-containing stable polymeric micelles for delivery of camptothecin. *J. Colloid Interface Sci.* **2014**, *430*, 305–314. [CrossRef] [PubMed]

© 2018 by the authors. Licensee MDPI, Basel, Switzerland. This article is an open access article distributed under the terms and conditions of the Creative Commons Attribution (CC BY) license (http://creativecommons.org/licenses/by/4.0/).

International Journal of
Molecular Sciences

MDPI

Article

Amphiphilic Polymeric Micelles Based on Deoxycholic Acid and Folic Acid Modified Chitosan for the Delivery of Paclitaxel

Liang Li [1], Na Liang [2], Danfeng Wang [1], Pengfei Yan [1], Yoshiaki Kawashima [3], Fude Cui [4] and Shaoping Sun [1,*]

1   Key Laboratory of Chemical Engineering Process & Technology for High-efficiency Conversion, College of Heilongjiang Province, School of Chemistry and Material Science, Heilongjiang University, Harbin 150080, China; lliang1991001@163.com (L.L.); dfwang626@163.com (D.W.); yanpf@vip.sina.com (P.Y.)
2   Key Laboratory of Photochemical Biomaterials and Energy Storage Materials, Heilongjiang Province, College of Chemistry & Chemical Engineering, Harbin Normal University, Harbin 150025, China; liangna528@163.com
3   Department of Pharmaceutical Engineering, School of Pharmacy, Aichi Gakuin University, Nagoya 464-8650, Japan; sykawa123@163.com
4   School of Pharmacy, Shenyang Pharmaceutical University, Shenyang 110016, China; syphucuifude@163.com
*   Correspondence: sunshaoping@hlju.edu.cn; Tel.: +86-451-8660-8616

Received: 14 September 2018; Accepted: 10 October 2018; Published: 12 October 2018

Abstract: The present investigation aimed to develop a tumor-targeting drug delivery system for paclitaxel (PTX). The hydrophobic deoxycholic acid (DA) and active targeting ligand folic acid (FA) were used to modify water-soluble chitosan (CS). As an amphiphilic polymer, the conjugate FA-CS-DA was synthesized and characterized by Proton nuclear magnetic resonance (^{1}H-NMR) and Fourier-transform infrared spectroscopy (FTIR) analysis. The degree of substitutions of DA and FA were calculated as 15.8% and 8.0%, respectively. In aqueous medium, the conjugate could self-assemble into micelles with the critical micelle concentration of 6.6×10^{-3} mg/mL. Under a transmission electron microscope (TEM), the PTX-loaded micelles exhibited a spherical shape. The particle size determined by dynamic light scattering was 126 nm, and the zeta potential was +19.3 mV. The drug loading efficiency and entrapment efficiency were 9.1% and 81.2%, respectively. X-Ray Diffraction (XRD) analysis showed that the PTX was encapsulated in the micelles in a molecular or amorphous state. In vitro and in vivo antitumor evaluations demonstrated the excellent antitumor activity of PTX-loaded micelles. It was suggested that FA-CS-DA was a safe and effective carrier for the intravenous delivery of paclitaxel.

Keywords: chitosan; deoxycholic acid; folic acid; amphiphilic polymer; micelles; paclitaxel

1. Introduction

Paclitaxel (PTX) is an important clinical chemotherapeutic drug that exhibits strong antitumour activity against a variety of cancer types. However, the low solubility of PTX due to its bulky polycyclic structure hampers its clinical application [1]. Many attempts have been made to find less toxic and better-tolerated carriers to increase the solubility of PTX for intravenous delivery, such as nanoparticles, dendrimers, liposomes and nanosuspensions [2–5]. In recent years, polymeric micelles have attracted growing interest due to their attractive characteristics, such as their excellent solubilization ability, small size, high stability, prolonged circulation time, low toxicity, ability to evade scavenging by the mononuclear phagocyte system (MPS), high biocompatibility and efficient accumulation in tumor tissues via an enhanced permeability and retention (EPR) effect [6,7]. The micelles have a unique core–shell structure with hydrophobic segments as the internal core and hydrophilic segments as the outer shell. The internal core provides a storeroom for poorly water-soluble drugs, and the outer shell allows the retention of the stability of micelles in aqueous medium and provides the opportunity to target the delivery of antitumor drugs to the tumor by further modification [8,9].

To date, numerous amphiphilic block or graft copolymers have been synthesized and applied as micellar drug delivery systems [10]. Among them, chitosan (CS) has been extensively studied for its biocompatibility, non-toxicity and biodegradability [11,12]. Moreover, the abundant active amine and hydroxyl groups in CS could offer many opportunities for chemical modification. In recent years, several chitosan-based PTX delivery systems have been developed, such as *N*-mercapto acetyl-*N'*-octyl-*O*, *N''*-glycol chitosan micelles [13], 3,6-*O*,*O'*-dimyristoyl chitosan micelles [14], folic acid–cholesterol–chitosan micelles [15], PTX conjugated trimethyl chitosan nanoparticles [16], palmitoyl chitosan nanoparticles [17], *N*-succinyl-chitosan nanoparticles [18] and PTX-loaded chitosan nanoparticles prepared by the nano-emulsion method [19]. However, chitosan with high molecular weight has poor solubility in aqueous medium at neutral pH, which limits its medical and pharmaceutical applications. In contrast, water-soluble chitosan with a low molecular weight and high degree of deacetylation is a superior candidate for amphiphilic copolymer synthesis [20]. In the present study, the water-soluble chitosan was used as the hydrophilic part of the copolymer to form a micellar system for PTX delivery.

Deoxycholic acid (DA) is a typical bile acid that is secreted from the gallbladder to emulsify fats and other hydrophobic compounds [21]. As an endogenous compound with a lipophilic nature, the introduction of DA to CS could adjust the hydrophilicity/hydrophobicity balance of the conjugate and would not lead to any serious toxicity [22]. DA has been approved as an excellent pharmaceutical additive for injection [23].

Molecular ligands were often grafted onto drug carriers to develop tumor-targeted drug delivery systems. It has been reported that folate receptors are over-expressed in many types of cancers, while almost undetectable in healthy tissues [24]. The folic acid-modified nanocarriers could improve therapeutic efficacy via folate receptor-mediated active targeting. The antitumor efficiency could be significantly enhanced by synergetic active and passive tumor targeting [25].

Based on the above, in present study, a biocompatible nanocarrier based on deoxycholic acid and folic acid-modified chitosan (FA-CS-DA) was designed for targeting the delivery of PTX. The synthesis, characterization and self-assembly of FA-CS-DA and the characterization and in vitro/in vivo antitumor activity of PTX-loaded micelles were studied in detail.

2. Results and Discussion

2.1. Preparation of FA-CS-DA

The synthesis of FA-CS-DA was performed via the amide bond formation between the amino groups of CS and the carboxyl groups of DA and FA. As shown in Figure 1, the FA-CS-DA was synthesized by a two-step reaction. First, the intermediate CS-DA was prepared by the conjugation of carboxylic groups of DA with the primary amino groups of CS. Then, an FA molecule was introduced by attaching the carboxyl groups of FA to the remaining terminal amino of CS-DA. EDC and NHS were used in both reactions to active the carboxyl groups [26]. The unreacted DA and FA, as well as any by-product, were removed by dialysis. The degree of substitutions (DS) of DA and FA were calculated as 15.8% and 8.0%, respectively.

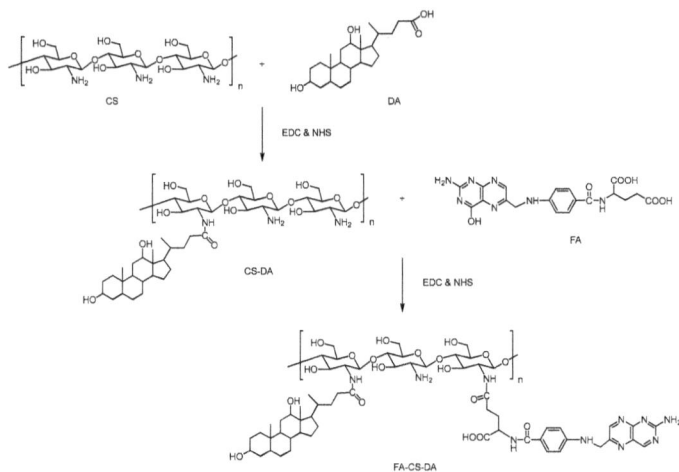

Figure 1. Scheme of the synthesis of FA-CS-DA (folic acid–chitosan–deoxycholic acid).

2.2. Characterization of FA-CS-DA

2.2.1. Proton Nuclear Magnetic Resonance (^{1}H-NMR) Characterization

To confirm the conjugate formation, the ^{1}H-NMR spectra of CS, CS-DA and FA-CS-DA are shown in Figure 2. Compared with CS, new peaks appeared in the range of 0.5–2.5 ppm in the spectrum of CS-DA and were assigned to the –CH$_3$ and –CH$_2$– protons of DA, which indicated the successful introduction of DA. Furthermore, FA-CS-DA showed characteristic signals attributed to the protons of FA at 7.20, 7.66 and 8.55 ppm. More specifically, the signal at 8.55 ppm was assigned to the proton of the pterin ring of FA, and signals at 7.66 and 7.20 ppm corresponded to the aromatic protons of FA. The aforementioned results revealed that both DA and FA were successfully grafted onto the backbone of CS.

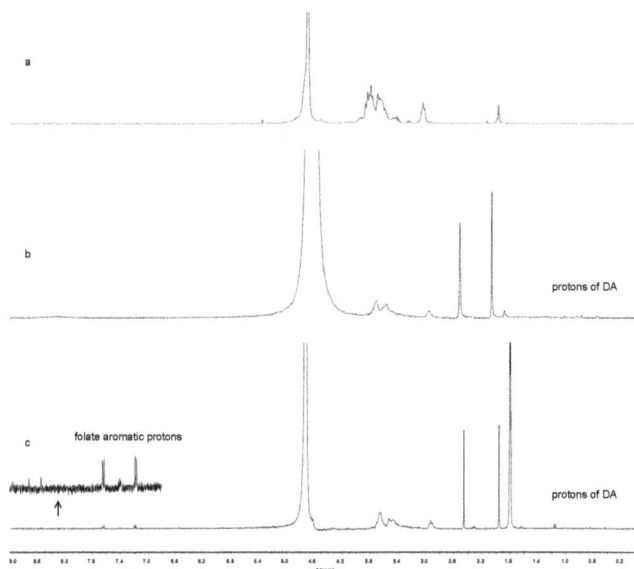

Figure 2. ^{1}H-NMR spectra of (**a**) CS, (**b**) CS-DA and (**c**) FA-CS-DA.

2.2.2. Fourier-Transform Infrared (FTIR) Characterization

FTIR analysis was used to further confirm the successful synthesis of FA-CS-DA. As presented in Figure 3, for CS, the signal at 1637 cm^{-1} was attributed to the C–O stretching vibration of C=O group of the amide I band, and the peak at 1517 cm^{-1} was assigned to the N–H bending vibration of the amide II band. In the spectrum of CS-DA, new peaks at 2925 and 2864 cm^{-1} were due to the C–H stretching vibration of methylene of DA. The enhancement of peak intensity at 3476 cm^{-1} suggested the increase of hydroxyl groups after the grafting of DA. For FA-CS-DA, the new signal at 1698 cm^{-1} was assigned to the unreacted carboxyl groups in FA, which implied the introduction of FA. All these differences indicated the formation of FA-CS-DA.

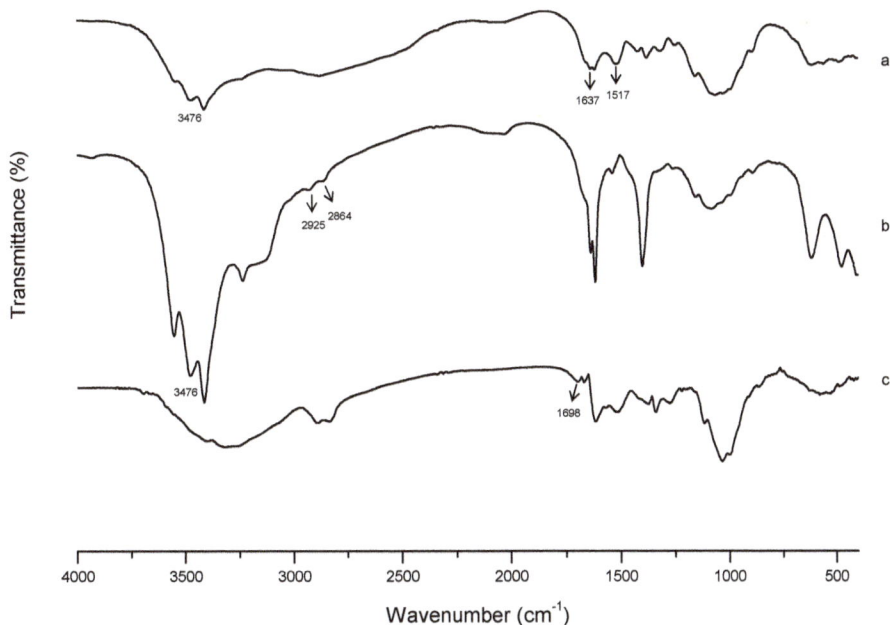

Figure 3. FTIR spectra of (**a**) CS, (**b**) CS-DA and (**c**) FA-CS-DA.

2.2.3. Critical Micelle Concentration (CMC) of FA-CS-DA

CMC is the lowest concentration for the amphiphilic polymer to form micelles in aqueous medium, and it is an important parameter that indicates the stability of micelles. In this study, pyrene was used as a fluorescence probe to measure the CMC of FA-CS-DA. At low concentrations, the polymer molecules existed in a single-stranded form, and the fluorescence intensity remained constant. Once the concentration was higher than CMC, the polymer molecules formed micelles, and pyrene was solubilized in the hydrophobic core of micelles. As a result, the fluorescence intensity increased significantly. Moreover, the intensity of the third energy peak (383 nm, I_3) increased more dramatically than the first peak (373 nm, I_1) [27]. The intensity ratio of I_1/I_3 was used as an indicator of the polarity of the environment, and the variation of I_1/I_3 against the logarithm of polymer concentration is shown in Figure 4. The CMC of FA-CS-DA could be determined from the point of inflection, and the value was calculated to be 6.6×10^{-3} mg/mL. The low value suggested the high stability of FA-CS-DA micelles.

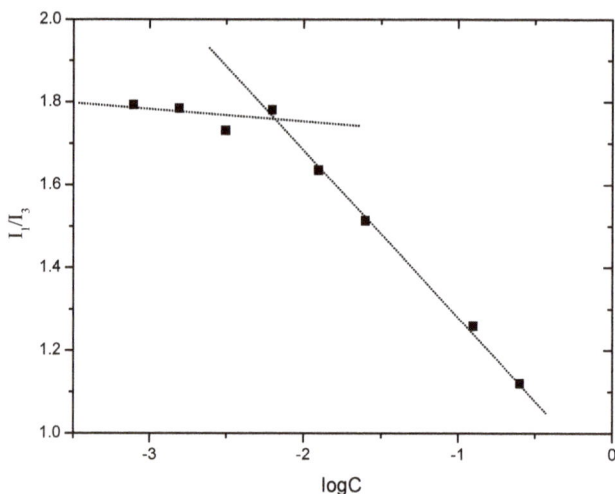

Figure 4. Variation of the fluorescence intensity ratio of I_1/I_3 against the logarithm of FA-CS-DA concentration.

2.3. Preparation of PTX-Loaded FA-CS-DA Micelles

As the polymer FA-CS-DA had an amphiphilic structure, including hydrophobic segments of DA, hydrophilic segments of CS and tumor targeting ligands of FA, in an aqueous environment, it could self-assemble into micelles. In this study, FA-CS-DA was dispersed in distilled water, and the micelles were prepared under ultrasonication without the addition of any emulsifier or stabilizer. The drug loading into the hydrophobic domains occurred simultaneously with the formation of micelles via either hydrophobic–hydrophobic interactions or Van der Waals interactions between PTX molecules and the hydrophobic groups of the polymer [28]. For the optimized drug-loaded micelles, the drug encapsulation efficiency and the drug loading capacity were calculated to be 81.2% and 9.1%, respectively.

2.4. Characterization of PTX-Loaded FA-CS-DA Micelles

2.4.1. Particle Size and Zeta Potential

It was well known that the particle size and size distribution of nanoparticles could dramatically affect the fate of the particles. Nanoparticles in the range of 10–200 nm could reduce the reticuloendothelial system (RES) uptake and enhance the endocytic uptake in tumors via the EPR effect [29]. The mean particle size of PTX-loaded FA-CS-DA micelles determined by the DLS method was 126 nm, with a polydispersity index (PDI) of 0.256, and this was larger than the bare ones (78 nm, PDI of 0.232), which indicated the encapsulation of PTX into the micelles.

Zeta potential is often used to indicate the stability of particle systems. With high zeta potential, the particles could repel each other and prevent aggregation, therefore enhancing the stability of the solution. The resultant zeta potential values of bare and PTX-loaded FA-CS-DA micelles were +29.1 mV and +19.3 mV, respectively. The relatively high positive potential was attributed to the ionized amino groups of CS. It was reported that the positively charged particles could enhance the endocytosis by cells [30].

2.4.2. Transmission Electron Microscopy (TEM) Observation

TEM was used to directly visualize the size and morphology of the micelles. The TEM micrograph of the PTX-loaded micelles presented in Figure 5 showed that the FA-CS-DA was capable of forming polymeric micelles, and the micelles had a near-spherical shape with narrow distribution. Furthermore, the size obtained by TEM was smaller than that measured by DLS, which was due to the different states of the particles in the measurements, i.e., the dried state and the hydrated state, respectively. More exactly, the outer shell of the micelles could be collapsed during the process in TEM experiment [31].

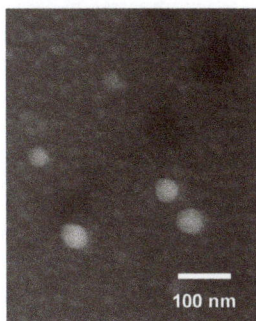

Figure 5. Transmission electron microscopy (TEM) image of paclitaxel (PTX)-loaded FA-CS-DA micelles.

2.4.3. X-Ray Diffraction (XRD) Analysis

XRD analysis was conducted to confirm the existence state of PTX in the polymeric micelles. As shown in Figure 6, the XRD diagram of PTX presented several peaks at 2θ of 5.53°, 8.87°, 10.04°, 11.14° and 12.53°, and there were a large number of small peaks in the range of 15° to 30°. For blank micelles, there were no typical crystal peaks in the pattern. The physical mixture of PTX and bare micelles still showed the typical crystal peaks of PTX with weaker intensity. However, the PTX-loaded micelles had a similar spectrum to the blank micelles and there were no PTX peaks. It was implied that PTX was entrapped in the FA-CS-DA micelles in an amorphous or molecular state, which might lead to better absorption of the drug.

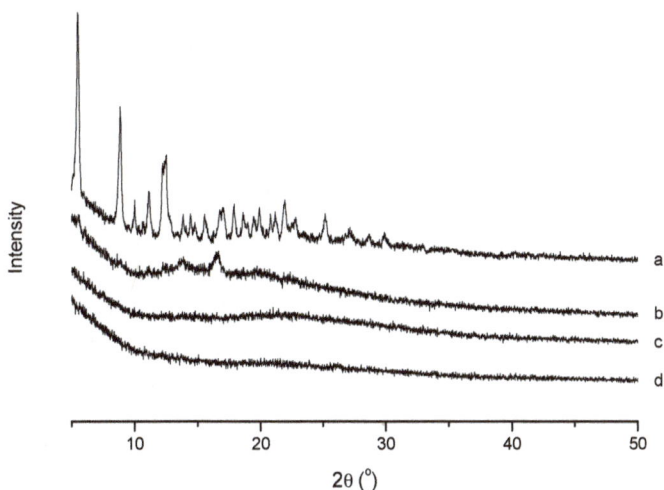

Figure 6. XRD spectra of (**a**) PTX, (**b**) a physical mixture of PTX and blank micelles, (**c**) PTX-loaded micelles and (**d**) blank micelles.

2.5. In Vitro Cytotoxicity Study

The in vitro cytotoxicity of the PTX-loaded micelles was evaluated by a standard MTT assay against MCF-7 cells. As illustrated in Figure 7, more than 99% of the cells were alive after the treatment of blank micelles even with high concentrations, which suggested the nontoxicity of the vehicle. For PTX formulations, the cytotoxicity was concentration-dependent. When the drug concentration increased, the cell viability decreased, which implied that a sufficient exposure level was important for the drug to kill the cells effectively. Moreover, it was exciting to see that the PTX-loaded micelles exhibited higher cytoxicity than the free PTX in dimethyl sulfoxide (DMSO). This might be explained by the effect of the FA-CS-DA micellar vehicle. For MCF-7 cells with over-expressed folate receptors on the surface [32], more FA-CS-DA micelles could be internalized into the cells via the receptor-mediated endocytosis. It could be speculated that FA-CS-DA might be a potential drug carrier for PTX.

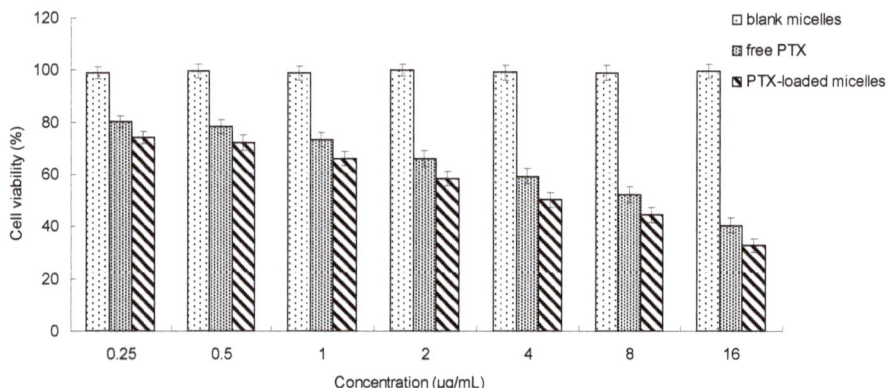

Figure 7. In vitro cytotoxicity of PTX-loaded FA-CS-DA micelles.

2.6. In Vivo Tumor Growth Inhibition Study

The tumor growth inhibition study was performed to further evaluate the in vivo antitumor activity of PTX-loaded FA-CS-DA micelles. As shown in Figure 8, the tumors excised from the mice in the PTX-loaded micelles group were significantly smaller than those from the normal saline group, and the TIR was as high as 78.1%. The outstanding antitumor efficacy could be explained by the following facts: first, there were a number of FA receptors on the surface of the tumor, and the PTX-loaded FA-CS-DA micelles could accumulate in the tumor tissues and then internalize into the tumor cells via the folate receptor-mediated active targeting. Second, via the electrostatic interaction, the positively charged micelles could efficiently bind to the tumor surface, which exhibited a high negative charge [33]. Moreover, the particles with positive charge were more likely to penetrate into the tumor cells [34]. Third, the particle size of <200 nm could facilitate the tumor accumulation of micelles by the EPR effect. In summary, all of these could increase the PTX concentration in the tumor, therefore obtaining excellent antitumor efficacy.

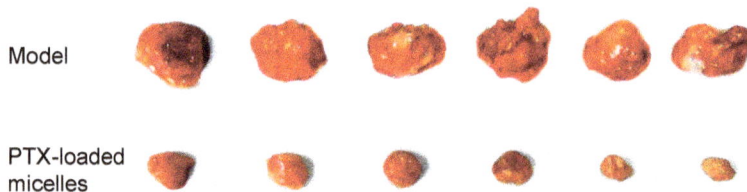

Figure 8. Tumors excised from the mice after intravenous injection treatment.

3. Materials and Methods

3.1. Materials

Chitosan (CS, Mw of 30 kDa, deacetylation degree > 90%) was obtained from Kittolife Co., Ltd., Seoul, Korea. Paclitaxel (PTX) was purchased from Natural Field Biological Technology Co., Ltd., Xi'an, China. Deoxycholic acid (DA), folic acid (FA), 1-Ethyl-3-(3-dimethylaminopropyl) carbodiimide hydrochloride (EDC) and *N*-hydroxysuccinimide (NHS) were supplied by Aladdin Industrial Co., Shanghai, China. Pyrene, 2,4,6-trinitrobenzene sulfonic acid (TNBS) and 3-(4,5-dimethylthiazol-2-yl)-2,5-diphenyl tetrazolium bromide (MTT) were supplied by Sigma Chemical Co., St. Louis, MO, USA. Dulbecco's modified Eagle's medium (DMEM), penicillin–streptomycin mixture and fetal bovine serum (FBS) were obtained from Gibco BRL, Carlsbad, CA, USA. All other chemicals and solvents were of analytical or chromatographic grade and used without further purification. Distilled water or Milli-Q water was used in all experiments.

3.2. Animals and Cell Lines

Specific pathogen-free mice weighing 20 ± 2 g were supplied by the Laboratory Animal Center of Harbin Medical University, Harbin, China. MCF-7 cells (human breast cancer cells) and H22 cells (mouse hepatocellular carcinoma cells) were kindly donated by the Department of Pharmacology, Harbin Medical University. All animal procedures were performed in compliance with the animal care protocols approved by the Animal Ethics Committee of Harbin Medical University (16 March 2018, No. 201803160028).

3.3. Synthesis of FA-CS-DA

FA-CS-DA was synthesized via the reaction of carboxyl groups of DA and FA with amino groups of chitosan under the catalyzation of EDC and NHS. The CS-DA was prepared as follows. Firstly, 1.0024 g of chitosan was dissolved in 50 mL of distilled water as solution A. The solution of DA (0.6012 g) with EDC (0.3426 g) and NHS (0.4113 g) in 50 mL of methanol was stirred for 1.5 h as solution B. Then solution B was added drop-wisely into solution A and stirred for 24 h at room temperature. After that, the reaction mixture was added to 50 mL of methanol. The resultant was filtered, and the filtrate was transferred into the mixture of methanol, water and ethanol at the ratio of 1:1:1 (*v/v*), then centrifuged at 6000 rpm for 20 min to remove unreacted DA and other impurities. The product of CS-DA was dried under vacuum at 40 °C.

For the synthesis of FA-CS-DA, a certain amount of CS-DA (0.14 g) was dissolved in 50 mL of acetate–acetate buffer solution (pH 4.7) as solution C. The solution of FA (0.04 g) with EDC (0.035 g) and NHS (0.042 g) in 20 mL of DMSO was stirred for 1 h as solution D. The solution D was added drop-wisely into solution C and stirred at an ambient temperature in dark condition. Twenty-four hours later, the resultant was dialyzed against an excess amount of distilled water (MWCO of 30 kDa) for 48 h for purification and then lyophilized to get the product of FA-CS-DA.

3.4. Characterization of FA-CS-DA

3.4.1. ^{1}H-NMR Characterization

^{1}H-NMR spectra were recorded on a Bruker Avance NMR spectrometer (AV-400, Bruker, Switzerland) operated at 400 MHz to analyze the structure of FA-CS-DA. The mixture of deuterated water (D_2O) and DMSO-d6 was used to prepared native and modified chitosan solution, respectively.

3.4.2. FTIR Characterization

The formation of FA-CS-DA was further verified by FTIR spectra that recorded with KBr pellets using FTIR spectrometer (Tensor II, Bruker, Switzerland) in the range from 4000 to 400 cm^{-1}, with a resolution of 2 cm^{-1}.

3.4.3. Measurement of the Degree of Substitution (DS)

The DS is defined as the number of DA (or FA) groups per 100 sugar units of CS. The DS of DA was determined by measuring the free amino groups of native chitosan and CS-DA via the TNBS method [35]. The DS of FA was measured by an ultraviolet-visible (UV-vis) spectrophotometer. Specifically, FA-CS-DA was dissolved in the mixture of water and DMSO at the volume ratio of 1:1, and the absorbance was determined by an UV-vis spectrophotometry (UV mini-1240, Shimadzu, Kyoto, Japan) at 363 nm. To get the standard curve for FA, a series of FA solutions with increasing amounts of FA were prepared.

3.4.4. Determination of Critical Micelle Concentration (CMC)

The CMC of FA-CS-DA was determined by measuring the fluorescence intensity of pyrene using a fluorescence spectrometer (F-2500 FL Spectrophotometer, Hitachi Ltd., Tokyo, Japan). In brief, 10 μL of pyrene in acetone solution was added into a series of 10 mL volumetric flasks, and the acetone was evaporated in dark. Then, the FA-CS-DA solutions with different concentrations were added to each flask, and the final concentration of pyrene was 6×10^{-7} mol/L. The samples were sonicated for 1 h, and then remained undisturbed and kept from light to reach equilibrium. The emission spectra of pyrene were recorded from 350 to 500 nm at the excitation wavelength of 337 nm. The slit widths for excitation and emission were set at 5 and 2.5 nm, respectively. The fluorescence intensity ratio of the first peak (I_1, 373 nm) to the third peak (I_3, 384 nm) was analyzed for determination of CMC.

3.5. Preparation of PTX-Loaded FA-CS-DA Micelles

The PTX-loaded FA-CS-DA micelles were prepared by ultrasonication using a probe-type sonicator (JY92-II, Ningbo Scientz Biotechnology Co., Ltd., Ningbo, China) [20]. Firstly, FA-CS-DA was dispersed in distilled water. Then the PTX in acetone solution was added quickly, and the mixture was sonicated at 224 W for 10 min with 2 s active/3 s duration. The mixture was dialyzed against distilled water for 5 h by using a dialysis bag (MWCO of 30 kDa) to remove unloaded PTX. The resultant was subsequently freeze-dried (Freeze Dryer Model FD-1A-50, Bioyikang Experimental Instrument Co., Ltd., Beijing, China) to get the PTX-loaded FA-CS-DA micelles powder.

3.6. Characterization of PTX-Loaded FA-CS-DA Micelles

3.6.1. Measurement of Particle Size and Zeta Potential

A dynamic light scattering method was applied to measure the particle size, zeta potential and size distribution of the samples using the Zetasizer® 3000 (Malvern Instruments, Southborough, MA, USA).

3.6.2. TEM Observation

TEM observation of the PTX-loaded micelles was performed using transmission electron microscopy (H-7650, Hitachi Ltd., Tokyo, Japan) operated at 60 kV. Before observation, an aqueous droplet of the sample was deposited on a copper grid coated with carbon. After 4 min, the grid was tapped with filter paper to remove surface water, followed by air drying and negatively stained with 2% phosphotangstic acid.

3.6.3. XRD Analysis

XRD analysis was used to study the existence state of PTX after loaded into micelles. Samples were measured by an X-ray diffractometer (Geigerflex, Rigaku Co., Tokyo, Japan) using Cu Kα radiation source at 30 kV and 30 mA. The relative intensity was recorded in the range of 5–50° (2θ) at scanning speed of 4°/min and step size of 0.02°.

3.7. Determination of Drug Loading and Drug Encapsulation Efficiency

The amount of PTX encapsulated into the micelles was determined as follows: a certain amount of lyophilized drug-loaded micelles was weighted, and then acetonitrile was added, followed by ultrasonication at 200 W for 10 min to extract PTX from the core of the micelles. After filtered through a 0.22 μm microporous filter, the PTX concentration was measured by HPLC method using a mobile phase delivery pump (LC-10ATVP, Shimadzu, Japan) and a Diamonsil^{TM} C_{18} reverse-phase column (200 mm × 4.6 mm, 5 μm, Dikma Technologies Inc., Beijing, China). The mobile phase was the mixture of acetonitrile and water at a volume ratio of 70:30. The column temperature was set at 30 °C. The detection wavelength was 227 nm, and the flow rate was 1 mL/min. The drug loading (DL%) and drug encapsulation efficiency (EE%) were calculated with the following formulas.

$$DL\% = W_{encapsulated}/W_{micelles} \times 100\% \tag{1}$$

$$EE\% = W_{encapsulated}/W_{fed} \times 100\% \tag{2}$$

where $W_{encapsulated}$, W_{fed} and $W_{micelles}$ represented the weight of PTX encapsulated in the micelles, the PTX fed initially, and the weight of PTX-loaded micelles, respectively.

3.8. In Vitro Cytotoxicity Study

The cytotoxicity of PTX-loaded micelles was investigated by MTT method. H22 cells in the logarithmic growth phase were seeded in a 96-well microtitre plate at the density of 1×10^5 cells/well and cultured in DMEM with 10% (*v/v*) fetal bovine serum and 1% penicillin–streptomycin at 37 °C in a humidified incubator with 5% CO_2. When the cells became 75% confluent, free PTX, blank FA-CS-DA micelles and PTX-loaded FA-CS-DA micelles with a PTX concentration ranging from 0.25 to 16 μg/mL was added, respectively. After incubation for 24 h, the cell viability was determined. At a predetermined time, the supernatant of each well was aspirated off and replaced with fresh medium, and then 10 μL of MTT solution was added. With the action of active mitochondrial dehydrogenase in live cells, the dissolved MTT could be converted to water-insoluble purple formazan crystals. After incubation for another 4 h, the unreacted MTT was discarded, followed by the addition of 100 μL of DMSO to dissolve the formazan crystals. The absorbance was analyzed by a BioRad microplate reader (Bio-Rad 680, Bio-Rad Laboratories, Hercules, CA, USA) at 490 nm. Untreated control cells were set as 100% viable. Cell viability was calculated by the following equation:

$$Cell\ viability\ (\%) = A_{sample}/A_{control} \times 100\% \tag{3}$$

where A_{sample} and $A_{control}$ were the absorbance of cells exposed to the sample and the absorbance of untreated cells, respectively.

3.9. In Vivo Antitumor Activity Study

The in vivo antitumor activity of PTX-loaded FA-CS-DA micelles was evaluated in H22 tumor-bearing mice. To establish the tumor model, mice were inoculated subcutaneously in the right armpit with 5×10^6 cells (0.2 mL) [36]. When the tumor xenografts became palpable, the mice were randomly divided into 2 groups (*n* = 6) and treated with physiological saline and PTX-loaded FA-CS-DA micelles (15 mg/kg), respectively. Samples were administered via the tail vein once every 3 days for 4 times. After 12 days of treatment, the mice were sacrificed and the tumors were harvested and weighted. The antitumor activity of the PTX-loaded micelles was expressed by the tumor inhibition rate (TIR), which could be calculated using Equation (4).

$$TIR = (1 - W_1/W_2) \times 100\% \tag{4}$$

where W_1 and W_2 represented the average tumor weight of PTX-loaded micelles group and normal saline group, respectively.

3.10. Statistical Analysis

Each experiment was performed in triplicate. Values were expressed as mean ± standard deviation. Statistical data analysis was performed using the Student's *t*-test with the significance level set at $p < 0.05$.

4. Conclusions

In the present study, an amphiphilic chitosan derivative of FA-CS-DA was synthesized via amidation reaction. The polymer could self-assemble into micelles in an aqueous milieu and showed good solubilization ability for PTX. The developed micellar system had excellent features, such as high stability upon dilution, high drug loading and encapsulation efficiency, small particle size and excellent cytotoxicity to tumor cells in vitro and in vivo. It could be concluded that the FA-CS-DA was a promising micellar carrier for PTX delivery, and further, more detailed studies will be performed.

Author Contributions: S.S. conceived and designed the experiments. L.L. performed the experiments, analyzed the data and wrote the paper. D.W. performed the in vitro cytotoxicity study. N.L. performed the in vivo tumor growth inhibition study and revised the paper. P.Y., Y.K. and F.C. gave us much useful advice and some pieces of guidance.

Funding: This work was funded by the National Natural Science Foundation of China (No. 51403057), Natural Science Foundation of Heilongjiang Province (No. E2018052), Research and Development Project of Scientific and Technological Achievements for Colleges and Universities of Heilongjiang Province (No. TSTAU-R2018023), Harbin Science and Technology Innovation Talents Special Fund Project (No. 2016RQQXJ097, No. 2016RQQXJ131), and the Doctoral Scientific Research Startup Foundation of Harbin Normal University (No. XKB201304).

Conflicts of Interest: The authors declare no conflict of interest.

References

1. Sofias, A.M.; Dunne, M.; Storm, G.; Allen, C. The battle of "nano" paclitaxel. *Adv. Drug Deliv. Rev.* **2017**, *122*, 20–30. [CrossRef] [PubMed]
2. Eloy, J.O.; Petrilli, R.; Topan, J.F.; Antonio, H.M.R.; Barcellos, J.P.A.; Chesca, D.L.; Serafini, L.N.; Tiezzi, D.G.; Lee, R.J.; Marchetti, J.M. Co-loaded paclitaxel/rapamycin liposomes: Development, characterization and in vitro and in vivo evaluation for breast cancer therapy. *Colloids Surf. B Biointerfaces* **2016**, *141*, 74–82. [CrossRef] [PubMed]
3. Tatiparti, K.; Sau, S.; Gawde, K.A.; Iyer, A.K. Copper-free 'click' chemistry-based synthesis and characterization of carbonic anhydrase-IX anchored albumin-paclitaxel nanoparticles for targeting tumor hypoxia. *Int. J. Mol. Sci.* **2018**, *19*, 838. [CrossRef] [PubMed]
4. Li, Y.; Zhao, X.; Zu, Y.; Zhang, Y. Preparation and characterization of paclitaxel nanosuspension using novel emulsification method by combining high speed homogenizer and high pressure homogenization. *Int. J. Pharm.* **2015**, *490*, 324–333. [CrossRef] [PubMed]
5. Yang, H. Targeted nanosystems: Advances in targeted dendrimers for cancer therapy. *Nanomedicine* **2016**, *12*, 309–316. [CrossRef] [PubMed]
6. Cagel, M.; Tesan, F.C.; Bernabeu, E.; Salgueiro, M.J.; Zubillaga, M.B.; Moretton, M.A.; Chiappetta, D.A. Polymeric mixed micelles as nanomedicines: Achievements and perspectives. *Eur. J. Pharm. Biopharm.* **2017**, *113*, 211–228. [CrossRef] [PubMed]
7. Liang, N.; Sun, S.; Gong, X.; Li, Q.; Yan, P.; Cui, F. Polymeric micelles based on modified glycol chitosan for paclitaxel delivery: Preparation, characterization and evaluation. *Int. J. Mol. Sci.* **2018**, *19*, 1550. [CrossRef] [PubMed]
8. Cong, Z.; Shi, Y.; Wang, Y.; Wang, Y.; Niu, J.E.; Chen, N.; Xue, H. A novel controlled drug delivery system based on alginate hydrogel/chitosan micelle composites. *Int. J. Biol. Macromol.* **2018**, *107*, 855–864. [CrossRef] [PubMed]

9. Shi, C.; Zhang, Z.; Wang, F.; Luan, Y. Active-targeting docetaxel-loaded mixed micelles for enhancing antitumor efficacy. *J. Mol. Liq.* **2018**, *264*, 172–178. [CrossRef]

10. Biswas, S.; Kumari, P.; Lakhani, P.M.; Ghosh, B. Recent advances in polymeric micelles for anti-cancer drug delivery. *Eur. J. Pharm. Sci.* **2016**, *83*, 184–202. [CrossRef] [PubMed]

11. Yang, Y.; Wang, S.; Wang, Y.; Wang, X.; Wang, Q.; Chen, M. Advances in self-assembled chitosan nanomaterials for drug delivery. *Biotechnol. Adv.* **2014**, *32*, 1301–1316. [CrossRef] [PubMed]

12. Ahsan, S.M.; Thomas, M.; Reddy, K.K.; Sooraparaju, S.G.; Asthana, A.; Bhatnagar, I. Chitosan as biomaterial in drug delivery and tissue engineering. *Int. J. Biol. Macromol.* **2018**, *110*, 97–109. [CrossRef] [PubMed]

13. Huo, M.; Fu, Y.; Liu, Y.; Chen, Q.; Mu, Y.; Zhou, J.; Li, L.; Xu, W.; Yin, T. N-mercapto acetyl-N'-octyl-O, N''-glycol chitosan as an efficiency oral delivery system of paclitaxel. *Carbohydr. Polym.* **2018**, *181*, 477–488. [CrossRef] [PubMed]

14. Silva, D.S.; Almeida, A.; Prezotti, F.; Cury, B.; Campana-Filho, S.P.; Sarmento, B. Synthesis and characterization of 3,6-O,O'- dimyristoyl chitosan micelles for oral delivery of paclitaxel. *Colloids Surf. B Biointerfaces* **2017**, *152*, 220–228. [CrossRef] [PubMed]

15. Cheng, L.C.; Jiang, Y.; Xie, Y.; Qiu, L.L.; Yang, Q.; Lu, H.Y. Novel amphiphilic folic acid-cholesterol-chitosan micelles for paclitaxel delivery. *Oncotarget* **2017**, *8*, 3315–3326. [CrossRef] [PubMed]

16. He, R.; Yin, C. Trimethyl chitosan based conjugates for oral and intravenous delivery of paclitaxel. *Acta Biomater.* **2017**, *53*, 355–366. [CrossRef] [PubMed]

17. Mansouri, M.; Nazarpak, M.H.; Solouk, A.; Akbari, S.; Hasani-Sadrabadi, M.M. Magnetic responsive of paclitaxel delivery system based on SPION and palmitoyl chitosan. *J. Magn. Magn. Mater.* **2017**, *421*, 316–325. [CrossRef]

18. Skorik, Y.A.; Golyshev, A.A.; Kritchenkov, A.S.; Gasilova, E.R.; Poshina, D.N.; Sivaram, A.J.; Jayakumar, R. Development of drug delivery systems for taxanes using ionic gelation of carboxyacyl derivatives of chitosan. *Carbohydr. Polym.* **2017**, *162*, 49–55. [CrossRef] [PubMed]

19. Gupta, U.; Sharma, S.; Khan, I.; Gothwal, A.; Sharma, A.K.; Singh, Y.; Chourasia, M.K.; Kumar, V. Enhanced apoptotic and anticancer potential of paclitaxel loaded biodegradable nanoparticles based on chitosan. *Int. J. Biol. Macromol.* **2017**, *98*, 810–819. [CrossRef] [PubMed]

20. Liang, N.; Sun, S.; Li, X.; Piao, H.; Piao, H.; Cui, F.; Fang, L. α-Tocopherol succinate-modified chitosan as a micellar delivery system for paclitaxel: Preparation, characterization and in vitro/in vivo evaluations. *Int. J. Pharm.* **2012**, *423*, 480–488. [CrossRef] [PubMed]

21. Heřmánková, E.; Žák, A.; Poláková, L.; Hobzová, R.; Hromádka, R.; Širc, J. Polymeric bile acid sequestrants: Review of design, in vitro binding activities, and hypocholesterolemic effects. *Eur. J. Med. Chem.* **2018**, *144*, 300–317. [CrossRef] [PubMed]

22. Hofmann, A.F.; Hagey, L.R. Bile acids: Chemistry, pathochemistry, biology, pathobiology, and therapeutics. *Cell. Mol. Life Sci.* **2008**, *65*, 2461–2483. [CrossRef] [PubMed]

23. Liu, M.; Du, H.; Zhai, G. Self-assembled nanoparticles based on chondroitin sulfate-deoxycholic acid conjugates for docetaxel delivery: Effect of degree of substitution of deoxycholic acid. *Colloids Surf. B Biointerfaces* **2016**, *146*, 235–244. [CrossRef] [PubMed]

24. Dhas, N.L.; Ige, P.P.; Kudarha, R.R. Design, optimization and in-vitro study of folic acid conjugated-chitosan functionalized PLGA nanoparticle for delivery of bicalutamide in prostate cancer. *Powder Technol.* **2015**, *283*, 234–245. [CrossRef]

25. Scomparin, A.; Salmaso, S.; Eldar-Boock, A.; Ben-Shushan, D.; Ferber, S.; Tiram, G.; Shmeeda, H.; Landa-Rouben, N.; Leor, J.; Caliceti, P.; et al. A comparative study of folate receptor-targeted doxorubicin delivery systems: Dosing regimens and therapeutic index. *J. Control. Release* **2015**, *208*, 106–120. [CrossRef] [PubMed]

26. Park, C.; Vo, C.L.N.; Kang, T.; Oh, E.; Lee, B.J. New method and characterization of self-assembled gelatin–oleic nanoparticles using a desolvation method via carbodiimide/N-hydroxysuccinimide (EDC/NHS) reaction. *Eur. J. Pharm. Biopharm.* **2015**, *89*, 365–373. [CrossRef] [PubMed]

27. Băran, A.; Stîngă, G.; Anghel, D.-F.; Iovescu, A.; Tudose, M. Comparing the spectral properties of pyrene as free molecule, label and derivative in some colloidal systems. *Sens. Actuators B Chem.* **2014**, *197*, 193–199. [CrossRef]

28. Zhang, C.; Qu, G.; Sun, Y.; Wu, X.; Yao, Z.; Guo, Q.; Ding, Q.; Yuan, S.; Shen, Z.; Ping, Q.; et al. Pharmacokinetics, biodistribution, efficacy and safety of N-octyl-O-sulfate chitosan micelles loaded with paclitaxel. *Biomaterials* **2008**, *29*, 1233–1241. [CrossRef] [PubMed]
29. Acharya, S.; Sahoo, S.K. PLGA nanoparticles containing various anticancer agents and tumour delivery by EPR effect. *Adv. Drug Deliv. Rev.* **2011**, *63*, 170–183. [CrossRef] [PubMed]
30. Wang, H.; Zuo, Z.; Du, J.; Wang, Y.; Sun, R.; Cao, Z.; Ye, X.; Wang, J.; Leong, K.W.; Wang, J. Surface charge critically affects tumor penetration and therapeutic efficacy of cancer nanomedicines. *Nano Today* **2016**, *11*, 133–144. [CrossRef]
31. Kim, C.; Lee, S.C.; Kang, S.W.; Kwon, I.C.; Kim, Y.H.; Jeong, S.Y. Synthesis and the micellar characteristics of poly(ethylene oxide)–deoxycholic acid conjugates. *Langmuir* **2000**, *16*, 4792–4797. [CrossRef]
32. Wang, F.; Chen, Y.; Zhang, D.; Zhang, Q.; Zheng, D.; Hao, L.; Liu, Y.; Duan, C.; Jia, L.; Liu, G. Folate-mediated targeted and intracellular delivery of paclitaxel using a novel deoxycholic acid-O-carboxymethylated chitosan–folic acid micelles. *Int. J. Nanomed.* **2012**, *7*, 325–337.
33. Yen, H.; Young, Y.; Tsai, T.; Cheng, K.; Chen, X.; Chen, Y.; Chen, C.; Young, J.; Hong, P. Positively charged gold nanoparticles capped with folate quaternary chitosan: Synthesis, cytotoxicity, and uptake by cancer cells. *Carbohydr. Polym.* **2018**, *183*, 140–150. [CrossRef] [PubMed]
34. Fröhlich, E. The role of surface charge in cellular uptake and cytotoxicity of medical nanoparticles. *Int. J. Nanomed.* **2012**, *7*, 5577–5591. [CrossRef] [PubMed]
35. Bernkop-Schnürch, A.; Krajicek, M.E. Mucoadhesive polymers as platforms for peroral peptide delivery and absorption: Synthesis and evaluation of different chitosan-EDTA conjugates. *J. Control. Release* **1998**, *50*, 215–223. [CrossRef]
36. Liang, N.; Sun, S.; Hong, J.; Tian, J.; Fang, L.; Cui, F. In vivo pharmacokinetics, biodistribution and antitumor effect of paclitaxel-loaded micelles based on α-tocopherol succinate-modified chitosan. *Drug Deliv.* **2016**, *23*, 2651–2660. [PubMed]

© 2018 by the authors. Licensee MDPI, Basel, Switzerland. This article is an open access article distributed under the terms and conditions of the Creative Commons Attribution (CC BY) license (http://creativecommons.org/licenses/by/4.0/).

International Journal of
Molecular Sciences

MDPI

Article

Study on Thermal Decomposition Behaviors of Terpolymers of Carbon Dioxide, Propylene Oxide, and Cyclohexene Oxide

Shaoyun Chen [1], Min Xiao [2], Luyi Sun [3] and Yuezhong Meng [2,*]

[1]   College of Chemical Engineering and Materials Science, Quanzhou Normal University, Quanzhou 362000, China; chshaoy@qztc.edu.cn
[2]   The Key Laboratory of Low-carbon Chemistry & Energy Conservation of Guangdong Province/State Key Laboratory of Optoelectronic Materials and Technologies, Sun Yat-Sen University, Guangzhou 510275, China; stsxm@mail.sysu.edu.cn
[3]   Department of Chemical & Biomolecular Engineering and Polymer Program, Institute of Materials Science, University of Connecticut, Storrs, CT 06269, USA; luyi.sun@uconn.edu
*   Correspondence: mengyzh@mail.sysu.edu.cn; Tel.: +86-20-8411-4113

Received: 20 September 2018; Accepted: 13 November 2018; Published: 23 November 2018

Abstract: The terpolymerization of carbon dioxide (CO_2), propylene oxide (PO), and cyclohexene oxide (CHO) were performed by both random polymerization and block polymerization to synthesize the random poly (propylene cyclohexene carbonate) (PPCHC), di-block polymers of poly (propylene carbonate–cyclohexyl carbonate) (PPC-PCHC), and tri-block polymers of poly (cyclohexyl carbonate–propylene carbonate–cyclohexyl carbonate) (PCHC-PPC-PCHC). The kinetics of the thermal degradation of the terpolymers was investigated by the multiple heating rate method (Kissinger-Akahira-Sunose (KAS) method), the single heating rate method (Coats-Redfern method), and the Isoconversional kinetic analysis method proposed by Vyazovkin with the data from thermogravimetric analysis under dynamic conditions. The values of ln k vs. T^{-1} for the thermal decomposition of four polymers demonstrate the thermal stability of PPC and PPC-PCHC are poorer than PPCHC and PCHC-PPC-PCHC. In addition, for PPCHC and PCHC-PPC-PCHC, there is an intersection between the two rate constant lines, which means that, for thermal stability of PPCHC, it is more stable than PCHC-PPC-PCHC at the temperature less than 309 °C and less stable when the decomposed temperature is more than 309 °C. Pyrolysis-gas chromatography/mass spectrometry (Py-GC/MS) and thermogravimetric analysis/infrared spectrometry (TG/FTIR) techniques were applied to investigate the thermal degradation behavior of the polymers. The results showed that unzipping was the main degradation mechanism of all polymers so the final pyrolysates were cyclic propylene carbonate and cyclic cyclohexene carbonate. For the block copolymers, the main chain scission reaction first occurs at PC-PC linkages initiating an unzipping reaction of PPC chain and then, at CHC–CHC linkages, initiating an unzipping reaction of the PCHC chain. That is why the $T_{-5\%}$ of di-block and tri-block polymers were not much higher than that of PPC while two maximum decomposition temperatures were observed for both the block copolymer and the second one were much higher than that of PPC. For PPCHC, the random arranged bulky cyclohexane groups in the polymer chain can effectively suppress the backbiting process and retard the unzipping reaction. Thus, it exhibited much higher $T_{-5\%}$ than that of PPC and block copolymers.

Keywords: polycarbonate; thermal decomposition kinetics; TG/FTIR; Py-GC/MS

1. Introduction

Carbon dioxide (CO_2) is a nontoxic, nonflammable material that exists naturally in abundance. The use of CO_2 has attracted increasing interest in recent years and has been considered as an alternative approach to reduce the release of this greenhouse gas [1–5]. One good approach is using CO_2 to produce biodegradable polymeric materials. In 1969, Inoue et al. first observed that the copolymerization of carbon dioxide with epoxides could form aliphatic polycarbonates [1]. Since then, much work has been done to make CO_2 copolymerize with other monomers [2–5]. Poly (propylene carbonate) (PPC) made from carbon dioxide and propylene oxide is the main kind of CO_2-based copolymer that has been widely investigated [6–10]. In previous work, the high molecular weight alternating PPC was synthesized in very high yield. The PPCs exhibit good biodegradability [11] but show inferior thermal stability due to the flexible carbonate linkage in the backbone. It has also been found that PPC is easily decomposed to cyclic carbonate by the unzipping reaction, which is initiated by the free hydroxyl terminal groups [12,13]. In order to enhance the thermal properties of PPC, a third monomer cyclohexene oxide (CHO) copolymerizing with CO_2 and PO is considered as a profitable mean [14,15]. However, the thermal degradation kinetics of the terpolymer has not been investigated until now.

Generally, TG analysis is an effective method for studying thermal decomposition kinetics and provides information on a frequency factor, activation energy, and overall reaction order. Due to the insufficient analytical capability of the evolved gas mixture analysis, it is still not possible to provide sufficient information on the mechanism of thermal degradation. Therefore, the direct analysis of gas composition by continuous monitoring with thermogravimetric analysis/Fourier transform infrared spectrometry (TG/FTIR) has attracted more attention in the identification of gaseous products to study the pyrolysis mechanism [16–18]. Pyrolysis gas chromatography/mass spectrometry (Py-GC/MS) is widely used to evaluate the thermal decomposition behavior of polymers because of its high sensitivity, rapidity, and effective separation ability of complex compounds containing a pyrolysis product of similar compositions [19–21]. Therefore, the combination of TG/FTIR and Py-GC/MS has been widely used to study the thermal decomposition mechanism [22–25].

In this paper, CHO was introduced to copolymerize with CO_2 and PO to get the random copolymer, the di-block copolymer, and the tri-block copolymer. The thermal decomposition behaviors of the resultant polymers with different sequence structures are studied by the combination of Py-GC/MS and TG/FTIR techniques. In addition, the thermal degradation kinetic parameters are obtained by using the multiple heating rate method (Kissinger-Akahira-Sunose (KAS) [26–28]), the single heating rate method (Coats-Redfern method) [29,30], and the Isoconversional kinetic analysis method proposed by Vyazovkin [31–33]. This used the data from thermogravimetric analysis under dynamic conditions such as activation energy E, the pre-exponential factor A, and the rate constant k. They will provide theoretical basis of thermal stability for further application.

2. Results and Discussion

2.1. Thermal Decomposition Behavior

The synthesized copolymers containing polycarbonates and polyesters are terpolymers as Scheme 1. These have proven by our previous paper [15]. ^{1}H NMR spectrum of the purified terpolymer shown in Figures S1–S4 demonstrates the formation of ester and carbonate linkages. The proton resonances 1.3 [3H, CH_3], 4.2 [2H, CH_2CH], and 5.0 [1H, CH_2CH] correspond to CH_3, CH_2, and CH groups in the polycarbonate sequence. The peaks of 1.7 ppm, 2.1 ppm, and 4.6 ppm representing CH_2 of M-, O-cycloalkane ring, and CH link to carbonate indicated the cyclohexene carbonate unit in the block copolymer. The basic properties of the resultant polymers are shown in Table 1.

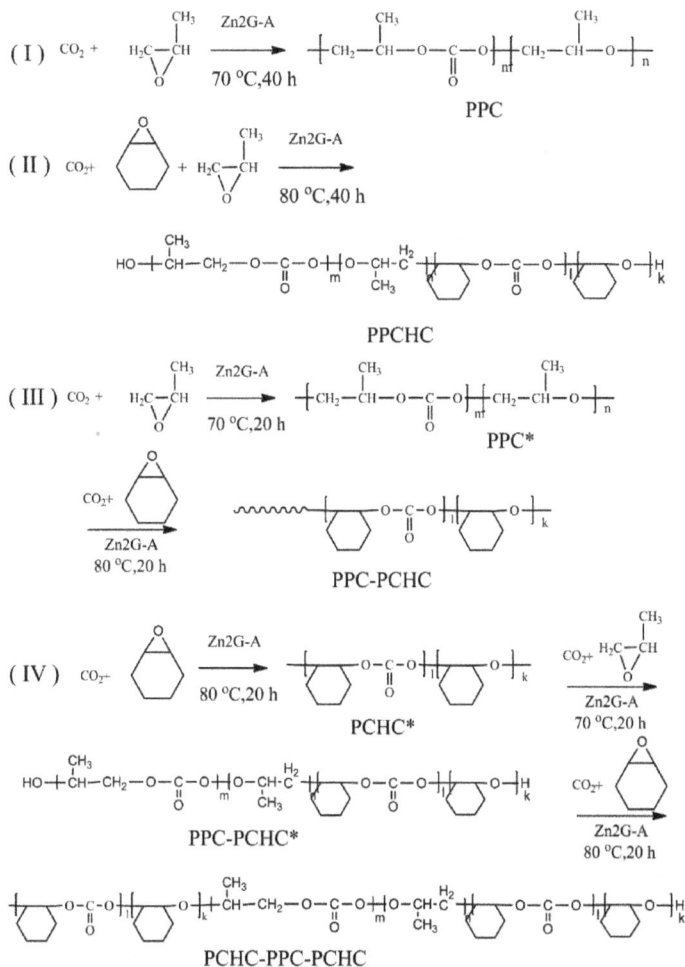

Scheme 1. The synthesized procedure of (I) PPC; (II) PPCHC; (III) PPC-PCHC; (IV) PCHC-PPC-PCHC copolymers.

Table 1. The properties of the resultant polymers.

Copolymer	Mn/Mw/PI [a]	$T_{-5\%}/T_{max}$ (°C) [c]	Composition (Molar Fraction %) [b]			
			f_{CO2}	f_{PC}	f_{CHC}	f_{PE}
PPC	$2.17 \times 10^5 / 3.78 \times 10^5 / 1.74$	255.7/278.2	48.9	48.9	-	2.2
PPCHC	$2.02 \times 10^5 / 7.13 \times 10^5 / 3.50$	281.0/313.4	47.9	36.2	11.7	4.2
PPC-PCHC	$2.97 \times 10^5 / 7.35 \times 10^5 / 2.47$	261.2/304.2, 342.7	48.0	35.1	12.9	4.0
PCHC-PPC-PCHC	$2.74 \times 10^5 / 7.88 \times 10^5 / 2.87$	275.2/305.8, 345.3	47.7	34.1	13.6	4.6

[a] Molecular weight was determined by GPC. [b] Determined by ^{1}H NMR spectroscopy. (see ESI Figure A1–A4) $f_{PC} = A_{4.2}/[(A_{4.2}+A_{4.6}) \times 2 + 0.8 \times A_{3.5}]$, $f_{CHC} = A_{4.6}/[(A_{4.2}+A_{4.6}) \times 2 + 0.8 \times A_{3.5}]$, $f_{CO2} = A_{4.2} + A_{4.6}/[(A_{4.2}+A_{4.6}) \times 2 + 0.8 \times A_{3.5}]$, $f_{PE} = 0.8 \times A_{3.5}/[(A_{4.2}+A_{4.6}) \times 2 + 0.8 \times A_{3.5}]$. [c] The heating rate is 10 °C/min.

In order to study the thermal degradation of the resultant polymers and to conclude the best way to improve the thermal stability of PPC, the non-isothermal kinetics of the thermal degradation of the resultant polymers were investigated with the Kissinger-Akahira-Sunose (KAS) method, the Coats-Redfern method, and the Isoconversional kinetic analysis method proposed by Vyazovkin.

TG-DTG curves of the polymers at different heating rates are presented in Figure 1. As the heating rate increased, thermal hysteresis became more and more evident. The thermal decomposition beginning temperature of the polymers also improved and the peak temperature moved to a higher temperature zone. Thermal decomposition of PPC and PPCHC were finished by one step and PPC-PCHC and PCHC-PPC-PCHC were mainly finished by two steps. Decomposition of four polymers at the heating rate 10 °C min^{-1} begin at 245.5 °C, 269.8 °C, 239.8 °C, and 257.2 °C, respectively. The maximum weight loss temperature from DTG curves for four polymers are 278.2 °C, 313.2 °C, 304.2 °C, and 305.8 °C, respectively.

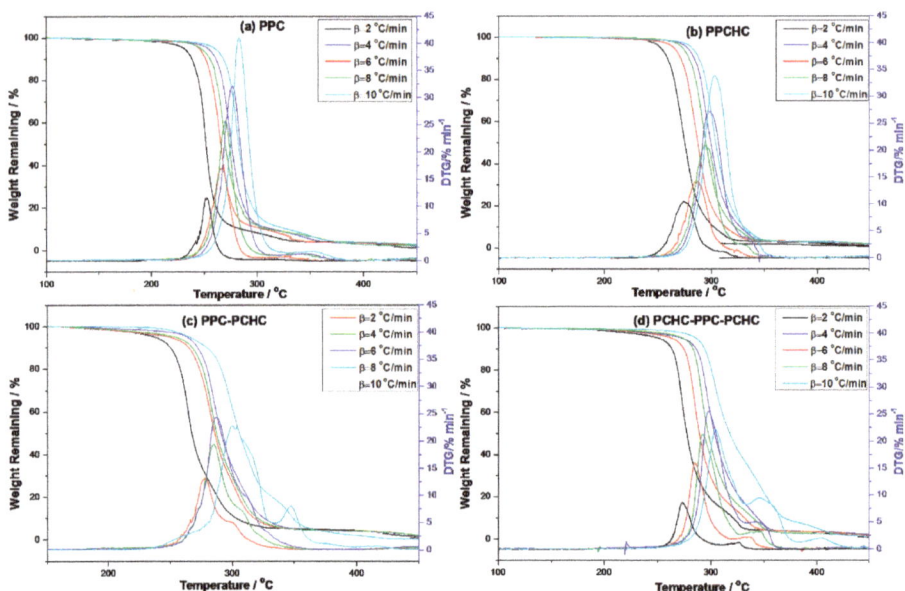

Figure 1. TG-DTG curves of the resultant polymers at different heating rates. (**a**) PPC; (**b**) PPCHC; (**c**) PPC-PCHC; (**d**) PCHC-PPC-PCHC.

In addition, the decomposition of four polymers are completed at ~350 °C. The KAS method has been employed to evaluate the activation energies of different polymers during thermal decomposition because of its good adaptability and validity for model-free approaches. At a constant value of conversion rate α, the plots of ln (β/T$_\alpha$²) versus 1/T$_\alpha$. As shown in Figure S5, the plots of ln (β/T$_\alpha$²) versus 1/T at several heating rates obtain well fitted straight lines whose slopes allow the evaluation of the apparent activation energy. The distribution of the activation energy (Eα) for different polymers is presented in Figure 2. It can be seen that the effective activation energy varies with conversion, which is an indication of a complex mechanism of the decomposition reaction. The process of PPC, PPC-PCHC, and PCHC-PPC-PCHC begins with Eα~110 kJ mol^{-1} and then slightly increases to 130 kJ mol^{-1}. In addition, Eα for block polymers PPC-PCHC and PCHC-PPC-PCHC decrease sharply to ~90 kJ mol^{-1} and 75 kJ mol^{-1}. For random polymer PPCHC, the Eα remains stable ~140–150 kJ mol^{-1}. The average activation energies for PPC, PPCHC, PPC-PCHC, and PCHC-PPC-PCHC are 124.6 ± 6.8 kJ mol^{-1}, 145.5 ± 3.0 kJ mol^{-1}, 109.9 ± 7.6 kJ mol^{-1}, and 104.9 ± 16.2 kJ mol^{-1}, respectively. The variation tendency tells us that the Eα of the block copolymer PPC-PCHC and PCHC-PPC-PCHC have a similar trend during the process of thermal decomposition while the variation tendency of PPC and PPCHC were relatively stable.

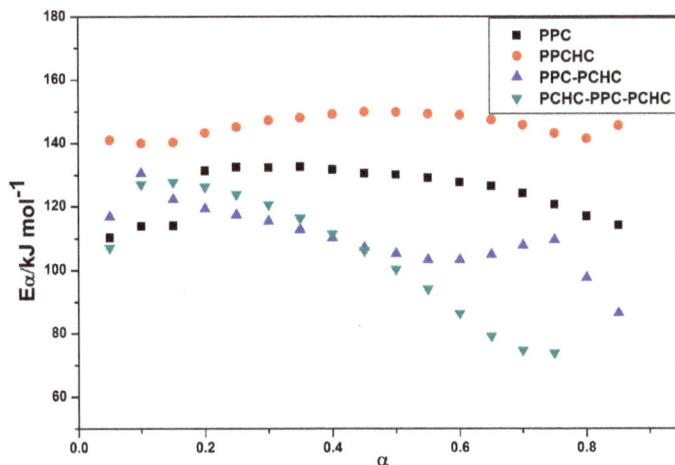

Figure 2. Values of the activation energy estimated by using the KAS method for the thermal decomposition of different polymers.

The model-fitting method (Coats-Redfern method) is suitable for the thermal decomposition of polymer during a certain pyrolysis stage in an entire thermal process. The Coats-Redfern method and 15 kinds of reaction models in solid-state reactions are adopted to focus on analyzing the pyrolysis kinetics of the polymers. The obtained kinetic parameters at different heating rates of polymers are presented in Table S1–S4. Then the iso-conversional pre-exponential factor values were evaluated by substituting the values of $E\alpha$ from the KAS method into the equation of the compensation effect in Equation (3) [31]. The compensation effect parameters a and b were determined by fitting the pairs of $\ln A_i$ and E_i from the Coats-Redfern method using 15 reaction models. The compensation effect parameters a and b were obtained by the average of five heating rates, which is shown in Figure S6. Therefore, the equations of compensation effect are $\ln A_\alpha = -3.23816 + 0.2244 E_\alpha$ for PPC, $\ln A_\alpha = -3.5574 + 0.21475 E_\alpha$ for PPCHC, $\ln A_\alpha = -2.9405 + 0.21602 E_\alpha$ for PPC-PCHC, and $\ln A_\alpha = -3.0112 + 0.211782 E_\alpha$ for PCHC-PPC-PCHC. The values of the pre-exponential factor as a function conversion for the thermal decomposition of different polymers are presented in Figure 3. The $\ln A_\alpha$ vs. α dependence exhibits similar facture as the one determined for the activation energy. The average preexponential factor for PPC, PPCHC, PPC-PCHC, and PCHC-PPC-PCHC are $24.7 \pm 1.5\,s^{-1}$, $27.7 \pm 0.6\,s^{-1}$, $20.8 \pm 1.6\,s^{-1}$, and $19.2 \pm 3.4\,s^{-1}$, respectively. Overall, the trend of activation energy values and pre-exponential factor values for decomposition is PCHC-PPC-PCHC < PPC-PCHC < PPC < PPCHC. However, the thermal stability of polymers determined by an increase activation energy values and a decrease pre-exponential factor values. In order to evaluate the overall effect of the activation energy and the pre-exponential factor on the kinetics of the process, the rate constant was calculated by using Equation (4) [32,33]. The values of $\ln k$ vs. T^{-1} for the thermal decomposition of four polymers are presented in Figure 4. From the results, decomposition of PPC, PPC-PCHC show higher values of the rate constant, so the thermal stability of PPC and PPC-PCHC are poorer than PPCHC and PCHC-PPC-PCHC. In addition, for PPCHC and PCHC-PPC-PCHC, there is an intersection between the two rate constant lines, which means that, for thermal stability, PPCHC is more stable than PCHC-PPC-PCHC at temperatures less than 309 °C and less stable when the temperature is more than 309 °C. The enhanced thermal stability of PPCHC is associated with an increase in the activation energy and the block polymer PPC-PCHC and PCHC-PPC-PCHC is mostly associated with a decrease in the pre-exponential factor. Above all, the random copolymerization of CHO, PO, and CO_2 is a better way to improve the thermal stability of PPC. We began with the idea that the tri-block polymer PCHC-PPC-PCHC had the same structure like SBS, which means its thermal stability was better than others. However, the results were not the same as predicted. In order to explain this, we studied the thermal degradation mechanisms of the polymers, which is discussed in the following part.

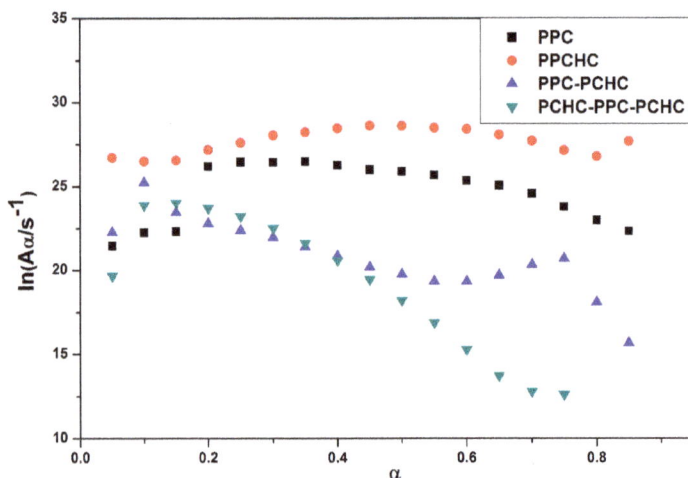

Figure 3. Pre-exponential factor as a function of conversion for the thermal decomposition of different polymers.

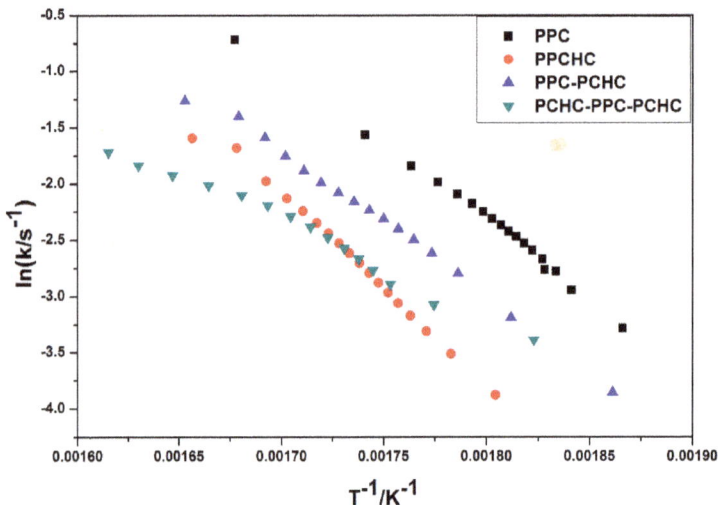

Figure 4. Rate constant as a function of reciprocal temperature for the thermal decomposition of different polymers.

2.2. Study the Thermal Decomposition Behavior Using Py-GC/MS and TG/FTIR

Thermal degradation kinetic parameters of the polymers were calculated by using thermogravimetric analysis. However, it could not provide enough information to analyze the thermal degradation mechanism. With the help of TG/IR technique, we can follow the dynamic process of the polymers decomposition. FTIR spectra (Figure S7) in three-dimensions for the decomposing dynamic process of polymers were obtained via this technique. The three-dimensional FTIR spectra of the pyrolysates derived from PPCHC, PPC-PCHC, and PCHC-PPC-PCHC are similar and the main IR absorbing peaks appear at 1109 cm^{-1}, 1863 cm^{-1}, and 2985 cm^{-1}, which indicates that the main pyrolysates of the block polymers are all cyclic carbonates.

According to previous research [34–37], the thermal decomposition behavior of PPC obeys two kinds of mechanism including the main chain scission reaction and the unzipping reaction. The unzipping reaction involves the backbiting of the terminal hydroxyl groups at the carbon of carbonate linkage leading to the formation of cyclic carbonate. From the FTIR spectra of pyrolysates at different temperatures shown in Figure 5, it can be seen that the peak stands for carbon dioxide is observed at 230 °C while the peaks stand for cyclic carbonates are observed at 250 °C for PPC, PPC-PCHC, and PCHC-PPC-PCHC and higher than 270 °C for PPCHC, which indicates the main chain scission reaction and the unzipping reaction during the decomposition process first occuring in PC-PC segments. In addition, the onset unzipping reaction temperature of the random polymer PPCHC is higher than that of PPC and block polymers. Comparing the intensity of the IR absorption peaks of the pyrolysates at a different temperature, we can see that large scale evolution of cyclic carbonates happens at a relatively lower temperature (about 300 °C) for PPC and higher temperature (about 350–400 °C) for PPCHC. While for the block copolymers, large scale evolution of cyclic carbonates happens at a wider temperature range (300–400 °C). It is supposed that the random polymer PPCHC probably has a large content of PC-CHC or CHC-PC linkages in the terpolymer, which means the steric hindrance of bulky cyclohexane groups can suppress the backbiting process and retard the unzipping reaction. Thus, the thermal stability of PPCHC is much better than PPC. While, for the block copolymers, the PC-PC block and CHC-CHC block exhibit similar thermal stability as PPC and PCHC, respectively, so the temperature range of large scale evolution of cyclic carbonate was wide and two maximum decomposition temperatures were observed, as shown in Table 1. The above supposition was further confirmed by Py-GC/MS, which provide a way to measure the thermal decomposition reactions in each stage by analyzing mixture-evolved gases. Based on the results of TG analysis, the pyrolysis temperatures were set at 200 °C, 250 °C, 300 °C, 350 °C, and 400 °C to obtain the formation curves of pyrolysates to analyze the degradation mechanisms. The chromatograms of pyrolysis products at different temperatures and their identifications are shown in Figure 6 and Table 2.

The pyrolysates of random copolymer PPCHC was not detected out at 300 °C and, at 350 °C, the pyrolysates are cyclic propylene carbonate (6.0–6.75 min) and cyclic cyclohexene carbonate (12.25–12.5 min). These phenomena confirms the large content of PC-CHC or CHC-PC linkages in the polymer chain. Due to the alternating PC and CHC main chain structure, the backbiting is suppressed by steric hindrance results in a retarded unzipping reaction and, once the unzipping reaction starts, the pyrolysates cyclic propylene carbonate and cyclic cyclohexene carbonate are produced at the same time.

For the di-block and tri-block copolymers, small peaks representing CO$_2$ (1.5–2.0 min) and cyclic propylene carbonate (6.0–6.75 min) appear when being pyrolyzed at 250 °C. The peak of cyclic cyclohexene carbonate appears at a pyrolysis temperature as high as 350 °C, which means that the main chain scission reaction first occurs at PC-PC linkages. This initiates the unzipping reaction of PC-PC linkages and then, at CHC-CHC linkages, initiates the unzipping reaction of the PCHC chain. In addition, this can explain why the degradation activation energies of the block polymers are lower at first and get higher with the decomposition conversion increasing. Based on above results and analysis, the possible decomposition pathways are given as Scheme 2. For the random copolymer PPCHC, the thermal decomposition behavior obeys the unzipping reaction mostly. In addition, for the block copolymers of PPC-PCHC and PCHC-PPC-PCHC, the thermal decomposition behavior obeys two kinds of mechanism including the main chain scission reaction and the unzipping reaction.

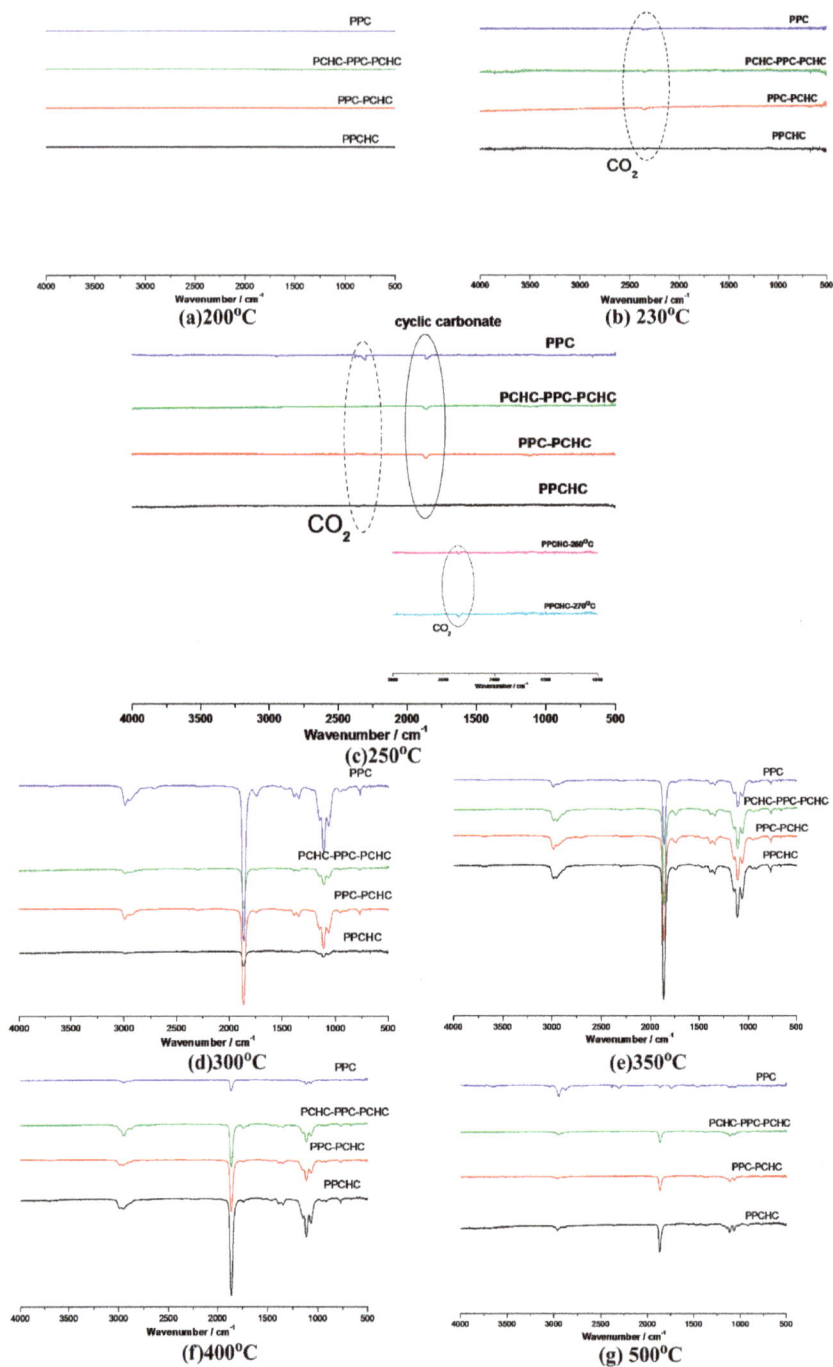

Figure 5. FTIR spectra of pyrolysates at (**a**) 200 °C; (**b**) 230 °C; (**c**)250 °C; (**d**) 300 °C; (**e**) 350 °C, (**f**) 400 °C, and (**g**) 500 °C.

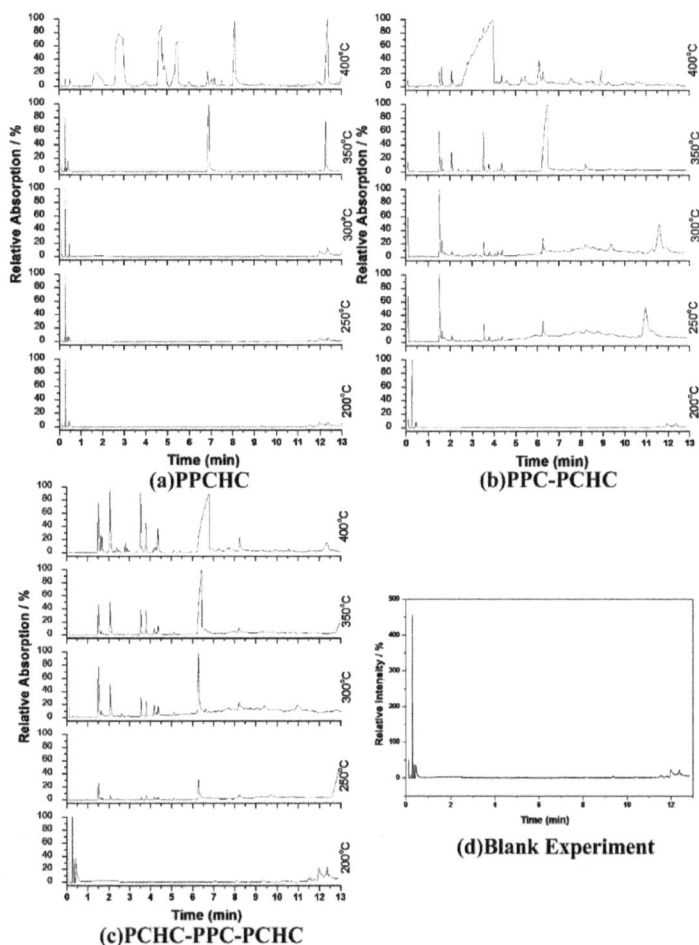

Figure 6. Chromatogram of pyrolysis products by Py-GC/MS at different temperatures :(**a**) PPCHC; (**b**)PPC-PCHC; (**c**)PCHC-PPC-PCHC; (**d**) Blank Experiment.

Table 2. Identification of pyrolysates in Py-GC/MS.

Compound	Retention Time (min)	Compound	Retention Time (min)
		CO_2	1.5–2 min
	4.0–4.5 min		4.5–5.0 min
	5.0–5.5 min		6.0–6.75 min
	7.5–8.0 min		12.25–12.5 min

(a)

(b)

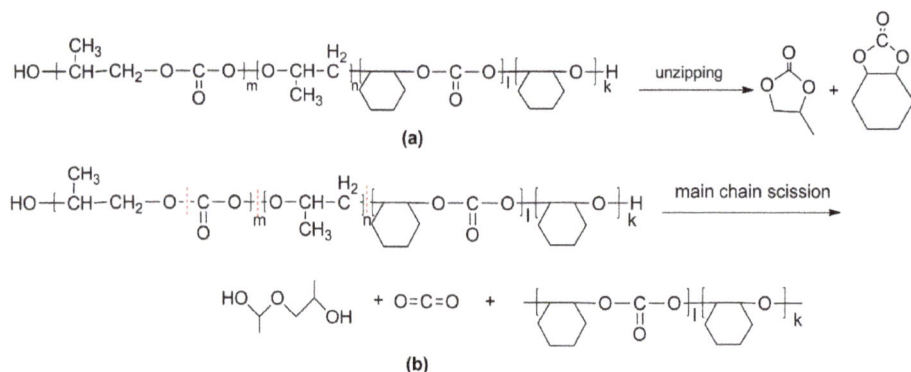

Scheme 2. Possible decomposition pathway of the polymers. (**a**) Unzipping of the polymer backbone and (**b**) main chain scission.

3. Materials and Methods

3.1. Materials

Carbon dioxide of a purity of 99.99% was used without further treatment. PO of a purity of 99.5% and CHO of a purity of 95.0% were refluxed over CaH_2 for 4 h and 24 h, respectively, and then distilled under dry nitrogen gas. Prior to use, they were stored over 4-Å molecular sieves. Other solvents and reagents such as ethanol and chloroform were of analytical grade and used without further purification.

Supported multi-component zinc dicarboxylate catalyst (Zn2G) was prepared according to previous work [38]. The catalyst was white powder with the Zn content of 11.6 wt %.

3.2. Preparation of Random Terpolymers and Block Copolymers

The di-block PPC-PCHC copolymerization was carried out using 'one pot, two steps' method in a 500 mL stainless steel autoclave equipped with a mechanical stirrer [15]. A multi-component catalyst (Zn2G) was introduced into the autoclave and the autoclave with the catalyst inside was dried for 24 h under vacuum at 80 °C and then cooled down to room temperature. Then the purified PO was immediately injected into the autoclave. The autoclave was pressurized to 5.2 MPa via a CO_2 cylinder and heated at 70 °C for 20 h. Following the evacuation of CO_2 and unreacted PO, CHO was introduced into the autoclave under an inert atmosphere. The autoclave was re-pressurized with 5.2 MPa of CO_2 and the reaction was performed at 80 °C for another 20 h. Then, the pressure in the autoclave was reduced to atmosphere to terminate the block copolymerization. Similarly, the tri-block polymer PCHC-PPC-PCHC was synthesized by using the 'one pot, three steps' method. The first step was the copolymerization of CO_2 with CHO, which is followed by the copolymerization of CO_2 with PO, and then the copolymerization of CO_2 with CHO again [15]. The random copolymer poly (propylene cyclohexene carbonate) (PPCHC) was prepared by terpolymerization of CO_2, PO, and CHO [14]. For the first 30 h, the temperature was set at 70 °C and, for the next 10 h, it was raised to 80 °C. PPC was made from PO and CO_2 while PCHC was a copolymer of CHO and CO_2. The resulting copolymers were separately dissolved in a proper volume of chloroform and 15 mL dilute HCl (5 wt %) was added to extract the catalyst residual from the product solution. The organic layer was then washed with distilled water three times. The viscous solution was concentrated to a proper concentration by using a rotary evaporator. Lastly, they were precipitated by being poured into vigorously stirred ethanol. The as-made copolymers were filtered and dried under vacuum at a temperature of 120 °C until a constant weight was obtained.

3.3. Determination of the Composition and the Molecular Weight of Polymers

^{1}H-NMR spectra of the block copolymer at room temperature using tetramethylsilane as an internal standard and D-chloroform ($CDCl_3$) as a solvent were recorded on a Bruker DRX-400 NMR spectrometer (Karlsruhe, Germany). Molecular weight (M_w and M_n) of the resultant polymer product was measured by a gel permeation chromatography (GPC) system (Waters 515 HPLC Pump, Waters 2414 detector) (Milford, MA, USA) with a set of three columns (Waters Styragel 500, 10,000, and 100,000 Å) and chloroform (HPLC grade) as eluent. The GPC system was calibrated by a series of polystyrene standards with polydisperisties of 1.02.

3.4. Py-GC/MS Measurement

The Py-GC/MS measurements were carried out using a PYR-2A micro-tube furnace pyrolyser (Shimadzu-PYR2A) (Kyoto, Japan) coupled to an HP 6890 gas chromatograph linking to an HP 5973 quadruple mass spectrometer. The sample was heated in the PYR-2A compact in an oxygen-free furnace. Sample aliquots of about 1.00 mg were pyrolyzed by using a platinum coil attachment. This furnace's temperatures were set at 200 °C, 250 °C, 300 °C, 350 °C, and 400 °C, respectively. A quartz capillary column (SE-30; 15 m × 0.2 mm) (Kyoto, Japan) was used in the GC. The column temperature was initially held at 45 °C for 1 min and then programmed to 250 °C at a heating rate of 10 °C/min. The GC/MS interface was set at 280 °C. The pyrolysed products of polymers were directly injected and separated by gas chromatography using helium as an eluent gas and then characterized by the mass spectrometer. The mass spectra were recorded under electron impact ionization energy at 70 eV. The MS detector scanned from 29 to 350 m/z at a scan rate of 1.8 scan/s.

3.5. TG/FTIR Measurement

Thermogravimetric analysis (TGA) measurements were performed in a PerkinElmer Pyris Diamond TG/DTA analyzer (Waltham, MA, USA) under a protective nitrogen atmosphere. The temperature ranged from 50–500 °C with a heating rate of 2, 4, 6, 8, 10 °C/min, respectively. TG/IR analysis was performed with a TG/IR system, which combined with a PerkinElmer Pyris Diamond TG/DTA analyzer and a PerkinElmer Spectrum 100 FTIR spectrometer (Waltham, MA, USA). Samples of about 10 mg were pyrolysed in the TG analyzer and the evolved gases were led to the FTIR spectrometer directly through a connected heated gas line to obtain three dimensional FTIR spectra. The flow rate of N_2 is 10 mL/min. The aluminum pans are used for the samples. The temperature of the heated transfer line is 200 °C. The heating rate for taking TGA/FTIR spectra is 10 °C /min. The operating conditions of the FTIR had a frequency range of 4000–400 cm^{-1}, a resolution of 2.0 cm^{-1}, and a scan rate of 1.0 scan/s.

3.6. Thermal Decomposition Kinetics

The "International Confederation for Thermal Analysis and Calorimetry (ICTAC)" committee recommended that utilizing multiple heating rate programs leads to more reliable kinetic parameters with respect to the single heating rate program [31]. The Kissinger-Akahira-Sunose (KAS) method is an integral method by which the E can be obtained through the conversion values of reactant [26–28], which is represented by Equation (1) below.

$$ln \frac{\beta}{T^2} = ln \left(\frac{AE}{Rg(\alpha)} \right) - \frac{E}{RT} \tag{1}$$

Definitions of all the variables and parameters are the same with previous equations. Since the value of $ln \left(\frac{AE}{Rg(\alpha)} \right)$ is approximately constant when the values of α are the same at different β, the plot $ln \frac{\beta}{T^2}$ versus $1/T$ is approximately linear. Thus, by plotting $ln \frac{\beta}{T^2}$ against $1/T$ at certain conversion rates, the slope $- \frac{E}{RT}$ is calculated E.

The parameters of the compensation effect of the polymers' pyrolysis were obtained by the Coats-Redfern method. Fifteen kinds of frequently used reaction mathematical models are substituted into the Coats-Redfern equation. The Coats-Redfern equation is expressed by Equation (2) [29,30].

$$ln\frac{g(\alpha)}{T^2} = ln\frac{AR}{\beta E} - \frac{E}{RT} \tag{2}$$

Substituting g(α) into Equation (2) and plotting $ln\frac{g(\alpha)}{T^2}$ versus $1/T$, E and lnA of the different mathematical models can be calculated based on the slope (-E/R) and intercept ($ln\frac{AR}{\beta E}$).

The iso-conversional pre-exponential factor lnA and rate constant *lnk* are obtained by using the iso-conversional kinetic analysis method proposed by Vyazovkin [31–33]. The ln A is determined by the reaction feature of the reactant and it is independent of temperature. Therefore, the calculation is important for understanding the reaction feature. Since the compensation effect exists in E and A, Equation (3) is usually used to calculate ln A [31].

$$ln A = aE + b \tag{3}$$

The compensation effect parameters a and b were determined by fitting the pairs of lnAi and Ei by 15 different models substituting into the Coats-Redfern method at each heating rate. Then, lnA can be determined for every conversion α by substituting the respective values of E from the KAS method into Equation (3) to obtain a dependence of lnA on α. LnA$_\alpha$ on α and Eα on α data were converted into the Arrhenius plot. Therefore, the rate constant for the thermal decomposition can be evaluated as Equation (4) [32,33].

$$ln k(T_\alpha) = ln A_\alpha - \frac{E_\alpha}{RT_\alpha} \tag{4}$$

4. Conclusions

Thermal decomposition behaviors and degradation kinetic parameters of terpolymers with different sequence structures derived from CO_2, PO, and CHO were studied by the combination of Py-GC/MS and TG/IR techniques. In addition, the thermal degradation kinetic parameters were calculated by the Kissinger-Akahira-Sunose (KAS) method, the Coats-Redfern method, and the iso-conversional kinetic analysis method proposed by Vyazovkin with the data from thermogravimetric analysis under dynamic conditions. The average degradation activation energies of the polymers are PCHC-PPC-PCHC (104.9 $\pm$ 16.2 kJ mol^{-1}) < PPC-PCHC (109.9 $\pm$ 7.6 kJ mol^{-1} L) < PPC (124.6 $\pm$ 6.8 kJ/mol)) < PPCHC (145.5 $\pm$ 3.0 kJ mol^{-1}). The average pre-exponential factor ln A$_\alpha$ for PPC, PPCHC, PPC-PCHC, and PCHC-PPC-PCHC are 24.7 $\pm$ 1.5 s^{-1}, 27.7 $\pm$ 0.6 s^{-1}, 20.8 $\pm$ 1.6 s^{-1}, and 19.2 $\pm$ 3.4 s^{-1}, respectively. The rate constant values ln k vs. T^{-1} for the thermal decomposition of four polymers demonstrated that the thermal stability of PPC and PPC-PCHC are poorer than PPCHC and PCHC-PPC-PCHC and for PPCHC and PCHC-PPC-PCHC. There is an intersection between the two rate constant lines, which means that, for thermal stability, PPCHC is more stable than PCHC-PPC-PCHC at temperatures less than 309 °C and less stable when the temperature is more than 309 °C. The thermal degradation mechanism of the polymers was elucidated by IR-TG and Py-GC/MS techniques to be the main chain scissor reaction followed by the unzipping reaction. Due to large content of PC-CHC or CHC-PC linkages in the PPCHC, the steric hindrance of bulky cyclohexane groups restricted the unzipping reaction to some extent. The random copolymer showed one step decomposition with a T$_{-5\%}$ and T$_{max}$ as high as 281.0 °C and 313.4 °C, respectively. For the block polymers, the chain scission and unzipping reaction occurred first at the PPC block and then at the PCHC block. The CHC-CHC linkages could not restrict the PC-PCs unzipping reaction well, so the T$_{-5\%}$ of di-block and tri-block polymers are 261.2 °C and 275.2 °C, respectively, while two maximum decomposition temperatures were observed at 304.2–342.7 °C, and 305.8–345.3 °C, respectively. Lastly, we can conclude that random copolymerization of CHO, PO, and CO$_2$ is a better way to improve the thermal stability of PPC than block copolymerization.

Supplementary Materials: Supplementary materials can be found at http://www.mdpi.com/1422-0067/19/12/3723/s1.

Author Contributions: Y.M. conceive and designed the experiments. S.C., M.X., and L.S. conducted the experiments. Y.M., S.C., and M.X. analyzed the results and wrote the manuscript. All authors read and approved the final manuscript.

Funding: The authors would like to thank the National Natural Science Foundation of China (51673131, 21643002, 51573215), the Special Project on the Integration of Industry, Education and Research of Guangdong Province (2017B090901003, 2016B010114004, 2016A050503001), the Natural Science Foundation of Guangdong Province (2016A030313354), and the Natural Science Foundation of Fujian Province (2015J05030) for the financial support of this work.

Conflicts of Interest: The authors declare no conflict of interest.

Abbreviations

TG/FTIR	Thermogravimetric analysis/ Fourier transform infrared spectrometry
Py-GC/MS	Pyrolysis-gas chromatography/mass spectrometry
PPC	poly (propylene carbonate)
PPCHC	Poly (propylene cyclohexene carbonate)
PPC-PCHC	Poly (propylene carbonate–cyclohexyl carbonate)
PCHC-PPC-PCHC	Poly (cyclohexyl carbonate–propylene carbonate–cyclohexyl carbonate)

References

1. Inoue, S.; Koinuma, H.; Tsuruta, T. Copolymerization of carbon dioxide and epoxide. *J. Polym. Sci. Part B Polym. Lett.* **1969**, *7*, 287–292. [CrossRef]
2. Darensbourg, D.J.; Holtcamp, M.W. Catalysts for the reactions of epoxides and carbon dioxide. *Coord. Chem. Rev.* **1996**, *153*, 155–174. [CrossRef]
3. Meng, Y.Z.; Du, L.C.; Tiong, S.C. Effects of the structure and morphology of zinc glutarate on the fixation of carbon dioxide into polymer. *J. Polym. Sci. Part A Polym. Chem.* **2002**, *40*, 3579–3591. [CrossRef]
4. Coates, G.W.; Moore, D.R. Discrete metal-based catalysts for the copolymerization of CO$_2$ and epoxides: Discovery, reactivity, optimization, and mechanism. *Angew. Chem. Int. Ed.* **2004**, *43*, 6618–6639. [CrossRef] [PubMed]
5. Lu, X.B.; Ren, W.M.; Wu, G.P. CO$_2$ copolymers from epoxides: Catalyst activity, product selectivity, and stereochemistry control. *Acc. Chem. Res.* **2012**, *45*, 1721–1735. [CrossRef] [PubMed]
6. Czaplewski, D.A.; Kameoka, J.; Mathers, R.; Coats, G.W.; Craighead, H.G. Nanofluidic channels with elliptical cross sections formed using a nonlithographic process. *Appl. Phys. Lett.* **2003**, *83*, 4836–4838. [CrossRef]
7. Cao, M.; Xiao, M.; Lu, Y.; Meng, Y. Novel in situ preparation of crosslinked ethylene-vinyl alcohol copolymer foams with propylene carbonate. *Mater. Lett.* **2006**, *60*, 3286–3291.
8. Zeng, S.; Wang, S.; Xiao, M.; Meng, Y. Preparation and properties of biodegradable blend containing poly (propylene carbonate) and starch acetate with different degrees of substitution. *Carbohydr. Polym.* **2011**, *86*, 1260–1265. [CrossRef]
9. Chen, W.; Pang, M.; Xiao, M.; Wen, L.; Meng, Y. Mechanical, thermal, and morphological properties of glass fiber-reinforced biodegradable poly (propylene carbonate) composites. *J. Reinf. Plastics Compos.* **2010**, *29*, 1545–1550. [CrossRef]
10. Thorat, S.D.; Phillips, P.J.; Semenov, V.; Gakh, A. Physical properties of aliphatic polycarbonates made from CO$_2$ and epoxides. *J. Appl. Polym. Sci.* **2003**, *89*, 1163–1176. [CrossRef]
11. Kember, M.R.; Buchard, A.; Williams, C.K. Catalysts for CO$_2$/epoxide copolymerization. *Chem. Commun.* **2011**, *47*, 141–163. [CrossRef] [PubMed]
12. Li, X.H.; Meng, Y.Z.; Zhu, Q.; Tjong, S.C. Thermal decomposition characteristics of poly (propylene carbonate) using TG/IR and Py-GC/MS techniques. *Polym. Degrad. STable* **2003**, *81*, 157–165. [CrossRef]

13. Lu, X.L.; Zhu, Q.; Meng, Y.Z. Kinetic analysis of thermal decomposition of poly (propylene carbonate). *Polym. Degrad. STable* **2005**, *89*, 282–288. [CrossRef]

14. Wu, J.S.; Xiao, M.; He, H.; Wang, S.; Han, D.M.; Meng, Y.Z. Synthesis and characterization of high molecular weight poly (1, 2-propylene carbonate-co-1, 2-cyclohexylene carbonate) using zinc complex catalyst. *Chin. J. Polym. Sci.* **2011**, *29*, 552–559. [CrossRef]

15. Chen, S.Y.; Xiao, M.; Wang, S.J.; Han, D.M.; Meng, Y.Z. Novel Ternary Block Copolymerization of Carbon Dioxide with Cyclohexene Oxide and Propylene Oxide Using Zinc Complex Catalyst. *J. Poly. Res.* **2012**, *19*. [CrossRef]

16. Bassilakis, R.; Carangelo, R.M.; Wojtowicz, M.A. TG-FTIR analysis of biomass pyrolysis. *Fuel* **2001**, *80*, 1765–1786. [CrossRef]

17. Jiao, L.; Xiao, H.; Wang, Q.; Sun, J. Thermal degradation characteristics of rigid polyurethane foam and the volatile products analysis with TG-FTIR-MS. *Polym. Degrad. STable* **2013**, *98*, 2687–2696. [CrossRef]

18. Bruno, S.S.; Ana Paula, D.M.; Ana Maria, R.F.T. TG-FTIR coupling to monitor the pyrolysis products from agricultural residues. *J. Therm. Anal. Calorim.* **2009**, *97*, 637–642.

19. Rio, J.C.D.; Gutierrez, A.; Hernando, M.; Landin, P.; Romero, J.; Martinez, A.T. Determining the influence of eucalypt lignin composition in paper pulp yield using Py-GC/MS. *J. Anal. Appl. Pyrolysis.* **2005**, *74*, 110–115.

20. Zhu, P.; Sui, S.; Wang, B.; Sun, K.; Sun, G. A study of pyrolysis and pyrolysis products of flame-retardant cotton fabrics by DSC, TGA, and PY-GC/MS. *J. Anal. Appl. Pyrolysis.* **2004**, *71*, 645–655. [CrossRef]

21. Lu, X.Q.; Hanna, J.V.; Johnson, W.D. Source indicators of humic substances: An elemental composition, solid state 13 C CP/MAS NMR and Py-GC/MS study. *Appl. Geochem.* **2000**, *15*, 1019–1033. [CrossRef]

22. Tsuge, S.; Ohtani, H. Structural characterization of polymeric materials by PyrolysisdGC/MS. *Polym. Degrad. STable* **1997**, *58*, 109–130. [CrossRef]

23. Gu, X.L.; Ma, X.; Li, L.X.; Liu, C.; Cheng, K.H.; Li, Z.Z. Pyrolysis of poplar wood sawdust by TG-FTIR and Py-GC/MS. *J. Anal. Appl. Pyrolysis.* **2013**, *102*, 16–23. [CrossRef]

24. Huang, G.; Zou, Y.; Xiao, M.; Wang, S.; Luo, W.; Han, D.; Meng, Y. Thermal degradation of poly (lactide-copropylene carbonate) measured by TG/FTIR and Py-GC/MS. *Polym. Degrad. STable* **2015**, *117*, 16–21. [CrossRef]

25. Luo, W.; Xiao, M.; Wang, S.; Ren, S.; Meng, Y. Thermal degradation behavior of Copoly (propylene carbonateε-caprolactone) investigated using TG/FTIR and Py-GC/MS methodologies. *Polymer Testing.* **2017**, *58*, 13–20. [CrossRef]

26. Kissinger, H.E. Variation of peak temperature with heating rate in differential thermal analysis. *J. Res. Natl. Bur. Stand.* **1956**, *57*, 217–221. [CrossRef]

27. Kissinger, H.E. Reaction kinetics in differential thermal analysis. *Anal. Chem.* **1957**, *29*, 1702–1706. [CrossRef]

28. Akahira, T.; Sunose, T. Method of determining activation deterioration constant of electrical insulating materials. *Res. Rep. Chiba Inst. Technol.* **1971**, *16*, 22–31.

29. Coats, A.W.; Redfern, J.P. Kinetic parameters from thermogravimetric data. *Nature* **1964**, *201*, 68–69. [CrossRef]

30. Yuan, J.J.; Tu, J.L.; Xu, Y.J.; Qin, F.G.F.; Li, B.; Wang, C.Z. Thermal stability and products chemical analysis of olive leaf extract after enzymolysis based on TG-FTIR and Py-GC-MS. *J. Therm. Anal. Calor.* **2018**, *132*, 1729–1740. [CrossRef]

31. Vyazovkin, S.; Burnham, A.K.; Criado, J.M.; Perez-Maqueda, L.A.; Popescu, C.; Sbirrazzuoli, N. ICTAC Kinetics Committee recommendations for performing kinetic computations on thermal analysis data. *Thermochim. Acta.* **2011**, *520*, 1–19. [CrossRef]

32. Liavitskaya, T.; Birx, L.; Vyazovkin, S. Thermal stability of Malonic Acid Dissolved in Pomy(vinylpyrrolidone) and Other Polymeric Matrices. *Ind. Eng. Chem. Res.* **2018**, *57*, 5228–5233. [CrossRef]

33. Osman, Y.B.; Liavitslaya, T.; Vyazovkin, S. Polyvinylpyrrolidone affects thermal stability of drugs in solid dispersions. *Int. J. Pharm.* **2018**, *551*, 111–120. [CrossRef] [PubMed]

34. Liu, M.; Teng, C.T.; Win, K.Y.; Chen, Y.; Zhang, X.; Yang, D.; Li, Z.; Ye, E. Polymeric Encapsulation of Turmeric Extract for Bioimaging and Antimicrobial Applications. *Macromol. Rapid Commun.* **2018**. [CrossRef] [PubMed]

35. Luinstra, G. Poly (propylene carbonate), old copolymers of propylene oxideand carbon dioxide with new interests: Catalysis and material properties. *Polym. Rev.* **2008**, *48*, 192–219. [CrossRef]

36. Chisholm, M.H.; Navarro-Llobet, D.; Zhou, Z. Poly (propylene carbonate). 1. More about poly (propylene carbonate) formed from the copolymerization of propylene oxide and carbon dioxide employing a zinc glutarate catalyst. *Macromolecules* **2002**, *35*, 6494–6504. [CrossRef]
37. Barreto, C.; Cannon, W.R.; Shanefield, D.J. Thermal decomposition behavior of poly (propylene carbonate): tailoring the composition and thermal properties of PPC. *Polym. Degrad. Stab.* **2012**, *97*, 893–904. [CrossRef]
38. Zhu, Q.; Meng, Y.; Tjong, S.; Zhao, X.; Chen, Y. Thermally stable and high molecular weight poly(propylene carbonate)s from carbon dioxide and propylene oxide. *Polym. Int.* **2002**, *51*, 1079–1085. [CrossRef]

© 2018 by the authors. Licensee MDPI, Basel, Switzerland. This article is an open access article distributed under the terms and conditions of the Creative Commons Attribution (CC BY) license (http://creativecommons.org/licenses/by/4.0/).

International Journal of
Molecular Sciences

MDPI

Article

Biodegradable and Toughened Composite of Poly(Propylene Carbonate)/Thermoplastic Polyurethane (PPC/TPU): Effect of Hydrogen Bonding

Dongmei Han [1,2], Guiji Chen [1,3], Min Xiao [1], Shuanjin Wang [1,*], Shou Chen [4],
Xiaohua Peng [4] and Yuezhong Meng [1,2,*]

[1] The Key Laboratory of Low-carbon Chemistry & Energy Conservation of Guangdong Province/State Key Laboratory of Optoelectronic Materials and Technologies, Sun Yat-Sen University, Guangzhou 510275, China; handongm@mail.sysu.edu.cn (D.H.); chenguiji@kingfa.com.cn (G.C.); stsxm@mail.sysu.edu.cn (M.X.)
[2] School of Chemical Engineering and Technology, Sun Yat-Sen University, Guangzhou 510275, China
[3] Shanghai Kingfa Science and Technology Development Co., Ltd., Shanghai 201714, China
[4] Shenzhen Beauty Star Co., Ltd., Shenzhen 518112, China; chens@beautystar.cn (S.C.); alice@beautystar.cn (X.P.)
[*] Correspondence: wangshj@mail.sysu.edu.cn (S.W.); mengyzh@mail.sysu.edu.cn (Y.M.); Tel.: +86-20-84114113 (Y.M.)

Received: 25 May 2018; Accepted: 30 June 2018; Published: 13 July 2018

Abstract: The blends of Poly(propylene carbonate) (PPC) and polyester-based thermoplastic polyurethane (TPU) were melt compounded in an internal mixer. The compatibility, thermal behaviors, mechanical properties and toughening mechanism of the blends were investigated using Fourier transform infrared spectra (FTIR), tensile tests, impact tests, differential scanning calorimetry (DSC), scanning electron microscopy (SEM) and dynamic mechanical analysis technologies. FTIR and SEM examination reveal strong interfacial adhesion between PPC matrix and suspended TPU particles. Dynamic mechanical analyzer (DMA) characterize the glass transition temperature, secondary motion and low temperature properties. By the incorporation of TPU, the thermal stabilities are greatly enhanced and the mechanical properties are obviously improved for the PPC/TPU blends. Moreover, PPC/TPU blends exhibit a brittle-ductile transition with the addition of 20 wt % TPU. It is considered that the enhanced toughness results in the shear yielding occurred in both PPC matrix and TPU particles of the blends.

Keywords: Poly(propylene carbonate); thermoplastic polyurethane; compatibility; toughness

---

1. Introduction

Poly(propylene carbonate) (PPC) is a biodegradable aliphatic polycarbonate derived from carbon dioxide and propylene oxide. It has been paid much attention because of its high value-added fixation of CO_2, biodegradability and excellent oxygen barrier performance [1–3]. However, because PPC exhibits amorphous nature and possess low glass-transition temperature (Tg), it becomes brittle at low temperature and quickly loses strength at elevated temperature, which severely limits its wide application [4]. Many efforts have been devoted to overcoming these drawbacks. Chemical methods such as terpolymerization of carbon dioxide and propylene oxide with another monomer—for example, cyclohexene oxide (CHO) [5], maleic anhydride [6], or caprolactone [7]—have been employed. Simultaneously, physical blending of PPC with other polymers remains attractive because it is simple and cost-effective. Several PPC-base alloys have been prepared, starch [8,9], poly(ε-hydroxybutyrate-co-ε-hydroxyvalerate) (PHBV) [10], poly(3-hydroxybutyrate) (PHB) [11], poly(lactic acid) (PLA) [12], poly(butylene succinate) (PBS) [13], poly(ethylene-co-vinyl alcohol) (EVOH) [14] or others have been used to modify PPC. Nevertheless, most of these blends exhibited low miscibility between each component due to relatively weak inter-molecular interactions. For this reason, toughening of PPC is rarely achieved.

Thermoplastic polyurethane (TPU) elastomer, a linear segmented block copolymer consist of alternating soft segments and hard segments, are extensively used due to its superior properties, including high strength, high toughness, abrasion resistance, low temperature flexibility, biocompatibility, durability, biostability and so forth. [15–17]. Several works have been reported that TPU can be used as flexibilizer in many brittle polymers, such as polypropylene [18], poly(lactic acid) [19,20], polyacetal [21], poly(butylene terephthalate) [22] and so on.

Because of the similarly chemical structure, it is believed that PPC and TPU have better miscibility. Therefore, according to the phenomena that PPC becomes brittle at low temperature and loses its strength rapidly at high temperature, thermoplastic polyurethanes (TPU) are used to toughen PPC, in this work. Furthermore, the urethane moiety in hard segment of TPU can form hydrogen bonding with the carbonyl groups in PPC (Scheme 1), which in turn increase the interaction of molecular chain between PPC and TPU. In these connections, PPC was blended with TPU is expected to improve its toughness and thermal stability. These improvements will broaden the practical application of the new biodegradable PPC.

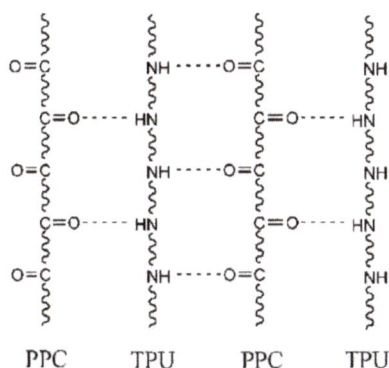

Scheme 1. Schematic illustration of hydrogen-bonding between the molecules of PPC and TPU.

2. Results and Discussion

2.1. FTIR Investigation

It is well known that the interaction between two components of the blend can be identified with FTIR technology. If two polymers are compatible, the absorption band shifts and broadening in the FTIR spectra indicates a chemical interaction between the molecular chains of one polymer and the other [23]. It was reported that there are the hydrogen-bonding interactions in their blends between Poly(propylene carbonate) (PPC) and other polymers, such as starch [8] and poly(ethylene-co-vinyl alcohol) (EVOH) [14]. Similarly, FTIR technique was used to confirm the existence of molecular chain interaction between the PPC and TPU in the blends. FTIR spectra of neat PPC, neat TPU and PPC/TPU blends with different TPU content are shown in Figure 1. The C=O absorption peak at 1749 cm^{-1} and N–H absorption peak at 3339 cm^{-1} were observed in the neat PPC and TPU respectively. In the blends, with the increase of PPC content, the N–H peak shifted towards lower wavenumber in the presence of PPC. It indicates a certain intermolecular hydrogen-bonding interaction between the N–H group of TPU and the carboxyl group of PPC. The peak for C=O group of blends become broader, which due to the shift of C=O groups in pure PPC and TPU is different. The wavenumber of C=O groups in TPU is lower than which in PPC.

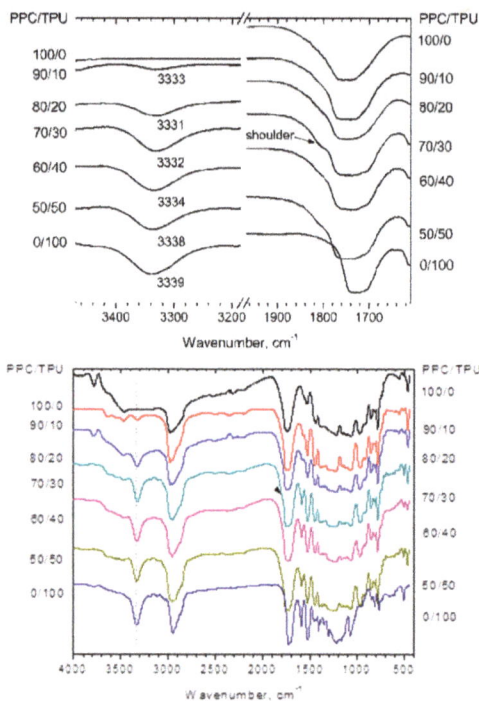

Figure 1. FTIR spectra of neat PPC, neat TPU and PPC/TPU blends.

2.2. Morphology Observation

Figure 2 represents the typical SEM micrographs of cryogenically fractured surfaces of the pure PPC and the PPC/TPU blends with various TPU contents. It can be seen that the pure PPC shows a smooth and uniform surface. For PPC/TPU blends, TPU particles are well dispersed in PPC matrix in fine droplets. The interfaces between PPC and TPU are fuzzy, demonstrating the good interfacial adhesion between suspended TPU particles and PPC matrix. This indicates the improved compatibility between PPC and TPU and which owe to the intermolecular hydrogen-bonding formation. This result is quite different from Huang's work and they reported that PLA is incompatible with TPU, although PLA also has carboxyl groups [24].

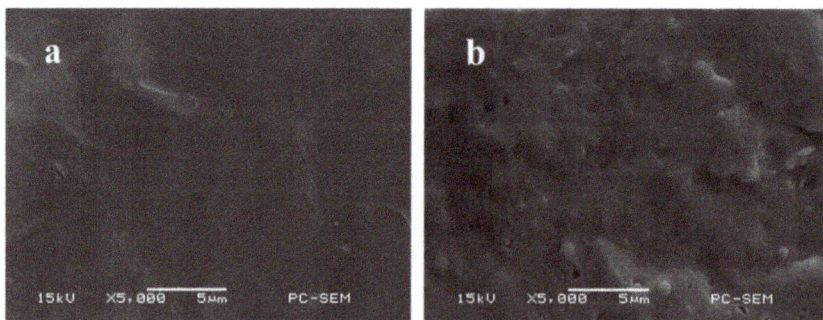

Figure 2. SEM micrographs of PPC and PPC/TPU blends. (**a**) Neat PPC, (**b**) PPC/10% TPU.

2.3. Thermal Behaviors

Figure 3 shows the DSC traces of the neat PPC, neat TPU and PPC/TPU blends recorded on the second heating step. The soft segment of TPU has a glass transition temperature (Tg) of −23.5 °C, while there is no obvious glass transition for the hard segment of TPU. The Tg of PPC increases with increasing TPU content. The value reaches 36.1 °C with the addition of 50 wt % TPU, while that of neat PPC is 31.7 °C. The increasing of PPC's Tg is presumably due to the hydrogen-bonding as indicated by FTIR investigation. Physical-crosslinking between PPC and TPU molecular chains greatly constrain the molecular movement of PPC matrix. On the contrary, the Tg value for soft segment of TPU decreases with increasing PPC content. It declines to about −34.2 °C with the addition of 50 wt % PPC. When PPC component increases to 90 wt %, the Tg value of TPU further decreases to as low as −45.6 °C. This is because the strong hydrogen-bonding between carbonyl of soft segment and the urothen groups of the TPU hard segment. The strong hydrogen-bonding decreases in turn constraining of soft segment of TPU, resulting in the decrease of Tg.

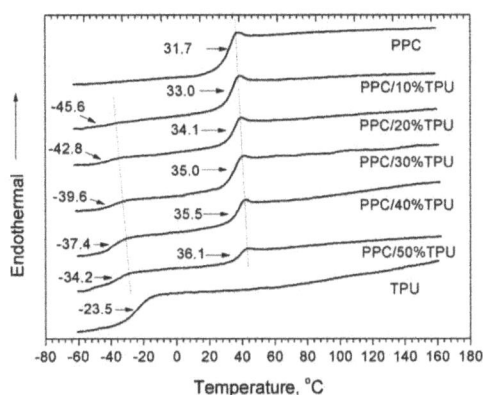

Figure 3. DSC curves of neat PPC, neat TPU and PPC/TPU blends.

The Vicat softening temperatures of the neat PPC, neat TPU and PPC/TPU blends are plotted in Figure 4. The Vicat temperatures shift to higher value with the incorporation of TPU. The value increases about 5 °C with every 10 wt % TPU addition. The improvement in thermal stability of PPC will certainly broaden the practical application of PPC.

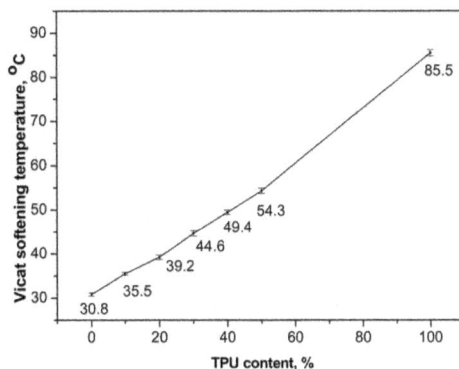

Figure 4. Effect of TPU content on Vicat softening temperatures of PPC/TPU blends.

2.4. Mechanical Properties

The stress-strain curves for the neat PPC, neat TPU and PPC/TPU blends are given in Figure 5. It is apparent that TPU has a low modulus but high strain and stress at break. All PPC/TPU blends exhibit yield characteristics, nevertheless, PPC shows the lowest elongation at break compared with other PPC/TPU blends. The yield strength for each PPC/TPU blend is higher than that of neat PPC, indicating the formation of the hydrogen-bonding between PPC matrix and TPU particles. The addition of TPU significantly changes the tensile behavior of the PPC, especially the enhancement of elongation at break.

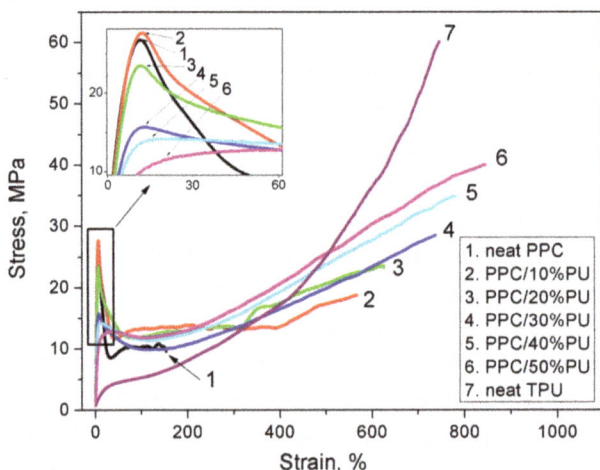

Figure 5. Stress-strain curves of PPC/TPU blends with different TPU contents.

The yield stress, stress at break, elongation at break and tensile energy to break values of neat PPC, neat TPU and PPC/TPU blends are listed in Table 1. It can be seen that the stress at break, elongation at break increase dramatically due to the hydrogen-bonding between PPC and TPU of the blends. Many reports have disclosed that the yield stress of blends decrease with the addition of TPU [20,22,25]. In this work, however, the yield stress of PPC/10%TPU blend is little higher than that of the neat PPC due to the hydrogen-bonding. Because of the lower modulus of TPU, the yield stress of PPC/TPU blends decreases with increasing TPU. Moreover, both elongation at break and tensile energy to break increase sharply with the addition of TPU. It is interesting to note that the PPC/TPU blend with only 10 wt % TPU exhibits a very high elongation at break of 566.2%, together with a little increase of yield stress. The tensile energy increases up to about 5 times for PPC/10%TPU blend and to more than 11 times for PPC/50%TPU blend. PPC/40%TPU and PPC/50%TPU blends even show higher tensile energy and strain to break than that of neat TPU. The results demonstrate that both toughness and tensile strength of PPC can be improved by the simply introduction of TPU.

The charpy impact strength of the neat PPC and PPC/TPU blends as a function of TPU content and is shown in Figure 6. The impact strength of the PPC/TPU blends increases slightly with the addition of 10 wt % TPU. The impact strength increases dramatically with further increasing TPU content. Some of the impact specimens cannot be broken completely during impact testing. The PPC/TPU blends exhibit a brittle-ductile transition behavior when 20 wt % TPU added because the maximum impact strength of our impact tester is 125 kJ/m² for standard specimen according to ASTM D256-05. These demonstrate the effective improvement of impact strength of PPC by the incorporation of TPU.

Table 1. Mechanical property of PPC/TPU blends with different TPU contents.

TPU Content, %	Yield Stress, MPa	Stress at Break, MPa	Elongation at Break, %	Tensile Energy to Break, MJ/m^3	Stress at 100%, MPa	Stress at 200%, MPa	Stress at 300%, MPa
0	26.5 ± 1.2	9.60 ± 0.6	180 ± 20	17.1 ± 3	10.3 ± 0.3	-	-
10	27.6 ± 0.9	18.7 ± 0.7	570 ± 30	82.2 ± 9	13.2 ± 0.2	13.8 ± 0.2	13.3 ± 0.3
20	23.5 ± 0.8	23.2 ± 0.4	620 ± 40	103 ± 7	11.9 ± 0.4	12.9 ± 0.3	13.5 ± 0.2
30	18.6 ± 0.5	27.5 ± 0.7	750 ± 50	124 ± 11	9.9 ± 0.2	10.7 ± 0.2	13.3 ± 0.2
40	14.3 ± 0.3	33.5 ± 2	780 ± 30	159 ± 9	11.4 ± 0.4	12.4 ± 0.3	15.2 ± 0.4
50	12.8 ± 0.5	40.0 ± 3	840 ± 50	195 ± 14	11.8 ± 0.2	12.9 ± 0.4	16.4 ± 0.4

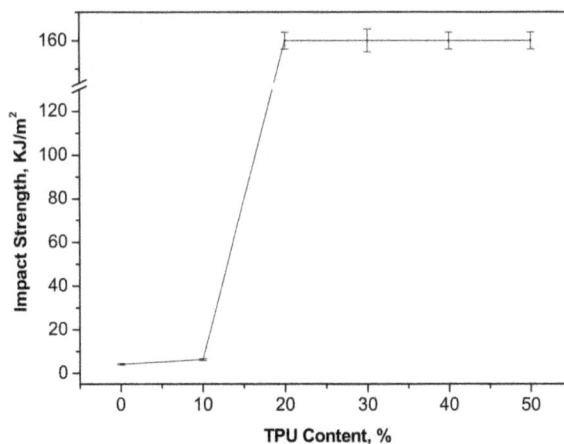

Figure 6. Impact strength of neat PPC and PPC/TPU blends as a function of TPU contents.

2.5. Toughening Mechanism

The results of tensile and impact experiments show that the highly effective toughness of TPU on PPC, especially for the PPC/TPU blends with more than 20 wt % TPU. Figure 7 shows the micrographs of the impact fractured surfaces of PPC/TPU blends. According to neat PPC (Figure 2a), with the addition of 10 wt % TPU, the fractured surface seems very roughness (Figure 7a). From the enlarged image (Figure 7b), the suspended TPU droplets can be clearly seen with observable gaps between PPC matrix and TPU particles. The dispersed TPU particles play a crucial role as stress concentration point within PPC matrix, where the deformation is much easier to happen than other areas within PPC/TPU blends. TPU particles serve as the plasticizing deformation phase because of its high elasticity. The deformation can absorb impact energy, therefore, the impact strength of PPC/10%TPU blend increases about 50 wt % compared with that of neat PPC. The PPC/TPU blends tend to become a tough failure with increasing TPU content. The dispersed TPU particles, as a large number of stress concentration points within PPC matrix, initiate crazing and shear banding, thus, absorb a large number of impact energy and impede crack growth under a strong external shock. PPC/30%TPU blend shows a typical tough fracture (Figure 7c). It can be seen that the cavitation and extensive plastic deformation of both PPC matrix and suspended TPU particles, implying that shear yielding is more predominant than crazing of PPC matrix and TPU particles.

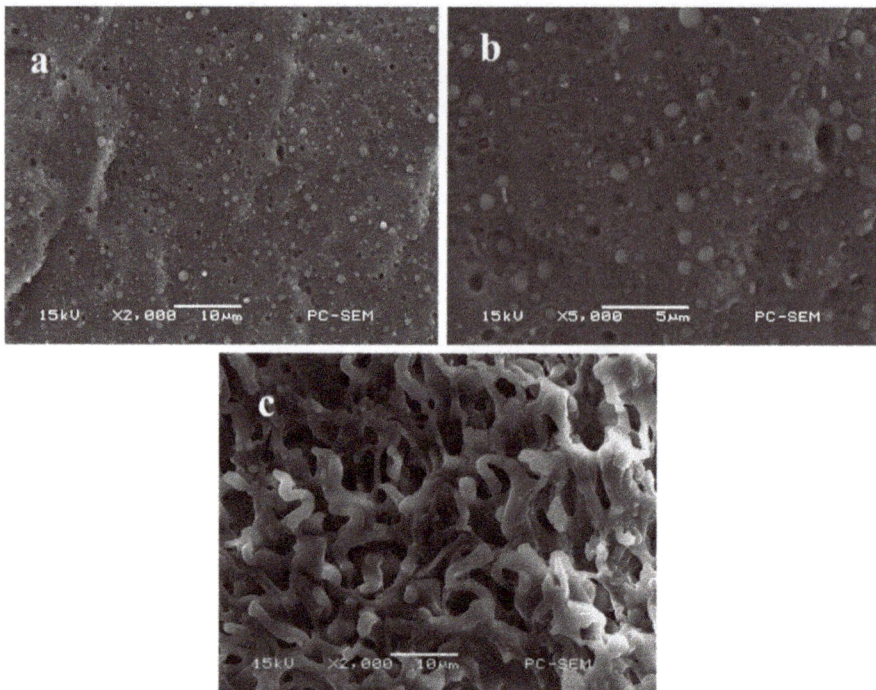

Figure 7. SEM micrographs of the impact fractured surfaces of PPC/TPU blends: (**a**) PPC/10%TPU, (**b**) enlarge view of a, (**c**) PPC/30%TPU.

In order to investigate the toughening mechanism, the different regions on the fractured surface of the impact fractured PPC/30%TPU blend sample is examined under SEM (Figure 8). The sample was not cracked completely and the cryogenically fractured surface along the impact direction (Figure 8a). Figure 8b–d shows different deformation states during impact testing. In the beginning of impact testing, the cracks were easily induced because of the high impact velocity and the stress concentration near the notch. The sample started to fracture immediately without large yielding as depicted in Figure 8b. Many voids around TPU particles were generated and some TPU particles showed small deformation and tended to be pulled out of PPC matrix. Thereafter, the impact energy was dissipated and the impact velocity decreased slightly. As shown in Figure 8c, further deformation of TPU particles can be seen clearly in point B. The PPC matrix around TPU particles deformed much obviously with shear yielding. This process absorbed the most energy of the impact fracture. As the impact velocity decreased sequentially, the sample showed again a tough failure in point C as shown in Figure 8d. The toughing mechanism can be explained as follows. First, TPU particles decrease the yield stress of PPC matrix following to generate shear yielding. Secondly, the PPC matrix deforms before cracking due to the decrease of impact velocity, which is attributed to the impact energy absorption from point A to point B. The micrograph of the freeze-fractured surface of the impact unfractured region is shown in Figure 8e. From this figure, it can be seen that the fractured surface is much smoother than other regions, only a few gaps between PPC matrix and TPU particles can be observed. Because the most of applied impact energy is dissipated before point D, there is not enough impact energy to destroy the blend sample with high TPU contents.

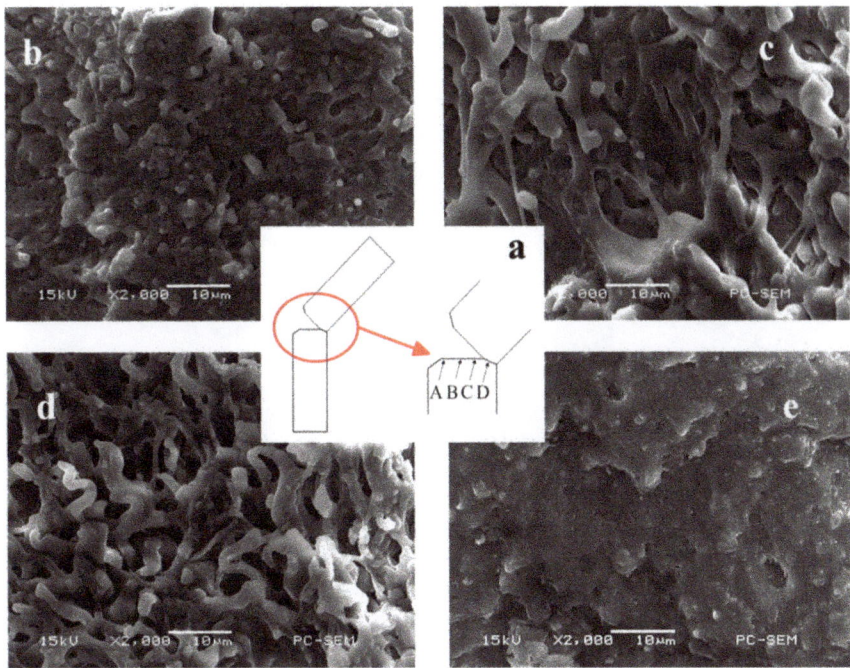

Figure 8. SEM micrographs of different regions on the impact fractured surface of the PPC/30%TPU sample. (**a**) Different regions of the impact fractured surface, (**b–e**) correspond to A–D in a, respectively.

2.6. Dynamic Mechanical Analysis

The dynamic mechanical analysis (DMA) can be used to characterize the secondary motion and properties of polymer molecules at low temperature [26], except for the characterization of the polymerized glass transition temperature.

The plots of storage modulus as a function of temperature for PPC, TPU and PPC/TPU are showed in Figure 9. It can be seen that in addition to a stress relaxation between 10 and 50 °C, pure PPC also has a secondary transition in the low temperature range of −100°C to −50 °C. Similarly, in addition to glass transition between −50 °C and 0 °C, pure TPU has a large secondary transition between −150 °C and −60 °C, which is important to the excellent flexibility and good performance at low temperature. Where When the TPU is in a highly elastic state, the storage modulus of the blends material gradually decreases as the TPU content increases, which indicates that the flexibility of the blends material is improved, which is one of the reasons why stretching and impact toughness is getting better with the increase of the TPU content.

Figure 10a shows the tg δ curves of PPC, TPU and their blends in the higher temperature region. The two peaks in the curve represent the glass transitions of PPC and TPU, respectively. The high temperature is according to the glass transition of PPC. The lower temperature is according to the glass transition of TPU. With the increase of TPU content, the corresponding peak of PPC becomes lower and lower and the corresponding peak of TPU becomes higher and higher. The corresponding value of the peak position is the glass transition temperature of the material and which is listed in Table 2. For PPC, as the content of TPU increases, the glass transition temperature of PPC increases gradually, from 44.7 °C of pure PPC to 53.5 °C when 50% TPU is added and 8.8 °C is increased. For TPU, as the content of PPC increases, the glass transition temperature of TPU gradually decreases, which is consistent with the results obtained by DSC analysis.

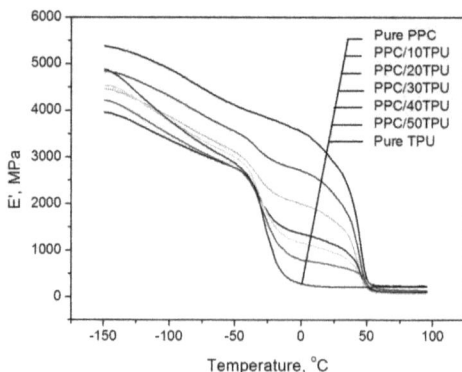

Figure 9. Storage modulus versus temperature for PPC, TPU and PPC/TPU blends.

It could be seen the tg δ curves of PPC, TPU and its blends in the lower temperature region in Figure 10b. It shows that both PPC and TPU have a β-transformation in the low temperature region and with the increase of TPU content, the β-transition peak of PPC in the blends material is becoming smaller, while the β-transition of TPU is more and more obvious. It indicates that as the TPU content increases, the high-temperature β-transition in the blends slowly disappears and replaced by a lower-temperature β-transition. The shift of β to low temperature means that the low temperature toughness of the material is improved. Therefore, the brittle-ductile transition temperature of the alloy material gradually decreases with the increase of the TPU content, showing better flexibility at low temperature.

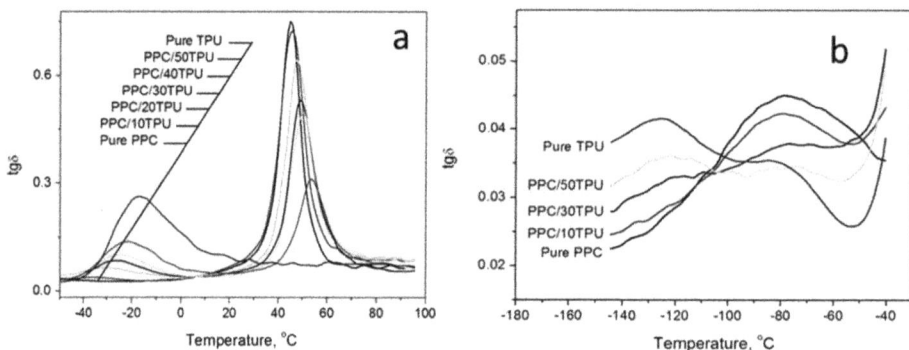

Figure 10. tg δ versus temperature for PPC, TPU and PPC/TPU blends: (**a**) higher temperature, (**b**) lower temperature.

Table 2. Glass transition temperature for PPC, TPU and their blends from dynamic mechanical analysis (DMA).

TPU Content, %	0	10	20	30	40	50	100
Tg (PPC), °C	44.7	45.4	47.5	48.9	50.2	53.5	-
Tg (TPU), °C	-	−35.9	−31.6	−26.3	−24.8	−22.1	−16.8

3. Materials and Methods

3.1. Materials

The high molecular weight Poly(propylene carbonate) (PPC), with a number-average molecular weight (Mn) of 109,000 Da and a polydispersity (PD) of 1.91, was provided by Tian'guan Enterprise Group Co. (Henan, China). Polyester-based TPU (grade S685AL, density = 1.2 g/cm^3), was obtained from Kin Join Co., Ltd. (Taiwan). Both PPC and TPU pellets were dried in a vacuum oven for 24 h at 80 °C before blending.

3.2. Preparation of PPC/TPU Blends

The PPC/TPU blends with weight ratios of 100/0, 90/10, 8/20, 70/30, 60/40, 50/50 and 0/100 were prepared in a Haake Rheomix RT 600 mixer (Haaker, Germany). The mixing was carried out at 160 °C with a rotary speed of 40 rpm for 7 min. The prepared blends were then melt pressed at 170 °C into standard dumbbell tensile bars (ASTM D638) and standard V-shaped notched impact bars (ASTM D256-05).

3.3. Fourier Transform Infrared (FTIR) Spectra

FTIR spectra were recorded on a Perkin-Elmer FTIR-100 spectrometer (Perkin-Elmer, Waltham, MA, USA). Samples were first melt pressed to thin film and then scanned with wavenumbers from 4000 to 400 cm^{-1} with a resolution of 4.0 cm^{-1}.

3.4. Scanning Electron Microscopy (SEM) Observation

The morphologies of the blends were observed using SEM (Jeol JSM-6380, JEOL, Tokyo, Japan). All of the Specimens were fractured in liquid nitrogen; the fracture surface was then coated with a thin layer of gold. The fracture surface after impact tests were also observed by the same SEM apparatus.

3.5. Differential Scanning Calorimetry (DSC) Investigation

The samples were scanned with differential scanning calorimetry (DSC, Netzsch 204, Selb, Bavaria, Germany). The temperature was initially heated from room temperature to 190 °C at a rate of 10 °C/min, maintained at 190 °C for a period of 3 min to eliminate previous thermal history, then cooled down to −100 °C at 10 °C/min, maintained at −100 °C for 3 min. The samples were subsequently scanned back to 190 °C with a heating rate of 10 °C/min. All the scanning processes were under a protective atmosphere of N2.

3.6. Vicat Softening Temperature

The Vicat softening temperatures of the blends were measured by a Vicat tester (New SANS, Shen Zhen, China) at 10 N load and heating rate of 50 °C/h according to ASTM D 1525. Six specimens have been examined for each blend and average value was reported accordingly.

3.7. Static Tensile Properties

The static tensile properties were investigated using a universal mechanical testing machine (New SANS) at 25 °C with a relative humidity of 50 ± 5%. The crosshead speed was set at 50 mm/min. Five specimens of each sample were measured and the average results were recorded. All the samples were conditioned at 25 °C and 50% RH for 24 h before testing.

3.8. Charpy Impact Strength

The Charpy impact strengths of notched specimens were determined using a Charpy impact tester (SANS ZBC-4B, MTSSANS, Eden Prairie, Minnesota, USA) at 25 °C and 50% RH according to ASTM D256-05. The maximum velocity of the pendulum is 2.9 m/s, generating impact energy of 4 J.

Each blend of five specimens was measured and the average results were recorded. All the samples were conditioned for 24 h at 25 °C and 50% RH before testing.

3.9. Dynamic Mechanical Analysis

The dynamic mechanical analysis of the material was carried out by a dynamic mechanical analyzer (DMA, model 242D, Netzsch, Selb, Germany). The test was conducted in a single cantilever vibration mode. The vibration frequency was 3.3 Hz and the amplitude was 100 μm. The size of the specimen is 35 mm × 10 mm × 1 mm and the temperature is set from −150 °C to 150 °C, with a heating rate of 5 °C/min.

4. Conclusions

PPC/TPU blends can be readily prepared by melt blending. Experimental results indicate the existence of intermolecular hydrogen-bonding between the carbonyl groups of PPC and the urethane groups of TPU. The SEM examination and thermal analysis of the PPC/TPU blends show a compatible blend system. The incorporation of TPU can obviously enhance the thermal stability and broaden the application temperature of PPC according to DSC and Vicat softening temperature investigation. Moreover, the toughness of the blends increases dramatically with increasing TPU content. The blend exhibits a brittle-ductile transition temperature at about 25 °C with the addition of 20 wt % TPU. Based on the experiment results, we have proposed a toughening mechanism of TPU for PPC matrix. The shear yielding occurs in both suspended TPU particles and PPC matrix, which accounts for the sharp increase in the mechanical properties of PPC.

Author Contributions: D.H., G.C., Y.M., S.W. and M.X. conceived and designed the experiments; D.H. and G.C. performed the experiments and analyzed the data; S.C. and X.P. contributed analysis tools. D.H. and G.C. wrote this paper.

Acknowledgments: This work was supported by funding from National Natural Science Foundation of China (No. 21376276, 51673131); Guangdong Province Sci & Tech Bureau (Nos. 2017B090901003, 2016B010114004, 2016A050503001); the Special-funded Program on National Key Scientific Instruments and Equipment Development of China (No. 2012YQ230043); and Fundamental Research Funds for the Central Universities (171gjc37).

Conflicts of Interest: The authors declare no conflict of interest.

References

1. Inoue, S.; Tsuruta, T.; Kobayash, M.; Koinuma, H. Reactivities of Some Organozinc Initiators for Copolymerization of Carbon Dioxide and Propylene Oxide. *Die. Makromol. Chem.* **1972**, *155*, 61–73. [CrossRef]
2. Darensbourg, D. Making Plastics from Carbon Dioxide: Salen Metal Complexes as Catalysts for the Production of Polycarbonates from Epoxides and CO_2. *J. Chem. Rev.* **2007**, *107*, 2388–2410. [CrossRef] [PubMed]
3. Luinstra, G.A. Poly(Propylene Carbonate), Old Copolymers of Propylene Oxide and Carbon Dioxide with New Interests: Catalysis and Material Properties. *Polym. Rev.* **2008**, *48*, 192–219. [CrossRef]
4. Inoue, S.; Tsuruta, T. Synthesis and thermal degradation of carbon dioxide-epoxide copolymer. *Appl. Polym. Symp.* **1975**, *26*, 257.
5. Ren, W.M.; Zhang, X.; Liu, Y.; Li, J.F.; Wang, H.; Lu, X.B. Highly Active, Bifunctional Co(III)-Salen Catalyst for Alternating Copolymerization of CO_2 with Cyclohexene Oxide and Terpolymerization with Aliphatic Epoxides. *Macromolecules* **2010**, *43*, 1396–1402. [CrossRef]
6. Song, P.F.; Xiao, M.; Du, F.G.; Wang, S.J.; Gan, L.Q.; Liu, G.Q.; Meng, Y.Z. Synthesis and properties of aliphatic polycarbonates derived from carbon dioxide, propylene oxide and maleic anhydride. *J. Appl. Polym. Sci.* **2008**, *109*, 4121–4129. [CrossRef]
7. Seong, J.E.; Na, S.J.; Cyriac, A.; Kim, B.W.; Lee, B.Y. Terpolymerization of CO_2 with Propylene Oxide and ε-Caprolactone Using Zinc Glutarate Catalyst. *Macromolecules* **2003**, *36*, 8210–8212. [CrossRef]
8. Zeng, S.S.; Wang, S.J.; Xiao, M.; Han, D.M.; Meng, Y.Z. Preparation and properties of biodegradable blend containing poly (propylene carbonate) and starch acetate with different degrees of substitution. *Carbohydr. Polym.* **2011**, *86*, 1260–1265. [CrossRef]

9. Ge, X.C.; Li, X.H.; Zhu, Q.; Li, L.; Meng, Y.Z. Preparation and properties of biodegradable poly(propylene carbonate)/starch composites. *Polym. Eng. Sci.* **2004**, *44*, 2134–2140. [CrossRef]
10. Tao, J.; Song, C.J.; Cao, M.F.; Hu, D.; Liu, L.; Liu, N.; Wang, S.F. Thermal properties and degradability of poly(propylene carbonate)/poly(β-hydroxybutyrate-*co*-β-hydroxyvalerate) (PPC/PHBV) blends. *Polym. Degrad. Stab.* **2009**, *94*, 575–583. [CrossRef]
11. Yang, D.Z.; Hu, P. Miscibility, crystallization, and mechanical properties of poly(3-hydroxybutyrate) and poly(propylene carbonate) biodegradable blends. *J. Appl. Polym. Sci.* **2008**, *109*, 1635–1642. [CrossRef]
12. Ma, X.F.; Yu, J.G.; Wang, N. Compatibility characterization of poly (lactic acid)/poly (propylene carbonate) blends. *J. Polym. Sci. Part B Polym. Phys.* **2006**, *44*, 94–101. [CrossRef]
13. Pang, M.Z.; Qiao, J.J.; Jiao, J.; Wang, S.J.; Xiao, M.; Meng, Y.Z. Miscibility and properties of completely biodegradable blends of poly(propylene carbonate) and poly(butylene succinate). *J. Appl. Polym. Sci.* **2008**, *107*, 2854–2860. [CrossRef]
14. Jiao, J.; Wang, S.J.; Xiao, M.; Xu, Y.; Meng, Y.Z. Processability, property, and morphology of biodegradable blends of poly (propylene carbonate) and poly (ethylene-*co*-vinyl alcohol). *Polym. Eng. Sci.* **2007**, *47*, 174–180. [CrossRef]
15. Simmons, A.; Hyvarinen, J.; Poole-Warren, L. The effect of sterilisation on a poly (dimethylsiloxane)/poly (hexamethylene oxide) mixed macrodiol-based polyurethane elastomer. *Biomaterials* **2006**, *27*, 4484–4497. [CrossRef] [PubMed]
16. Lebedev, E.V.; Ishchenko, S.S.; Denisenko, V.D.; Dupanov, V.O.; Privalko, E.G.; Usenko, A.A.; Privalko, V.P. Physical characterization of polyurethanes reinforced with the in situ-generated silica-polyphosphate nano-phase. *Compos. Sci. Technol.* **2006**, *66*, 3132–3137. [CrossRef]
17. Xian, W.Q.; Song, L.N.; Liu, B.H.; Ding, H.L.; Li, Z.; Cheng, M.P.; Ma, L. Rheological and mechanical properties of thermoplastic polyurethaneelastomer derived from CO_2 copolymer diol. *J. Appl. Polym. Sci.* **2018**, *135*, 45974. [CrossRef]
18. Lu, Q.W.; Macosko, C.W.; Lu, Q.W.; Macosko, C.W. Comparing the compatibility of various functionalized polypropylenes with thermoplastic polyurethane (TPU). *Polymer* **2004**, *45*, 1981–1991. [CrossRef]
19. Feng, F.; Ye, L. Morphologies and mechanical properties of polylactide/thermoplastic polyurethane elastomer blends. *J. Appl. Polym. Sci.* **2011**, *119*, 2778–2783. [CrossRef]
20. Li, Y.J.; Shimizu, H. Toughening of polylactide by melt blending with a biodegradable poly (ether) urethane elastomer. *Macromol. Biosci.* **2007**, *7*, 921–928. [CrossRef] [PubMed]
21. Palanivelu, K.; Balakrishnan, S.; Rengasamy, P. Thermoplastic polyurethane toughened polyacetal blends. *Polym. Test.* **2000**, *19*, 75–83. [CrossRef]
22. Palanivelu, K.; Sivaraman, P.; Reddy, M.D. Studies on thermoplastic polyurethane toughened poly (butylene terephthalate) blends. *Polym. Test.* **2002**, *21*, 345–351. [CrossRef]
23. Peng, S.W.; Wang, X.Y.; Dong, L.S. Special interaction between poly (propylene carbonate) and corn starch. *Polym. Compos.* **2005**, *26*, 37–41. [CrossRef]
24. Han, J.J.; Huang, H.X. Preparation and characterization of biodegradable polylactide/thermoplastic polyurethane elastomer blends. *J. Appl. Polym. Sci.* **2011**, *120*, 3217–3223. [CrossRef]
25. Chang, F.C.; Yang, M.Y. Mechanical fracture behavior of polyacetal and thermoplastic polyurethane elastomer toughened polyacetal. *Polym. Eng. Sci.* **1990**, *30*, 543–552. [CrossRef]
26. Borah, J.S.; Chaki, T.K. Dynamic mechanical, thermal, physico-mechanical and morphological properties of LLDPE/EMA blends. *J. Polym. Res.* **2011**, *18*, 569–578. [CrossRef]

© 2018 by the authors. Licensee MDPI, Basel, Switzerland. This article is an open access article distributed under the terms and conditions of the Creative Commons Attribution (CC BY) license (http://creativecommons.org/licenses/by/4.0/).

International Journal of
Molecular Sciences

MDPI

Article

Chitin Nanofibrils and Nanolignin as Functional Agents in Skin Regeneration

Serena Danti [1,2,*], Luisa Trombi [2], Alessandra Fusco [2,3], Bahareh Azimi [2], Andrea Lazzeri [1,2], Pierfrancesco Morganti [3], Maria-Beatrice Coltelli [1,2] and Giovanna Donnarumma [2,3]

[1]   Department of Civil and Industrial Engineering, University of Pisa, 56122 Pisa, Italy;
       andrea.lazzeri@unipi.it (A.L.), maria.beatrice.coltelli@unipi.it (M.-B.C.)
[2]   Consorzio Interuniversitario Nazionale per la Scienza e Tecnologia dei Materiali (INSTM),
       50121 Florence, Italy; l.trombi@yahoo.it (L.T.), alessandra.fusco@unicampania.it (A.F.),
       b.azimi@ing.unipi.it (B.A.), giovanna.donnarumma@unicampania.it (G.D.)
[3]   Department of Experimental Medicine, University of Campania "Luigi Vanvitelli", 80138 Naples, Italy;
       pierfrancesco.morganti@mavicosmetics.it
*   Correspondence: serena.danti@unipi.it; Tel.: +39-050-2217874

Received: 28 April 2019; Accepted: 28 May 2019; Published: 30 May 2019

Abstract: Chitin and lignin, by-products of fishery and plant biomass, can be converted to innovative high value bio- and eco-compatible materials. On the nanoscale, high antibacterial, anti-inflammatory, cicatrizing and anti-aging activity is obtained by controlling their crystalline structure and purity. Moreover, electropositive chitin nanofibrlis (CN) can be combined with electronegative nanolignin (NL) leading to microcapsule-like systems suitable for entrapping both hydrophilic and lipophilic molecules. The aim of this study was to provide morphological, physico-chemical, thermogravimetric and biological characterization of CN, NL, and CN-NL complexes, which were also loaded with glycyrrhetinic acid (GA) as a model of a bioactive molecule. CN-NL and CN-NL/GA were thermally stable up to 114 °C and 127 °C, respectively. The compounds were administered to in vitro cultures of human keratinocytes (HaCaT cells) and human mesenchymal stromal cells (hMSCs) for potential use in skin contact applications. Cell viability, cytokine expression and effects on hMSC multipotency were studied. For each component, CN, NL, CN-NL and CN-NL/GA, non-toxic concentrations towards HaCaT cells were identified. In the keratinocyte model, the proinflammatory cytokines IL-1α, IL-1 β, IL-6, IL-8 and TNF-α that resulted were downregulated, whereas the antimicrobial peptide human β defensin-2 was upregulated by CN-LN. The hMSCs were viable, and the use of these complexes did not modify the osteo-differentiation capability of these cells. The obtained findings demonstrate that these biocomponents are cytocompatible, show anti-inflammatory activity and may serve for the delivery of biomolecules for skin care and regeneration.

Keywords: chitin; lignin; anti-inflammatory; immunomodulation; keratinocytes; mesenchymal stem cells; cosmetics; glycyrrhetinic acid; skin regeneration; bio-based; nanomaterials

1. Introduction

In recent years, the development of natural biopolymer-based products suitable for cosmetic, skin care and regeneration have received considerable attention [1]. The key specifications of these products are biocompatibility and biodegradation, both in the environment and in the human body. The constituent polymers and their decomposition products are thus expected to be safe and without side effects. Chitin and lignin, by-products of fishery and plant biomass, can be reused and converted to novel high value materials for biomedical and cosmetic applications, which are bio- and eco-compatible [2,3]. On the nanoscale, chitin and lignin crystalline structure and purity can be controlled, resulting in several interesting properties, such as antibacterial [4], anti-inflammatory [5], cicatrizing and anti-aging effectiveness [6,7]. Chitin nanofibrils (CN) represent the purest crystal form of chitin and show positive surface charges. As such, CN are able to combine with electronegative compounds entrapping different ingredients for skin-friendly applications, such as innovative cosmetics for aged [6,8], as well as problematic and sensitive skin [9]. Because of these properties, CN have been used in combination with biopolymers [3,10], such as with poly (lactic acid) (PLA) [11–15], by several researchers. Morganti et al. [6] developed CN-Hyaluronan block copolymeric nanoparticles, entrapping several active ingredients in order to evaluate their efficiency and safety as biologically active rejuvenation treatments able to accelerate skin regeneration. They concluded that these biodegradable polymer particles have some advantages over other colloidal carrier systems thanks to their higher stability and versatility in tailoring the ingredient load and its release rate. In another attempt [7], the same authors investigated CN-hyaluronan block-copolymeric nanoparticles for their ability to load lutein, as an active anti-wrinkle agent, and demonstrated their effectiveness as anti-aging cosmetic emulsion compounds. Combining electropositive CN and electronegative nanolignin (NL) also gives rise to microcapsules able to entrap both hydrophilic and lipophilic molecules such as vitamins, microelements, anti-inflammatory drugs, antioxidants, anti-ageing substances, immunomodulating agents and enzymes [3,16].

In view of the abovementioned first studies, since cosmetic product consumers are becoming more and more aware of the environmental impact of products and increasingly trust the efficacy of green ingredients, the interest in such biopolymeric nanomaterials is expected to grow in the coming years. Therefore, it is significantly important to assess the specific properties of CN and NL and CN-NL complexes, including their material characterization, in order to improve or enable efficient processing technologies and biological characterization, including usable concentrations and the disclosure of their advantages in skin contact products as well as potential side effects, if any. To this purpose, advanced in vitro platforms can be efficiently exploited as ethical models for material screening. The HaCaT cell line represents an established model of human keratinocytes (i.e., epidermis) which can give preliminary information about new materials with respect to their inflammatory reactions through the expression of several cytokines, and of antimicrobial response through the expression of antimicrobial peptides of the innate immunity, such as defensins.

The aim of this study was to test in vitro the efficacy and safety of CN, NL and CN-NL complexes, obtained in powder by spray drying, with human keratinocytes (HaCaT cells) and human mesenchymal stromal cells (hMSCs). CN-LN was also investigated for the delivery of glycyrrhetinic acid (GA), as a biomolecule with anti-inflammatory activity deriving from licorice plants (Glycyrrhiza). GA (also known as enoxolone or glycyrrhetic acid) is a pentacyclic triterpenoid obtained from the hydrolysis of glycyrrhizic acid. GA and its salts and esters are cosmetic ingredients which act as flavoring or skin-conditioning agents and have excellent antimicrobial, anti-inflammatory, antioxidant, anti-ulceration, antiviral and analgesic properties [17,18]. In this context, hMSCs represent a model of immature cells present in the connective tissue (e.g., dermis), which may be affected by the addition of nanocomponents, as they can penetrate across skin layers. The hMSC multipotency (i.e., capability of multiple lineage differentiation proper of stem cells) must be preserved for a correct tissue turn over.

To our knowledge, this is the first study investigating the efficacy of CN-NL complexes as carriers of GA. The successful fabrication of innovative complexes entrapping antibacterial and anti-inflammatory ingredients would enable the introduction of novel products and their industrial upscale for potential use in dermatologic and cosmetic applications.

2. Results

2.1. Morphological Characterization of CN, NL and CN-NL Complexes

CN was spray-dried in the presence of 2% (*w/w*) of poly(ethylene glycol) (PEG) to obtain CN samples and consisted of scraps of micrometric dimensions (Figure 1A). The use of PEG can be advantageous to avoid CN aggregation when the solution is concentrated [15]. In fact, this aggregation, by giving rise to particles of micrometric size, would result in a lower effectiveness of CNs because of the reduction of their active surface. The spray-dried CN-NL complex in powder, pretreated with PEG, showed a specific morphology, as it consisted of almost round particles of micrometric dimensions (Figure 1B). The supplied lignin in powder, NL, appeared as nanostructured micrometric aggregates (Figure 1C,D).

Figure 1. FE-SEM micrographs of the powders: (**A**) Spray dried CN; (**B**) Spray dried CN-NL complex; (**C**) Morphology of NL powder taken from the pristine NL sample; (**D**) Zoomed-in magnification of NL (16,000×) in back scattered modality to observe the NL nanostructure.

For better investigating the morphology of CN, pure CN suspensions at 2% (*w/w*) were diluted 1:1000 in distilled water and the suspension was deposited on a glass window on which the sample was dried. CN appeared as "whiskers" having a nanometric thickness and a micrometric length (Figure 2A). The spray dried complex CN-NL in powder (Figure 1B) was suspended in water (at 80 ppm) and analyzed after deposition on a glass window followed by drying. This sample showed a completely different morphology, consisting in micrometric disks having a round or ellipsoidal shape (Figure 2B). At higher magnification, the presence of a nanostructured system was highlighted (Figure 2C,D), in which the presence of both CN and NL particles could be observed.

The spray-dried CN-NL complex was easily suspended in water to obtain flat micrometric nanostructured agglomerates that can deposit onto a surface, and be used to modify its properties thanks to CN and NL functionalities.

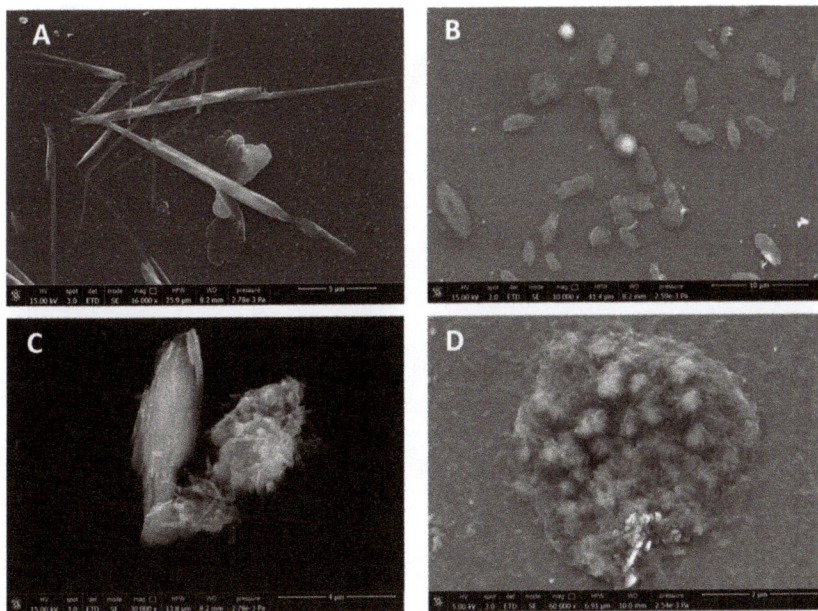

Figure 2. FE-SEM micrographs of (**A**) pure CN from deposition from diluted water suspension; (**B**) CN-NL complex; (**C**) and (**D**) CN-NL complex at higher magnification.

2.2. Chemical Structure and Thermal Stability of CN, NL, GA and CN-NL/GA Complexes

The starting materials and complexes CN-NL and CN-NL/GA were characterized by infrared spectroscopy. Spray-dried CN powder showed a spectrum (Figure 3A) with characteristic amide I and Amide II bands at 1619 and 1552 cm^{-1}, respectively. The Amide I band is split into two components at 1656 cm^{-1} and 1619 cm^{-1}. This is typical of α-chitin [19]. The most intense bands at 1010 cm^{-1} and 1070 cm^{-1} are typical of C-O stretching and agree with the polysaccharidic nature of this biopolymer. The band at 3439 cm^{-1} is attributable to O-H stretching, the bands at 3256 cm^{-1} and 3102 cm^{-1} to N-H stretching of amine and amide groups, respectively. The band at 2874 cm^{-1} indicates C-H stretching.

NL shows a spectrum (Figure 3B) characterized by the presence of strong band s at 1030 cm^{-1}, 1120 cm^{-1} and 1222 cm^{-1}. The band at 1222 cm^{-1} can be associated with C–C plus C–O plus C=O stretching.

The band at 1030 cm^{-1} can be attributed to aromatic C–H deformation that appears as a complex vibration associated with the C–O, C–C stretching and C–OH bending in polysaccharides. Carbohydrate originating vibrations are associated to the band at 1120 cm^{-1} [20]. In the carbonyl/carboxyl region, medium bands are found at 1705–1720 cm^{-1}, attributable to unconjugated carbonyl/carboxyl stretching. The bands at 1601 cm^{-1} and 1511 cm^{-1} can be associated with aromatic skeleton vibrations.

The GA spectrum (Figure 3C) is characterized by a band at 3430 cm^{-1} typical of OH stretching and a band at 2945 cm^{-1} attributable to CH stretching; both the bands at 1700 cm^{-1} and 1660 cm^{-1} are assigned to the C=O stretching (not conjugated and conjugated with the double bond, respectively) and the band at 1460 cm^{-1} to CH bending; the band at 1025 cm^{-1} can be attributed to C-O stretching and the band at 990 cm^{-1} can be attributed to the rocking of methyl groups [21].

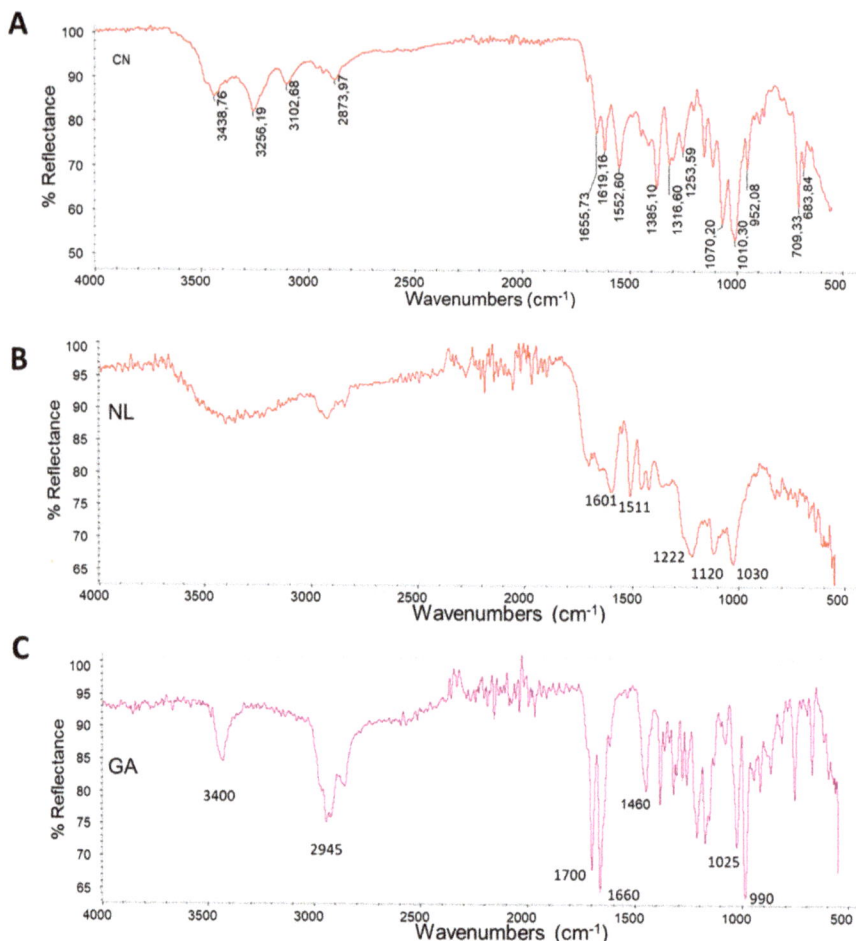

Figure 3. Infrared ATR spectra of (**A**): CN; (**B**) NL; (**C**) GA.

The complex CN-NL and the same complex entrapping GA were characterized by ATR spectroscopy (Figure 4). The obtained spectra were quite similar and they showed the bands already observed in the previous spectra of CN and NL, but slightly shifted because of the interactions occurring between CN and NL. The presence of GA can be revealed despite of its low concentration in the sample (0.2% w/w). As it can be observed in Figure 4B where the spectrum of GA, the spectrum of CN-NL and the spectrum of CN-NL/GA are compared in the region 2400–4000 cm^{-1}, the higher intensity of the CN-NL/GA with respect to CN-NL at 3400 cm^{-1} is in agreement with the presence of GA in the CN-NL/GA complex. A similar comparison is proposed in Figure 4C for the region 1100-2000 cm^{-1}. The increase of intensity of the band at 1650 cm^{-1} and 1470 cm^{-1} clearly observed in the spectrum of CN-NL/GA complex, can be attributed to the presence of GA in the powder. Interestingly the band at 1700 cm^{-1} of the carboxylic group of GA is not much evident in Figure 4C. In agreement with the work of Cheng et al. [22] this can be attributed to the interactions or reactions of the carboxylic groups with the functional group of CN and NL.

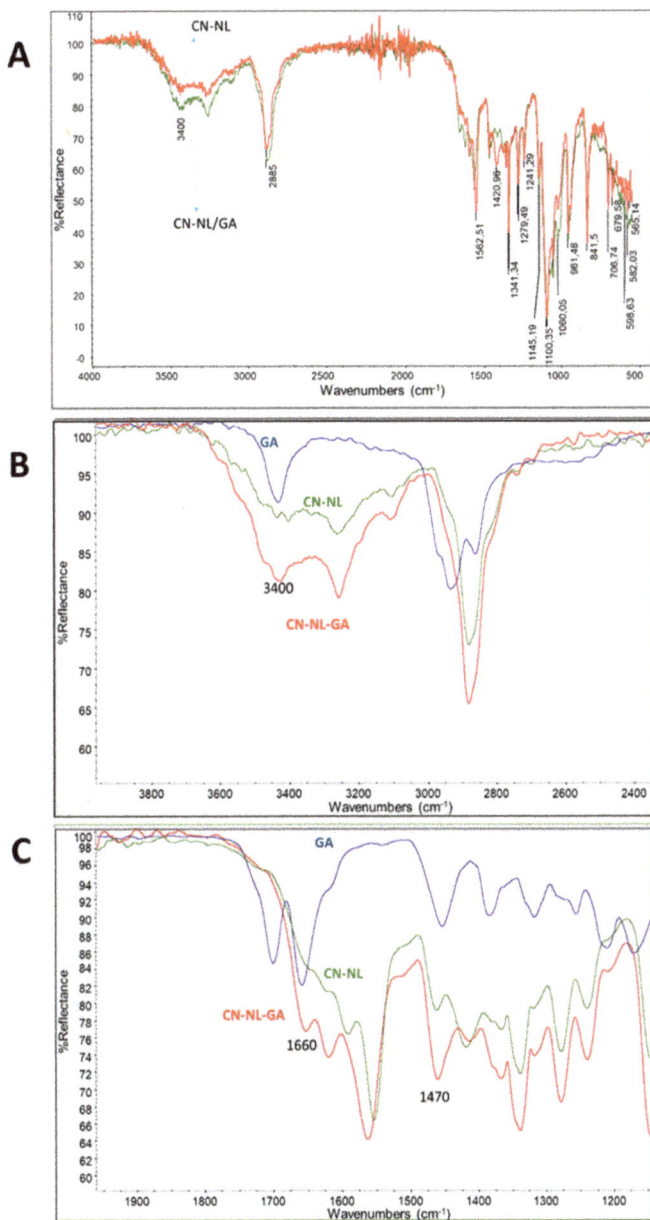

Figure 4. Infrared ATR spectra of (**A**) CN-NL and CN-NL/GA powders; (**B**) CN-NL, CN-LN/GA and GA in the region 2400-4000 cm^{-1}; (**C**) CN-NL, CN-LN/GA and GA in the region 1100-2000 cm^{-1}.

The thermogravimetric analysis (TGA) in nitrogen flow of pure components and of the two complexes was investigated and it was observed that GA, having an onset temperature at 387 °C, has a higher thermal stability than CN and NL. This different behavior is attributable to its action as a radical scavenger [23]. GA decomposition occurred in a single step and resulted in an almost complete decomposition. As observed in Figure 5A and in Table 1, the final residue is only 1.97% by weight.

The behavior of CN and NL was different, as several mass loss steps were observed (Figure 5A) and the final residue was 38.9% for NL and 17.4% for CN. These results evidenced that NL and CN formed a high amount of carbonaceous residue during thermal decomposition. Interestingly, NL showed a not negligible mass loss at 40 °C, probably because of partial decomposition, resulting in volatile fragments. The thermogravimetric trend of the complex did not include this mass loss. Hence, NL appeared more stable thanks to the formation of the complex with CN. However, the thermal stability of CN was reduced due to the presence of NL (the onset 1 and onset 2 temperatures were both decreased comparing those of CN-NL with CN) (Table 1). By the addition of GA, the stability of the complex was improved. This could be reasonably due to the radical scavenger activity of GA. Although the thermogravimetric trends of CN-NL and CN-NL/GA were similar, the mass loss recorded for CN-NL/GA was clearly lower than the one of CN-NL. This evidence corroborated the improved stability of CN-NL/GA thanks to the presence of GA.

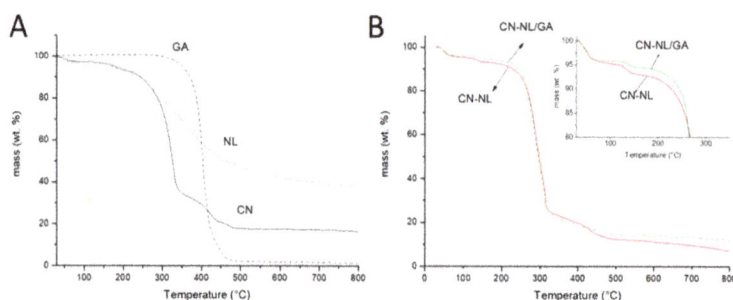

Figure 5. Thermogravimetric trends related to: (**A**) GA, NL and CN; (**B**) CN-NL and CN-NL/GA.

Table 1. Results of TGA analysis.

	Onset 1 (°C)	Mass Loss 1 (%)	Onset 2 (°C)	Mass Loss 2 (%)	Onset 3 (°C)	Mass Loss 3 (%)	Residue (%)
GA	386.95	−98.92	-	-	-	-	1.97
NL	40	−5.26	262	−55.63	-	-	38.93
CN	159	−5.35	300	−57. 54	387	−17.85	17.42
CN-NL	114	−2.60	266	−69.25	406	−10.39	12.37
CN-NL/GA	127	−1.92	251	−68.53	472	−11.63	13.84

CN, CN-NL and CN-NL/GA represent the spray-dried samples containing 2% (*w/w*) of PEG.

2.3. HaCaT Cell Viability

HaCaT cells were seeded at a density of $1·10^3$/well in 96-well culture plates. After 24 hours, the cells were treated with CN, CN-NL and CN-NL/GA at the following concentrations: 10 μg/mL, 5 μg/mL, μg/mL, 1 μg/mL, 0.5 μg/mL, 0.2 μg/mL, 0.1 μg/mL, 75 ng/mL, 50 ng/mL, 25 ng/mL.

After 24 h, the cells were incubated with MTT at 0.5 mg/mL at 37 °C for four 4 h and, subsequently, with DMSO at room temperature for 15 min. The spectrophotometric absorbance of the samples was determined by using an Ultra Multifunctional Microplate Reader (Biorad) at 570–655 nm [21]. Results, expressed as percentage of viable treated cell to respect with untreated control, are shown in Figure 6. In vitro studies performed on HaCaT cells enabled the establishment of the safe concentration of these green nano- and micro-compounds to be used. In particular, CN and CN-NL, did not decrease cell viability at 10 μg/mL and 0.2 μg/mL, respectively. Furthermore, CN-NL/GA complexes did not decrease cell viability at 0.5 μg/mL.

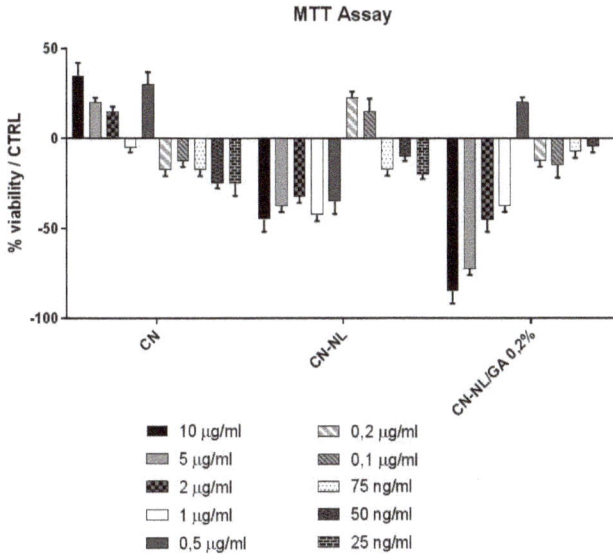

Figure 6. Results of MTT assay for different concentrations of CN, CN-NL, CN-NL/GA (0.2%), using HaCaT cells. The results were normalized by the viability of untreated cells as control.

2.4. HMSC Morphology, Viability and Osteodifferentiative Potential

Basing on the findings reported in Figure 6, hMSCs were treated with the above reported concentrations employed for HaCaT cells and the cell culture results are shown in Figures 7 and 8.

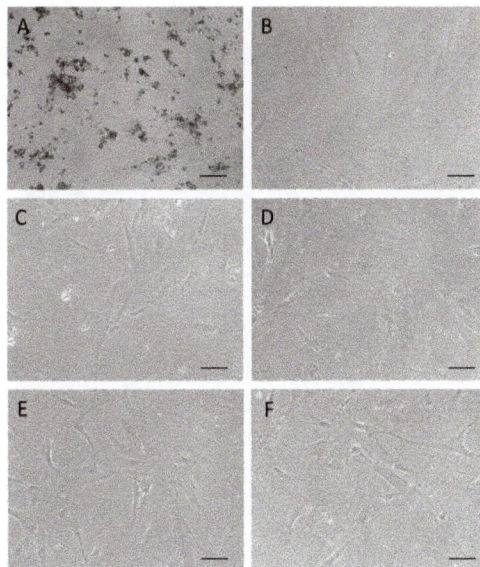

Figure 7. Micrographs at the inverted optical microscope of hMSCs in culture after 24 h of treatment with: (**A**) CN at 10 μg/mL; (**B**) NL at 10 μg/mL; (**C**) CN-NL at 0.2 μg/mL; (**D**) CN-NL/GA at 0.5 μg/mL; (**E**) GA at 0.5 μg/mL; (**F**) control. Micrographs are representative of each compound administrated at the intermediate concentration in the tested ranges. Original magnification 100×, scale bar = 100 μm.

Lower and higher concentrations of CN-NL (0.1 µg/mL and 0.5 µg/mL) and both CN-NL/GA and GA alone (0.2 µg/mL and 1 µg/mL) were tested with hMSCs. CN and NL were assayed at the maximum concentration. Treatments with CN, LN, CN-LN, CN-LN/GA complexes and GA did not affect the morphology of hMSCs at all the tested concentrations (Figure 7). CN treated samples showed micrometric aggregations that did not alter cell adhesion and growth (Figure 7A). Beside using pure CN, the other samples appeared clean and the spindle-like shape of hMSCs was clearly visible. AlamarBlue®test was performed on undifferentiated hMSCs in growth conditions in order to assess the cytocompatibility of the CN-NL complex and GA separately, and the combined CN-NL/GA complex. The obtained results are shown in Figure 8.

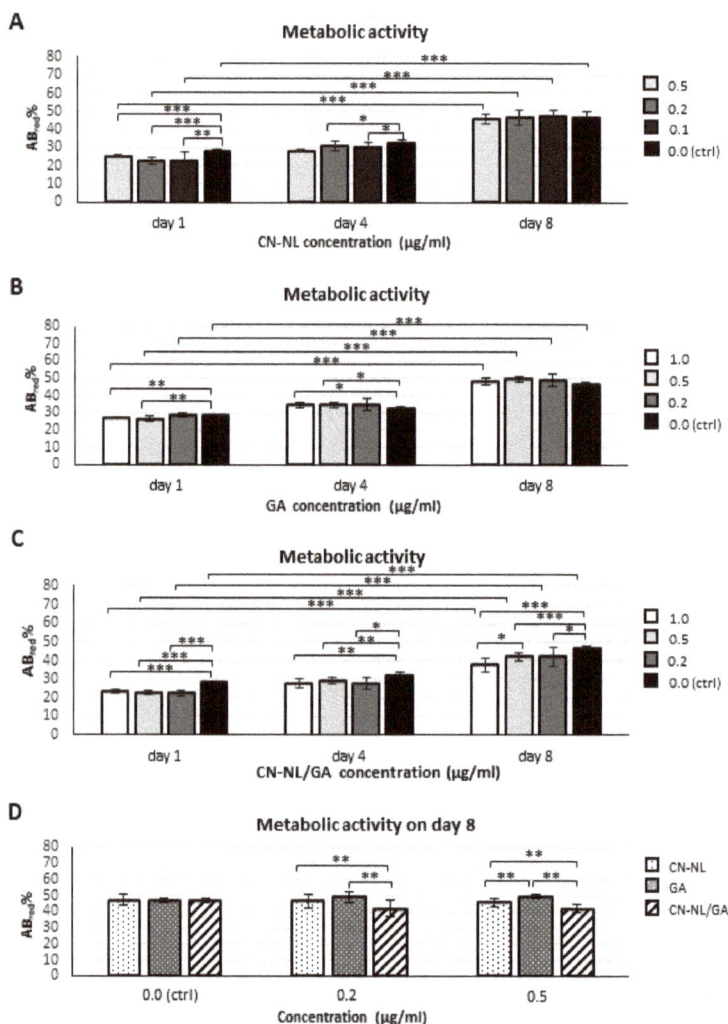

Figure 8. Bar graphs showing metabolic activity, as obtained by alamarBlue®test, performed on undifferentiated hMSCS using CN-NL, GA and CN-LN/GA complexes: (**A–C**) at different time points (1, 4 and 8 days in culture) and selected concentrations for each compound; (**D**) on day 8, comparing the three compounds within homogeneous concentration groups (* $p < 0.01$, ** $p < 0.001$, *** $p < 0.0001$).

In all the samples, the metabolic activity significantly increased over time ($p < 0.0001$) and was not significantly different from that of control samples (p = n.s), but using CN-NL/GA. In these samples, at the endpoint, the metabolic activity statistically decreased with respect to controls by increasing CN-NL/GA concentration ($p < 0.01$ at 0.2 µg/mL; $p < 0.0001$ at 0.5 µg/mL and 1.0 µg/mL) (Figure 8C). The highest GA concentrations (0.5 µg/mL and 1.0 µg/mL) increased hMSC metabolic activity with statistical significance only at day 4 ($p < 0.01$) (Figure 8B). On day 8, GA resulted the most performing treatment over CN-NL and CN-NL/GA both at 0.2 µg/mL and 0.5 µg/mL; instead, CN-NL/GA reduced hMSC metabolic activity ($p < 0.001$) (Figure 8D). To assess if the multipotency of hMSCs was maintained by the administration of these green compounds, the hMSCs were differentiated towards the osteogenic lineage in presence of all the additives and compared to undifferentiated hMSCs as a control. Mineral matrix deposition was investigated as a proof of osteo-differentiation using von Kossa staining (Figure 9).

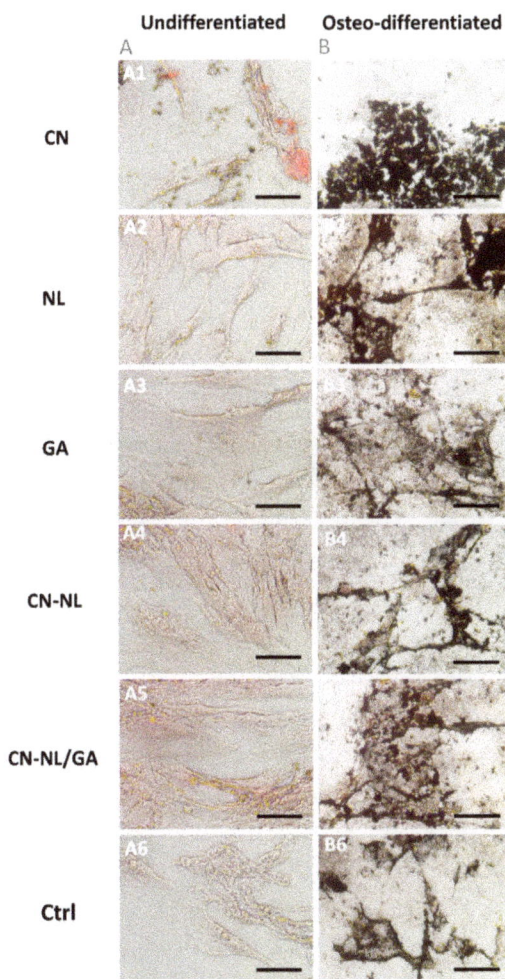

Figure 9. Light micrographs of von Kossa staining performed on (**A**) undifferentiated (control) and (**B**) osteodifferentiated hMSCs after incubation with all the nanocomponents (CN, NL, GA, CN-NL, CN-NL/GA) for 15 days at the highest concentrations in the tested ranges. Original magnification 200×; scale bar = 100 µm.

The obtained findings highlighted that both treated and untreated hMSCs, cultured in osteodifferentiation medium, produced mineral granules, visible in black, with no appreciable differences according to morphology and extension (Figure 9).

2.5. Anti-Inflammatory and Immune Responses of HaCaT Cells

Our data indicated that CN, CN-NL and GA markedly down-regulated the expression of pro-inflammatory cytokines interleukin-1 alpha (IL-1α), IL-1β, IL-6, IL-8 and tumor necrosis factor alpha (TNF-α) at mRNA level, especially after 24 h of treatment. The observed panel is in line with modulation of inflammatory activity exerted by all these components on human keratinocytes (Figure 10). The CN-NL/GA complex was the most effective in modulating the expression of the largest number of pro-inflammatory cytokines: IL-1α, IL-6, and IL-8. CN-NL was able to up-regulate the expression of an antimicrobial peptide, the human β-defensin (HBD-2), which is involved in innate immune response (Figure 10). Differently, transforming growth factor beta (TGF-β) was not modulated (data not shown).

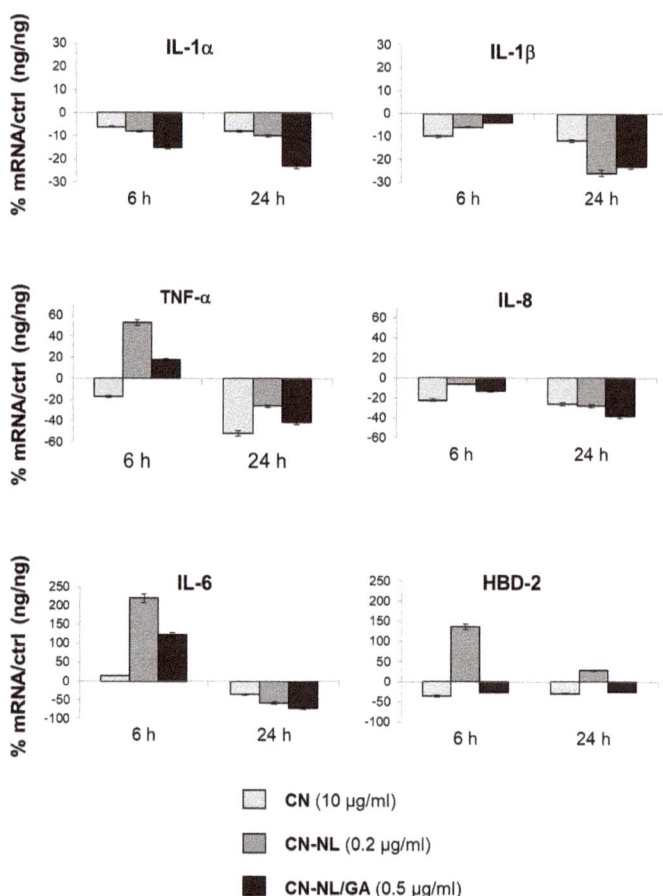

Figure 10. Bar graphs showing the results of RT-PCR performed on HaCaT cells exposed to the biopolymer nanomaterials al 6 h and 24 h for different cytokines involved in inflammatory response and HBD-2 and an antimicrobial peptide. The results are normalized by the expression in untreated cells as control.

These results are very encouraging for the use of these natural products for cosmetic applications and tissue repair, by acting on the modulation of cytokines and antimicrobial peptide HBD-2. On the other hand, these composites did not modify the proliferative and differentiative potential of hMSCs, thus demonstrating a good biocompatibility and cell affinity.

3. Discussion

The spray-dried CN-NL complex in powder was found to be water-suspendable and applicable on a surface in the form of micrometric flat nanostructured ellipsoids. Regulating the concentration of the water suspension can be fundamental to treat and confer to a surface some functional properties, such as anti-microbial and antioxidant properties. The resultant NL thermal stability was lower than that of CN. However, the CN-NL complex displayed an improved thermal stability with respect to pure NL. In the presence of GA, despite of its low amount in the powder (2% by weight), the thermal stability was further improved, in agreement with the radical scavenging properties of this substance. On the whole the complex CN-NL in powder, also in its version loaded with GA, can be considered a functionalizing agent much useful for functionalizing various substrates.

Interaction with skin cells is fundamental to assess the functional properties and demonstrate safety in skin contact applications, as they may include cosmetics, personal care as well as skin regeneration. It is a fact that nanosized objects can migrate into tissues [7], potentially leading to unknown reactions. The nanomaterials investigated in this study are biodegradable and therefore their interaction with the human body is temporary, so they do not pose harmful risks due to organ accumulation. The initial interactions with skin cells was investigated to understand their safety and disclose possible inflammatory reactions. In our study, we used HaCaT cells, as a model of human keratinocytes (i.e., epithelial cells), to select the optimal concentrations to be used for each of these biocompounds, which were 10 µg/mL for CN and NL alone, 0.2 µg/mL for CN-NL and 0.5 µg/mL for CN-NL/GA complexes. The higher concentration usable for CN and NL in comparison with those of the complexes may depend on the larger size (micrometric) of the CN-NL.

HMSCs were used as a model of mesenchymal (stem) cells present in the innermost layers of the skin, such as dermis and hypodermis (where they originate fibroblasts, among other cell types), as well as connective tissue in general [24], in order to assess the cytocompatibility and safety. Starting from the optimal concentrations chosen for HaCaT cells, each component was diluted at higher and lower concentrations to investigate the sensitivity of hMSCs to the same component. In the selected concentration ranges, the hMSCs were viable, with no appreciable differences in their morphology and osteo-differentiation capability with respect to untreated samples [25], thus demonstrating good biocompatibility and cell affinity of these green components, even at two times higher concentrations than the optimal ones selected using HaCaT cells. The wide range of cytocompatibility with hMSCs ensures a safe use of these bionanomaterials in the selected ranges. The significant reduction in hMSC metabolic activity following the administration of CN-NL/GA samples highlighted at the endpoint (day 8) with respect to GA and CN-LN administrated singularly can be a consequence of the synergistic activity exerted by the two components in the complex, and in particular by CN-NL, which alone largely decreased the metabolic activity of HaCaT cells at 1 µg/mL concentration. GA is known to affect some cellular functions in hepatocytes [17], and in our study showed a positive effect on hMSC metabolism. It is thus possible that the size and the neutral charge of the complex can affect cell metabolic activity by promoting other cellular functions.

HaCaT cells were also used to understand the reaction of the epidermis layer by investigating an array of cytokines involved in inflammation and immune response. HBD-2 is an inducible antimicrobial peptide with a molecular mass of 4–6 kDa acting as an endogenous antibiotic in the defense of host against Gram-positive and Gram-negative bacteria, fungi and the envelope of some viruses, and is involved in the innate immune response because its release is induced by pro-inflammatory cytokines, endogenous stimuli, infections or wounds [26]. Cytokines are multi-functional biological molecules that play a key role in the innate immune response, and that are involved in important biological activities such as hematopoiesis, infectious diseases, tumorigenesis, homeostasis and tissue repair, growth and cell development [27]. In particular, IL-1 promotes local inflammation and coagulation, increases the expression of adhesion molecules, and causes the release of chemokines and recruitment of leukocytes to the site of inflammation. In addition, it is an endogenous pyrogen, stimulates the synthesis of acute phase proteins and induces cachexia [28]. TNF-α is essential mediator in inflammation [29]. Its release in the site of the inflammation involves a localized vascular endothelial activation and vasodilation with increased vascular permeability. TNF-α also acts on platelet adhesiveness, favoring the formation of thrombus and occlusion of blood vessels and, therefore, reducing infection but also the tissue necrosis. IL-6 is a pleiotropic molecule that stimulates hepatocytes to synthetize many plasma proteins, such as fibrinogen, which ultimately contribute to the inflammatory acute phase response. Finally, IL-8 is a chemokine with many functions, including the attraction and activation of polymorphonuclear leukocytes, chemotaxis of basophils and a role in angiogenesis [30]. Remarkably, CN-NL and CN-NL/GA upregulated the mRNA expression of TNF-α and IL-6 at 6 h, which resulted subsequently downregulated after 24 h. This behavior can be explained by giving the nanocomposites a role in the wound repair process. In fact, wound healing is characterized a series of phases, and in particular in a first phase a number of overlapping events occur, including the production of pro-inflammatory cytokines. TNF-α represents the primary cytokine for pro-inflammatory responses; the direct effect of its release is the upregulation of IL-6 [31]. The obtained results showed that all the selected components were able to modulate the inflammatory response by decreasing the expression of proinflammatory cytokines. This fact was particularly evident in the CN-NL/GA complex, which downregulated the mRNA expression of the largest number of pro-inflammatory cytokines tested: IL-1α, IL-6, and IL-8. The unloaded complex, CN-NL alone, was able to up-regulate the expression of HBD-2. The reason why CN-NL/GA does not maintain the same functional activity towards HBD-2 may depend on specific cellular interactions with the GA-loaded complex. Indeed, GA shows different antibacterial mechanisms that are exerted directly towards bacteria and are not cell-mediated [32]. The obtained findings showed that CN, NL and specifically CN-NL complexes are very promising for skin contact applications, such as for coating biomedical, personal care and cosmetic products. Further investigation using primary human keratinocytes will provide a comprehensive validation of these compounds.

4. Materials and Methods

4.1. Materials

CN, NL, and GA were supplied by Mavi Sud, Aprilia (LT), (Milan, Italy). CN, CN-NL and CN-NL/GA were prepared in powder by using a Buchi Mini B-190 spray drier (Flawil, Switzerland) [33] by adding 2% with respect to CN-NL of PEG8000 from Sigma-Aldrich (Milan, Italy). Complexes CN-NL and CN-NL/GA were thus prepared in accordance with previous works [34,35]. The ratio between CN and NL is 2:1 by weight. The content of GA in CN-NL/GA is 0.2% by weight. HMSCs were supplied from Merck Millipore S.A.S., (Burlington, MA, USA). HaCaT, Dulbecco's Minimal Essential Medium (DMEM), L-glutamine, penicillin, streptomycin and fetal calf serum were purchased from Invitrogen, (Carlsbad, CA, USA). Fetal Bovine Serum and AlamarBlue®were purchased from Thermo Fisher Scientific, (Waltham, MA, USA). The HMSC Differentiation Osteogenic Bulletkit was purchased from Lonza, (Basel, Switzerland). Dulbecco's phosphate-buffered saline (DPBS), silver nitrate, pyrogallol, sodium thiosulphate, nuclear fast red, dimethyl sulfoxide (DMSO), $MgCl_2$ and MTT were purchased from Sigma-Aldrich (Milan, Italy). Aluminum sulphate was purchased from Carlo Erba (Milan, Italy).

4.2. Morphological Characterization of CN, NL and CN-NL Complexes

The morphology of the materials samples was investigated by field emission scanning electron microscopy (FESEM) using a FEI FEG-Quanta 450 instrument (Field Electron and Ion Company, Hillsboro, OR, USA). The samples were sputtered with Gold (Gold Edwards SP150B, England) before analysis. Inverted optical microscope (Nikon Ti, Nikon Instruments, Amsterdam, The Netherlands) was used to evaluate the morphology of hMSCs on CN, NL and CN-NL and CN-NL/GA complexes.

4.3. Chemical Structure and Thermal Stability Charachterisation of CN, NL and CN-NL Complexes

The powders as provided by MAVI were characterized by infrared spectroscopy using a Nicolet T380 Thermo Scientific instrument equipped with a Smart ITX ATR accessory with diamond plate.

Thermogravimetric tests onto the powders were performed on 4–10 mg of sample using a Mettler-Toledo Thermogravimetric Analysis/Scanning Differential Thermal Analysis (TGA/SDTA) 851 instrument operating with nitrogen as the purge gas (60 mL/min) at 10 °C/min heating rate in the 25–800 °C temperature range.

4.4. In vitro Culture of hMSCs and HaCaT

CN, NL, CN-NL, 0.2%GA and CN-NL/0.2%GA were solubilized in DPBS at 0.1 mg/mL and then diluted in the culture media to be tested with the cells. HMSCs isolated from the bone marrow were cultured in 24 well plate at 20,000 cells/well in DMEM Low Glucose with 2 mM L-glutamine, 100 IU/mL penicillin, 100 mg/mL streptomycin and 10% heat-inactivated FBS, with the above-mentioned components at different concentrations for 1 week to select the best concentration to be used. Immortalized human keratinocytes (HaCaT cell line) were cultured in DMEM supplemented with 1% Pen-Strep, 1% glutamine and 10% fetal calf serum at 37 °C in air and 5% CO_2. Subsequently, cells were dispensed into 6-well culture plates and left to grow until 80% of confluence.

4.5. MTT Assay

To establish the optimal non-toxic concentrations to be used in subsequent treatments, HaCaT cells were seeded at a density of 1×10^3/well in 96-well culture plates ($n = 2$). After 24 h, the cells were treated with CN, CN-NL and CN-NL/GA at different concentrations (from 10 µg/mL to 25 ng/mL) for 24 h and then incubated with MTT (0.5 mg/mL) at 37 °C for 4 h and, subsequently, with DMSO at room temperature for 5 min. The spectrophotometric absorbance of the samples was determined by using Ultra Multifunctional Microplate Reader (Bio-Rad, Hercules, California, USA) at 570–655 nm [36]. Results were given as average in Table 2.

Table 2. Concentrations of compounds tested with hMSCs.

CN-NL	0.5 µg/mL	0.2 µg/mL	0.1 µg/mL
GA	1 µg/mL	0.5 µg/mL	0.2 µg/mL
CN-NL/GA	1 µg/mL	0.5 µg/mL	0.2 µg/mL

4.6. Evaluation of hMSCs Viability and Differentiation Potential

AlamarBlue®test was performed at days 1, 4 and 8 to monitor hMSC viability in 24 well plates ($n = 6$) following the manufacturer's instructions. HMSCs were cultured for one week with the following concentrations of the components (Table 2):

Subsequently, hMSCs were osteoinduced using the differentiation Bulletkit osteogenic medium for one week in order to test hMSC regeneration potential. The samples plates ($n = 3$) were fixed with 1% *w/v* neutral buffered formalin. Mineralized matrix production was detected by von Kossa staining: cells were incubated with 1% *w/v* silver nitrate exposed to light for 15 min, 0.5% *w/v* Pyrogallol for 2 minutes and 5% *w/v* sodium thiosulphate for 2 min. All the solutions were in distilled water. The counterstaining was performed incubating cells with 0.1% *w/v* nuclear fast red diluted in a distilled water solution containing 5% *w/v* aluminum sulphate, for 5 min and washing in tap water for 5 min in order to reveal the reaction. The staining was observed with a Nikon Eclipse Ci microscope (Nikon Instruments, Amsterdam, The Netherlands) equipped with a digital camera by three independent observers for a qualitative analysis.

4.7. Anti-Inflammatory and Immune Responses Evaluation of HaCaT Cells

HaCaT cells, seeded in 6-well culture plates ($n = 3$) until 80% of confluence, were treated with CN, CN-NL and CN-NL/GA at selected concentrations (i.e., 10 μg/mL, 0.2 μg/mL and 0.5 μg/mL, respectively) for 6 h and 24 h. At the end of the experiment, total ribonucleic acid (RNA) was isolated and one microgram of this were reverse-transcribed into complementary deoxyribonucleic acid (cDNA) using random hexamer primers (Promega, Italy) at 42 °C for 45 min, according to the manufacturer's instructions. The anti-inflammatory and immune responses of HaCaT cells were evaluated by assaying the expression of pro-inflammatory cytokines IL-1α, IL-1β, IL-6, IL-8 and TNF-α, anti-inflammatory cytokine TGF-β, and antimicrobial peptide HBD-2 by Real time PCR with the LC Fast Start DNA Master SYBR Green kit from Roche Applied Science (Euroclone S.p.A., Pero, Italy) using 2 μl cDNA, corresponding to 10 ng of total RNA in a 20 Mm μl final volume, 3 mM Magnesium Chloride ($MgCl_2$) and 0.5 μM of sense primer and antisense primers (Table 3).

Table 3. Primer sequences and RT-PCR conditions.

Gene	Primers Sequence	Conditions	Product Size (bp)
IL-1α	5'-CATGTCAAATTTCACTGCTTCATCC -3' 5'-GTCTCTGAATCAGAAATCCTTCTATC -3'	5 s at 95 °C, 8 s at 55 °C, 17 s at 72 °C for 45 cycles	421
TNF-α	5'-CAGAGGGAAGAGTTCCCCAG -3' 5'-CCTTGGTCTGGTAGGAGACG -3'	5 s at 95°C, 6 s at 57°C, 13 s at 72°C for 40 cycles	324
IL-6	5'-ATGAACTCCTTCTCCACAAGCGC-3' 5'-GAAGAGCCCTCAGGCTGGACTG-3'	5 s at 95°C, 13 s at 56°C, 25 s at 72°C for 40 cycles	628
IL-8	5-ATGACTTCCAAGCTGGCCGTG -3' 5-TGAATTCTCAGCCCTCTTCAAAAACTTCTC-3'	5 s at 94°C, 6 s at 55°C, 12 s at 72°C for 40 cycles	297
TGF-β	5'-CCGACTACTACGCCAAGGAGGTCAC-3' 5'-AGGCCGGTTCATGCCATGAATGGTG-3'	5 s at 94°C, 9 s at 60°C, 18 s at 72°C for 40 cycles	439
IL-1β	5'-GCATCCAGCTACGAATCTCC-3' 5'-CCACATTCAGCACAGGACTC-3'	5 s at 95°C, 14 s at 58°C, 28 s at 72°C for 40 cycles	708

At the end of each run, the melting curve profiles were achieved by cooling the sample to 65° C for 15 s and then heating it slowly at 0.20 °C/s up to 95 °C with continuous measurement of fluorescence to confirm the amplification of specific transcripts. Cycle-to-cycle fluorescence emission readings were monitored and analyzed using LightCycler®software (Roche Diagnostics GmbH). Melting curves were generated after each run to confirm the amplification of specific transcripts. We used the b-actin coding gene, one of the most commonly used housekeeping genes, as an internal control gene. All reactions were carried out in triplicate, and the relative expression of a specific mRNA was determined by calculating the fold change relative to the b-actin control. The fold change of the tested gene mRNA was obtained with LightCycler®software by using the amplification efficiency of each primer, as calculated by the dilution curve. The specificity of the amplification products was verified by subjecting the amplification products to electrophoresis on 1.5% agarose gel and visualization by ethidium bromide staining [37].

4.8. Statistical Analysis

Statistical analyses were carried out by SPSS (SPSS v.16.0; IBM). All data were analyzed using a one-way analysis of variance (ANOVA) and post hoc test (Duncan) for multiple comparisons. Probability (p) values < 0.05 were considered as statistically significant differences.

5. Conclusions

CN-NL and CN-NL containing GA, obtained in powder by a spray drier technology can be optimal agents to form nanostructured and functional surfaces through a simple deposition by water suspension. The thermal stability of the complexes CN-NL and CN-NL/GA were investigated, and it was found that they can be considered thermally stable up to 114 °C and 127 °C, respectively. The performed characterization can give indications for a suitable application of CN-NL complexes as coating for material surfaces. All the investigated biopolymeric components were cytocompatible with HaCaT cells and hMSCs and optimal concentrations were selected for each of them. These findings demonstrate that CN-NL/GA complexes are able to downregulate a panel of anti-inflammatory cytokines in human keratinocytes and do not modify the proliferative and osteo-differentiative capacity of hMSC. Thus, these materials are very promising for skin contact applications, such as in biomedical, personal care and cosmetic products.

Author Contributions: Conceptualization, P.M., M.-B.C., G.D. and S.D.; methodology, L.T., A.F. and G.D.; formal analysis, B.A.; validation, L.T., G.D. and M.-B.C.; investigation, L.T., A.F.; resources, M.-B.C., A.L., P.M.; writing—original draft preparation, S.D.; writing—review and editing, M.B.-C., G.D., P.M.; visualization, S.D.; supervision, S.D.; project administration, M.-B.C. and S.D.; funding acquisition, M.-B.C., A.L, P.M.

Funding: This research was funded by the Bio-Based Industries Joint Undertaking under the European Union Horizon 2020 research program (BBI-H2020), PolyBioSkin project, grant number G.A 745839

Acknowledgments: The authors gratefully acknowledge Delfo D'Alessandro, University of Pisa, for his precious technical contribution to histology and Prof. Stefano Berrettini, University of Pisa, for providing access to equipment and facilities. Sabrina Bianchi, of the Department of Chemistry and Industrial Chemistry of the University of Pisa, and Randa Ishak, of the Department of Civil and Industrial Engineering of the University of Pisa, are kindly acknowledged for the support in thermogravimetry and FESEM investigations respectively.

Conflicts of Interest: The authors declare no conflict of interest.

Abbreviations

CN	Chitin nanofibrlis
NL	Nanolignin
GA	Glycyrrhetinic acid
HaCaT	Human keratinocytes
HMSCs	Human mesenchymal stromal cells
IL	Interleukin
TNF-α	Tumor necrosis factor alpha
PEG	Poly(ethylene glycol)
HBD-2	Human beta-defensin 2
TGF-β	Transforming growth factor beta
DMEM	Dulbecco's Minimal Essential Medium
DPBS	Dulbecco's phosphate-buffered saline
RT-PCR	Reverse transcription polymerase chain reaction
DMSO	Dimethyl sulfoxide
RNA	Ribonucleic acid
cDNA	Complementary deoxyribonucleic acid
MgCl2	Magnesium chloride
FESEM	Field Emission Scanning Electron Microscopy
ATR	Attenuated Total Reflection
TGA/SDTA	Thermogravimetric Analysis/Scanning Differential Thermal Analysis

References

1. Babu, R.P.; O'connor, K.; Seeram, R. Current progress on bio-based polymers and their future trends. *Prog. Biomate.* **2013**, *2*, 8–24. [CrossRef] [PubMed]
2. Morganti, P.; Danti, S.; Coltelli, M.B. Chitin and lignin to produce biocompatible tissues. *Res. Clin. Dermatol.* **2018**, *1*, 5–11.
3. Morganti, P.; Febo, P.; Cardillo, M.; Donnarumma, G.; Baroni, A. Chitin nanofibril and nanolignin: Natural polymers of biomedical interest. *J. Clin. Cosmet. Dermatol.* **2017**, *1*. [CrossRef]
4. Beisl, S.; Friedl, A.; Miltner, A. Lignin from micro-to nanosize: Applications. *Int. J. Mol. Sci.* **2017**, *18*, 2367. [CrossRef] [PubMed]
5. Morganti, P.; Del Ciotto, P.; Stoller, M.; Chianese, A. Antibacterial and anti-inflammatory green nanocomposites. *Chem. Eng. Trans.* **2016**, *47*, 61–66. [CrossRef]

6.   Morganti, P.; Svolacchia, F.; Del Ciotto, P.; Carezzi, F. New insights on anti-aging activity of chitin nanofibril-hyaluronan block copolymers entrapping active ingredients: *In vitro* and *in vivo* study. *J. Appl. Cosmetol.* **2013**, *31*, 1–29.

7.   Morganti, P.; Palombo, M.; Tishchenko, G.; Yudin, V.E.; Guarneri, F.; Cardillo, M.; Del Ciotto, P.; Carezzi, F.; Morganti, G.; Fabrizi, G. Chitin-hyaluronan nanoparticles: A multifunctional carrier to deliver anti-aging active ingredients through the skin. *Cosmet* **2014**, *1*, 140–158. [CrossRef]

8.   Morganti, P.; Carezzi, F.; Del Ciotto, P.; Tishchenco, G.; Chianese, A.; Yudin, V.E. A green multifunctional polymer from discarded material: Chitin Nanofibrils. *Br. J. Appl. Sci. Technol.* **2014**, *4*, 4175–4190. [CrossRef]

9.   Morganti, P.; Tishchenko, G.; Palombo, M.; Kelnar, I.; Brozova, L.; Spirkova, M.; Pavlova, H.; Kobera, L.; Carezzi, F. Chitin nanofibrils for biomimetic products: nanoparticles and nanocomposite chitosan films in health care. In *Marine Biomaterials: Isolation, Characterization and Application*, 1st ed.; Kim, S., Ed.; CRC Press: Boca Raton, FL, USA, 2013; pp. 681–715. ISBN 9781138076389.

10. Morganti, P.; Coltelli, M.B.; Danti, S. Biobased Tissues for innovative Cosmetic products: Polybioskin as an EU Research project. *Glob. J. Nano.* **2018**, *3*, 555620. [CrossRef]

11. Rizvi, R.; Cochrane, B.; Naguib, H.; Lee, P.C. Fabrication and characterization of melt-blended polylactide-chitin composites and their foams. *J. Cell. Plast.* **2011**, *47*, 283–300. [CrossRef]

12. Herrera, N.; Singh, A.A.; Salaberria, A.M.; Labidi, J.; Mathew, A.P.; Oksman, K. Triethyl Citrate (TEC) as a Dispersing Aid in Polylactic Acid/Chitin Nanocomposites Prepared via Liquid-Assisted Extrusion. *Polymers* **2017**, *9*, 406. [CrossRef] [PubMed]

13. Guan, Q.; Naguib, H.E. Fabrication and Characterization of PLA/PHBV-Chitin Nanocomposites and Their Foams. *J. Polym. Environ.* **2014**, *22*, 119–130. [CrossRef]

14. Herrera, N.; Roch, H.; Salaberria, A.M.; Pino-Orellana, M.A.; Labidi, J.; Fernandes, S.C.M.; Radic, D.; Leiva, A.; Oksman, K. Functionalized blown films of plasticized polylactic acid/chitin nanocomposite: Preparation and characterization. *Mater. Des.* **2016**, *92*, 846–852. [CrossRef]

15. Coltelli, M.B.; Cinelli, P.; Gigante, V.; Aliotta, L.; Morganti, P.; Panariello, L.; Lazzeri, A. Chitin Nanofibrils in Poly(Lactic Acid) (PLA) Nanocomposites: Dispersion and Thermo-Mechanical Properties. *Int. J. Mol. Sci.* **2019**, *20*, 504. [CrossRef]

16. Morganti, P. Composition and material comprising chitin nanofibrils, lignin and a co-polymer and their uses. Patent WO 2016/042474 Al, 2015.

17. Andersen, F.A. Final report on the safety assessment of Glycyrrhetinic Acid, Potassium Glycyrrhetinate, Disodium Succinoyl Glycyrrhetinate, Glyceryl Glycyrrhetinate, Glycyrrhetinyl Stearate, Stearyl Glycyrrhetinate, Glycyrrhizic Acid, Ammonium Glycyrrhizate, Dipotassium Glycyrrhizate, Disodium Glycyrrhizate, Trisodium Glycyrrhizate, Methyl Glycyrrhizate, and Potassium Glycyrrhizinate. *Int. J. Toxicol.* **2007**, *26*, 79–112. [CrossRef]

18. Hussain, H.; Green, I.R.; Shamraiz, U.; Saleem, M.; Badshah, A.; Abbas, G.; Rehman, N.U.; Irshad, M. Therapeutic potential of glycyrrhetinic acids: a patent review (2010-2017). *Expert. Opin. Ther. Pat.* **2018**, *28*, 383–398. [CrossRef] [PubMed]

19. Kumirska, J.; Czerwicka, M.; Kaczyński, Z.; Bychowska, A.; Brzozowski, K.; Thöming, J.; Stepnowski, P. Application of Spectroscopic Methods for Structural Analysis of Chitin and Chitosan. *Mar. Drugs* **2010**, *8*, 1567–1636. [CrossRef]

20. Boeriu, C.G.; Bravo, D.; Gosselink, R.J.A.; Van Dam, J.E.G. Characterisation of structure-dependent functional properties of lignin with infrared spectroscopy. *Ind. Crops Prod.* **2004**, *20*, 205–218. [CrossRef]

21. Zu, Y.; Meng, L.; Zhao, X.; Ge, Y.; Yu, X.; Zhang, Y.; Deng, Y. Preparation of 10-hydroxycamptothecin- loaded glycyrrhizic acid-conjugated bovine serum albumin nanoparticles for hepatocellular carcinoma-targeted drug delivery. *Int. J. Nanomed.* **2013**, *8*, 1207–1222. [CrossRef]

22. Cheng, M.; Gao, X.; Wang, Y.; Chen, H.; He, B.; Xu, H.; Li, Y.; Han, J.; Zhang, Z. Synthesis of Glycyrrhetinic Acid-Modified Chitosan, 5-Fluorouracil Nanoparticles and Its Inhibition of Liver Cancer Characteristics in Vitro and in Vivo. *Mar. Drugs* **2013**, *11*, 3517–3536. [CrossRef]

23. Ablis, M.; Leininger-Muller, B.; Wong, C.D.; Siest, G.; Loppinet, V.; Visvikis, S. Synthesis and *in Vitro* Antioxidant Activity of Glycyrrhetinic Acid Derivatives Tested with the Cytochrome P450/NADPH System. *Chem. Pharm. Bull.* **2004**, *52*, 1436–1439. [CrossRef]

24. Nilforoushzadeh, M.A.; Ahmadi Ashtiani, H.R.; Jaffary, F.; Jahangiri, F.; Nikkhah, N.; Mahmoudbeyk, M.; Fard, M.; Ansaria, Z.; Zare, S. Dermal Fibroblast Cells: Biology and Function in Skin Regeneration. *J. Skin Stem Cell* **2017**, *4*, e69080. [CrossRef]

25. Danti, S.; D'Acunto, M.; Trombi, L.; Berrettini, S.; Pietrabissa, A. A micro/nanoscale surface mechanical study on morpho-functional changes occurred to multilineage-differentiated human mesenchymal stem cells. *Macromol. Biosci.* **2007**, *7*, 589–598. [CrossRef] [PubMed]

26. Donnarumma, G.; Paoletti, I.; Fusco, A.; Perfetto, B.; Buommino, E.; de Gregorio, V.; Baroni, A. β-Defensins: Work in Progress. *Adv. Exp. Med. Biol.* **2016**, *901*, 59–76. [CrossRef] [PubMed]

27. Mantovani, A.; Sica, A.; Sozzani, S.; Allavena, P.; Vecchi, A.; Locati, M. The chemokine system in diverse forms of macrophage activation and polarization. *Trends Immunol.* **2004**, *25*, 677–686. [CrossRef] [PubMed]

28. Dinarello, C.A. Interleukin-1. *Cytokine Growth Factor Rev.* **1997**, *8*, 253–265. [CrossRef]

29. Esposito, E.; Cuzzocrea, S. TNF-alpha as a therapeutic target in inflammatory diseases, ischemia-reperfusion injury and trauma. *Curr. Med. Chem.* **2009**, *16*, 3152–3167. [CrossRef] [PubMed]

30. Koch, A.E.; Polverini, P.J.; Kunkel, S.L.; Harlow, L.A.; DiPietro, L.A.; Elner, V.M.; Elner, S.G.; Strieter, R.M. Interleukin-8 as a macrophage-derived mediator of angiogenesis. *Science* **1992**, *258*, 1798–1801. [CrossRef]

31. Morganti, P.; Fusco, A.; Paoletti, I.; Perfetto, B.; Del Ciotto, P.; Palombo, M.; Chianese, A.; Baroni, A.; Donnarumma, G. Anti-inflammatory, immunomodulatory, and tissue repair activity on human keratinocytes by green innovative nanocomposites. *Materials* **2017**, *10*, 843. [CrossRef] [PubMed]

32. Oyama, K.; Kawada-Matsuo, M.; Oogai, Y.; Hayashi, T.; Nakamura, N.; Komatsuzawa, H. Antibacterial effects of glycyrrhetinic acid and its derivatives on staphylococcus aureus. *PLoS ONE* **2016**, *11*, e0165831. [CrossRef]

33. Morganti, P.; Muzzarelli, C. Spray-dried chitin nanofibrils, method for production and uses thereof. Patent WO2007060628A1, 23 November 2005.

34. Morganti, P. Method of preparation of chitin and active principles complexes and the so obtained complexes. Patent WO2012143875A1, 19 April 2011.

35. Brooker, A.D.M.; Vaccaro, M.; Scialla, S.; Walker, S.J.; Morganti, P.; Carezzi, F.; Benjelloun-Mlayah, B.; Crestini, C.; Lange, H.; Bartzoka, E. A consumer goods product comprising chitin nanofibrils, lignin and a polymer or co-polymer. Patent EP2995321B1, 15 April 2014.

36. Mosmann, T. Rapid colorimetric assay for cellular growth and survival: application to proliferation and cytotoxicity assays. *J. Immunol. Methods* **1983**, *1*, 55–63. [CrossRef]

37. Fusco, A.; Coretti, L.; Savio, V.; Buommino, E.; Lembo, F.; Donnarumma, G. Biofilm formation and immunomodulatory activity of Proteus mirabilis clinically isolated strains. *Int. J. Mol. Sci.* **2017**, *18*, 414. [CrossRef] [PubMed]

© 2019 by the authors. Licensee MDPI, Basel, Switzerland. This article is an open access article distributed under the terms and conditions of the Creative Commons Attribution (CC BY) license (http://creativecommons.org/licenses/by/4.0/).

MDPI
St. Alban-Anlage 66
4052 Basel
Switzerland
Tel. +41 61 683 77 34
Fax +41 61 302 89 18
www.mdpi.com

International Journal of Molecular Sciences Editorial Office
E-mail: ijms@mdpi.com
www.mdpi.com/journal/ijms

www.ingramcontent.com/pod-product-compliance
Lightning Source LLC
Chambersburg PA
CBHW051716210326
41597CB00032B/5498

* 9 7 8 3 0 3 9 2 1 1 3 2 6 *